continued on back

ELEMENTS OF APPLIED
STOCHASTIC PROCESSES

To Girija, Girish, and Gouri

PREFACE

Retaining the objective, approach, level, and structure of the first edition, this edition attempts to correct deficiencies and to provide a broader exposure to the state of application of stochastic processes in various fields.

As before, the book can be divided roughly into two parts. Chapters 1–9 deal with the theory of Markov, renewal and stationary processes. Chapters 10–21 are on application and include six new chapters. The first group of chapters have seen the following major changes. The separate chapter on two-state Markov process has been eliminated. The two-state chain is now a section under Finite Markov Chains and the continuous parameter process is a section under Simple Markov Processes. A new section on graph theoretic methods has been added to the chapter on the classification of states of Markov chains. With the addition of the two-state chain, special topics such as higher-order, lumpable, and reversible chains, and additional material on inference, the chapter on finite Markov chains became too long. Consequently it has been divided into two chapters—one on time-dependent and limiting behavior and the other on special topics and inference. Chapter 9, on Stationary Processes, has been expanded to include topics such as ergodic theorems, spectral density, and linear filters. Needless to say, new material has been incorporated throughout other chapters in order to make the treatment comprehensive and up-to-date. A major deficiency of the first edition has been the scarcity of exercises. This new edition includes more than 200 exercises in the first nine chapters. Answers to these exercises are provided at the end of the book.

The second part of the book has seen considerable change in the form of new chapters covering either new application areas or expanded discussions of topics included in the earlier edition. Specifically the following changes may be noted. New chapters include Queueing Networks (Chapter 12), Communication and Information Systems (Chapter 13), Inventory and Storage Processes (Chapter 14), Combat Models (Chapter 16), Markov Models in Biological Sciences (Chapter 18), and Stochastic Models in Traffic Flow Theory and Geological Sciences (Chapter 20). The chapter on Time Series Analysis has been expanded considerably to make it up to date and comprehensive. It has been moved to the end of the book to maintain the continuity of treatment of similar models.

While going through these chapters on applications, the reader should bear in mind two key observations. It is not the intention of the author to provide a "state-of-the-art" survey of an application area. The purpose of the discussion is the illustration of an applied stochastic process as appearing in the literature on the topic. Consequently the discussion is not comprehensive and includes only sample problems from the published literature. We do not claim unbiasedness in the sample either. In many cases the literature is vast, and the sample is likely to reflect the interests of the author and the accessibility of the literature. Nevertheless, efforts have been made to direct the reader toward relevant literature in the form of a list of articles and books under the heading "For Further Reading." Using this list, an interested reader should be able to construct a comprehensive bibliography in the area of interest. Incidentally, it should be noted that this list plays the role of exercises incorporated in the first nine chapters.

The second observation to be noted is the predominance of models based on Markov processes. This is the result of the stage in the growth of use of probability models in a particular field and our objective of keeping the level of mathematical sophistication low. Nevertheless, it is not an oversimplification to state that in all areas of application of stochastic processes, simpler models are mostly Markovian.

Another major change is in the Appendix. Sections on Probability and Probability Distributions (Appendix A) and Sampling Statistics and Statistical Inference (Appendix B) of the earlier edition have been dropped. This book cannot be used without a basic course in probability and some exposure to statistical inference. For persons with that background it is unnecessary to have a compendium of results from those topics. Therefore while expanding the book in other areas, eliminating sections which are not totally necessary seemed prudent.

The author has taught a course out of this text several times during the past 12 years. The revision carried out for this edition is the result of cumulative experience as well as the comments and criticisms received over these years. I have been gratified by the many positive comments that have come from unexpected sources, and I am indebted to all the criticisms that have helped in improving the book. Particular mention should be made of colleagues Randy L. Eubank and Henry L. Gray in helping me revise chapters on stationary processes and time series analysis through comments, criticisms, and discussions.

Special acknowledgment of indebtedness is made of the constructive criticisms made by Professors J. Stuart Hunter and Carl M. Harris who reviewed earlier drafts of this edition. Professor Hunter was the reviewer of the first edition as well. It is no exaggeration to say that this revision could not have been so comprehensive without their timely contribution.

Finally acknowledgment is also made of the help given me by Kim Batchelor, Sheila Crain, and Janis Ekanem in typing the manuscript.

U. NARAYAN BHAT

Dallas, Texas
June 1984

PREFACE TO THE FIRST EDITION

With the expanding interest in stochastic models, courses in probability and stochastic processes are being created by educational units other than the traditional mathematics and statistics departments. Students in applied areas derive motivation through real systems, and unfortunately most available introductory textbooks on stochastic processes lack sufficient motivation and require an advanced level of mathematical sophistication. As a result, many applied scientists view a course in stochastic processes as a necessary burden rather than a stimulating experience. From a pedagogical point of view, therefore, a book designed to motivate and cater to the needs of such students as well as to broaden their horizons of knowledge seems to be essential—this is the major objective of this volume.

This book has grown out of a senior-level and an introductory graduate-level course given in the Department of Operations Research at Case Western Reserve University, Cleveland, Ohio, and the Department of Statistics and the Computer Science/Operations Research Center at Southern Methodist University, Dallas, Texas. Assuming a one-semester course in probability and statistics as background, this book is suitable for a following one-semester course in applied stochastic processes. A basic course in calculus, familiarity with some simple combinatorial mathematics, and matrix algebra are assumed to be prerequisites. These requirements are not unreasonable for most graduate students. Consequently, the book is viewed as suitable for a graduate course in applied stochastic processes in business, economics, and some fields of engineering, or for a masters-level course in specialized areas such as statistics, operations research, and computer science. It is the experience of the author that students motivated by a course of this nature derive a good deal from their advanced course in the theory of stochastic processes.

The book can be divided roughly into two parts. Chapters 1–9 deal with the theory of Markov and renewal processes, and Chapters 10–15 treat their applications. Keeping in mind the lack of analytical sophistication of the applied scientist, the theory is built by relating the particular to the general.

For instance, an entire chapter is devoted to the study of two-state Markov processes, and a detailed discussion is given on finite Markov chains before discussing the properties of Markov chains with infinitely many states. Because of their simplicity and applicability, more theoretical discussion is allotted to Markov processes. Nevertheless, some simple concepts of renewal theory and stationary processes have also been included. Special mention may also be made of the inclusion of inference problems in Markov processes which are finding increasing applications in investigations in applied probability.

In the chapters on applications, five specific areas are considered in some detail: Markov decision processes, congestion processes, reliability theory, time series analysis, and social and behavioral processes. Chapter 15 mentions various topics such as marketing, financial management, and the sport of fencing. In these chapters, the aim is to familiarize the reader with the use of stochastic models in different areas. The intent is to indicate the potential for further work rather than give an exhaustive survey. The key word is motivation. Recommended and advanced reading lists are provided for the benefit of readers who would like to pursue these problems further.

A four-part appendix gives some fundamental concepts of probability theory, distributions, sampling theory, statistical inference, matrix algebra, and other miscellaneous theorems and results which would be required by the applied scientist in working with stochastic models. It should be noted that the text of the book does not depend on all items included in the appendix, nor is any claim made for completeness of the appendix with respect to the mathematical and statistical techniques needed by the applied scientist in his work. The appendix only lists the most useful results and provides references to advanced techniques of possible benefit to the applied scientist.

This book can serve as a text for a one-semester (or a two-quarter or two-trimester) course in the following manner:

1. *An introduction to stochastic processes:* Follow the sequence of topics as in the book, and spend more time on Chapters 1–9.

2. *Applied stochastic processes:* Discuss applications from Chapters 10–15 while covering theory topics from Chapters 1–9, as follows:

Theory (Chapter)	Applications (Section)
1, 2	Discuss general problems from Chapters 10–15
3	10.2–10.4
4, 5	10.5; 11.10; 14.2–14.4; 15.2–15.4

6	11.7; 11.9; 11.10
7	11.2–11.6; 11.8; 12.3–12.5; 14.5; 14.6; 15.5
8	11.10; 12.2
9	13.2–13.5

3. *Stochastic systems:* Focus attention on the analysis of stochastic systems from Chapters 10–15, and discuss theory topics from the earlier chapters as providing analysis techniques.

A concentrated effort has been made to give credit to the proper author in all cases. Nevertheless, for well-known results, techniques, and/or formulations credits may be unintentionally omitted or misplaced. Apologies are submitted in advance for any errors of this nature.

The author is indebted to the authors of the many books and articles he has read. In the course of his writing he has benefited from his discussions with Professor Richard V. Evans of the University of Illinois and Professor Richard E. Nance of Southern Methodist University. This contribution is gratefully acknowledged. Acknowledgement is also made of the contribution of the author's students, who used and tested earlier drafts of this text. Particular mention should be made of the help received from Martin J. Fischer, who read an earlier draft of the manuscript, and of the excellent typing of the manuscript by Mrs. Mary Kateley of the Computer Science/Operations Research Center at Southern Methodist University.

U. NARAYAN BHAT

Dallas, Texas
December 1971

CONTENTS

ELEMENTS OF APPLIED STOCHASTIC PROCESSES

Chapter 1

INTRODUCTION

Mathematical models can be categorized broadly as being probabilistic or deterministic. Among situations where probabilistic models are more suitable, very often a better representation is given by considering a collection or a family of random variables instead of a single one. Collections of random variables that are indexed by a parameter such as time and space are known as *stochastic processes* (or "random" or "chance" processes).

In applied statistics after the collection of empirical data, a theoretical probability distribution is fitted in order to extract more information from the data. If the fit is good, the properties of the set of data can be approximated by the properties of the distribution. In a similar way, suppose that a real-life process has been observed to have the characteristics of a stochastic process. The knowledge of the behavior of the stochastic process in question is then highly desirable in understanding the real-life situation. This is especially true when the system to deal with is complex. Here are a few examples where stochastic processes are better representations than deterministic models.

Example 1.1 A Brand-Switching Model for Consumer Behavior

Before introducing a new brand of coffee, a manufacturer wants to study consumer behavior relative to the brands already available. Suppose there are three brands on sale, say A, B, and C. The consumers either buy the same brand for a few months or change their brands every now and then. There is also a strong possibility that when a superior brand is introduced, some of the old brands will be left with only a few customers. To gauge consumer behavior, a sample survey of the stores is necessary–it would be more useful if such a survey is conducted before and after the introduction of a new brand.

In such a survey, conducted over a period of time, suppose the estimates obtained for the consumer brand-switching behavior are as follows: Out of those who buy brand A in one month, during the next month 60% buy A

1

again, 30% switch to brand B, and 10% switch to brand C. For brands B and C these figures are, B to A 50%, B to B 30%, B to C 20%, and C to A 40%, C to B 40%, C to C 20%.

If we are interested in the number of people who buy a certain brand of coffee, then that number could be represented as a stochastic process. The behavior of the consumer can also be considered as a stochastic process that can enter three different states A, B, or C. Some of the questions that arise are: What is the expected number of months that a consumer stays with one specific brand? What are the mean and variance of the number using a particular brand after a certain number of months? Which is the product preferred most by the customers in the long run?

Example 1.2 An Insurance Risk Problem

An insurance company starts with a fixed capital x_0. The state of the company is considered after every period, say a month. During the nth period suppose its income is I_n and total claims are C_n. The state of the company after n such periods is given by

$$X_n = x_0 + (I_1 - C_1) + (I_2 - C_2) + \cdots + (I_n - C_n)$$

Clearly, whenever $X_n \leq 0$, the company is said to be ruined.

This is a simplified version of the following general problem: let $\{ X(t), t \geq 0 \}$ be the state of the company at time t. Some of the uncertainties involved in this process are the arrival and magnitude of claims and income from policyholders and other sources. All these can be modeled in terms of probability distributions, making $X(t)$ a stochastic process with time parameter t. With simplifying assumptions we can use some standard stochastic processes to study its behavior. Some questions that could be asked in this problem are: What are the possibilities of the insurance company getting ruined given a certain policy rate? Given a cost structure and profit level, what should be the premium rates for the effective operation of the company?

Example 1.3 A Dam Problem

The problem of a dam used for storing water has some characteristics similar to the insurance risk problem. Water flows into the dam from external sources, and it is stored in the dam for use at proper times. The input and output are subject to uncertainties, and hence the content of the dam could very well be described as a random variable indexed by time. The dam has a maximum capacity beyond which water will overflow and will either be wasted or added to the output of a subsequent reservoir. Also there is no guarantee that the dam will not dry up. From the point of view of the behavior of the process, the periods to first emptiness, first overflow, interoverflow time, the content at any time, and so forth, are interesting problems. From the operational point of view the problems of interest are the capacity of the dam and the control of the input and output rate such that some optimality conditions are satisfied.

Example 1.4 An Inventory Problem

A manufacturing firm produces a single product (for simplicity) for which the demand in a single time period may be considered to be a random variable with a standard distribution. The firm is vitally interested in deciding the amount of the product to be manufactured in a time period such that its profit is maximized under a certain cost structure. In view of this the stochastic process of interest is the amount of inventory on hand at the end of the time period. Short-term as well as long-term behavior of the inventory process need to be studied in this problem. Several variations and generalizations of problems of this nature are studied in inventory theory.

Example 1.5 A Queueing Problem

A bus taking students back and forth between the dormitory complex and the student union arrives at the student union several times during the day. The bus has a capacity to seat K persons. If there are K or less waiting persons when the bus arrives, it departs with the available number. If there are more then K waiting, the driver allows the first K to board, and the remaining persons must wait for the next bus. The university administrators would like to know at each time period during the day how many students are left behind when a bus departs.

The number waiting at the bus stop is a stochastic process dependent on the arrival process (e.g., with some probability distribution). Some desirable characteristics of the system are reduction in the number waiting at any time, minimizing the time a student has to wait for a ride, and minimum operational cost.

Example 1.6 Population Growth Problems

It is more realistic to consider the growth of a population as stochastic rather than deterministic. The external factors that influence the growth of animals, such as weather conditions, disease, and availability of food, are too varied and uncertain for deterministic models. When these factors are identified and accounted for, the population size at any time can be considered a stochastic process. In problems of this nature we would not only be interested in the behavior of the process but also in using such information in the control of the growth or decline of the population. (Lack of such action could result in the species becoming extinct.)

Example 1.7   Recovery, Relapse, and Death Due to a Disease

The process of recovery, relapse, and death in the case of some major diseases such as cancer is governed by several random causes, and therefore stochastic process models have been found useful in the study of hospital data related to such cases. For instance, four different states of the patient can be identified: (1) the initial state of being under treatment, (2) the state of being dead immediately following treatment, (3) the state of recovery, and (4) the state of being lost after recovery (not being able to trace the patient). From a model like this, problems related to the effectiveness of treatment can be studied.

Example 1.8   Acceptance Sampling

In a quality control problem samples are taken from lots of manufactured items to make sure that the quality is maintained in the accepted lots. One of the sampling methods used is a sequential scheme where items are selected one at a time. At every step a decision is made to examine one more item on the basis of information available from the sampled items. The number of defectives in the sampled items can be considered as a stochastic process, and its behavioral properties can be used to determine optimum stopping rules, actual outgoing quality, and other operating characteristics of the system.

Example 1.9   A Time-Sharing Computer System

Jobs of varied length come to a computing center from various sources. The number of jobs arriving, as well as their length, can be said to follow certain distributions. Under these conditions the number of jobs waiting at any time and the time a job has to spend in the system can be represented by stochastic processes. Under a strictly "first-come, first-served" policy there is a good chance of a long job delaying a much more important shorter job over a long period of time. For the efficient operation of the system, in addition to minimizing the number of jobs waiting and the total delay, it is necessary to adopt a different service policy. A round-robin policy in which the service is performed on a single job only for a certain length of time, say 3 or 5 sec and those jobs that need more service are put back in the queue, is one of the common practices adopted under these conditions.

Example 1.10   A Data Storage and Retrieval System

As computers get faster and more efficient, the problems arising from job execution will not be as significant as problems arising from job handling. A major portion of the job-handling problem pertains to the retrieval of necessary data from storage units. A simplified system can be described as follows: suppose there are two storage units, one for faster jobs and the other for slower ones. Attached to each is a computer that executes the job. Each unit has access to the other so that a job is not turned away for lack of material. For the sake of simplicity assume that all congestion is due to job handling, not to job execution. Connected with each unit, therefore, there are two waiting lines of jobs: one that requests data from the same unit and one that wishes to have access to the second storage unit. The arrivals to the waiting lines are at least of three types: requests coming from the second unit, requests generated by the execution of the computer units, and requests of jobs new to the system. Due to these characteristics an efficient operation of the system should take into account the dependence between the two units (this can also be generalized to several units). The processes involved are mainly stochastic processes which are dependent in nature.

Example 1.11   A Congestion Problem at a Road Intersection

Consider an intersection where a minor road and a major road intersect. Vehicles coming from the minor road should give way to the vehicles on the major road. Therefore associated with each vehicle on the minor road is a delay period whose length is dependent on the gaps created by the vehicles on the major road. The arrival of vehicles on both roads is better described by stochastic processes, and therefore the delay is also a stochastic process. An understanding of the nature of this process is essential if a traffic engineer wants to look into the problem of installing signals at the intersection. This is only a simplified model of a congestion problem in traffic theory. More realistic models should include complex characteristics such as road intersections in series, arrival of vehicles not independent of each other, effects of merging, and lane changing.

Example 1.12   A Reliability Problem

The reliability of a system (or a component) is defined as the probability that it performs its assigned mission. If the mission is to accomplish a task with no reference to time, we can consider it as the probability of success at a time epoch. However, when the success has to be measured over a period of time, with several parts to the system, one has to consider the life distributions of different parts and the system structure in determining its reliability.

Consider a traffic light system with red, green, and amber lights mounted at a north–south–east–west intersection. The system is run by a time clock. Assuming that the electrical connections are perfect, there are 12 bulbs and the time clock to be considered. Life distributions of all these items are now relevant. Furthermore the light system is such that north–south and east–west lights are separately synchronized. The mission of the light system can be defined in several ways, for instance, the entire system is trouble free, or a specified number of light bulbs are out, but the system is not hazardous. Depending on the mission, suppose one is interested in determining the probability of accomplishing that mission during a specified length of time. Because of the probabilistic nature of the life lengths of the components, the underlying process is a stochastic process, and the reliability of the system can be determined from the properties of this process.

Example 1.13   A Problem in Genetics

The phenomenon of brother-sister mating has received considerable attention in genetics. Initially two individuals are mated, and in successive generations mating occurs between two individuals of the opposite sex among their direct descendents. Identifying three genotypes AA, Aa, and aa for each parent, parents can be identified as belonging to the following six possible distinct combinations. AA × AA, AA × Aa, AA × aa, Aa × Aa, Aa × aa, aa × aa. Assuming that genes are transmitted in a random fashion, the parental combinations can be considered to be a stochastic process that can enter any one of the six states. Some of the relevant questions one may be interested in the study of this process are: What fraction of the parents at the nth generation can one expect to belong to each of the parental combinations? In the long run how many generations would be needed for the parental combinations to become purely homozygotic (e.g., AA × AA and aa × aa)? It can be shown that in a set up like this ultimately individuals do end up being homozygotes.

Example 1.14   The Gambler's Ruin Problem

A gambler wins or loses a dollar (or any other unit of currency) based on the outcome of a game. Let p be the probability of a win and $q(=1-p)$ be the probability of loss. Let \$$a$ be the gambler's initial capital and \$$(m-a)$ be the capital of an adversary at that time. Thus they play with a total capital of \$$m$, and the play stops when one of them acquires all the capital they have between them. When the gambler's capital reduces to zero, this individual is said to be ruined.

This is one of the classical problems in random walk in which a particle moves along a path back and forth, up or down, and so forth, in a random fashion, and the underlying process is stochastic. In problems such as these one would be interested in the probability of the gambler's ruin and the probability distribution of the duration of the game. Several variations of this problem can be obtained be changing the nature and location of the barriers 0 and m, by considering the walk in more than two dimensions and by making the time parameter continuous. Even if we retain the random walk terminology, these models apply to many physical phenomena such as generalized sampling schemes and diffusion processes.

The systems described here are only a few examples where stochastic processes are more realistic as mathematical models. A study of such systems has three essential aspects. These can be referred to as behavioral, statistical, and operational. In order to obtain the maximum benefit, the study is complete only if all three aspects of the system are investigated. A behavioral study is aimed at understanding the general as well as specific details of the behavior of the system under real-life situations. This can be carried out in several ways. One way is to build an exact model and deduce the behavior of the system from the model. A miniature replica of an intended irrigation system, the architect's model of a building, and simulation models of complex systems are examples of such exact models. The information that can be derived from these models is limited by the complexities of the system and difficulties in their modification.

Another method of study is to build an idealized model for the complex situation and deduce some of the general properties from this model. This is where the importance of an analytical model lies. An idealized model helps us to weed out the unnecessary details and to look into only the essential features of the system. Once the general properties are known, the specifics can be deduced by other means. An analytical model has maneuverability and scope limited only by the techniques available at the time. (For instance, difference differential equations for the behavior of some queueing systems were not solved when they were derived during 1920–1940. They were solved only during the period 1950–1960.)

Even if the behavior of a wide variety of stochastic processes is known, selection of the right model can only be made by statistical means. Needed data have to be collected and analyzed, and the essential characteristics of the system have to be estimated. Hypotheses made about the features of the system need to be tested through reliable means.

Identification of the correct model does not complete the analysis. For the study to be useful, the results of the analysis should be used for improving the system operation. One way of doing this is to manipulate the exact or the idealized model until we arrive at the desirable features of the model. But a more efficient way would be to use the behavioral results derived from the analytical model in yet another analytical model for the operational aspects of the system. A simple example would be that of minimizing a cost equation for the system.

In the preceding paragraphs we have tried to draw the attention of the reader to the use of stochastic processes as models for real-life situations and to the mode of their analyses. But it should be clearly understood that it is neither the intention nor within the scope of this book to discuss all these aspects. Also we do not intend to give all the necessary techniques to analyze the systems discussed. Our intention is to demonstrate how simple some processes are and how useful they can be when handled with the right spirit. We visualize this look only as an elementary introduction to the theory of stochastic processes with due emphasis on their applications.

FOR FURTHER READING

Bartholomew, D. J. (1973). *Stochastic Models for Social Processes*. 2nd ed. New York: Wiley, Chap. 1.

Bartlett, M. S. (1975). *Probability, Statistics and Time*. New York: Chapman and Hall.

Cramer, H. (1961). "Model Building with the Aid of Stochastic Processes." *Bull. Inter. Stat. Inst.* **39**(2), 1–30. Reprinted in *Technometrics* **6**(2), 133–159.

Kendall, M. G., ed. (1970). *Mathematical Model Building in Economics and Industry*. New York: Hafner.

Markov, A. A. (1981). *The Correspondence between A. A. Markov and A. A. Chuprov on the Theory of Probability and Mathematical Statistics* (Ed: Kh. O. Ondar, Eng. Trans.: Charles and Margaret Stein). New York: Springer-Verlag.

Marshak, J. (1954). Probability in the Social Sciences. *Mathematical Thinking in the Social Sciences*. Glencoe, Ill.: The Free Press, pp. 166–215.

Chapter 2
STOCHASTIC PROCESSES: DESCRIPTION AND DEFINITION

2.1 DESCRIPTION

In this section we shall give an informal description of a stochastic process with the help of examples. Other expressions used as synonyms for "stochastic process" in the literature are "chance process" and "random process."

In the brand-switching model for consumer behavior (Example 1.1) the consumer preference at any time t would be any one of the three brands of coffee. This set of preferences $\{A, B, C\}$ is the sample space for a random experiment performed in time: the observation of consumer preferences. Associating numerical values to the elements of this space, we may define a family of random variables $\{X(t), t \geq 0\}$ indexed by the time parameter t. The process $X(t)$ is called a *stochastic process*. The values assumed by the process are called the *states*, and the set of possible values is called the *state space*. Clearly in this experiment the state space is discrete (comprised of a countable set of points). In examples such as dam models (Example 1.3) a stochastic process with a continuous state space would be a better representation.

There are examples of describing a stochastic process with a nonnumerical state space. The brand-switching model is one such case. This practice is just a convenience, and even then an unambiguous assignment of numerical values is always possible for the purpose of rigor.

The set of possible values of the indexing parameter is called the *parameter space*, which can be either continuous or discrete. For the sake of convenience, when the indexing parameter is discrete, we shall denote it by n and represent the process as $\{X_n, n = 0, 1, 2, \ldots\}$. A word about the parameter space may also be added. Even though in most physical problems time is a natural index parameter, other kinds of parameters such as space may also arise. For instance, the number of defects on a sheet of metal can be considered a stochastic process with area as the index parameter, or the number of fibers at a point on yarn can be considered a stochastic process with the length of the yarn as the parameter. But since time is the parameter used in the majority of problems, in our discussion we shall use the expression "time parameter" in a generic sense.

In identifying the nature of stochastic processes, as a first step it helps to classify them on the basis of the nature of their parameter space and state space:

1. Stochastic processes with discrete parameter and state spaces.

   Example 1.1 Consumer preferences observed on a monthly basis.

   Example 1.8 Number of defective items in an acceptance sampling scheme, with the number of items inspected as the indexing parameter.

2. Stochastic processes with continuous parameter space and discrete state space.

   Example 1.4 Amount of inventory on hand (when they are discrete items) over a period of time.

   Example 1.5 Number of students waiting for the bus at any time of day.

3. Stochastic processes with discrete parameter space and continuous state space.

   Example 1.4 Product is not denumerable, such as cloth, gasoline, and the inventory on hand is observed only at discrete epochs of time.

   Example 1.5 Waiting time of the nth student arriving at the bus stop.

4. Stochastic processes with continuous parameter and state spaces.

   Example 1.3 Content of a dam observed over an interval of time.

   Example 1.9 Waiting time of an arriving job until it gets into service, with the arriving time of the job now the parameter.

In arriving at appropriate stochastic processes in these examples, we had to idealize the experiments to some extent. Thus we should bear in mind that a study of the behavior of these stochastic processes can give us the behavior of the actual process only to a certain degree of approximation. When it is possible to identify the stochastic process without being able to study its behavior due to its complexity, its sample function behavior can be obtained through simulation techniques. This is done by simulating the process several times and averaging the sample characteristics so obtained. In the absence of analytical results simulation results also can be used.

For the purpose of illustration we give three realizations of a stochastic process with discrete state space and discrete parameter space. For this specific example it is also possible to arrive at the behavior of the underlying process through analytical means.

Example 2.1.1

Peter and John play a coin-tossing game. Peter agrees to pay John 1¢ whenever the coin falls "heads," and John agrees to pay Peter 1¢ whenever it is "tails." Let S_n be the amount earned by Peter in n tosses of the coin. Clearly $\{S_n, n = 0, 1, 2, \ldots\}$ is a stochastic process characterized by a discrete parameter (number of tosses) and a discrete state space $(0, \pm 1, \pm 2, \ldots)$.

Peter's earnings S_n in n tosses of the coin can be represented as the sum of n random variables $X_r, r = 1, 2, \ldots, n$, where X_r is the amount of his earnings at the rth toss. Assuming that the coin used is unbiased, the probability distribution of X_r is given by

$$
\begin{array}{ccc}
X_r & 1 & -1 \\
\text{Probability} & \tfrac{1}{2} & \tfrac{1}{2}
\end{array}
\qquad (2.1.1)
$$

The mean and variance of X_r are given, respectively, as

$$E(X_r) = \tfrac{1}{2} - \tfrac{1}{2} = 0. \qquad (2.1.2)$$

$$V(X_r) = \tfrac{1}{2} + \tfrac{1}{2} = 1 \qquad (2.1.3)$$

Further the series of tosses can be treated as a series of repeated independent trials with the same probability distribution for the outcomes. Therefore $\{X_r, r = 1, 2, \ldots\}$ forms a sequence of independent and identically distributed random variables. Then, for $S_n = X_1 + X_2 + \cdots + X_n$, we get

$$E(S_n) = \sum_{r=1}^{n} E(X_r) = 0 \qquad (2.1.4)$$

$$V(S_n) = \sum_{r=1}^{n} V(X_r) = n \qquad (2.1.5)$$

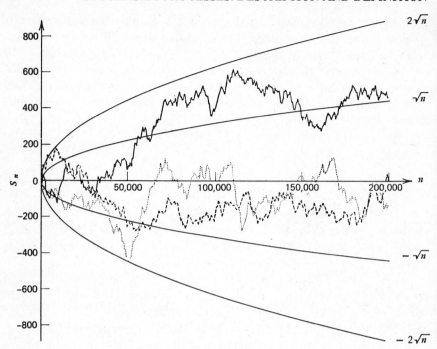

Figure 2.1.1   Typical realizations of the stochastic process $\{S_n, n = 0, 1, 2, \ldots\}$ (Wold, 1965, p. 8).

Three typical realizations (sample functions) of the stochastic process $\{S_n, n = 0, 1, 2, \ldots\}$ are given in Figure 2.1.1.* It may be noted from the figure, that even though the mean $E(S_n) = 0$ for $n = 1, 2, \ldots$, the actual path lies mostly on one side of the mean, and the side depends on the results of the first few tosses.

2.2   PROBABILITY DISTRIBUTIONS

For a given value of the time parameter t the stochastic process $X(t)$ is a simple random variable, and its probability distribution can be obtained as for any other random variable. However, when t varies in a space T, the information on the process $X(t)$ is not provided by a simple distribution for a given t. For the complete information on the process we need the joint distribution of the basic random variables of the family $\{X(t), t \in T\}$.

---

*Figure 2.1.1 is one of several such graphs that can be found in Wold (1965), which is highly recommended for preliminary reading.

When t is continuous, obtaining such a joint distribution is an impossibility, as the membership of the family is infinite in number. Under these circumstances it seems legitimate to assume that the behavior of the process can be obtained by studying it at discrete sets of points, and a joint distribution function defined at these points seems appropriate. Let $(t_1, t_2, \ldots, t_n)$, with $t_1 < t_2 < t_3 < \cdots < t_n$, be such a discrete set of points within T. The joint distribution for the process $X(t)$ at these points can now be defined as

$$P[X(t_1) \le x_1, X(t_2) \le x_2, \ldots, X(t_n) \le x_n] \qquad (2.2.1)$$

This distribution has the simplest form when the random variables are independent. Then it is given by the product of individual distributions, and therefore the study of the stochastic process is reduced to the study of a single random variable. In most practical situations, however, this is not the case, and we are faced with the problem of assuming that some sort of dependence exists among these random variables.

Even though we need joint distributions of the type defined in (2.2.1) for the complete description of the process, most of the information needed in practice can be derived through transition distribution functions. These are conditional probability distribution functions based on some information of the stochastic process available for a specific value of the time parameter.

Let t_0 and t_1 be two points in T such that $t_0 \le t_1$. Then we may define the conditional transition distribution function as

$$F(x_0, x_1; t_0, t_1) = P[X(t_1) \le x_1 | X(t_0) = x_0] \qquad (2.2.2)$$

When the stochastic process has discrete parameter and state spaces, we may define the transition probabilities as

$$P_{ij}^{(m, n)} = P(X_n = j | X_m = i) \qquad n \ge m \qquad (2.2.3)$$

A stochastic process $\{X(t), t \in T\}$ is said to be *time-homogeneous* ("parameter-homogeneous," to be exact) if the transition distribution function given by (2.2.2) depends only on the difference $t_1 - t_0$ instead of t_0 and t_1. Then we have

$$F(x_0, x; t_0, t_0 + t) = F(x_0, x; 0, t) \qquad (2.2.4)$$

for $t_0 \in T$. For convenience we can write (2.2.4) as $F(x_0, x; t)$. The corresponding expression for the process $\{X_n, n = 0, 1, 2, \ldots\}$ would then be $P_{ij}^{(n)}$.

Let $F(x, t)$ be the unconditional distribution of the process $X(t)$ defined as

$$F(x, t) = P[X(t) \le x]$$

Also let $f(x_0)$ be the probability density of $X(0)$. Now $F(x, t)$ can be determined from $F(x_0, x; t)$ from the simple relation

$$F(x, t) = \int_{x_0 \in S.} F(x_0, x; t) f(x_o) \, dx_0 \qquad (2.2.5)$$

where S is used to represent the state space. In the discrete case the corresponding relation takes the form

$$p_j^{(n)} = \sum_{i \in S} p_i^{(0)} P_{ij}^{(n)} \qquad (2.2.6)$$

where

$$p_j^{(n)} = P[X_n = j]$$

2.3 MARKOV PROCESSES AND RENEWAL PROCESSES

Stochastic processes occurring in most real-life situations are such that for a discrete set of parameters $t_1, t_2, \ldots, t_n \in T$, the random variables $X(t_1), X(t_2), \ldots, X(t_n)$ exhibit some sort of dependence. For instance, in Example 1.2 the state of the insurance company after n periods (X_n) may conceivably depend on the state of the company during the preceding $n - 1$ periods. In Example 2.1.1. Peter's accumulated earnings after n coin tosses depend on his earnings at the end of $n - 1$ coin tosses. Clearly the analysis of the process gets complicated as the dependence structure becomes complex. The simplest type of dependence is the first-order dependence underlying the stochastic process in Example 2.1.1. This is called *Markov-dependence*, which may be defined as follows: Consider a finite (or countably infinite) set of points $(t_0, t_1, \ldots, t_n, t)$, $t_0 < t_1 < t_2 < \cdots t_n < t$ and $t, t_r \in T (r = 0, 1, \ldots, n)$ where T is the parameter space of the process $\{X(t)\}$. The dependence exhibited by the process $\{X(t), t \in T\}$ is called *Markov-dependence* if the conditional distribution of $X(t)$ for given values of $X(t_1), X(t_2), \ldots, X(t_n)$ depends only on $X(t_n)$ which is the most recent known value of the process, that is, if

$$P[X(t) \le x | X(t_n) = x_n, X(t_{n-1}) = x_{n-1}, \ldots, X(t_0) = x_0]$$

$$= P[X(t) \le x | X(t_n) = x_n]$$

$$= F(x_n, x; t_n, t) \qquad (2.3.1)$$

The stochastic process exhibiting this property is called a *Markov process*. In a Markov process, therefore, if the state is known for any specific value of the time parameter t, that information is sufficient to predict the behavior of the process beyond that point.

As a consequence of the property given by (2.3.1), we have the following relation:

$$F(x_0, x; t_0, t) = \int_{y \in S} F(y, x; \tau, t) \, dF(x_0, y; t_0, \tau) \qquad (2.3.2)$$

where $t_0 < \tau < t$ and S is the state space of the process $\{X(t)\}$.

When the stochastic process has a discrete state space and a discrete parameter space, (2.3.1) and (2.3.2) take the following forms: for $n > n_1 > n_2 > \cdots > n_k$ and n and $n_1, \ldots, n_k$ belonging to the parameter space,

$$P\left(X_n = j | X_{n_1} = i_1, X_{n_2} = i_2, \ldots, X_{n_k} = i_k\right) = P\left(X_n = j | X_{n_1} = i_1\right)$$

$$= P_{i_1 j}^{(n_1, n)} \qquad (2.3.3)$$

Using this property, for $m < r < n$ we get

$$P_{ij}^{(m, n)} = P(X_n = j | X_m = i)$$

$$= \sum_{k \in S} P(X_n = j | X_r = k) P(X_r = k | X_m = i)$$

$$= \sum_{k \in S} P_{ik}^{(m, r)} P_{kj}^{(r, n)} \qquad (2.3.4)$$

where we have again used S as the state space of the process.

Equations (2.3.2) and (2.3.4) are called the *Chapman-Kolmogorov equations* for the process—these are basic equations in the study of Markov processes. They enable us to build convenient relationships for the transition probabilities between any points in T at which the processes exhibit the property of Markov-dependence. Later we shall see that it is possible to extract such Markov points even from processes that are seemingly non-Markovian.

Depending on the nature of the state space and the parameter space, we can divide Markov processes into four different classes, which are given here in the form of a table. Whenever the parameter and state spaces are discrete, we shall call such Markov processes Markov chains. In this book we shall be concerned mostly with Markov chains, and Markov processes with discrete state spaces. When the state space is continuous, we need more sophisticated tools for its analysis.

	State Space	
Parameter Space	Discrete	Continuous
Discrete	×	×
	(Markov chain)	
Continuous	×	×̇

In practice we shall also encounter events occurring in time. Accidents on roads, arrival of customers into a queueing system, and the occurrence of demands in an inventory system are examples of such processes. Let $N(t)$ be the number of events occurring during the time interval $(0, t)$. Considering the occurrence of an event as the outcome of an experiment, $\{N(t), t \geq 0\}$ is a family of random variables defined on the sample space of such experiments, and hence it is a stochastic process. It is called a renewal counting process if the time intervals between consecutive occurrences of events are independent and identically distributed random variables.

Let $t_0, t_1, t_2, \ldots, t_n, \ldots$ be the epochs of time at which these events occur. When $t_1 - t_0, t_2 - t_1, \ldots, t_n - t_{n-1}, \ldots$ are independent and identically distributed random variables, we have a process that renews itself at these points in time, and $N(t)$ counts the number of renewals that have taken place during that time. Therefore, if we are able to identify renewal points in a given stochastic process, the study of its behavior is greatly simplified; the bulk of the information we then need would be for the part of the process beyond the last renewal point.

2.4 SOME COMMON STOCHASTIC PROCESSES

We shall close this chapter with the discussion of some commonly occurring stochastic processes.

The Bernoulli Process

This is a process with discrete state and parameter space. Consider a series of independent repeated trials with two outcomes, say "success" and "failure." Let p and q $(p + q = 1)$ be the probabilities of success and failure, respectively, at each trial. Denote by S_n the number of successes in n trials. Clearly $\{S_n\}$ is a stochastic process with state space $\{0, 1, 2, \ldots\}$. The probability distribution of S_n for a given n is

$$P(S_n = k) = \binom{n}{k} p^k q^{n-k} \qquad k = 0, 1, 2, \ldots, n \qquad (2.4.1)$$

The process $\{S_n\}$ is called the Bernoulli process. After the occurrence of the $(k-1)$th success the number of trials W_k needed to produce the kth success (the length of the discrete time interval between consecutive successes) has a geometric distribution given by

$$P(W_k = r) = q^{r-1}p \qquad r = 1, 2, \ldots \quad \text{for all } k \qquad (2.4.2)$$

The random variables W_k are independent and identically distributed, and therefore the Bernoulli process is a simple discrete time renewal counting process. For the reasons given in Section 2.3, it is also a Markov chain with a discrete state space.

Example 2.4.1

Due to some random causes a manufacturing process does not turn out all good items. Suppose a percentage $100p$ of the items produced are defective. Let D_n be the number of defectives found in n manufactured items. Then $\{ D_n, n = 0, 1, 2, \ldots \}$ is a Bernoulli process with parameter p.

The Poisson Process

This is a process with discrete state space and continuous parameter space. Consider the events occurring under the following postulates:

1. Events occurring in nonoverlapping intervals of time are independent of each other.
2. There is a constant λ such that the probabilities of occurrence of events in a small interval of length Δt are given as follows:
   a. $P\{$Number of events occurring in

   $$(t, t + \Delta t] = 0\} = 1 - \lambda \Delta t + o(\Delta t)$$

   b. $P\{$Number of events occurring in

   $$(t, t + \Delta t] = 1\} = \lambda \Delta t + o(\Delta t)$$

   c. $P\{$Number of events occurring in

   $$(t, t + \Delta t] > 1\} = o(\Delta t)$$

   where $o(\Delta t)$ is such that $o(\Delta t)/\Delta t \to 0$ as $\Delta t \to 0$.

Under assumptions a, b, and c, λ is the mean number of events per unit time.

Let $X(t)$ be the number of events occurring in time $(0, t]$. The process $\{X(t)\}$ is stochastic with state space $\{S:0, 1, 2, \ldots\}$ and parameter space $\{t \geq 0\}$. As will be seen in Chapter 7, for a given t, $X(t)$ has the Poisson distribution with mean λt:

$$P[X(t) = k] = e^{-\lambda t}\frac{(\lambda t)^k}{k!} \qquad k = 0, 1, 2, \ldots \qquad (2.4.3)$$

The process $\{X(t)\}$ is therefore called the Poisson process.

It will be shown later that the time intervals between consecutive occurrences of events in a Poisson process are independent random variables identically distributed with the probability density function $f(x)$ given by

$$f(x) = \lambda e^{-\lambda x} \qquad x > 0 \qquad (2.4.4)$$

Therefore the Poisson process is also a renewal counting process. Due to postulates 1 and 2 the process $\{X(t), t \geq 0\}$ has the Markov property; consequently, the Poisson process is also a Markov process.

Example 2.4.2

The number of telephone calls at a switchboard can be modeled reasonably well as a Poisson process. In view of the fact that the rate of occurrence of calls varies depending on the time of day, to make the process time-homogeneous, the day must be split up into several periods of constant rate of occurrences. Within these periods the number of calls would then have the distribution given by (2.4.3) with suitable values for λ.

The Gaussian Process

This is a process with continuous state and parameter spaces. Consider a stochastic process $\{X(t)\}$ with the property that for an arbitrary set of n time points $\{t_1, t_2, \ldots, t_n\}$ the joint distribution of $X(t_r)$ $(r = 1, 2, \ldots, n)$ is n-variate normal. Then the process is called Gaussian. If, in addition, for any finite set of points $\{t_r\}$ $(r = 1, 2, \ldots)$ the random variables $X(t_r)$ are mutually independent and $X(t)$ has a normal distribution for all t, then it is called a *purely random* Gaussian process.

Example 2.4.3

Most common examples of Gaussian processes occur in communications theory. The autocorrelation function of a stochastic process with lag h is the correlation of process values separated by h units in the time parameter.

The Fourier cosine transform of the autocorrelation function is known as the spectral density function for the process. In communications theory signals made up of the superposition of a very large number of independent random effects of brief duration (e.g., the impact of a stream of electrons on the anode of a vacuum tube) can be modeled as a stochastic process with a constant spectral density function which is also called a "white noise process." Suppose a white noise process is subjected to a linear transformation (identified in engineering as a linear filter). Then in the limit the output will have the property of a Gaussian process.

The Wiener Process (Brownian Motion Process)

This is a process with continuous state and parameter spaces. Consider a stochastic process $\{X(t)\}$ with the following properties:

1. The process $\{X(t), t \geq 0\}$ has stationary independent increments. This means that for $t_1, t_2 \in T$ (parameter space) and $t_1 < t_2$, the distribution of $X(t_2) - X(t_1)$ is the same as $X(t_2 + h) - X(t_1 + h)$ for any $h > 0$, and for any nonoverlapping time intervals (t_1, t_2) and (t_3, t_4), with $t_1 < t_2 < t_3 < t_4$, the random variables $X(t_2) - X(t_1)$ and $X(t_4) - X(t_3)$ are independent.

2. For any given time interval (t_1, t_2), $X(t_2) - X(t_1)$ is normally distributed with mean zero and variance $\sigma^2(t_2 - t_1)$.

Then the process $\{X(t)\}$ is called the Wiener process.

If the mean of $X(t_2) - X(t_1)$ is $\mu(t_2 - t_1) \neq 0$, then we have a Wiener process with drift μ.

Example 2.4.4

The displacement of a particle suspended in a fluid and moving under rapid, successive random impacts of neighboring particles is the classical example of a Wiener process. The physical phenomenon of this process was first noticed by the botanist Robert Brown in 1827, and hence the process is also called the Brownian motion process. The theory of the behavior of this process is due to Einstein (1956) and Wiener (1923).

REFERENCES

Einstein, A. (1956). *Investigations on the Theory of the Brownian Movement.* New York: Dover (trans. of Einstein's 1905 papers).

Wiener, N. (1923). "Differential Space." *J. Math. Phys.* **2**, 131–174.

Wold, H. O. A. (1965). "A Graphic Introduction to Stochastic Processes." *International Statistical Institute Bibliography of Time Series and Stochastic Processes.* Cambridge, Mass.: The MIT Press.

FOR FURTHER READING

Cox, D. R., and Miller, H. D. (1965). *The Theory of Stochastic Processes.* New York: Wiley, Chap. 1.

Haight, F. A. (1981). *Applied Probability,* New York: Plenum Press.

Parzen E. (1962). *Stochastic Processes.* San Francisco: Holden-Day, Chap. 1.

Ross, S. M. (1980). *Introduction to Probability Models.* 2nd ed. New York: Academic Press.

Thompson, W. A., Jr. (1969). *Applied Probability.* New York: Holt, Rinehart and Winston.

MODELING EXERCISES

1. Describe the state and parameter spaces of the stochastic process representing consumer behavior in Example 1.1, and discuss the suitability of a Markov process model for it. How can one get the distribution of the number of people buying a certain brand of coffee using the model for consumer behavior?

2. Describe the state and parameter spaces of the stochastic processes representing the state of the insurance company in both the simplified and the general version of Example 1.2. Using the definition of Markov processes given in this chapter, determine the conditions in general terms under which a Markov process model can be used for the state of the company.

3. We are interested in representing the amount of inventory on hand by a stochastic process model. Identify the elements that need characterization for this model.

4. In the sequential sampling scheme of Example 1.8, let D_n be the number of defective items when the sample size is n. Let p be the probability that an item is defective and $q = 1 - p$. What is the distribution of D_n when n is known? Give a probability relationship between D_n and D_{n+1}.

5. Suppose we are interested in representing the number of jobs waiting at time t at a computer center (Example 1.9) as a stochastic process. Describe the appropriate state space and parameter space for the

process. Also identify the elements that need characterization in order to develop this model.

6. In a road intersection without traffic lights, a minor road vehicle can move across the major road only if a gap of acceptable length is available in the major road traffic. Let T be the minimum acceptable length of the gap, and let $f(x)$ be the probability density function of the gap length. For convenience, assume that the gaps include the length of vehicles as well. If a minor road vehicle has to wait until the $(n + 1)$st gap to cross the major road, give the conditional probability density of this waiting time. Hence determine the unconditional distribution of the waiting time in terms of the probability density of the gap length.

   Hint First obtain the probability density of the unacceptable gap length.

7. The size of the population in a region changes due to births, deaths, immigration, and emigration. Suppose we are interested in setting up a stochastic process model for the population size. Give a set of elements whose characteristics should be defined in order to set up the model.

8. Let $f(x)$ be the probability density of the length of life of a component. Assume that N components have been put in operation at time 0, and let $X(t)$ be the number of surviving components at time t. Assuming that they operate independently of each other, determine the distribution of $X(t)$. What is the state space for the process $X(t)$?

   Hint The cumulative distribution $F(x)$ of the lifetime density $f(x)$ gives the probability that the component has failed at or before x.

9. The life time of a light bulb has a probability density $f(x)$. Assume that a bulb is replaced as soon as it burns out and that a new bulb was placed in operation at time 0. What is the distribution of time until the bulb is replaced n times? What is the probability that exactly n replacements have been made during a period of length T? Specialize this problem where $f(x) = \lambda e^{-\lambda x}$ $(x > 0)$.

10. Suppose in Exercise 9 the bulb replacement takes an amount of time after burnout that has a probability density $g(x)$. Determine the distribution of time until the nth replacement with this modification, again assuming that a new bulb was placed in operation at time 0. What is the probability that exactly n replacements have been made during a period of length T?

11. a. Generalize the Bernoulli process of Section 2.4 to the hypergeometric case as follows: N_1 items in a population of size N have been identified as "successes" and the remaining $N - N_1$ as "failures." Let $S_n^{(N, N_1)}$ be the number of successes in a random sample of size n from the population, selected without replacement. Give the probability distribution of $S_n^{(N, N_1)}$.

b. Suppose the kth success was chosen at the nth selection. Let W_{k+1} be the number of selections required for the next success. Show that the probability distribution of W_{k+1} is given by

$$P(W_{k+1} = r | k\text{th success at the } n\text{th selection})$$

$$= \frac{\binom{N - n - r}{N_1 - k - 1}}{\binom{N - n}{N_1 - k}}$$

12. a. In a Bernoulli process determine the probability

$$P(S_{n+1} = j | S_n = i) \qquad j = i - 1, i, i + 1, \ldots$$

b. In a hypergeometric process, as defined in Exercise 11a, determine the probability

$$P\left(S_{n+1}^{(N, N_1)} = j | S_n^{(N, N_1)} = i\right) \qquad j = \cdots i - 1, i, i + 1, \ldots$$

ELEMENTARY REVIEW EXERCISES

1. A sample space S consists of n mutually exclusive and collectively exhaustive events $A_1, A_2, \ldots, A_n$. An event E is defined on this space. Then show that

a. $P(E) = \sum_{r=1}^{n} P(A_r) P(E | A_r)$

b. $P(A_i | E) = \dfrac{P(A_i) P(E | A_i)}{\sum_{r=1}^{n} P(A_r) P(E | A_r)}$

2. Coefficient of variation is defined as the ratio of standard deviation to the mean. Determine the coefficient of variation for the following distributions:

a. Bernoulli: $p_0 = 1 - p$; $p_1 = p$.

b. Binomial: $p_k = \binom{n}{k} p^k (1 - p)^{n-k}$, $k = 0, 1, 2, \ldots, n$.

c. Geometric: $p_n = p(1 - p)^{n-1}$, $n = 1, 2, \ldots$.

d. Negative binomial: $p_n = \binom{n + k - 1}{k} p^n (1 - p)^k$, $n = 1, 2, 3 \ldots$.

e. Poisson: $p_k = e^{-\lambda} \dfrac{\lambda^k}{k!}$, $k = 0, 1, 2, \ldots$.

f. Exponential: $f(x) = \lambda e^{-\lambda x}$, $x > 0$.

g. Gamma: $f(x) = e^{-\lambda x} \dfrac{\lambda^n x^{n-1}}{\lceil n}$, $x > 0$.

3. Let $b(k; n, p)$ be the probability mass function of a binomial distribution (with n as the number of trials and p as the probability of success) given by

$$b(k; n, p) = \binom{n}{k} p^k (1 - p)^{n-k} \qquad k = 0, 1, , \ldots, n$$

Show that

$$b(k; n, p) = \left(\frac{n}{n - k} \right) \left(\frac{p}{1 - p} \right) b(k - 1; n, p) \qquad k = 1, 2, \ldots, n$$

4. Let $p(k, \lambda)$ be the probability mass function of a Poisson distribution given by

$$p(k, \lambda) = e^{-\lambda} \frac{\lambda^k}{k!} \qquad k = 0, 1, 2, \ldots$$

Show that

$$p(k, \lambda) = \frac{\lambda}{k} p(k - 1, \lambda) \qquad k = 1, 2, \ldots$$

5. Derive the Poisson distribution as a limit to the binomial.

   Hint Write $\lambda = np$, and let $n \to \infty$ and $p \to 0$, keeping λ constant.

6. A discrete random variable X has a geometric distribution

$$g(k) = p(1 - p)^{k-1} \qquad k = 1, 2, \ldots$$

Show that

$$P(X > n) = (1 - p)^n$$

Use this probability to derive the exponential distribution as the limit of the geometric.

Hint First write $t = n\Delta$ and $p = \lambda\Delta$, and then let $n \to \infty$ and $\Delta \to 0$, keeping t constant.

7. Random variables $X_1, X_2, \ldots, X_n$ have independent Poisson distributions with means $\lambda_1, \lambda_2, \ldots, \lambda_n$, respectively. Show that the sum $S_n = X_1 + X_2 + \ldots + X_n$ has a Poisson distribution with mean $\lambda_1 + \lambda_2 + \cdots + \lambda_n$.

8. In a garage automobiles brought in for service generate two classes of jobs: short jobs such as oil changes and lubrication and long jobs such as major repairs. Service times for both job categories have been found to follow exponential distributions with means $1/\mu_s$ and $1/\mu_1$ ($\mu_s > \mu_1$). The proportion of short jobs is p, with $1 - p$ being the proportion of long jobs. What is the distribution of service time for a job selected at random? Also determine its mean, variance, and coefficient of variation. (*Note*: The resulting distribution is known as a hyperexponential distribution.)

9. Generalize Exercise 8 to include K job categories with $1/\mu_i$ as the mean service time for a job in category i and p_i as the proportion of jobs in that category ($i = 1, 2, \ldots, K$ and $\sum_{i=1}^{K} p_i = 1$).

10. Obtain the probability generating function $P(z) = \sum_k p_k z^k$ ($|z| \le 1$) for the discrete distributions identified in Exercise 2.

11. Obtain the Laplace transform

$$\psi(\theta) = \int_0^\infty e^{-\theta x} f(x)\, dx \qquad \text{Re}(\theta) > 0$$

for the continuous distributions:

a. Exponential: $f(x) = \lambda e^{-\lambda x}$, $x > 0$.
b. Uniform: $f(x) = 1/(a - b)$, $a < x < b$.
c. Deterministic:

$$F(x) = \begin{cases} 0 & x < b \\ 1 & x \ge b \end{cases}$$

12. Statistically, the reliability of a component can be defined as the probability that the unit will perform its mission. Say a system has n components arranged in series. For system success, all components must be working; the probability that a component works is p $(0 < p \le 1)$. Determine the reliability of the system.

13. Suppose that in Exercise 12 the n components are arranged in parallel. Now, for system success, at least k components must be working. Determine the reliability of the system.

14. Suppose the system has five components arranged as shown in the following figure. In the parallel structure, at least one component must be working for system success. Determine the reliability of the system with the reliabilities of the components as indicated in the figure.

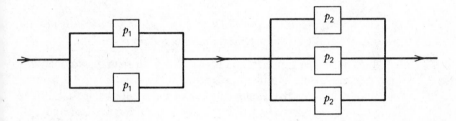

15. Determine the reliability of the system whose structure, along with the reliabilities of individual components, is given as follows:

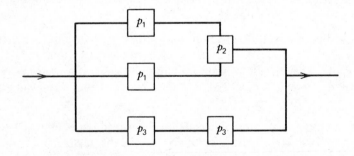

16. A triple modular redundant (TMR) system is a common element of many communication systems. In such a system there are three components, two of which must be working for system success. The input entering the system passes through the components and then a voter, which indicates whether the majority of the components are function-

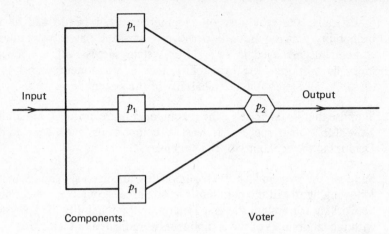

ing. The above diagram of a TMR shows both the reliabilities of the components and the voter. Determine the system reliability.

17. A retailer has three suppliers, A, B and C, for an item of merchandise. Forty percent of the supplies come from A, 55 from B, and 5 from C. Percentages of defective items vary among the suppliers, that is, 2%, 3%, and 0.5%, respectively, for the three of them. Suppose an item sold is found to be defective. What are the probabilities that it was supplied by A, by B, and by C?

18. Suppose the retailer in Exercise 17 can eliminate supplier A as being the source of the defective items. Now what are the probabilities that it was supplied by B and by C?

19. Based on national television ratings, Monday Night Football is viewed by 60% of adults. A Monday night birthday party is attended by 20 people. They decide to watch the football game if a majority of them are regular viewers. What is the probability that they will end up watching the game?

20. A sampling inspection plan calls for the inspection of 50 items out of a lot of 1000. If there are no defectives in the sample, it is accepted. Otherwise, it is rejected. If a lot of 1% defective items is submitted for inspection, find the probability of acceptance using a Poisson approximation for the probability.

21. In a garage car repairs have been found to take an average of 40 minutes, with the repair times following an exponential distribution. Say when a customer brings his car for repair, he finds the mechanic

busy on another car. The mechanic, however, promises to work on his car as soon as the other one is done. Instead of waiting for his car, the customer decides to come back after 40 min. What is the probability that on his arrival, he finds that the mechanic has not yet started work on his car?

ADVANCED REVIEW EXERCISES

22. Let X be a continuous random variable with probability density $f(x)$ $(0 < x < \infty)$ and cumulative distribution function $F(x)$. Show that

$$E(X) = \int_0^\infty [1 - F(x)]\, dx$$

Also show that if X takes values in the range $(-\infty, \infty)$, then

$$E(X) = \int_0^\infty [1 - F(x)]\, dx - \int_{-\infty}^0 F(x)\, dx$$

23. The joint probability density of two random variables X and Y is given by

$$f_{XY}(x, y) = xye^{-(x^2+y^2)/2}$$

Are X and Y independent? Are they identically distributed?

$$\text{Let } Z = \max(X, Y).$$

Show that

$$P(Z \le z) = \left(1 - e^{-z^2/2}\right)^2$$

24. Random variables X and Y have the joint probability density

$$f_{XY}(x, y) = e^{-x-y-\theta xy}\{(1 + \theta x)(1 + \theta y) - \theta\}$$

$$x, y > 0, 0 \le \theta \le 1$$

Obtain the marginal densities for X and Y. Are they independent? Show that the conditional density is

$$f_{Y|X}(y|x) = e^{-y(1+\theta x)}\{(1 + \theta x)(1 + \theta y) - \theta\}$$

and that

$$E[Y|X = x] = \frac{1 + \theta + \theta x}{(1 + \theta x)^2}$$

$$V[Y|X = x] = \frac{(1 + \theta + \theta x)^2 - 2\theta^2}{(1 + \theta x)^4}$$

25. X_1 and X_2 are independent random variables with Poisson distributions and parameters λ_1 and λ_2, respectively. When $X_1 + X_2 = n$, show that, the distribution of X_1 is given by

$$P(X_1 = k|X_1 + X_2 = n) = \binom{n}{k}\left(\frac{\lambda_1}{\lambda_1 + \lambda_2}\right)^k\left(\frac{\lambda_2}{\lambda_1 + \lambda_2}\right)^{n-k}$$

26. Consider two discrete random variables X and Y. Let $E(X|Y = y)$ be the conditional expectation of X when $Y = y$. Show that

$$E(X) = \sum_y E(X|Y = y)\, P(Y = y)$$

Note: Writing $E[X|Y]$ for the function of the random variable Y whose value at $Y = y$ is $E[X|Y = y]$, this relation is usually written in a compact form as

$$E(X) = E[E(X|Y)]$$

27. Let $S_N = X_1 + X_2 + \cdots + X_N$, where X_i, $i = 1, 2, \ldots, N$ are independent and identically distributed random variables with mean $E(X)$ and variance $V(X)$, and N is a nonnegative integer-valued random variable independent of the X's. Then, using the result from Exercise 26, show that

$$E(S_N) = E(X)E(N)$$

and

$$V(S_N) = E(N)V(X) + V(N)[E(X)]^2$$

28. $X_1, X_2, \ldots, X_n$ are independent and identically distributed random variables with the exponential distribution

$$f(x) = \lambda e^{-\lambda x} \qquad x > 0$$

Show that the probability density function of $S_n = X_1 + X_2 + \cdots + X_n$ is given by

$$g(x) = e^{-\lambda x}\frac{\lambda^n x^{n-1}}{\lceil n} \qquad x > 0$$

where $\lceil n = (n - 1)!$ Note that this is a gamma distribution. For integral values of n this distribution is identified often as an Erlangian distribution in the literature.

29. Establish the identity

$$\int_0^x e^{-\lambda y}\frac{\lambda^n y^{n-1}}{(n-1)!}\, dy = e^{-\lambda x}\sum_{r=n}^{\infty}\frac{(\lambda x)^r}{r!}$$

30. Show that the Laplace transform of

$$f(x) = e^{-k\lambda x}\frac{(k\lambda)^k x^{k-1}}{\lceil k}$$

is obtained as

$$\psi(\theta) = \int_0^{\infty} e^{-\theta x}f(x)\, dx = \left(\frac{k\lambda}{\theta + k\lambda}\right)^k$$

Hence show that, as $k \to \infty$, $f(x)$ tends to a distribution whose cumulative distribution function $F(x)$ is given by

$$F(x) = \int_0^x f(y)\, dy = \begin{cases} 0 & x < \dfrac{1}{\lambda} \\[2mm] 1 & x \geq \dfrac{1}{\lambda} \end{cases}$$

31. The probability of success in repeated independent trials is p ($q = 1 - p$). Show that the probability that n trials will be needed to obtain k successes is given by

$$f(k; n, p) = \binom{n-1}{k-1}p^k q^{n-k} \qquad n \geq k$$

Using the definition

$$\binom{-n}{x} = \frac{(-n)(-n-1)\ldots(-n-x+1)}{x!}$$

show that $f(k; n, p)$ can be expressed in the negative binomial form

$$f(k; n, p) = \binom{-k}{n-k} p^k (-q)^{n-k}$$

Hint   Note that the kth success should occur at the nth trial.

32.   a.   Random variables X_i, $i = 1, 2, \ldots$ are independent and identically distributed with cumulative distribution function F. Consider a random sum of N random variables $X_1, X_2, \ldots, X_N$, and let $S_N = X_1 + X_2 + \cdots + X_N$. Define the probability generating function $\eta(z)$ of N and the Laplace transform $\psi(\theta)$ of F as

$$\eta(z) = \sum_{n=0}^{\infty} P(N = n) z^n \quad \text{and} \quad \psi(\theta) = \int_0^{\infty} e^{-\theta x} dF(x)$$

and show that the transform of the joint distribution of S_N and N is given by

$$\sum_{n=0}^{\infty} z^n \int_0^{\infty} e^{-\theta x} d_x P(N = n, S_N \le x)$$

Hence show that

$$E[S_N] = E(N) E(X)$$

and

$$V[S_N] = E(N) V(X) + V(N) [E(X)]^2$$

b.   In a city, the number of residential fires per week has been found to have a Poisson distribution with mean 3. Fire damage estimates average \$20,000 per property, and data indicate that damage estimates can be represented by an exponential distribution. Determine the mean and variance of weekly fire damage in the city.

33.   a.   The number of events occurring in a unit time has a Poisson distribution with parameter λ. Each event has a probability p of getting registered on a counter. Derive the distribution of events registered in a unit time.

Hint   Use a combination of Poisson and binomial distributions in an appropriate manner and simplify. Alternately, use the results of

Exercise 32a with a Bernoulli distribution for the X's. Note that from a joint probability-generating function (p.g.f.) for two distributions, the p.g.f. for a marginal distribution is obtained by setting the other argument equal to one.

b. The number of traffic tickets issued per day by a police department of a city for moving violations has a Poisson distribution with mean 55. It has been found that only 40% of such cases end up in court. What is the probability that on a typical day the number of traffic tickets issued that require court action is at least 20?

34. A hyperexponential distribution with two job classes has p_i and $1/2p_i\lambda$ as job proportion and mean service time, respectively, for the ith class ($i = 1, 2$). Determine the mean, variance, and the coefficient of variation of a randomly chosen job from the mix. Also show that the coefficient of variation is the minimum when $p_1 = p_2 = \frac{1}{2}$.
   Derive the coefficient of variation of the gamma distribution

$$f(x) = e^{-k\lambda x}\frac{(k\lambda)^k x^{k-1}}{\lceil k}$$

and show that together with the hyperexponential given earlier this gamma distribution makes up a class of distributions with the same mean and coefficient of variation ranging from 0 to ∞.

35. Mixed distributions are obtained by considering the parameter of a distribution as a random variable and attaching a distribution to it. Derive the two mixed distributions that follow.

a. *Beta-binomial:* In the binomial distribution

$$b(k; n, p) = \binom{n}{k} p^k (1 - p)^{n-k}$$

$$0 < p < 1, \; k = 0, 1, 2, \ldots, n$$

let p be a random variable with a beta distribution

$$\beta(p; a, b) = \frac{\lceil a + b}{\lceil a \lceil b} p^{a-1}(1 - p)^{b-1} \qquad a, b > 0, 0 < p < 1$$

Then

$$f(k) = \int_0^1 b(k; n, p)\beta(p; a, b)\, dp$$

$$= \frac{\lceil n + 1}{\lceil k + 1 \lceil n - k + 1} \cdot \frac{\lceil a + b}{\lceil a \lceil b} \cdot \frac{\lceil k + a \lceil n - k + b}{\lceil n + a + b}$$

Note: $\lceil a = (a - 1)!$

$$\int_0^1 x^\alpha (1 - x)^\gamma dx = \frac{\lceil \alpha + 1 \lceil \gamma + 1}{\lceil \alpha + \gamma + 2}$$

b. *Pólya distribution:* In the Poisson distribution

$$p(n, \lambda t) = e^{-\lambda t} \frac{(\lambda t)^n}{n!} \qquad \lambda > 0,\ n = 0, 1, 2, \ldots$$

let λ be a random variable with a gamma distribution. Also let

$$g(\lambda; \mu, k) = e^{-\mu \lambda} \frac{\mu^k \lambda^{k-1}}{\lceil k} \qquad \mu > 0,\ k \text{ an integer}$$

Then

$$f(n) = \int_0^\infty p(n, \lambda t) g(\lambda; \mu, k)\, d\lambda$$

$$= \binom{n + k - 1}{k - 1} \left(\frac{\mu}{t + \mu}\right) \left(\frac{t}{t + \mu}\right)^n$$

Chapter 3
MARKOV CHAINS: CLASSIFICATION OF STATES

3.1 INTRODUCTION AND EXAMPLES

In this chapter we shall give some basic definitions and properties of Markov chains (i.e., Markov processes with discrete time and parameter spaces) whose state space could be finite or countably infinite. Chapters 4 and 5 will be devoted to specific properties of finite Markov chains, and Chapter 6 will discuss Markov chains with countably infinite state spaces. The following notations are necessary for introducing examples.

Let $\{ X_n, n = 0, 1, 2, \ldots \}$ be a Markov chain with a state space $S \subseteq \mathscr{S} = \{0, 1, 2, \ldots\}$. While discussing a finite m-state chain, we shall identify the state space S to be given by the set $\{1, 2, \ldots, m\}$. Generally, we shall discuss only time-homogeneous Markov chains, and therefore the n-step transition probabilities defined in (2.2.3) will be denoted as

$$P_{ij}^{(n)} = P(X_n = j | X_0 = i) \qquad (3.1.1)$$

It should be noted that the term "n-step" refers to the time interval between observations. When $n = 1$, we shall write $P_{ij}^{(1)} = P_{ij}$. Because of the dual subscripts it is convenient to arrange these transition probabilities in a matrix form. Thus we shall write

$$\boldsymbol{P} = \| P_{ij} \| = \begin{bmatrix} P_{00} & P_{01} & P_{02} & \cdots \\ P_{10} & P_{11} & P_{12} & \cdots \\ \vdots & \vdots & \vdots & \end{bmatrix}$$

and

$$\boldsymbol{P}^{(n)} = \| P_{ij}^{(n)} \| = \begin{bmatrix} P_{00}^{(n)} & P_{01}^{(n)} & P_{02}^{(n)} & \cdots \\ P_{10}^{(n)} & P_{11}^{(n)} & P_{12}^{(n)} & \cdots \\ \vdots & \vdots & \vdots & \end{bmatrix} \qquad (3.1.2)$$

Clearly we have

$$\sum_{j \in S} P_{ij}^{(n)} = 1 \qquad n = 1, 2, \ldots$$

Consider the following examples.

Example 3.1.1

A computer center has an old computer which seems to be developing too many minor problems to go unnoticed. The center's director has instituted a study of its operation and collected data over a considerable period of time. At every minute the state of the computer has been noted down as working or not working (if it is not working at one observation point, repair is initiated immediately, which may take one or more observation periods). The analysis of such data over a long period of time has revealed the following pattern. If the computer is not working at an observation epoch, the probability that it will still be under repair at the following observation epoch (after 1 min) is 0.8. On the other hand, if it is working at an observation epoch, the probability is 0.95 that it will continue to be working at the next observation epoch. Assuming that these probabilities of changes of state remain constant each time, the director would like to get answers for the following questions:

1. Given that the computer was working at 8:00 A.M., what are the chances that it would be working after 15 min? What are the chances that it would not be working after one hour?
2. Suppose the computer is observed not working at an observation epoch. How long is the repair expected to continue?

If the state of the computer is modeled as a stochastic process X_n with states 0 (working) and 1 (not working), the transition probability matrix can be given as

$$P = \begin{bmatrix} 0.95 & 0.05 \\ 0.20 & 0.80 \end{bmatrix}$$

Example 3.1.2

A barbershop has three chairs for waiting customers and one barber in attendance. Customers who arrive when all the chairs are occupied leave without a haircut. The barber takes exactly 15 min for a haircut, and as long as there is somebody waiting, she does not take any free time. The number of customers arriving in a 15 min interval is found to have the following distribution:

Number of arrivals	0	1	2	3	4	5 and above
Probability	0.20	0.70	0.07	0.02	0.01	0.00

Also it has been found that this distribution remains the same regardless of the starting point of the observation period. (As will be seen later, this assumption implies a Poisson assumption for the arrival process.)

Suppose the barbershop is observed every 15 min; that is, let 0, σ, 2σ, $3\sigma,\ldots, n\sigma$ be the observation epochs, where $\sigma = 15$ min. Let Q_n be the number of customers in the barbershop at $n\sigma$. Possible values of Q_n are $\{0, 1, 2, 3, 4\}$, with one customer getting a haircut (if $Q_n > 0$) and $Q_n - 1$ customers waiting.

Since the length of the observation period is the same as the time needed for a haircut, the following relationship between Q_n and Q_{n+1} can be obtained:

$$Q_{n+1} = \begin{cases} \min(4, X_{n+1}) & \text{if } Q_n = 0 \\ \min(4, Q_n - 1 + X_{n+1}) & \text{if } Q_n > 0 \end{cases} \qquad (3.1.3)$$

where X_{n+1} is the number of customers arriving during $[n\sigma, (n + 1)\sigma]$.

If $Q_n = 0$, Q_{n+1} is equal to the number of customers who arrive during the 15 min period following $n\sigma$, since nobody would have finished a haircut. If $Q_n > 0$, one person would have finished the haircut during the observation period. Relation (3.1.3) shows that Q_{n+1} depends only on Q_n and X_{n+1} (whose distribution is known and external to the system), and not on any of the Q_r ($r < n$). Thus $\{Q_n, n = 0, 1, 2, \ldots\}$ is a Markov chain whose transition probability matrix is obtained as follows:

If $Q_n = 0$,

then $P(Q_{n+1} = j) = P(X_{n+1} = j)$ $\qquad j = 0, 1, 2, \ldots, 4$

giving $P_{00} = 0.20$, $P_{01} = 0.70$, $P_{02} = 0.07$, $P_{03} = 0.02$, $P_{04} = 0.01$.
If $Q_n = i > 0$,
$i = 1$

then $P(Q_{n+1} = j) = P[\min(4, X_{n+1}) = j] = P(X_{n+1} = j)$

giving $P_{10} = 0.20$, $P_{11} = 0.70$, $P_{12} = 0.07$, $P_{13} = 0.02$, $P_{14} = 0.01$
$i = 2$

then $P(Q_{n+1} = j) = P[\min(4, X_{n+1} + 1) = j]$

giving $P_{20} = 0$, $P_{21} = 0.20$, $P_{22} = 0.70$, $P_{23} = 0.07$, $P_{24} = 0.02 +$ 0.01.

$i = 3$

$$\text{then} \quad P(Q_{n+1} = j) = P[\min(4, X_{n+1} + 2) = j]$$

giving $P_{30} = 0$, $P_{31} = 0$, $P_{32} = 0.20$, $P_{33} = 0.70$, $P_{34} = 0.07 + 0.02 + 0.01$.

$i = 4$

$$\text{then} \quad P(Q_{n+1} = j) = P[\min(4, X_{n+1} + 3) = j]$$

giving $P_{40} = 0$, $P_{41} = 0$, $P_{42} = 0$, $P_{43} = 0.20$, $P_{44} = 0.70 + 0.07 + 0.02 + 0.01$.

Summarizing these results, the transition probability matrix can be given as

$$
P =
\begin{matrix}
 & \begin{matrix} 0 & \quad 1 & \quad 2 & \quad 3 & \quad 4 \end{matrix} \\
\begin{matrix} 0 \\ 1 \\ 2 \\ 3 \\ 4 \end{matrix} &
\begin{bmatrix}
0.20 & 0.70 & 0.07 & 0.02 & 0.01 \\
0.20 & 0.70 & 0.07 & 0.02 & 0.01 \\
0 & 0.20 & 0.70 & 0.07 & 0.03 \\
0 & 0 & 0.20 & 0.70 & 0.10 \\
0 & 0 & 0 & 0.20 & 0.80
\end{bmatrix}
\end{matrix}
\qquad (3.1.4)
$$

Example 3.1.3

Stores eventually must classify some charge accounts as bad debts. In this context it is helpful for a manager to have a "loss expectancy" rate to estimate the ultimate loss. Use of a Markov model for accounts is one way of getting such an estimate. A simplified model* is as follows: a store allows charge accounts only if they are less than 3 months behind payment. An account is classified as 2 months old if it has an unpaid bill (at least partially) which is 2 months old starting from the billing date. One and zero month old accounts are classified similarly. When the account becomes 3 months old, it is either paid up or reclassified as a bad debt, and other procedures are adopted for its recovery. Let A_n be the state of the account at the nth month (since the opening of the account). Possible values of A_n are P (paid up), B (bad debt), and 0, 1, and 2. Experience shows that the transitions from one state to another depend only on the present state, not

*Model based on Cyert et al. (1962).

the previous one. Therefore it is reasonable to consider $\{A_n\}$ as a Markov chain with five states. The transitions are such that when once the process enters P or B, it does not leave that state. Further the possible transitions are $(0 \to 0)$, $(0 \to 1)$, $(0 \to P)$, $(1 \to 0)$, $(1 \to 1)$, $(1 \to 2)$, $(1 \to P)$, $(2 \to 0)$, $(2 \to 1)$, $(2 \to 2)$, $(2 \to P)$, $(2 \to B)$, $(P \to P)$, and $(B \to B)$. By analyzing the charge accounts at a certain store, these transitions are found to have the probabilities given by

$$
P = \begin{array}{c} \\ 0 \\ 1 \\ 2 \\ P \\ B \end{array}
\begin{array}{ccccc}
0 & 1 & 2 & P & B \\
\left[\begin{array}{ccccc}
0.6 & 0.1 & 0 & 0.3 & 0 \\
0.2 & 0.5 & 0.1 & 0.2 & 0 \\
0.2 & 0.2 & 0.4 & 0.1 & 0.1 \\
0 & 0 & 0 & 1 & 0 \\
0 & 0 & 0 & 0 & 1
\end{array}\right]
\end{array}
\qquad (3.1.5)
$$

Example 3.1.4

Two persons A and B start a game with \$3 each as their capital. At the end of every game the loser pays \$1 to the winner. With every game the probability of A winning is 0.6 and B winning is 0.4. They quit playing when one of them either loses or wins all. Let C_n be the capital of A after n games. Then

$$C_{n+1} = C_n - 1 \qquad \text{with probability } 0.4$$

$$\phantom{C_{n+1}} = C_n + 1 \qquad \text{with probability } 0.6 \qquad (3.1.6)$$

Clearly C_{n+1} depends only on C_n; thus we have a Markov chain. Note that $C_{n+1} = C_n$ only if they have stopped playing. The possible transitions are $(0 \to 0)$, $(1 \to 0)$, $(1 \to 2)$, $(2 \to 1)$, $(2 \to 3)$, $(3 \to 2)$, $(3 \to 4)$, $(4 \to 3)$, $(4 \to 5)$, $(5 \to 4)$, $(5 \to 6)$, and $(6 \to 6)$. The probabilities of these transitions are given by

$$
P = \begin{array}{c} \\ 0 \\ 1 \\ 2 \\ 3 \\ 4 \\ 5 \\ 6 \end{array}
\begin{array}{ccccccc}
0 & 1 & 2 & 3 & 4 & 5 & 6 \\
\left[\begin{array}{ccccccc}
1 & 0 & 0 & 0 & 0 & 0 & 0 \\
0.4 & 0 & 0.6 & 0 & 0 & 0 & 0 \\
0 & 0.4 & 0 & 0.6 & 0 & 0 & 0 \\
0 & 0 & 0.4 & 0 & 0.6 & 0 & 0 \\
0 & 0 & 0 & 0.4 & 0 & 0.6 & 0 \\
0 & 0 & 0 & 0 & 0.4 & 0 & 0.6 \\
0 & 0 & 0 & 0 & 0 & 0 & 1
\end{array}\right]
\end{array}
\qquad (3.1.7)
$$

Example 3.1.5

In a modification of the gambling game of Example 3.1.4, suppose A plays against an infinitely rich adversary (e.g., a casino). Now the state space for A's capital C_n is countably infinite with $S = \{0, 1, 2, \ldots\}$. With the same probability structure as given in (3.1.6) for the outcomes of a game, we have the following matrix for the transition probabilities:

$$
P = \begin{array}{c} \\ 0 \\ 1 \\ 2 \\ 3 \\ \\ \\ \\ \end{array}
\begin{array}{cccccc}
0 & 1 & 2 & 3 & 4 & 5 & \cdot & \cdot \\
\left[\begin{array}{cccccccc}
1 & 0 & 0 & 0 & 0 & 0 & \cdot & \cdot \\
0.4 & 0 & 0.6 & 0 & 0 & 0 & \cdot & \cdot \\
0 & 0.4 & 0 & 0.6 & 0 & 0 & \cdot & \cdot \\
0 & 0 & 0.4 & 0 & 0.6 & 0 & \cdot & \cdot \\
\cdot & \cdot & \cdot & \cdot & \cdot & \cdot & \cdot & \cdot \\
\cdot & \cdot & \cdot & \cdot & \cdot & \cdot & \cdot & \cdot \\
\cdot & \cdot & \cdot & \cdot & \cdot & \cdot & \cdot & \cdot \\
\end{array}\right]
\end{array} \qquad (3.1.8)
$$

In the foregoing examples we have used Markov chain models to represent the state of the process. Sections 3.2–3.5 provide the broad framework needed for the study of such chains.

3.2   THE n-STEP TRANSITION PROBABILITY MATRIX

Let P be the transition probability matrix of a Markov chain $\{X_n, n = 0, 1, 2, \ldots\}$ as defined in (3.1.2). Let $p_j^{(n)}$ be the probability that the process is in state j after n transitions [this is the unconditional probability defined in (2.2.6)]. Denote by the row vector $p^{(n)}$, the vector of probabilities $p_j^{(n)}, j \in S$.

The n-step transition probabilities $P_{ij}^{(n)}$ and the unconditional probabilities $p_j^{(n)}, i, j \in S$ are determined by Theorem 3.2.1.

Theorem 3.2.1

$$P^{(n)} = P^n \qquad (3.2.1)$$

and

$$p^{(n)} = p^{(0)} P^n \qquad (3.2.2)$$

Proof   From the Chapman-Kolmogorov equation (2.3.4) we have

$$P_{ij}^{(r+s)} = \sum_{k \in S} P_{ik}^{(r)} P_{kj}^{(s)} \qquad \text{for given } r \text{ and } s \qquad (3.2.3)$$

Set $r = 1$, $s = 1$ in (3.2.3) so that

$$P_{ij}^{(2)} = \sum_{k \in S} P_{ik} P_{kj}$$

Clearly $P_{ij}^{(2)}$ is the (i, j)th element of the matrix product $P \cdot P = P^2$. Based on this result, assume that

$$P^{(r)} = P^r \qquad r = 1, 2, \ldots, n - 1$$

Setting $r = n - 1$, $s = 1$ in (3.2.3), we get

$$P_{ij}^{(n)} = \sum_{k \in S} P_{ik}^{(n-k)} P_{kj}$$

which again can be seen as the (i, j)th element of the matrix product $P^{n-1} \cdot P = P^n$, which proves (3.2.1). The result (3.2.2) is obtained by noting that

$$P(X_n = j) = \sum_{i \in S} P(X_n = j | X_0 = i) P(X_0 = i) \qquad \square$$

From this theorem it is clear that when the size of the state space is small (actual size would depend on the power of the computer used), the n-step transition probabilities can be easily obtained by simple matrix multiplication. For larger state spaces efficient methods for the calculation of P^n are needed. We shall discuss this problem later for special classes of Markov chains. (See Section 4.3 for the two-state Markov chain).

3.3   CLASSIFICATION OF STATES

Let us consider Examples 3.1.1 through 3.1.5 in more detail. In all examples the stochastic process model is a Markov chain. In Example 3.1.1 the state of the computer alternates between 0 and 1, and hence each state is accessible from the other.

Looking at P in Example 3.1.2, it is clear that at a given time the number of customers in the barbershop can be any one of the five values. Also, given a sufficient number of transitions, the process can reach any state from any other state (e.g., transitions from states 0, 1 to any other state are possible in one step, but a transition from state 2 to state 0 needs at least two steps). So all the states in this example can be said to belong to one class.

In Example 3.1.3 the states of the system are given by the classification of the charge account—namely paid up (P), 0 month old (0), 1 month old (1), 2 months old (2), and bad debt (B). Looking at P of equation (3.1.5), it is clear that out of the five states, states 0, 1, and 2 belong to one class and states P and B belong to another. Further once the system reaches either P or B, it cannot go to 0, 1, or 2. From this point of view perhaps the study of the chain in this example has to be different from the studies of the chain of Examples 4.1.1 and 4.1.2.

In Example 3.1.4 the states of the Markov chains are $(0, 1, 2, \ldots, 6)$, the capital of A. The game continues until either A's capital is 0 or 6, the total capital both A and B have. Clearly P of (3.1.7) has two states, 0 and 6, which belong to a different class than the rest. In addition to this, it can be seen that it takes an even number of steps to come back to the original state $(1, 2, \ldots, 5)$, if it does not visit states 0 and 6 in the meantime (e.g., suppose that the chain is in state 3 originally; after one step it is either in state 2—probability 0.4—or in state 4—probability 0.6).

Finally, in Example 3.1.5 the state space is countably infinite, and there is no state comparable to state 6 of Example 3.1.4. Nevertheless, it is not hard to visualize that however long it may take, the process has to enter state 0 eventually. Also states $\{1, 2, \ldots\}$ have the periodic character described for the states of Example 3.1.4.

The examples discussed previously illustrate the need to classify the states of a Markov chain according to certain basic properties. Such a classification can be based on the definitions that follow. The definitions are given without reference to the number of states, as they are common for Markov chains with finite as well as countably infinite number of states.

DEFINITION   *State j is said to be* accessible *from state i if j can be reached from i in a finite number of steps. If two states i and j are accessible to each other, then they are said to* communicate. *Probabilistically these definitions imply*

$$i \to j \ (\,j \ accessible \ from \ i\,) \quad if \ for \ some \quad n \geq 0, \quad P_{ij}^{(n)} > 0$$

$$j \to i \ (\,i \ accessible \ from \ j\,) \quad if \ for \ some \quad n \geq 0, \quad P_{ji}^{(n)} > 0$$

$$i \leftrightarrow j \ (\,i \ and \ j \ communicate\,) \quad if \ for \ some \quad n \geq 0, \quad P_{ij}^{(n)} > 0$$

$$and \ for \ some \quad m \geq 0, \quad P_{ji}^{(m)} > 0$$

$$(3.3.1)$$

Conversely,

$i \nrightarrow j$ (*j not accessible from i*)	*if for all* $n \geq 0$,	$P_{ij}^{(n)} = 0$
$j \nrightarrow i$ (*i not accessible from j*)	*if for all* $n \geq 0$,	$P_{ji}^{(n)} = 0$
$i \nleftrightarrow j$ (*i and j do not communicate*)	*if* $P_{ij}^{(n)} = 0$ *for all* $n \geq 0$	

$$\text{or } P_{ji}^{(n)} = 0 \text{ for all } n \geq 0$$

$$\text{or } P_{ij}^{(n)} = 0 \text{ and } P_{ji}^{(n)} = 0$$

$$\text{for all } n \geq 0$$

As a consequence of the definition we have the following properties of this communication relation:

1. *Reflexivity:* $i \leftrightarrow i$, for

$$P_{ij}^{(0)} = \delta_{ij} = \begin{cases} 1 & \text{if} \quad i = j \\ 0 & \text{if} \quad i \neq j \end{cases} \tag{3.3.2}$$

2. *Symmetry: if* $i \leftrightarrow j$, *then* $j \leftrightarrow i$.
3. *Transitivity: if* $i \leftrightarrow j$ *and* $j \leftrightarrow k$, *then* $i \leftrightarrow k$.

For there exist two integers r and s such that

$$P_{ij}^{(r)} > 0 \quad \text{and} \quad P_{jk}^{(s)} > 0$$

But we have

$$P_{ik}^{(r+s)} = \sum_{l} P_{il}^{(r)} P_{lk}^{(s)}$$

$$\geq P_{ij}^{(r)} P_{jk}^{(s)} > 0 \tag{3.3.3}$$

Therefore $i \rightarrow k$. Similarly, it can be shown that there exists an integer n such that

$$P_{ki}^{(n)} > 0$$

and hence $k \rightarrow i$. Combining these two results, we have $i \leftrightarrow k$.

Incidentally, it may be mentioned that because of the properties of reflexivity, symmetry, and transitivity exhibited by the communication relation as just illustrated, it is an *equivalence relation*. (A relation exhibiting these properties is an equivalence relation.) The set of all states of a Markov chain that communicate (with each other) can therefore be grouped into a single *equivalence class*. A Markov chain may have more than one such equivalence class. If there are more than one, then it is not possible to have communicating states in different equivalence classes. However, it is possible to have states in one class that are accessible from another class.

DEFINITION *If a Markov chain has all its states belonging to one equivalence class, it is said to be* irreducible.

Clearly, in an irreducible chain all states communicate. To illustrate this concept consider Examples 3.1.1–3.1.5:

Example 3.1.1 Irreducible chain since both states communicate.

Example 3.1.2 Irreducible chain since all states $(0, 1, 2, 3, 4)$ communicate.

Example 3.1.3 The chain has three equivalence classes $\{0, 1, 2\}$, $\{P\}$, and $\{B\}$. Classes $\{P\}$ and $\{B\}$ can be entered from $\{0, 1, 2\}$, but not vice versa. $\{P\}$ and $\{B\}$ are not even accessible in any direction.

Example 3.1.4 The chain has three equivalence classes $\{0\}$, $\{1, 2, 3, 4, 5\}$ and $\{6\}$. $\{0\}$ and $\{6\}$ are accessible from the second equivalence class, but not vice versa.

Example 3.1.5 The chain has two equivalence classes $\{0\}$ and $\{1, 2, 3, \dots\}$. $\{0\}$ is accessible from the states in the other class, but not vice versa.

The equivalence classification of states takes into account the external relationship between the states. Another closely related classification is based on the internal nature of each state.

In the three examples considered we have come across four different types of states. For instance, in Example 3.1.2, since the chain can go to any of the five states without any restriction, the probability is 1 that, starting from any state, the process will return to that state in finite time. But the states $\{0, 1, 2\}$ of Example 3.1.3 do not have this property. Clearly there is a positive probability that the process will go out of this class. Once it gets out, it may go either to $\{P\}$ or $\{B\}$, and in either case it stays in that state thereafter. Finally, the states $\{1, 2, 3, 4, 5\}$ of Example 3.1.4, in addition to exhibiting the second property of allowing the process to get out of their class with a positive probability, are such that the return to a state is possible only with an even number of steps.

The following definitions identify the states with respect to their internal nature: Let

$$f_{ii}^{(n)} = P[X_n = i, X_r \neq i(r = 1, 2, \ldots, n - 1)|X_0 = i] \qquad (3.3.4)$$

and

$$f_{ii}^* = \sum_{n=1}^{\infty} f_{ii}^{(n)} \qquad (3.3.5)$$

In other words, $f_{ii}^{(n)}$ is the probability that, starting from i, the process returns to i for the first time in n steps, and f_{ii}^* is the probability that, starting from i, the return to the initial state occurs in finite time.

DEFINITION *A state i is said to be* recurrent *if and only if, starting from state i, eventual return to this state is certain.*

In terms of the probabilities f_{ii}^* this implies that the state i is recurrent if and only if $f_{ii}^* = 1$.

When state i is recurrent, define

$$\mu_i = \sum_{n=1}^{\infty} n f_{ii}^{(n)} \qquad (3.3.6)$$

which is the mathematical expectation of the number of steps required for the first return to state i. The number of steps required for the first return to the same state is called the *recurrence time*; μ_i can then be called the *mean recurrence time* of the state i.

Using μ_i, a recurrent state can be further classified either as null recurrent or positive recurrent:

1. A recurrent state i is said to be *null recurrent* if and only if the mean recurrence time is ∞; that is, if $\mu_i = \infty$.
2. A recurrent state is said to be *positive recurrent* if and only if the mean recurrence time is finite; that is, $\mu_i < \infty$.

For a finite Markov chain, μ_i, $i \in S$ is always finite. Therefore null recurrence is possible only when the state space is countably infinite. *Note* In these definitions, limiting operations are implied, and therefore the concepts of recurrence and $\mu_i = \infty$ are not contradictory.

DEFINITION *A state i is said to be* transient *if and only if, starting from state i, there is a positive probability that the process may not eventually return to this state. This implies $f_{ii}^* < 1$.*

Another criterion for classification of states as recurrent or transient is given in terms of the probabilities $P_{ii}^{(n)}$, the probability that the process occupies state i after n steps, given that the initial state was also i (note that this return to state i is not necessarily for the first time), whereas $f_{ii}^{(n)}$ refers to the probability of the first return to i.

Theorem 3.3.1 *A state i is recurrent or transient according as*

$$\sum_{n=1}^{\infty} P_{ii}^{(n)} = \infty \quad or \quad < \infty \tag{3.3.7}$$

A rigorous proof of this theorem is beyond the scope of this book. We give here only an outline of one of the proofs of this theorem. Combinatorially we can write

$$P_{ii}^{(0)} = 1$$

$$P_{ii}^{(n)} = \sum_{r=1}^{n} f_{ii}^{(r)} P_{ii}^{(n-r)} \quad n > 0 \tag{3.3.8}$$

Let

$$P_{ii}(z) = \sum_{n=0}^{\infty} P_{ii}^{(n)} z^n$$

and

$$F_{ii}(z) = \sum_{n=1}^{\infty} f_{ii}^{(n)} z^n \quad |z| < 1$$

Multiplying both sides of (3.3.8) with appropriate powers of z and summing over all values of n, we get

$$P_{ii}(z) = 1 + F_{ii}(z) P_{ii}(z)$$

that is,

$$P_{ii}(z) = \frac{1}{1 - F_{ii}(z)} \quad |z| < 1 \tag{3.3.9}$$

Letting $z \to 1-$ on both sides of (4.3.9), we have

$$\lim_{z \to 1-} P_{ii}(z) = \sum_{n=0}^{\infty} P_{ii}^{(n)}$$

$$\lim_{z \to 1-} \frac{1}{1 - F_{ii}(z)} = \frac{1}{1 - f_{ii}^*}$$

The theorem now follows by appealing to Theorem B.3.3 of Appendix B and the definitions for recurrent and transient states.

The preceding theorem is used in obtaining the following important property.

Theorem 3.3.2   *If $i \leftrightarrow j$ and if i is recurrent, then j is also recurrent.*

Proof   Because $i \leftrightarrow j$, there exist integers n and m such that

$$P_{ij}^{(n)} > 0 \quad \text{and} \quad P_{ji}^{(m)} > 0$$

Consider

$$P_{jj}^{(n+s+m)} \geq P_{ji}^{(m)} P_{ii}^{(s)} P_{ij}^{(n)} \qquad s \geq 0 \tag{3.3.10}$$

This gives

$$\sum_{s=0}^{\infty} P_{jj}^{(n+s+m)} \geq P_{ji}^{(m)} P_{ij}^{(n)} \sum_{s=0}^{\infty} P_{ii}^{(s)} \tag{3.3.11}$$

If

$$\sum_{s=0}^{\infty} P_{ii}^{(s)} = \infty$$

then

$$\sum_{s} P_{jj}^{(n+s+m)} = \infty$$

The theorem now follows from Theorem 3.3.1 □

The importance of Theorem 3.3.2 lies in the fact that it shows recurrence as a class property. Since all states in an equivalence class communicate, they are all either recurrent or transient. Therefore we can now consider the class as a whole as being either recurrent or transient. We shall use this property in the classification of Markov chains.

More definitions are given as follows.

DEFINITION   *A state i is said to be an* absorbing state *if and only if $P_{ii} = 1$.*

Clearly when i is absorbing, $f_{ii}^{(1)} = P_{ii} = 1$, and hence $f_{ii}^* = 1$ and $\mu_i = 1$, showing that i is positive recurrent.

DEFINITION   *The* period *of a state i is defined as the greatest common divisor of all integers $n \geq 1$, for which $P_{ii}^{(n)} > 0$. When the period is 1, the state is referred to as* aperiodic.

Periodic or aperiodic states exhibit the following important property.

Theorem 3.3.3 *If state i has a period d_i (≥ 1), then there exists an integer N such that for all integers $n \geq N$, $P_{ii}^{(nd_i)} > 0$.*

Proof The proof of this theorem depends on a result in number theory (Section B.4) which states that if $n_1, n_2, \ldots, n_k$ are positive integers with the greatest common divisor d, there exists a positive integer N such that for all $n \geq N$ we can find c_i (nonnegative integers) satisfying the relation

$$nd = \sum_{i=1}^{k} c_i n_i \qquad (3.3.12)$$

Now for state i let $n_1, n_2, \ldots, n_k$ be the integers for which $P_{ii}^{(n_r)} > 0$ ($r = 1, 2, \ldots, k$), and let d_i be their greatest common divisor. Without loss of generality we assume c_i, $i = 1, 2, \ldots, k$, as positive integers. From (3.3.12) we have

$$P_{ii}^{(nd_i)} = P_{ii}^{(\Sigma_1^k c_r n_r)}$$

$$\geq \left[P_{ii}^{(n_1)} \right]^{c_1} \left[P_{ii}^{(n_2)} \right]^{c_2} \cdots \left[P_{ii}^{(n_k)} \right]^{c_k} > 0 \qquad (3.3.13)$$

Hence the theorem. □

COROLLARY   *If $P_{ji}^{(m)} > 0$, then there exists a positive integer N such that, for $n \geq N$,*

$$P_{ji}^{(m+nd_i)} > 0 \qquad (3.3.14)$$

This result follows directly from (3.3.13) *if we write*

$$P_{ji}^{(m+nd_i)} \geq P_{ji}^{(m)} P_{ii}^{(nd_i)} \qquad (3.3.15)$$

Theorem 3.3.4 *If $i \leftrightarrow j$, then i and j have the same period.*

Proof Let d_i be the period of i, and let d_j be the period of j. From the definition of the period and the property given in Theorem 3.3.3, with $d_i = s$,

$$P_{ii}^{(Ns)} > 0; \; P_{ii}^{[(N+1)s]} > 0; \ldots$$

Clearly d_i divides $(N + 1)s - Ns$, $(N + 2)s - Ns, \ldots$. Since $i \leftrightarrow j$, there exist integers n and m such that

$$P_{ji}^{(n)} > 0 \quad \text{and} \quad P_{ij}^{(m)} > 0$$

Now

$$P_{jj}^{(n+Ns+m)} \geq P_{ji}^{(n)}P_{ii}^{(Ns)}P_{ij}^{(m)} > 0 \qquad (3.3.16)$$

Similarly,

$$P_{jj}^{[n+(N+1)s+m]} > 0$$

Therefore the period of j (d_j) divides $[n + (N + 1)s + m] - (n + Ns + m) = s = d_j$. Similarly, starting with $P_{ii}^{(m+Nr+n)}$, where $d_j = r$, we can show that d_i divides d_j. This is possible only if $d_i = d_j$.   □

Theorem 3.3.4 demonstrates that periodicity is also a class property. Therefore we can speak of an equivalence class as having a certain period. A class having states of period one is called *aperiodic*. For irreducible chains with aperiodic states we have Theorem 3.3.5.

Theorem 3.3.5   *Let P be the transition probability matrix of an irreducible, aperiodic, finite Markov chain. Then there exists an N such that for all $n \geq N$, the n-step transition probability matrix P^n has no zero elements.*

Proof   Let S be the state space of the Markov chain. All states communicate with each other, and therefore for each pair of states (i, j), $i, j \in S$, there exist nonnegative integers $m(i, j)$ such that

$$P_{ij}^{[m(i,j)]} > 0$$

From (3.3.14) we know that for each pair (i, j) there exists a positive integer $N(i, j)$ such that for $n \geq N(i, j)$

$$P_{ij}^{[m(i,j)+n]} > 0$$

Now, if we take

$$N = \max_{i,j \in S} [m(i, j) + N(i, j)]$$

then clearly for all $n \geq N$,

$$P_{ij}^{(n)} > 0 \qquad \text{for all } i, j \in S \qquad □$$

DEFINITION   *For the sake of convenience we shall call a positive recurrent aperiodic state an* ergodic *state.*

In general terms the reason for calling these states ergodic can be given as follows: To determine mean value characteristics (e.g., mean and variance) of a stochastic process at any time point, one should be able to observe a sufficient number of realizations of the process at that time. (This is similar to having a random sample of sufficient size.) However, normally one is able to observe only one realization of the process over a period of time. Therefore, if one can use a time average (average of the process value over time) in place of an ensemble average (average of the process values obtained at the same time from several realizations), collecting information on stochastic processes will be much simpler. As will be discussed later in Chapter 9, some stochastic processes in fact have the property that their time averages converge to ensemble averages and such processes are considered to be ergodic. Irreducible Markov chains with positive current aperiodic states belong to the class of ergodic stochastic processes.

Going back to the examples considered earlier, we can see that the two states of Markov chain in Example 3.1.1 and all states of the Markov chain in Example 3.1.2 are recurrent and the states P and B of Example 3.1.3 are absorbing, whereas states $\{0, 1, 2\}$ are transient. In Example 3.1.4 the states 0 and 6 are absorbing, and states $\{1, 2, 3, 4, 5\}$ are transient and have a period 2. Finally in Example 3.1.5 state 0 is absorbing, and all other states are transient and have a period 2.

3.4 A CANONICAL REPRESENTATION OF THE TRANSITION PROBABILITY MATRIX

For the sake of convenience renumber the states in such a way that all those belonging to the same equivalence class have consecutive numbers. Then arrange these equivalence classes in such a way that the process can go from a given state to another in the same or in a preceding class, but not a state in a following class. This is done by placing recurrent classes before transient classes. If there are more than one transient classes, they could be arranged with the same restriction.

Let $C_1, C_2, \ldots, C_k, C_{k+1}, \ldots, C_n$ be the equivalence classes arranged in this order such that $C_1, C_2, \ldots, C_k$ are equivalence classes with recurrent states and $C_{k+1}, C_{k+2}, \ldots, C_n$ are equivalence classes with transient states. Let $P_1, P_2, \ldots, P_k$ and $Q_{k+1}, Q_{k+2}, \ldots, Q_n$ be the corresponding transition probability matrices within the equivalence classes. Also let

$$[R_s] = [R_{s1}, R_{s2}, \ldots, R_{s,s-1}]$$

be the matrix of transition probabilities of going from a state in the sth equivalence class to another state in an earlier class. With these definitions we can represent the transition probability matrix P as

$$
P = \begin{bmatrix}
P_1 & & & & & & & \\
& P_2 & & & & & \bigcirc & \\
& \bigcirc & & \ddots & & & & \\
& & & & P_k & & & \\
R_{k+1,1} & & R_{k+1,2} & \cdots & R_{k+1,k} & Q_{k+1} & & \\
\vdots & & \vdots & & \vdots & & \ddots & \\
R_{n1} & & R_{n2} & \cdots & R_{nk} & R_{n,k+1} & \cdots & Q_n
\end{bmatrix}
$$

$$(3.4.1)$$

The regions with 0 consist entirely of zeros. A transition probability matrix arranged in this manner is said to be in the canonical form. Usefulness of this representation will be evident from Chapter 4.

The transition probability matrix of Example 3.1.3 has the following representation:

$$
P = \begin{array}{c}
\begin{array}{ccccc} P & B & 0 & 1 & 2 \end{array} \\
\begin{array}{c} P \\ B \\ 0 \\ 1 \\ 2 \end{array}
\left[\begin{array}{cc|ccc}
1 & 0 & 0 & 0 & 0 \\
0 & 1 & 0 & 0 & 0 \\
\hline
0.3 & 0 & 0.6 & 0.1 & 0 \\
0.2 & 0 & 0.2 & 0.5 & 0.1 \\
0.1 & 0.1 & 0.2 & 0.2 & 0.4
\end{array}\right]
\end{array}
\qquad (3.4.2)
$$

In the case of the Markov chain of Example 3.1.4 we get

$$
P = \begin{array}{c}
\begin{array}{ccccccc} 0 & 6 & 1 & 2 & 3 & 4 & 5 \end{array} \\
\begin{array}{c} 0 \\ 6 \\ 1 \\ 2 \\ 3 \\ 4 \\ 5 \end{array}
\left[\begin{array}{cc|ccccc}
1 & 0 & 0 & 0 & 0 & 0 & 0 \\
0 & 1 & 0 & 0 & 0 & 0 & 0 \\
\hline
0.4 & 0 & 0 & 0.6 & 0 & 0 & 0 \\
0 & 0 & 0.4 & 0 & 0.6 & 0 & 0 \\
0 & 0 & 0 & 0.4 & 0 & 0.6 & 0 \\
0 & 0 & 0 & 0 & 0.4 & 0 & 0.6 \\
0 & 0.6 & 0 & 0 & 0 & 0.4 & 0
\end{array}\right]
\end{array}
\qquad (3.4.3)
$$

The definitions and the classifications given so far in this chapter are common for Markov chains with a finite number of states as well as for Markov chains with a countable infinity of states (infinite Markov chains). However, while studying their properties, it is more convenient to treat them separately. Then the special nature of the states due to their being finite or infinite in number can be used to give certain special properties of Markov chains of their class. For instance, Theorem 3.4.1 and its corollaries give one such property.

Consider the canonical representation of a transition probability matrix given by

$$P = \begin{bmatrix} P_1 & 0 \\ R & Q \end{bmatrix} \tag{3.4.4}$$

Taking powers of P, we get

$$P^n = \begin{bmatrix} P_1^n & 0 \\ R_n & Q^n \end{bmatrix} \tag{3.4.5}$$

where $R_1 = R$ and $R_n = R_{n-1}P_1 + Q^{n-1}R$. As $n \to \infty$, we may write

$$\lim_{n \to \infty} P^n = \begin{bmatrix} P_1^\infty & 0 \\ R_\infty & Q^\infty \end{bmatrix} \tag{3.4.6}$$

Of the three elements P_1^∞, R_∞, and Q^∞ of matrix (3.4.6), Q^∞ is shown to be a null matrix in Theorem 3.4.1, which follows.

Theorem 3.4.1 *In a finite Markov chain, as the number of steps tends to infinity, the probability that the process is in a transient state tends to zero irrespective of the state at which the process starts.*

Proof If the process is initially in a recurrent state, there is no possibility of getting into a transient equivalence class, and therefore the probability of finding the process in a transient state is zero. Suppose the process is initially in state i, $i \in T$, where T is the set of transient states. By definition, given a sufficient number of steps, there is a positive probability that the process will get out of the transient class T. Let n be the maximum of the least number of steps required for this to happen, starting from any of the states belonging to T. In terms of the transition probability matrix, n is the least power of P, such that there is at least one nonzero element in each row of the submatrix whose elements are the probabilities of transition from a state in T to a state outside T.

Since we are considering the maximum number n, we can also say there is a positive number p, such that the probability of the process leaving the set T of transient states in at most n steps is greater than or equal to p ($p < 1$). This implies that the probability of the process not leaving the set T of transient states in n steps is less than or equal to $1 - p$. This can be represented by the inequality

$$\sum_{r \in T} P_{ir}^{(n)} \leq 1 - p \qquad \text{for all } i \in T$$

Further

$$P_{ir}^{(nk)} = \sum_{r_1} \cdots \sum_{r_{k-1}} P_{ir_1}^{(n)} P_{r_1 r_2}^{(n)} \cdots P_{r_{k-1}r}^{(n)}$$

This gives

$$\sum_{r \in T} P_{ir}^{(nk)} \leq (1 - p)^k \qquad k = 0, 1, 2, \ldots$$

and

$$\sum_{r \in T} P_{ir}^{(nk+l)} \leq (1 - p)^k \qquad 0 \leq l < n$$

that is,

$$P_{ij}^{(nk+l)} \leq (1 - p)^k \qquad j \in T \tag{3.4.7}$$

Clearly, as $k \to \infty$, the right-hand side of (3.4.7) tends to zero, and hence the probability on the left-hand side also tends to zero. $\qquad \square$

COROLLARY 1 *For i and $j \in T$, the n-step transition probability $P_{ij}^{(n)}$ approaches zero geometrically. Alternately, there exist numbers $c > 0$ and $0 < r < 1$, such that*

$$P_{ij}^{(n)} \leq cr^n \tag{3.4.8}$$

Proof   From (3.4.7) we have

$$P_{ij}^{(nk+l)} \le (1-p)^k \qquad k = 0,1,2,\dots;\ 0 \le l < n$$

Set $nk + l = m$; then we get

$$P_{ij}^{(m)} \le (1-p)^{(m-l)/n}$$

$$\le (1-p)^{(m/n)-1} = (1-p)^{-1}\left(\sqrt[n]{1-p}\right)^m \tag{3.4.9}$$

The corollary now follows by putting $(1-p)^{-1} = c$ and $\sqrt[n]{1-p} = r$.   □

COROLLARY 2   *In a finite Markov chain not all states can be transient.*

This result follows directly from Theorem 3.4.1 when we observe that in an m-state Markov chain

$$\sum_{j=1}^{m} P_{ij}^{(n)} = 1 \qquad \text{for all } n$$

In practice there are situations where the Markov chain to be studied has several recurrent as well as transient classes. Because of the noncommunicability of the states in different recurrent classes, once the process enters a recurrent class, it can then be studied in isolation. Similarly, a transient class has to be studied only in relationship with another recurrent class, because if there are two transient classes in the same chain, the states of one do not communicate with the states of the other, and for all practical purposes, the class whose states are accessible acts as a recurrent class. In view of this we shall consider in detail only the following Markov chains:

1. *Finite Markov chains*:
    a. With recurrent and transient states.
    b. With only ergodic states.
2. *Infinite Markov chains* (number of states is countably infinite):
    c. Aperiodic irreducible Markov chains.

Markov chains with states whose periods are greater than one do occur in practice, as illustrated by Examples 3.1.4 and 3.1.5. For a study of such chains the reader is referred to advanced texts on the subject.

3.5   CLASSIFICATION OF STATES USING GRAPH ALGORITHMS

The last section uses the properties of states to give a canonical representation of the transition probability matrix in order to organize and simplify the analysis of a Markov chain. There are three major tasks in representing the transition probability matrix in the canonical form: (1) identification of irreducible equivalence classes, (2) classification of equivalence classes as being recurrent or transient, and (3) recognition of the hierarchy among equivalence classes. When the number of states is small, these three tasks can be accomplished either by observation or by simply noting the communication relations between the states. However, when the number of states get large, there is a need for some algorithm that can accomplish these tasks in an orderly fashion. In this section we outline procedures by which available algorithms in graph theory can be used for this purpose.*

Basic Definitions from Graph Theory[†]

Identify the states of a Markov chain as vertexes (or nodes) of a graph and the possible one-step transitions by directed arcs. Thus an arc (i, j) between states i and j exists if the probability of the transition $i \rightarrow j$ is greater than zero. The graph defined through this procedure is known as a Markov graph or a stochastic graph. Now consider the following definitions from graph theory.

Let $x: \{x_1, x_2, \ldots, x_m\}$ be the vertexes (or nodes), and let a be the set of directed arcs between these vertexes. The graph $G = \{x, a\}$ is a *directed graph* (also called digraph). A *path* in a directed graph is any sequence of arcs where the final vertex of one is the initial vertex of the next one. A path $a_1, a_2, \ldots, a_k$ in which the initial vertex of a_1 is the same as the final vertex of a_k is called a *circuit*.

The *outdegree* of a vertex x_i is defined as the number of arcs that have x_i as their initial vertex. The *indegree* of a vertex x_i is the number of arcs that have x_i as their final vertex.

A graph is said to be *strongly connected* or *strong* if for any two distinct vertexes x_i and x_j there is at least one path going from x_i to x_j. A *maximal subgraph* of G is a strong subgraph of G which is not contained in any other strong subgraph of G. Such a subgraph is called a *strong component* of G.

---

*Preliminary research for material covered in this section was done by Krishna M. Kavipurapu (1979).
[†]Our treatment follows generally that of Christofides (1975) and Harary (1969).

Suppose $y = \{\, y_1, y_2, \ldots, y_k \,\}$ are the strong components of G, where y_i is a subgraph with vertexes $\{\, x_{i1}, x_{i2}, \ldots, x_{ir_i} \,\}$, $i = 1, 2, \ldots, k$, and identify a vertex corresponding to each y_i. Let c be the set of arcs between these vertexes. The resulting graph $G^* = (\, y, c)$ is called the *condensed graph* of G.

Given a graph G, the *adjacency matrix* $A = \|a_{ij}\|$ is given by

$$a_{ij} = 1 \quad \text{if arc } (x_i, x_j) \text{ exists in } G$$

$$= 0 \quad \text{otherwise}$$

Thus the elements of the adjacency matrix indicate the possibility of moving between any two vertexes in one step (moving along one arc is called a step). Extending this definition to more than one step, we can define the *reachability matrix* $R = \|r_{ij}\|$ as

$$r_{ii} = 1 \quad \text{if } x_j \text{ can be reached from } x_i \text{ with no restrictions} \\ \text{on the number of steps taken.}$$

$$= 0 \quad \text{otherwise}$$

Conceptually the ith row of the reachability matrix R (i.e., if one starts from vertex x_i) gives the union of vertexes that can be reached in $1, 2, 3, \ldots, m - 1$ steps. Clearly, if the total number of vertexes is m, the maximum number of steps to be considered is $m - 1$. Symbolically we may write this as follows.

Let $R^k(x_i)$ be the set of vertexes that can be reached from x_i in exactly k steps and $R(x_i)$, the set of vertexes that can be eventually reached from x_i. We have

$$R(x_i) = \{\, x_i \,\} \cup R^1(x_i) \cup R^2(x_i) \cup \cdots \cup R^{p_i}(x_i)$$

where p_i is the largest number of steps needed for this operation. (The operation stops when the total set stops increasing in size.) Let $p = \max_i(p_i)$; using Boolean operations on the adjacency matrix A ($0 + 1 = 1$, $1 + 1 = 1$, $0 \cdot 1 = 0$, $1 \cdot 1 = 1$, etc.), it can be shown that the nonzero elements in the ith row of A^k identify the vertexes in the set $R^k(x_i)$. Thus we can write

$$R = I + A + A^2 + \cdots + A^p \tag{3.5.1}$$

(The objectives of the Boolean operation are accomplished if all nonzero elements of R are replaced by 1, after carrying out the matrix multiplications and additions in the usual manner. But computationally, Boolean operations are more efficient.)

An algorithm given by Warshall (1962) simplifies this procedure considerably. The algorithm determines the reachability matrix $R = R^{(p)}$ in an iterative manner. Let $R^{(k)} = \left\| r_{ij}^{(k)} \right\|$. Determine $R^{(k)}$, $k = 1, 2, 3, 4, \ldots, p$ iteratively as

$$r_{ij}^{(1)} = a_{ij}$$

$$r_{ij}^{(k)} = r_{ij}^{(k-1)} \vee \left[r_{ik}^{(k-1)} \wedge r_{kj}^{(k-1)} \right] \qquad \text{for } k > 1 \qquad (3.5.2)$$

where $\vee$ and $\wedge$ are used to denote the Boolean operations of addition and multiplication, respectively. For hand calculations the second equation in (3.5.2) can be implemented by writing down a supplementary matrix $S^{(k)}$ representing the Boolean operation within braces on the right-hand side. If $r_{ik}^{(k-1)} = 0$, the ith row of $S^{(k)}$ will have all elements zero, and if $r_{ik}^{(k-1)} = 1$, the ith row of $S^{(k)}$ will be the same as the kth row of $R^{(k-1)}$. Now $R^{(k)}$ is obtained by carrying out the element-by-element Boolean addition of matrices $R^{(k-1)}$ and $S^{(k)}$. For a computer algorithm the matrices $S^{(k)}$ need not be determined. The matrix $R^{(k)}$ can be directly obtained from $R^{(k-1)}$ using the relation

$$r_{ij}^{(k)} = \begin{cases} r_{ij}^{(k-1)} & \text{if } r_{ik}^{(k-1)} = 0 \\ r_{ij}^{(k-1)} \vee r_{kj}^{(k-1)} & \text{if } r_{ik}^{(k-1)} = 1 \end{cases} \qquad (3.5.3)$$

(See Reingold et al., 1977, where (3.5.3) is identified as the transitive closure algorithm.)

We have defined $R(x_i)$ as the set of vertexes of the graph G that can be reached from x_i. Now we define $Q(x_i)$ as the set of vertexes from which x_i can be reached. The matrix $Q = \|q_{ij}\|$, where $q_{ij} = 1$, if x_i can be reached from x_j and $= 0$ otherwise is called the *reaching matrix*. Clearly we have $Q = R^T$ (which is the transpose of R). Thus, if one is interested in identifying the vertexes that can be mutually reached from one another (i.e., vertexes that correspond to the states belonging to the same equivalence class), one has to carry out the element-by-element Boolean multiplication of the matrices R and Q. We shall denote this product by $R \otimes Q$. Vertexes corresponding to the nonzero elements in the distinct rows of this matrix now give the strong components of graph G.

Example 3.5.1

Consider the graphs represented in Figures 3.5.1*a* and *b*. The properties of these graphs can be identified as follows:

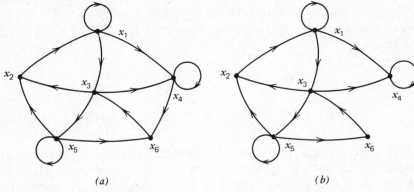

<table>
<tr><td>(a)</td><td>(b)</td></tr>
<tr><td>Figure 3.5.1a</td><td>Figure 3.5.1b</td></tr>
</table>

---

1. Outdegree of $x_4 = 1$; indegree of $x_4 = 2$.	Outdegree of $x_4 = 0$; indegree of $x_4 = 2$.
2. Strongly connected	Not strongly connected, but $\{x_1, x_2, x_3, x_5, x_6\}$ is a strong component.

3. Adjacency matrix:

$$
\begin{array}{c c c c c c c}
 & x_1 & x_2 & x_3 & x_4 & x_5 & x_6 \\
x_1 & 1 & & 1 & 1 & & \\
x_2 & 1 & & & & & \\
x_3 & & 1 & & 1 & 1 & \\
x_4 & & & & 1 & & 1 \\
x_5 & & 1 & & & 1 & 1 \\
x_6 & & & 1 & & &
\end{array}
$$

Adjacency matrix:

$$
\begin{array}{c c c c c c c}
 & x_1 & x_2 & x_3 & x_4 & x_5 & x_6 \\
x_1 & 1 & & 1 & 1 & & \\
x_2 & 1 & & & & & \\
x_3 & & 1 & & 1 & 1 & \\
x_4 & & & & 1 & & \\
x_5 & & 1 & & & 1 & 1 \\
x_6 & & & 1 & & &
\end{array}
$$

4. Reachability matrix:

$$
\begin{array}{c c c c c c c}
 & x_1 & x_2 & x_3 & x_4 & x_5 & x_6 \\
x_1 & 1 & 1 & 1 & 1 & 1 & 1 \\
x_2 & 1 & 1 & 1 & 1 & 1 & 1 \\
x_3 & 1 & 1 & 1 & 1 & 1 & 1 \\
x_4 & 1 & 1 & 1 & 1 & 1 & 1 \\
x_5 & 1 & 1 & 1 & 1 & 1 & 1 \\
x_6 & 1 & 1 & 1 & 1 & 1 & 1
\end{array}
$$

Reachability matrix:

$$
\begin{array}{c c c c c c c}
 & x_1 & x_2 & x_3 & x_4 & x_5 & x_6 \\
x_1 & 1 & 1 & 1 & 1 & 1 & 1 \\
x_2 & 1 & 1 & 1 & 1 & 1 & 1 \\
x_3 & 1 & 1 & 1 & 1 & 1 & 1 \\
x_4 & & & & 1 & & \\
x_5 & 1 & 1 & 1 & 1 & 1 & 1 \\
x_6 & 1 & 1 & 1 & 1 & 1 & 1
\end{array}
$$

Markov Chains as Graphs

As mentioned earlier, a Markov chain can be easily represented by a Markov graph in which vertexes represent states and directed arcs represent one-step transitions with nonzero probabilities. Using this representation, equivalent properties of the Markov chain and the Markov graph can be identified as follows:

1. *Markov chain*: States x_i and x_j communicate.

   Markov graph: Vertexes x_i and x_j are mutually reachable.

2. *Markov chain*: The set of states that mutually communicate with each other form an equivalence class.

   Markov graph: A maximal strong subgraph is called a strong component.

3. *Markov chain*: An equivalence class is recurrent if and only if a transition from a state within the class to a state outside the class is not possible. If such a transition is possible it is transient.

   Markov graph: Strong components of a graph can be treated as vertexes of a condensed graph. States corresponding to the vertexes of a strong component y_i can be classified as recurrent or transient using the following property: an equivalence class of a Markov chain is recurrent if, in the condensed graph of the Markov graph, the outdegree of the vertex y_i representing the corresponding strong component is zero. The equivalence class is transient if the outdegree is greater than zero. Also condensed graphs cannot have circuits.

Based on these properties, it is clear that states of a Markov chain can be grouped into equivalence classes by identifying the strong components of the corresponding Markov graph. As described earlier, vertexes corresponding to the nonzero elements in the distinct rows of the Boolean matrix $R \otimes Q$, where R is the reachability matrix and Q is the reaching matrix, give strong components of the Markov graph. If the equivalence class is recurrent, the corresponding vertex in the condensed graph has an outdegree 0. Otherwise, the class is transient; that is, if x_i, x_j, x_k belong to the equivalence class, the ith, jth and kth rows of the reachability matrix should have nonzero elements only at the ith, jth, and kth columns.

The next step in our analysis is arranging the transition probability matrix in the canonical form, that is, in the corresponding Markov graph establishing a hierarchy for the strong components.

A simple algorithm is obtained for this purpose if we consider the level of hierarchy as being dependent on the "distance" of a transient class from other recurrent equivalence classes. The "distance" used here is the number of steps corresponding to the longest path (critical path) from the vertex to the vertexes representing recurrent classes.

It has been noted earlier that (1) the condensed graph, which has strong components for vertexes, has no circuit and (2) vertexes corresponding to a recurrent class have an outdegree 0. In a canonical form recurrent equivalence classes occupy higher levels in the hierarchy than the transient classes. Based on the outdegree information, levels can be identified for each vertex of the condensed graph (i.e., for each equivalence class in the Markov chain) as follows.

Let $L(y_i)$ be the level of vertex y_i. Further define $\lambda(y_i)$ as the set of vertexes y_j of the condensed graph, such that the arc (y_i, y_j) exists. In terms of the Markov chain, $\lambda(y_i)$ includes the set of equivalence classes that are accessible from the class corresponding to y_i. Now the levels of the vertexes of the condensed graph can be determined in the following manner:

1. All vertexes with zero outdegree are at level zero; that is,

$$L(y_i) = 0 \qquad \text{if } \lambda(y_i) = \phi \qquad (3.5.4)$$

2. For vertexes with outdegree greater than zero, the level is obtained as

$$L(y_i) = \left[\min_{y_j \in \lambda(y_i)} \{ L(y_j) \} \right] - 1 \qquad (3.5.5)$$

It should be noted that in order to carry out the operations in equation (3.5.5), levels of y_j for $y_j \in \lambda(y_i)$ must be known. It is therefore necessary to start with the vertexes with outdegree zero and then proceed to other neighboring vertexes in an orderly fashion. If complete information is not available for a neighboring vertex, another vertex should be chosen. By considering all vertexes, it is possible to come up with a sequence of y_i's that allows us to use the algorithm effectively.

3. Arrange the vertexes of the condensed graph (the equivalence classes of the Markov chain) in the decreasing order of their levels, with level zero at the top of the hierarchy. When there are more than one recurrence classes in a Markov chain, it helps to arrange the absorbing states at the top of the equivalence class hierarchy; among the vertexes of the condensed graph the corresponding

vertexes should be given higher ranking among those with level zero. (A vertex y_i corresponds to an absorbing state if in the reachability matrix only the diagonal element is nonzero in the ith row.)

4. Other vertexes with the same level can be arranged in any order among themselves.

It is clear from the foregoing algorithm that the level of a vertex is obtained as the negative of the longest path from vertexes at level zero.

A Markov chain with a relatively small state space is the subject of the next example. When the state space is large, the use of the computer is necessary. Warshall's (1962) algorithm described earlier is simple and is based on the reachability matrix (also see Reingold et. al., 1977). A more efficient algorithm is the depth first search algorithm known since the 19th century, widely recognized after the appearance of a paper by Tarjan (1972). In this algorithm, starting from a vertex, all vertexes are visited one after another backtracking whenever one of those previously visited is reached. Because of the labeling needed to identify different classes of vertexes according to the nature of visits made, the algorithm is complicated in structure; nevertheless, for the determination of strong components it seems to be the most efficient method available at this time (see Even, 1979).

For the determination of the hierarchical levels of strong components Dijkstra's (1959) algorithm can be used after appropriate modifications. Dijkstra's algorithm is meant for finding the shortest path between a pair of vertexes in a directed graph.

Another algorithm for identifying recurrent classes and transient states in a Markov chain is given by Fox and Landi (1968; also see references cited in this paper). This algorithm uses an efficient iterative row-search approach in which the adjacency matrix is collapsed with the identification of states in the same recurrent class.

Isaacson and Madsen (1976) employ a different approach from the one described here for the analysis of a Markov chain using a computer. First, they determine whether the entire chain is a single recurrent class. Then, if the answer to the first question is negative, they proceed to identify the different equivalence classes in the chain. In all these determinations the adjacency matrix defined earlier plays a significant role.

Example 3.5.2

A Markov chain has six states $\{1, 2, 3, 4, 5, 6\}$ with the transition probability matrix that follows. Classify the states and arrange the matrix in a canonical form. (All zero elements in the matrices are omitted.)

$$
P = \begin{array}{c} \\ 1 \\ 2 \\ 3 \\ 4 \\ 5 \\ 6 \end{array}
\begin{array}{c}
\begin{array}{cccccc} 1 & \;2 & \;3 & \;4 & \;5 & \;6 \end{array} \\
\left[\begin{array}{cccccc}
0.5 & & & & 0.5 & \\
 & 0.2 & 0.2 & 0.5 & 0.1 & \\
 & & 1 & & & \\
 & 0.6 & & 0.4 & & \\
0.3 & & & & 0.7 & \\
0.2 & 0.2 & 0.1 & & 0.2 & 0.3
\end{array} \right]
\end{array}
$$

The adjacency matrix A is obtained as

$$
A = \begin{array}{c} \\ 1 \\ 2 \\ 3 \\ 4 \\ 5 \\ 6 \end{array}
\begin{array}{c}
\begin{array}{cccccc} 1 & \;2 & \;3 & \;4 & \;5 & \;6 \end{array} \\
\left[\begin{array}{cccccc}
1 & & & & 1 & \\
 & 1 & 1 & 1 & 1 & \\
 & & 1 & & & \\
 & 1 & & 1 & & \\
1 & & & & 1 & \\
1 & 1 & 1 & & 1 & 1
\end{array} \right]
\end{array}
$$

Using the recursive relations (3.5.2), the matrices $R^{(k)}$ are obtained as

$$
\begin{array}{c}
R^{(1)} = A \\
\begin{array}{c} \\ 1 \\ 2 \\ 3 \\ 4 \\ 5 \\ 6 \end{array}
\begin{array}{c}
\begin{array}{cccccc} 1 & 2 & 3 & 4 & 5 & 6 \end{array} \\
\left[\begin{array}{cccccc}
1 & & & & 1 & \\
 & 1 & 1 & 1 & 1 & \\
 & & 1 & & & \\
 & 1 & & 1 & & \\
1 & & & & 1 & \\
1 & 1 & 1 & & 1 & 1
\end{array} \right]
\end{array}
\end{array}
\qquad
\begin{array}{c}
S^{(2)} \\
\begin{array}{c} \\ 1 \\ 2 \\ 3 \\ 4 \\ 5 \\ 6 \end{array}
\begin{array}{c}
\begin{array}{cccccc} 1 & 2 & 3 & 4 & 5 & 6 \end{array} \\
\left[\begin{array}{cccccc}
 & & & & & \\
 & 1 & 1 & 1 & 1 & \\
 & & & & & \\
 & 1 & 1 & 1 & 1 & \\
 & & & & & \\
 & 1 & 1 & 1 & 1 &
\end{array} \right]
\end{array}
\end{array}
$$

$$
\begin{array}{c}
R^{(2)} \\
\begin{array}{c} \\ 1 \\ 2 \\ 3 \\ 4 \\ 5 \\ 6 \end{array}
\begin{array}{c}
\begin{array}{cccccc} 1 & 2 & 3 & 4 & 5 & 6 \end{array} \\
\left[\begin{array}{cccccc}
1 & & & & 1 & \\
 & 1 & 1 & 1 & 1 & \\
 & & 1 & & & \\
 & 1 & 1 & 1 & 1 & \\
1 & & & & 1 & \\
1 & 1 & 1 & 1 & 1 & 1
\end{array} \right]
\end{array}
\end{array}
\qquad
\begin{array}{c}
S^{(3)} \\
\begin{array}{c} \\ 1 \\ 2 \\ 3 \\ 4 \\ 5 \\ 6 \end{array}
\begin{array}{c}
\begin{array}{cccccc} 1 & 2 & 3 & 4 & 5 & 6 \end{array} \\
\left[\begin{array}{cccccc}
 & & & & & \\
 & & 1 & & & \\
 & & 1 & & & \\
 & & 1 & & & \\
 & & & & & \\
 & & 1 & & &
\end{array} \right]
\end{array}
\end{array}
$$

$R^{(3)}$

	1	2	3	4	5	6
1	1				1	
2		1	1	1	1	
3			1			
4		1	1	1	1	
5	1				1	
6	1	1	1	1	1	1

$S^{(4)}$

	1	2	3	4	5	6
1						
2		1	1	1	1	
3						
4		1	1	1	1	
5						
6		1	1	1	1	

$R^{(4)}$

	1	2	3	4	5	6
1	1				1	
2		1	1	1	1	
3			1			
4		1	1	1	1	
5	1				1	
6	1	1	1	1	1	1

$S^{(5)}$

	1	2	3	4	5	6
1	1				1	
2	1				1	
3						
4	1				1	
5	1				1	
6	1				1	

$R^{(5)}$

	1	2	3	4	5	6
1	1				1	
2	1	1	1	1	1	
3			1			
4	1	1	1	1	1	
5	1				1	
6	1	1	1	1	1	1

$S^{(6)}$

	1	2	3	4	5	6
1						
2						
3						
4						
5						
6	1	1	1	1	1	1

$R^{(6)} = R$

	1	2	3	4	5	6
1	1				1	
2	1	1	1	1	1	
3			1			
4	1	1	1	1	1	
5	1				1	
6	1	1	1	1	1	1

$Q = R^{T}$

	1	2	3	4	5	6
1	1	1		1	1	1
2		1		1		1
3		1	1	1		1
4		1		1		1
5	1	1		1	1	1
6						1

$R \otimes Q$

	1	2	3	4	5	6
1	1				1	
2		1		1		
3			1			
4		1		1		
5	1				1	
6						1

Comparing $R \otimes Q$ and R, we have the equivalence classes as follows:

   $\{3\}$:   recurrent (absorbing state).
$\{1,5\}$:   recurrent.
$\{2,4\}$:   transient.
   $\{6\}$:   transient.

Denote these equivalence classes as

$$y_1 = \{3\} \qquad y_2 = \{1,5\} \qquad y_3 = \{2,4\} \qquad y_4 = \{6\}$$

From the reachability matrix for the condensed graph, $y = \{\, y_1, y_2, y_3, y_4 \,\}$, we have the following information (od $=$ outdegree).

$$\text{od}(\, y_1) = 0 \qquad \text{od}(\, y_2) = 0 \qquad \text{od}(\, y_3) = 2 \qquad \text{od}(\, y_4) = 3$$

Using equations (3.5.4) and (3.5.5), we have

$$L(\, y_1) = 0 \qquad L(\, y_2) = 0$$
$$L(\, y_3) = \min\{\, L(\, y_1), L(\, y_2)\} - 1 = -1$$
$$L(\, y_4) = \min\{\, L(\, y_1), L(\, y_2), L(\, y_3)\} - 1$$
$$= -1 - 1 = -2$$

Hence the class hierarchy for the canonical form is given as $\{\, y_1, y_2, y_3, y_4 \,\}$. Now the matrix P can be arranged in the following canonical form:

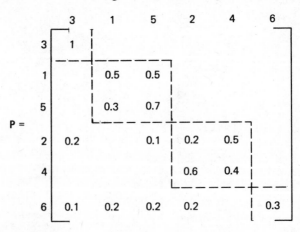

Answer

REFERENCES

Christofides, N. (1975). *Graph Theory: An Algorithmic Approach*. New York: Academic Press.

Cyert, R. M., Davidson, H. J., and Thompson, G. L. (1962). "Estimation of the Allowance for Doubtful Accounts by Markov Chains." *Management Sci.*, 287–303.

Dijkstra, E. W. (1959). "A Note on Two Problems in Connection with Graphs." *Numerische Mathematic*, 269.

Even, S. (1979). *Graph Algorithms*. Potomac, Md.: Computer Science Press.

Fox, B. L., and Landi, D. M. (1968). "An Algorithm for Identifying the Ergodic Subchains and Transient States of a Stochastic Matrix." *Comm. ACM* **11**, 619–621.

Harary, F. (1969). *Graph Theory*. Reading, Mass.: Addison-Wesley.

Isaacson, D. L., and Madsen, R. W. (1976). *Markov Chains: Theory and Applications*. New York: Wiley.

Kavipurapu, K. M. (1979). "Classification of Markov Chains Using Graph Algorithms." Tech. Rep. CSE7912, Department of Computer Science and Engineering, Southern Methodist University, Dallas, Tex.

Reingold, E. M., Nievergelt, J., and Deo, N. (1977). *Combinatorial Algorithms: Theory and Practice*. Englewood Cliffs, N.J.; Prentice-Hall.

Tarjan, R. E. (1972). "Depth First Search and Linear Graph Algorithms." *SIAM J. Comp.*, 146–160.

Warshall, S. (1962). "A Theorem on Boolean Matrices." *J. ACM*, 11–12.

FOR FURTHER READING

Chung, K. L. (1967). *Markov Chains*. New York: Springer-Verlag, pp. 1–28.

Çinlar, E. (1975). *Introduction to Stochastic Processes*. Englewood Cliffs, N.J.: Prentice-Hall.

Cox, D. R., and Miller, H. D. (1965). *The Theory of Stochastic Processes*. London: Methuen. Reprinted 1972: Chapman and Hall.

Feller, W. (1968). *An Introduction to Probability Theory and Its Applications*. 3rd ed. New York: Wiley, Chap. 15.

Isaacson, D. L., and Madsen, R. W. (1976). *Markov Chain Theory and Applications*. New York: John Wiley.

Karlin, S., and Taylor, H. M. (1975). *A First Course in Stochastic Processes*. New York: Academic Press, Chap. 2.

Kemeny, J. G., and Snell, J. L. (1960). *Finite Markov Chains*. New York: Van Nostrand, Chap. 2.

Parzen, E. (1962). *Stochastic Processes*, San Francisco: Holden-Day, Chap. 6.

Prabhu, N. U. (1965). *Stochastic Processes*. New York: Macmillan, Chap. 2.

EXERCISES

1. Let P be the transition probability matrix of an m-state Markov chain, and let

$$|P - \lambda I| = 0$$

be the characteristic equation of the square matrix P. Using the property of square matrices that the matrix satisfies its own characteristic equation, show that

$$P^n = A_{m-1} P^{n-1} + \cdots + A_0 P^{n-m} \qquad n \geq m$$

where $A_i = -A_i^*/A_m^* \, (i = 0, 1, \ldots, m-1)$ satisfy the relation

$$|P - \lambda I| = \sum_{i=0}^{m} A_i^* \lambda^i$$

2. A sequential sampling procedure known as Bartky's sampling inspection scheme uses successive samples of size N. If in the initial sample the number of defectives is zero, the lot is accepted. If the number of defectives exceeds a predetermined number a, the lot is rejected. From the second sample onward one defective per sample is allowed. Thus after n such samples, the lot will be accepted if the total number of defectives is $\leq n$ and rejected if the number of defectives is $> a + n$. Let S_n be the total excess number of defectives (i.e., after allowing for one defective per sample) after taking n samples. Show that $\{S_n\}$ is a Markov chain, and identify its state space. Let X_k be the excess number at the kth sample, that is, $X_k = $ (number of defectives out of N) $- 1$. Assuming p to be the fraction defective in the lot, determine the distribution of X_k. Obtain the transition probability matrix for $\{S_n\}$ in terms of the distribution of X_k. Classify the states of $\{S_n\}$.

3. The Ehrenfest model for heat exchange is described as follows: N particles are distributed between two containers A and B. At every step a particle is chosen at random from one container and moved to the other. This procedure is repeated. The number of particles in A is the state of the process. Note that when there are n particles in a container, the probability that a particle from that container will be moved to the other container is n/N. Show that the process is Markovian.

   Obtain the transition probability matrix for the number of particles in container A, and classify the states of the process.

4. a. A coin with $P(\text{heads}) = p$ and $P(\text{tails}) = q = 1 - p$ is tossed repeatedly. A process $\{X_n\}$ is defined with states $1, 2, 3, 4$ according as the $(n-1)$st and nth tosses give HH, HT, TH, and TT, respectively. Show that $\{X_n\}$ is a Markov chain, and obtain its transition probability matrix.

   b. Suppose, in the preceding problem $\{X_n\}$ is defined with states $(2, 1, 0)$ based on the number of heads in the $(n-1)$st and nth tosses. Is $\{X_n\}$ a Markov chain? If yes, give the transition probability matrix. If no, justify your answer.

5. In the coin-tossing example of Exercise 4a, define a process $\{X_n\}$ based on the outcomes of the $(n-2)$nd, $(n-1)$st and nth tosses, and

show that it is a Markov chain. Determine its transition probability matrix.

6. A run is defined as the successive occurrence of the same event. In repeated trials with S(success) and F(failure) as the outcomes, a sequence of outcomes FSSFFFSFSSSSFS has 4 success runs (2 of length 1, 1 of length 2, and 1 of length 4) and 4 failure runs (3 of length 1 and 1 of length 3). Suppose the probability of success is p, with $q = 1 - p$. Let X_n be the length of the success run after the nth trial ($X_n = 0$, if the nth trial is a failure). Show that $\{ X_n, n = 1, 2, \ldots \}$ is a Markov chain, and determine its transition probability matrix. Determine the distribution of recurrence time to state 0 and its mean. Using these results, classify the states of the Markov chain.

7. The following method of estimating the size of the fish population in a pond has been suggested. Fish are caught and returned to the pond after tagging them, say with a dye. Fishing is continued until a tagged fish is caught. Noting that the number of fish caught until stopping is related to the success run of Exercise 6, the estimation procedure would depend on the distribution properties of the success run of untagged fish.

   Let $\{ X_n \}$ be the success run of untagged fish. Show that it is a Markov chain, and determine its transition probability matrix.

8. It is a common practice to have standby redundant units in mechanical and electrical systems so as to attain a high degree of reliability. Suppose two machines are available, one in use and one on a standby basis. The probability that a machine that is in use fails during a unit time period is $p(q = 1 - p)$. It takes three units of time to repair a failed machine. Define a process with the states identified by different combinations of two elements: number of machines in a working condition; expended repair time. Thus the states are $20, 10, 11,$ $12, 01, 02$. Show that this process is a Markov chain, and determine its transition probability matrix.

9. In a discrete time queue, customers arrive or leave the system just before discrete epochs $0, \sigma, 2\sigma, 3\sigma, \ldots$. The probability of an arrival is $p(q = 1 - p)$, and the probability of service completion is $r(s = 1 - r)$. Let X_n be the number of customers in the system. Show that $\{ X_n \}$ is a Markov chain, and determine its transition probability matrix.

10. A service agency gets help from a self-employed repairperson in its routine repair service. The maximum number of jobs assigned to her, in addition to the one she is working on at any time is $N(\geq 1)$. If she cannot finish the assigned jobs on a given day, she starts with the

remaining ones the following day. However, if at any time of the day she finishes all the jobs assigned to her, she returns to her own work and becomes unavailable for any more agency jobs that day. Let a_j be the probability that j ($j \geq 0$) new jobs arrive during a service period. Let X_n be the number of jobs assigned to her at the end of the nth service. Under what conditions is $\{X_n\}$ a Markov chain? Determine its transition probability matrix, and classify its states.

11. A gambler starts a game with an initial capital of $100 with him. The game requires the player to bet one dollar in every game. If he wins he gets back two dollars. If he loses, he forfeits the bet. As an additional incentive, the gambling house lets him continue the game even when his capital is zero with the provision that if he wins at that time, he gets back only one dollar. Let p be the probability of his win and q ($= 1 - p$), the probability of his loss in a game. Determine the transition probability matrix for his capital.

12. In trying to assess her chances for the nomination, a presidential candidate classifies her own candidacy during primaries as follows:

    a. Sure winner (W).

    b. Developing support (D).

    c. Sinking nomination chances (S).

    d. Sure loser (L).

    She further considers that the states W and L are terminal and that if she has developed support in one primary, for the following primary that chances of her moving to the winner state as 30%, staying in the same state as 40%, and slipping into the sinking states as 30%. On the other hand, if her chances are slipping and is in the sinking state in one primary, for the following primary she considers her chances of developing support as 20%, staying in the same sinking state as 50%, and becoming a sure loser as 30%.

    Model her performance in the primaries as a Markov chain, and determine its transition probability matrix. Classify the states of the process.

13. In a college 65% of the freshmen advance to the sophomore level, 80% of the sophomores advance to the junior level, 92% of the juniors advance to the senior level, and 95% of the seniors graduate. It is also known that the percentages of dropouts (and transfers to other colleges) for each class are: freshmen 25%, sophomores 10%, juniors 3%, and seniors 1%; other students remain at the same level the following year. Set up a Markov model, and determine its transition probability matrix. Classify the states of the process.

14. A discount store records indicate that the daily demands for a certain item can be represented by the following distribution:

Demand	0	1	2	3	4	5	6	7	8
Probability	0.1	0.1	0.2	0.1	0.1	0.1	0.1	0.1	0.1

Based on this information, the storekeeper has set up an inventory system as follows. The store starts the day with at least 3 items, but not more than 7. If during the day the inventory level falls below 3, new items are ordered, and the following day starts with a level of 7 again (that means the storekeeper orders 5 or more depending on the day's demand. Backorders are filled in this manner). It is assumed that ordered items arrive by the following morning. Orders are not placed for items if the inventory level does not go below 3.

Describe the state space for the inventory level at the beginning of the day, and show that it is a Markov chain. Determine its transition probability matrix.

15. Six boys (Dick, Harry, Joe, Mark, Sam, and Tom) play catch. If Dick has the ball, he is equally likely to throw it to Harry, Mark, Sam, and Tom. If Harry gets the ball, he is equally likely to throw it to Dick, Joe, Sam, and Tom. If Sam has the ball, he is equally likely to throw it to Dick, Harry, Mark, and Tom. If either Joe or Tom gets the ball, they keep throwing it to each other. If Mark gets the ball, he runs away with it. Obtain the transition probability matrix and classify the states of the chain.

16. Three tanks A, B, and C fight a duel. Tank A hits its target with probability 2/5, tank B with probability 2/3, and tank C with probability 3/5. Shots are fired simultaneously, and once a tank is hit it is out of action. When all three tanks are in action, each one picks its target at random. Let the set of tanks still in action be the states of the system. Show that the transition probability matrix is given by the following:

	E	A	B	C	AB	AC	BC	ABC
E	1	0	0	0	0	0	0	0
A	0	1	0	0	0	0	0	0
B	0	0	1	0	0	0	0	0
C	0	0	0	1	0	0	0	0
AB	$\frac{4}{15}$	$\frac{2}{15}$	$\frac{6}{15}$	0	$\frac{3}{15}$	0	0	0
AC	$\frac{6}{25}$	$\frac{4}{25}$	0	$\frac{9}{25}$	0	$\frac{6}{25}$	0	0
BC	$\frac{6}{15}$	0	$\frac{4}{15}$	$\frac{3}{15}$	0	0	$\frac{2}{15}$	0
ABC	$\frac{3}{75}$	$\frac{11}{75}$	$\frac{11}{75}$	$\frac{11}{75}$	$\frac{10}{75}$	$\frac{8}{75}$	$\frac{15}{75}$	$\frac{6}{75}$

Classify the states of the Markov chain.

17. Modify the tank duel problem of Exercise 16 by changing target selection procedure when all the tanks are in action as follows: When all the tanks are in action, A shoots at B, B at C, and C at A. Obtain the transition probability matrix in this case.

18. You have just been assigned by a consulting firm to investigate some marketing problems for a manufacturer. This particular manufacturer markets two brands of a specific product type. The manufacturer has been collecting data on the market situation and thinks that a Markov chain analysis is appropriate. The manufacturer knows that there is only one other brand (from another manufacturer) that ever has a significant share of the market.

    a. The manufacturing representative asks you, as a consultant, to comment on the appropriateness of a Markov chain analysis for this situation. Give your comments.

    b. Given that the manufacturer will use a Markov chain approach whether or not it is appropriate, she is concerned over the one-step transition probability matrices below which might result from her analysis. The manufacturer's brands are A and B.

    (i)
    $$\begin{array}{c} \\ A \\ B \\ C \end{array} \begin{array}{ccc} A & B & C \\ \begin{bmatrix} 1 & 0 & 0 \\ 0 & 1 & 0 \\ 0 & 0 & 1 \end{bmatrix} \end{array}$$

    (ii)
    $$\begin{array}{c} \\ A \\ B \\ C \end{array} \begin{array}{ccc} A & B & C \\ \begin{bmatrix} 0 & \frac{1}{2} & \frac{1}{2} \\ \frac{1}{2} & 0 & \frac{1}{2} \\ \frac{1}{2} & \frac{1}{2} & 0 \end{bmatrix} \end{array}$$

    (iii)
    $$\begin{array}{c} \\ A \\ B \\ C \end{array} \begin{array}{ccc} A & B & C \\ \begin{bmatrix} \frac{1}{2} & 0 & \frac{1}{2} \\ \frac{1}{3} & \frac{1}{3} & \frac{1}{3} \\ \frac{1}{2} & 0 & \frac{1}{2} \end{bmatrix} \end{array}$$

    (iv)
    $$\begin{array}{c} \\ A \\ B \\ C \end{array} \begin{array}{ccc} A & B & C \\ \begin{bmatrix} 0 & 1 & 0 \\ 0 & 0 & 1 \\ 1 & 0 & 0 \end{bmatrix} \end{array}$$

    For each of the matrices (i)–(iv), describe the situation in terms of the market behavior exhibited, and make recommendations for the manufacturer of brands A and B.

19. Classify the states of the Markov chains with the following transition probability matrices. Also arrange the transition probability matrices in canonical forms.

    a.

	1	2	3	4	5	6
1	0	0	0	0	0	1
2	0	0.3	0.7	0	0	0
3	0	0	1	0	0	0
4	0.2	0	0	0.5	0.3	0
5	0	0	0	0.3	0.3	0.4
6	1	0	0	0	0	0

b.

	1	2	3	4	5
1	0.3	0	0	0.6	0.1
2	0	0.6	0.4	0	0
3	0.2	0.5	0.3	0	0
4	0	0	0	0.8	0.2
5	0.5	0	0	0.2	0.3

20. Using graph algorithms, classify the states of the Markov chains with the following transition probability matrices. Also arrange the matrices in canonical forms.

a.

	1	2	3	4	5	6
1	0.2	0	0	0.8	0	0
2	0	0.3	0	0	0.7	0
3	0	0	0.5	0	0	0.5
4	0.6	0	0	0.4	0	0
5	0	1	0	0	0	0
6	0	0	0.3	0	0.3	0.4

b.

	1	2	3	4	5	6	7	8	9	10
1	0.2	0	0	0.3	0	0	0.3	0.2	0	0
2	0	1	0	0	0	0	0	0	0	0
3	0	0.2	0.2	0	0	0	0	0	0.3	0.3
4	0	0	0	0.3	0	0	0.6	0.1	0	0
5	0	0	0	0	0.5	0.5	0	0	0	0
6	0	0	0	0	0	1	0	0	0	0
7	0.4	0	0	0.5	0	0	0.1	0	0	0
8	0.2	0	0	0.4	0	0	0.2	0.2	0	0
9	0.3	0	0.1	0	0	0	0	0	0.4	0.2
10	0	0	0.4	0	0	0	0	0	0.3	0.3

Chapter 4

FINITE MARKOV CHAINS: TIME-DEPENDENT AND LIMITING BEHAVIOR

In this chapter we shall study m-state Markov chains ($m < \infty$) in some detail and derive results of practical interest. Results given in Theorems 4.1.2–4.1.5, 4.2.1, and 4.2.4 are due to Kemeny and Snell (1960) and are now an integral part of the theory of stochastic processes. Section 4.3 deals with a two-state Markov chain which is in fact a special case of the m-state chain discussed in Section 4.2. Since it is amenable for analysis through simpler techniques and the process has a significant place in probability modeling, a separate section is devoted for that process.

4.1 FINITE MARKOV CHAINS WITH RECURRENT AND TRANSIENT STATES

In this section our main interest is the behavior of the stochastic process as related to the transient states. As shown in Corollary 2 to Theorem 3.4.1, it is not possible to have only transient states in a finite Markov chain. Therefore for the purpose of our study we consider a chain with one transient class and a set of recurrent states that may be grouped into one or more equivalence classes. Some of these recurrent states may even be absorbing. A convenient way of representing the transition probability matrix is the canonical form described earlier.

Let the m-state Markov chain consist of r recurrent states and $(m - r)$ transient states belonging to a single equivalence class. Let T be the set of these transient states and T^c the set of recurrent states. The transition probability matrix P can now be put in the form

$$P = \left[\begin{array}{c|c} P_1 & 0 \\ \hline R & Q \end{array} \right] \tag{4.1.1}$$

where P_1 is an $r \times r$ matrix with transition probabilities among the recurrent states for its elements; Q is an $(m - r) \times (m - r)$ substochastic matrix (with at least one row sum less than 1) with probabilities of transition only among the transient states for its elements; and R is an $(m - r) \times r$ matrix whose elements are the probabilities of the one-step transition from the $(m - r)$ transient states to the r recurrent states.

In the study of Markov chains with recurrent and transient states we are interested in answering some of the following questions: Given that the process is in a transient state i initially, what is the average number of visits it makes to another transient state j (i.e., $i, j \in T$) before it eventually enters any one of the recurrent states? What is the variance of the number of visits to j from i? What are the mean and variance of the number of steps required to leave the transient class? Starting from a state $i \in T$, what are the probabilities of eventually entering a given recurrent state? The relevance of these questions can be easily understood if we consider the problems arising in Examples 3.1.3 and 3.1.4.

The Fundamental Matrix M

The matrix $M = (I - Q)^{-1}$ plays a useful role in the determination of the means and variances mentioned previously. We shall call M the *fundamental matrix*. The existence of the inverse of the matrix $(I - Q)$ can be established by Theorem 4.1.1.

Theorem 4.1.1 *For a Markov chain with transition probability matrix partitioned as in* (4.1.1), *the inverse* $(I - Q)^{-1}$ *exists, and*

$$(I - Q)^{-1} = I + Q + Q^2 + \cdots = \sum_{r=0}^{\infty} Q^r \qquad (4.1.2)$$

Proof   From direct multiplication we have

$$(I - Q)(I + Q + Q^2 + \cdots + Q^{n-1}) = I - Q^n \qquad (4.1.3)$$

Due to Theorem 3.4.1 and Corollary 1 Q^n tends to a matrix with zero elements as $n \to \infty$. Hence the determinant

$$|I - Q^n| \to 1 \qquad \text{as } n \to \infty$$

Therefore, for a sufficiently large n, $|I - Q^n| \neq 0$; whence

$$|I - Q||I + Q + Q^2 + \cdots + Q^{n-1}| \neq 0 \qquad (4.1.4)$$

This is possible only if both the determinants on the left-hand side of (4.1.4) are nonzero; that is, $|I - Q| \neq 0$, which is a sufficient condition for the existence of its inverse. Now multiplying both sides of (4.1.3) by $(I - Q)^{-1}$ and making $n \to \infty$, we have

$$I + Q + Q^2 + \cdots = (I - Q)^{-1} \tag{4.1.5}$$

$\square$

We shall now try to answer some of the questions posed earlier. All results are given in terms of the matrix $M = (I - Q)^{-1}$.

Let $N_{ij}(i, j \in T)$ be the random variable denoting the number of times the process visits j before it eventually enters a recurrent state, having initially started from state i. Let $\mu_{ij} = E[N_{ij}]$.

Theorem 4.1.2   *For $i, j \in T$*

$$\|\mu_{ij}\| = M \tag{4.1.6}$$

Proof   Initially the Markov chain is in state $i \in T$. If in one step it enters a recurrent state (with probability $\Sigma_{k \in T^c} P_{ik}$), the number of visits to j is zero unless $j = i$. If δ_{ij} is the Kronecker δ function such that $\delta_{ij} = 1$ if $i = j$ and 0 if $i \neq j$, then we can write $N_{ij} = \delta_{ij}$ with probability $\Sigma_{k \in T^c} P_{ik}$. On the other hand, suppose the Markov chain moves to a state $k \in T$ at the first step (with probability P_{ik}). From that position onward the number of visits to j is N_{kj}. However, if $i = j$, then the total number of visits to j would be $N_{kj} + \delta_{ij}$. Thus we have

$$N_{ij} = \begin{cases} \delta_{ij} & \text{with probability } \displaystyle\sum_{k \in T^c} P_{ik} \\[2mm] N_{kj} + \delta_{ij} & \text{with probability } P_{ik}\ k \in T \end{cases} \tag{4.1.7}$$

Taking expectations, we get

$$E(N_{ij}) = \sum_{k \in T^c} P_{ik}\delta_{ij} + \sum_{k \in T} P_{ik} E(N_{kj} + \delta_{ij}) \tag{4.1.8}$$

which gives

$$\mu_{ij} = \delta_{ij} + \sum_{k \in T} P_{ik}\mu_{kj}$$

Using all the elements of the matrix $\|\mu_{ij}\|$, we have

$$\|\mu_{ij}\| = I + Q\|\mu_{ij}\| \tag{4.1.9}$$

Hence

$$\|\mu_{ij}\| = (I - Q)^{-1} = M \qquad \qquad \square$$

Related to the matrix M, define the matrices that follow. Let $M = \|\mu_{ij}\|$ ($i, j \in T$ and $i, j = r + 1, r + 2, \ldots, m$):

$$M_D = \begin{bmatrix} \mu_{r+1,r+1} & & & & \bigcirc \\ & \mu_{r+2,r+2} & & & \\ & & \ddots & & \\ \bigcirc & & & & \\ & & & & \mu_{m,m} \end{bmatrix} \qquad (4.1.10)$$

and

$$M_2 = \begin{bmatrix} \mu_{r+1,r+1}^2 & \mu_{r+1,r+2}^2 & \cdots & \mu_{r+1,m}^2 \\ \mu_{r+2,r+1}^2 & \mu_{r+2,r+2}^2 & \cdots & \mu_{r+2,m}^2 \\ \vdots & \vdots & & \vdots \\ \mu_{m,r+1}^2 & \mu_{m,r+2}^2 & \cdots & \mu_{m,m}^2 \end{bmatrix} \qquad (4.1.11)$$

Let $\sigma_{ij}^2 = V(N_{ij})$.

Theorem 4.1.3

$$\|\sigma_{ij}^2\| = M(2M_D - I) - M_2 \qquad i, j \in T \qquad (4.1.12)$$

Proof By definition

$$V(N_{ij}) = E(N_{ij}^2) - \left[E(N_{ij}) \right]^2$$

From Theorem 4.1.2 we know that

$$\left[E(N_{ij}) \right]^2 = \mu_{ij}^2$$

and hence

$$\left\| \left[E(N_{ij}) \right]^2 \right\| = M_2 \qquad (4.1.13)$$

Before we derive an expression for $E(N_{ij}^2)$, we have to show that it is finite. This result is proved as follows: Let

$$Y_{ij}^{(n)} = \begin{cases} 1 & \text{if the system is in state } j \text{ at the } n\text{th step} \\ 0 & \text{otherwise} \end{cases}$$

Clearly we have

$$N_{ij} = \sum_{n=0}^{\infty} Y_{ij}^{(n)} \tag{4.1.14}$$

and therefore

$$E\left(N_{ij}^2\right) = E\left[\sum_{n=0}^{\infty} Y_{ij}^{(n)}\right]^2$$

$$= E\left[\sum_{n_1=0}^{\infty} \sum_{n_2=0}^{\infty} Y_{ij}^{(n_1)} Y_{ij}^{(n_2)}\right]$$

$$= \sum_{n_1=0}^{\infty} \sum_{n_2=0}^{\infty} E\left[Y_{ij}^{(n_1)} Y_{ij}^{(n_2)}\right] \tag{4.1.15}$$

However,

$$Y_{ij}^{(n_1)} Y_{ij}^{(n_2)} = \begin{cases} 1 & \text{if the system is in } j \text{ at the } n_1\text{th and } n_2\text{th steps} \\ 0 & \text{otherwise} \end{cases}$$

and hence

$$P\left(Y_{ij}^{(n_1)} Y_{ij}^{(n_2)} = 1\right) = P\left(Y_{ij}^{(n_1)} = 1, Y_{ij}^{(n_2)} = 1\right) \tag{4.1.16}$$

Let $p = \min(n_1, n_2)$ and $q = |n_1 - n_2|$. Then we have

$$P\left(Y_{ij}^{(n_1)} Y_{ij}^{(n_2)} = 1\right) = P\left(Y_{ij}^{(p)} = 1\right)\left[P\left(Y_{ij}^{(p+q)} = 1 | Y_{ij}^{(p)} = 1\right)\right]$$

$$= P_{ij}^{(p)} P_{jj}^{(q)} \tag{4.1.17}$$

Equation (4.1.15) can now be written as

$$E\left(N_{ij}^2\right) = \sum_{n_1=0}^{\infty} \sum_{n_2=0}^{\infty} P_{ij}^{(p)} P_{jj}^{(q)} \tag{4.1.18}$$

From Corollary 1 to Theorem 3.4.1 we further know that there exist $c > 0$ and $0 < r < 1$, such that

$$P_{ij}^{(p)} \leq cr^p \qquad P_{jj}^{(q)} \leq cr^q \qquad (4.1.19)$$

Hence

$$E\left(N_{ij}^2\right) \leq \sum_{n_1=0}^{\infty} \sum_{n_2=0}^{\infty} (cr^p)(cr^q)$$

$$= c^2 \sum_{n_1=0}^{\infty} \sum_{n_2=0}^{\infty} r^{p+q} \qquad (4.1.20)$$

But

$$p + q = \min(n_1, n_2) + |n_1 - n_2|$$

$$= \max(n_1, n_2)$$

Thus, if $n_1 > n_2$, then the term on the right-hand side of (4.1.20)

$$= c^2 \sum_{n_1=0}^{\infty} \sum_{n_2=0}^{n_1-1} r^{n_1}$$

$$= c^2 \sum_{n_1=0}^{\infty} n_1 r^{n_1}$$

Similarly, if $n_2 > n_1$,

$$= c^2 \sum_{n_2=0}^{\infty} n_2 r^{n_2}$$

Finally, if $n_1 = n_2$,

$$= c^2 \sum_{n_1=0}^{\infty} r^{n_1}$$

Adding all these terms, we get

$$E\left(N_{ij}^2\right) \leq c^2 \left[\sum_{n_1=0}^{\infty} n_1 r^{n_1} + \sum_{n_2=0}^{\infty} n_2 r^{n_2} + \sum_{n_1=0}^{\infty} r^{n_1} \right]$$

$$= c^2 \left[\frac{2r}{(1-r)^2} + \frac{1}{1-r} \right] < \infty \qquad (4.1.21)$$

In simplifying (4.1.21), we have used the results

$$\sum_{n=0}^{\infty} nr^n = \frac{r}{(1-r)^2} \qquad |r| < 1 \qquad (4.1.22)$$

and

$$\sum_{n=0}^{\infty} r^n = \frac{1}{1-r} \qquad |r| < 1 \qquad (4.1.23)$$

Now that we have shown that $E[N_{ij}^2]$ is finite, we proceed to derive it as follows: From (4.1.7) we have

$$N_{ij}^2 = \begin{cases} \delta_{ij}^2 & \text{with probability } \sum_{k \in T^c} P_{ik} \\ (N_{kj} + \delta_{ij})^2 & \text{with probability } P_{ik}, k \in T \end{cases} \qquad (4.1.24)$$

We also have $\delta_{ij}^2 = \delta_{ij}$.

Taking expectations, we get

$$E(N_{ij}^2) = \sum_{k \in T^c} P_{ik}\delta_{ij}^2 + \sum_{k \in T} P_{ik}E(N_{kj} + \delta_{ij})^2$$

$$= \sum_{k \in T^c} P_{ik}\delta_{ij}^2 + \sum_{k \in T} P_{ik}E(N_{kj}^2) + 2\sum_{k \in T} P_{ik}E(N_{kj})\delta_{ij} + \sum_{k \in T} P_{ik}\delta_{ij}^2$$

$$= \sum_{k \in T} P_{ik}E(N_{kj}^2) + 2\sum_{k \in T} P_{ik}E(N_{kj})\delta_{ij} + \delta_{ij}$$

giving

$$\|E(N_{ij}^2)\| = Q\|E(N_{ij}^2)\| + 2(QM)_D + I \qquad (4.1.25)$$

But we have

$$QM = Q(I + Q + Q^2 + \cdots)$$

$$= Q + Q^2 + Q^3 + \cdots \qquad (4.1.26)$$

$$= M - I$$

and

$$(QM)_D = (M - I)_D$$

$$= M_D - I$$

These give

$$\|E\big(N_{ij}^2\big)\| = (I - Q)^{-1}[2(M_D - I) + I] \qquad (4.1.27)$$

and

$$\|\sigma_{ij}^2\| = M(2M_D - I) - M_2 \qquad i, j \in T \qquad \square$$

The random variable N_{ij} gives the number of visits to a transient state j before entering a recurrent state. If we are interested in the total number of steps spent in transient states before entering a recurrent state, the variable to be considered is $N_i = \sum_{j \in T} N_{ij}$.

Using the results given by Theorems 4.1.2 and 4.1.3, we can get the mean and variance of $N_i (i \in T)$. For the sake of convenience define

$$M_\rho = \left\| \sum_{j \in T} \mu_{ij} \right\| \qquad (4.1.28)$$

a column vector where the kth component is the sum of the elements in the kth row of M, and

$$M_{\rho 2} = \left\| \left(\sum_{j \in T} \mu_{ij} \right)^2 \right\| \qquad (4.1.29)$$

a column vector whose kth component is the square of the kth component in M_ρ.

Theorem 4.1.4

$$\|E(N_i)\| = M_\rho \qquad i \in T \qquad (4.1.30)$$

Proof   Clearly,

$$E(N_i) = E\left(\sum_{j \in T} N_{ij} \right)$$

$$= \sum_{j \in T} E(N_{ij}) \qquad \square$$

Theorem 4.1.5

$$\|V(N_i)\| = (2M - I)M_\rho - M_{\rho 2} \qquad (4.1.31)$$

Proof Proceeding as in Theorem 4.1.3, we have

$$E(N_i^2) = \sum_{k \in T^c} P_{ik} \cdot 1 + \sum_{k \in T} P_{ik} E(N_k + 1)^2 \qquad (4.1.32)$$

Note that, since we are interested in the total number of steps, in place of δ_{ij} we have $\sum_{k \in T} \delta_{ij} = 1$. We get

$$E(N_i^2) = \sum_{k \in T} P_{ik} E(N_k^2) + 2 \sum_{k \in T} P_{ik} E(N_k) + 1$$

giving

$$\|E(N_i^2)\| = Q\|E(N_i^2)\| + 2QM_\rho + e \qquad (4.1.33)$$

where e is a column vector consisting only of ones. Thus we obtain

$$\|E(N_i^2)\| = (I - Q)^{-1}(2QM_\rho + e)$$

$$= 2MQM_\rho + M_\rho$$

$$= 2(M - I)M_\rho + M_\rho$$

$$= (2M - I)M_\rho \qquad (4.1.34)$$

Proceeding as in Theorem 4.1.4, we also get

$$\|[E(N_i)]^2\| = M_{\rho 2} \qquad (4.1.35)$$

The theorem now follows by

$$\|V(N_i)\| = \|E(N_i^2)\| - \|[E(N_i)]^2\| \qquad \square$$

Finally, we shall consider the probability of entering a particular recurrent state when the process leaves the transient class. It should be noted that if the recurrent state entered by the process is absorbing, it does not leave that state.

Let $f_{ij}^{(n)}$ be the probability that starting from transient state i, the process enters a recurrent state j in n steps. Noting that when $i \in T$ and $j \in T^c$, a $j \to i$ transition is not possible, we can consider the number of steps as the time needed for a first passage transition $i \to j$. If we define T_{ij}

to represent this random variable, we may write $f_{ij}^{(n)}$ as its distribution given by

$$P\left(T_{ij} = n\right) = f_{ij}^{(n)} \qquad i \in T, \, j \in T^c \qquad (4.1.36)$$

Let

$$f_{ij} = \sum_{n=1}^{\infty} f_{ij}^{(n)}$$

the probability of eventual passage to j. In matrix notations also let

$$\boldsymbol{F}^{(n)} = \|f_{ij}^{(n)}\| \quad \text{and} \quad \boldsymbol{F} = \|f_{ij}\|$$

We have

Theorem 4.1.6

$$\boldsymbol{F}^{(n)} = \boldsymbol{Q}^{n-1}\boldsymbol{R} \qquad (4.1.37)$$

and

$$\boldsymbol{F} = \boldsymbol{M}\boldsymbol{R} \qquad (4.1.38)$$

where we have used notations introduced in (4.1.1) and (4.1.6).

Proof   We have

$$f_{ij}^{(1)} = P_{ij} \qquad i \in T, \, j \in T^c$$

$$f_{ij}^{(n)} = \sum_{k \in T} P_{ik} f_{kj}^{(n-1)} \qquad (4.1.39)$$

that is,

$$\boldsymbol{F}^{(1)} = \boldsymbol{R}$$

$$\boldsymbol{F}^{(n)} = \boldsymbol{Q}\boldsymbol{F}^{(n-1)} \qquad (4.1.40)$$

which, on iteration on $n = 1, 2, \ldots$ gives

$$\boldsymbol{F}^{(n)} = \boldsymbol{Q}^{n-1}\boldsymbol{R}$$

Also

$$F = \sum_{n=1}^{\infty} F^{(n)}$$

$$= R + \sum_{n=2}^{\infty} Q^{n-1}R$$

$$= (I + Q + Q^2 + \cdots)R$$

$$= MR$$

This completes the proof. □

COROLLARY 1   Let R_j be the jth column of R. Then the probabilities that, starting from any transient state which is an element of T, the process enters state j, $j \in T^c$, is given by MR_j.

COROLLARY 2   If $p_T^{(0)}$ is an $(m - r)$ element row vector, with probabilities of initially finding the process in any one of the $(m - r)$ transient states as elements, then the probability that the process enters the recurrent state j, $j \in T^c$, when it leaves the class T, is given by $p_T^{(0)}MR_j$.

If the state j in the preceding corollaries is absorbing, the probabilities mentioned may be called *absorption probabilities*. Otherwise, they are merely *first passage probabilities*.

Example 4.1.1   (Example 3.1.3 Continued)

The transition probability matrix of the accounting problem is given by

$$P = \begin{bmatrix} 1 & 0 & 0 & 0 & 0 \\ 0 & 1 & 0 & 0 & 0 \\ 0.3 & 0 & 0.6 & 0.1 & 0 \\ 0.2 & 0 & 0.2 & 0.5 & 0.1 \\ 0.1 & 0.1 & 0.2 & 0.2 & 0.4 \end{bmatrix} \tag{4.1.41}$$

so that

$$Q = \begin{bmatrix} 0.6 & 0.1 & 0 \\ 0.2 & 0.5 & 0.1 \\ 0.2 & 0.2 & 0.4 \end{bmatrix} \tag{4.1.42}$$

and

$$M = (I - Q)^{-1} = \begin{bmatrix} 0.4 & -0.1 & 0 \\ -0.2 & 0.5 & -0.1 \\ -0.2 & -0.2 & 0.6 \end{bmatrix}^{-1}$$

$$= \begin{bmatrix} \frac{140}{49} & \frac{30}{49} & \frac{5}{49} \\ \frac{70}{49} & \frac{120}{49} & \frac{20}{49} \\ \frac{70}{49} & \frac{50}{49} & \frac{90}{49} \end{bmatrix} \qquad (4.1.43)$$

From M we get

$$M_\rho = \begin{bmatrix} \frac{175}{49} \\ \frac{210}{49} \\ \frac{210}{49} \end{bmatrix} \quad M_{\rho 2} = \begin{bmatrix} 12.76 \\ 18.37 \\ 18.37 \end{bmatrix}$$

$$M_D = \begin{bmatrix} \frac{140}{49} & & \bigcirc \\ & \frac{120}{49} & \\ \bigcirc & & \frac{90}{49} \end{bmatrix}$$

and

$$M_2 = \begin{bmatrix} 8.16 & 0.38 & 0.01 \\ 2.04 & 6.00 & 0.17 \\ 2.04 & 1.04 & 3.37 \end{bmatrix}$$

Using (4.1.12), we get

$$\|\sigma_{ij}^2\| = M(2M_D - I) - M_2$$

$$= \begin{bmatrix} 5.31 & 2.01 & 0.27 \\ 4.70 & 3.55 & 0.92 \\ 4.70 & 2.94 & 1.54 \end{bmatrix}$$

and from (4.1.31) we get

$$\|V(N_i)\| = \begin{bmatrix} 10.19 \\ 12.03 \\ 12.03 \end{bmatrix}$$

Some of the conclusions we may draw from these results are as follows: suppose an account is now zero month old. Before it is finally classified either as a paid up account or as a bad debt, the expected number of months it will be zero month old is 140/49, with a variance of 5.31; the expected number of months it will be a one month old account is 30/49, with variance 2.01; and the expected number of months it will be a two month old account is 5/49, with a variance 0.27. Further the total time the account will be kept alive (before it is finally classified as paid up or bad debt) has an expected value of 175/49 months, with variance 10.19. Similar interpretations can be given to the results regarding accounts that start at one and two months old.

Now consider the "loss expectancy ratio," or the proportion of accounts that get classified eventually as bad debt. From (4.1.38) we have

$$F = MR$$

For this problem we get

$$F = \begin{bmatrix} \frac{48.5}{49} & \frac{0.5}{49} \\ \frac{47}{49} & \frac{2}{49} \\ \frac{40}{49} & \frac{9}{49} \end{bmatrix}$$

The conclusions we may draw are the following:

1. Of accounts currently classified as zero month old (50/49%), approximately 1% will end up being bad debts.
2. Of accounts currently classified as one month old (200/49%), approximately 4% will end up as bad debts.
3. Of accounts currently classified as two month old (900/49%), approximately 18% will end up as bad debts.

4.2 IRREDUCIBLE FINITE MARKOV CHAINS WITH ERGODIC STATES

We have defined an ergodic state as one that is aperiodic and positive recurrent. Clearly, when a finite Markov chain is irreducible, it has to be positive recurrent. As a first step in the study of irreducible finite Markov chains with ergodic states, we shall describe a method of determining the n-step transition probabilities when their transition probability matrices admit distinct eigenvalues. This method can be extended to Markov chains with more general characteristics. Basic results from matrix algebra required in this procedure are given in Appendix A. These results relate to eigenvalues and eigenvectors.

If P is an $m \times m$ matrix, we define *eigenvalues* of P as those numbers λ for which the characteristic equation

$$PX = \lambda X \tag{4.2.1}$$

has a solution $X \neq 0$. The column vector X is called the *right eigenvector* belonging to the *eigenvalue* λ. Similarly we can define a row vector Y as a *left eigenvector* belonging to the eigenvalue λ if there exists $Y \neq 0$ such that

$$YP = \lambda Y \tag{4.2.2}$$

Incidentally, it may be noted that the eigenvalues are obtained by solving the determinant equation

$$|\lambda I - P| = 0 \tag{4.2.3}$$

When the m eigenvalues of P are all distinct, we can get m linearly independent right eigenvectors and m linearly independent left eigenvectors belonging to these distinct eigenvalues.

Let Q be the nonsingular matrix $(X_1, \ldots, X_m)$, where X_i is the right eigenvector (column) belonging to the eigenvalue $\lambda_i (i = 1, 2, \ldots, m)$. Then we can write

$$PX_j = \lambda_j X_j \tag{4.2.4}$$

so that

$$PQ = Q\Lambda \tag{4.2.5}$$

and

$$Q^{-1}PQ = \Lambda \tag{4.2.6}$$

where Λ is the diagonal matrix

$$\Lambda = \begin{bmatrix} \lambda_1 & & & \\ & \lambda_2 & & \bigcirc \\ & & \ddots & \\ & \bigcirc & & \lambda_m \end{bmatrix} \tag{4.2.7}$$

In (4.2.6) Q^{-1} turns out to be the nonsingular matrix composed of the m linearly independent left eigenvectors $Y_i (i = 1, 2, \ldots, m$, a row vector) as row vectors.

When (4.2.6) holds, the matrix P is said to be *diagonalized*. From this relation we also obtain

$$P = Q\Lambda Q^{-1}$$

$$P^2 = (Q\Lambda Q^{-1})(Q\Lambda Q^{-1}) = Q\Lambda^2 Q^{-1} \qquad (4.2.8)$$

$$\vdots$$

$$P^n = Q\Lambda^n Q^{-1} \qquad (4.2.9)$$

where

$$\Lambda^n = \begin{bmatrix} \lambda_1^n & & & \\ & \lambda_2^n & & \bigcirc \\ & & \ddots & \\ & \bigcirc & & \lambda_m^n \end{bmatrix} \qquad (4.2.10)$$

Thus (4.2.9) gives a convenient form of P^n for computational purposes. The various steps in this procedure can be summarized as follows:

1. Determine the eigenvalues of the transition probability matrix P by solving the determinant equation

$$|\lambda I - P| = 0$$

   If the eigenvalues $\lambda_i (i = 1, 2, \ldots, m)$ are distinct, proceed; otherwise, stop. (Then one must use the Jordan canonical form. Refer to advanced books on the subject.)

2. If λ_i's are distinct, obtain m column vectors $(X_1, X_2, \ldots, X_m)$ corresponding to the m eigenvalues by solving equations of the type (4.2.1). Thus determine

$$Q = \|X_1, X_2, \ldots, X_m\|$$

3. Obtain Q^{-1}. This can also be determined by solving equations of the type (4.2.2), but then the multiplicative constant should be chosen so as to make

$$QQ^{-1} = I$$

4. Obtain P^n from the relation

$$P^n = Q\Lambda^n Q^{-1}$$

We illustrate this method by solving a simple example of a two-state chain.

Example 4.2.1

Let

$$P = \begin{bmatrix} 0 & 1 \\ \frac{1}{2} & \frac{1}{2} \end{bmatrix} \tag{4.2.11}$$

We have the characteristic equation as

$$\left| \begin{bmatrix} \lambda & 0 \\ 0 & \lambda \end{bmatrix} - \begin{bmatrix} 0 & 1 \\ \frac{1}{2} & \frac{1}{2} \end{bmatrix} \right| = 0$$

$$2\lambda^2 - \lambda - 1 = 0$$

giving

$$\lambda_1 = 1 \qquad \lambda_2 = -\tfrac{1}{2} \tag{4.2.12}$$

For the right eigenvectors

$$X_1 = \begin{bmatrix} x_{11} \\ x_{12} \end{bmatrix} \quad \text{and} \quad X_2 = \begin{bmatrix} x_{21} \\ x_{22} \end{bmatrix}$$

we have

$$\begin{bmatrix} 0 & 1 \\ \frac{1}{2} & \frac{1}{2} \end{bmatrix} \begin{bmatrix} x_{11} \\ x_{12} \end{bmatrix} = 1 \cdot \begin{bmatrix} x_{11} \\ x_{12} \end{bmatrix}$$

$$x_{12} = x_{11}$$

$$\tfrac{1}{2}(x_{11} + x_{12}) = x_{12} \tag{4.2.13}$$

and

$$\begin{bmatrix} 0 & 1 \\ \frac{1}{2} & \frac{1}{2} \end{bmatrix} \begin{bmatrix} x_{21} \\ x_{22} \end{bmatrix} = -\tfrac{1}{2} \begin{bmatrix} x_{21} \\ x_{22} \end{bmatrix}$$

$$x_{22} = -\tfrac{1}{2} x_{21} \tag{4.2.14}$$

$$\tfrac{1}{2}(x_{21} + x_{22}) = -\tfrac{1}{2} x_{22}$$

The two equations in (4.2.13) are not independent. Hence the solution is unique only up to a multiplicative constant. Let $x_{11} = 1$; then we get $x_{12} = 1$. Similarly assuming $x_{21} = 1$, we get $x_{22} = -\frac{1}{2}$. Therefore we get

$$Q = \begin{bmatrix} x_{11} & x_{21} \\ x_{12} & x_{22} \end{bmatrix} = \begin{bmatrix} 1 & 1 \\ 1 & -\frac{1}{2} \end{bmatrix} \tag{4.2.15}$$

Let

$$Q^{-1} = \begin{bmatrix} z_{11} & z_{21} \\ z_{12} & z_{22} \end{bmatrix}$$

Using the result

$$QQ^{-1} = I$$

we get

$$Q^{-1} = \begin{bmatrix} \frac{1}{3} & \frac{2}{3} \\ \frac{2}{3} & -\frac{2}{3} \end{bmatrix} \tag{4.2.16}$$

Finally, we write

$$P^n = \begin{bmatrix} 1 & 1 \\ 1 & -\frac{1}{2} \end{bmatrix} \begin{bmatrix} 1 & 0 \\ 0 & \left(-\frac{1}{2}\right)^n \end{bmatrix} \begin{bmatrix} \frac{1}{3} & \frac{2}{3} \\ \frac{2}{3} & -\frac{2}{3} \end{bmatrix}$$

$$= \begin{bmatrix} \frac{1}{3} + \frac{2}{3}\left(-\frac{1}{2}\right)^n & \frac{2}{3} - \frac{2}{3}\left(-\frac{1}{2}\right)^n \\ \frac{1}{3} - \frac{1}{3}\left(-\frac{1}{2}\right)^n & \frac{2}{3} + \frac{1}{3}\left(-\frac{1}{2}\right)^n \end{bmatrix} \tag{4.2.17}$$

Answer

A great deal of information regarding the behavior of finite Markov chains can be derived from the nature of eigenvalues of the associated transition probability matrices. For instance, it is easy to see that 1 is always an eigenvalue of the transition probability matrix. From the Perron-Frobenius theorem of matrix theory (Appendix A), we find that for an irreducible chain all other eigenvalues are less than or equal to 1 in modulus. It can also be shown that the rate of approach of the n-step transition probabilities to their limiting values is determined by the eigenvalue of largest modulus less than unity. Further in the case of periodic chains the periodicity of the chain is given by the number of eigenvalues of unit modulus. A discussion of these and other results based on matrix theory requires a mathematical sophistication beyond the level assumed for this book. We shall therefore restrict ourselves to the study of the limiting behavior of the transition probabilities through other simpler means. First consider the following lemma.

LEMMA 4.2.1 *Let P be an $(m \times m)$ stochastic matrix with no zero elements. Let ε be the smallest entry of P. Let X be any m-component column vector with minimum component a_0 and maximum component b_0. Let a_1 and b_1 be the minimum and maximum components of PX. Then*

$$a_1 \geq a_0 \qquad b_1 \leq b_0 \tag{4.2.18}$$

and

$$b_1 - a_1 \leq (1 - 2\varepsilon)(b_0 - a_0) \tag{4.2.19}$$

Proof Let X' be the column vector obtained by replacing all components of X except the minimum component by the maximum component b_0. We have

$$X = \begin{bmatrix} x_1 \\ x_2 \\ \vdots \\ x_m \end{bmatrix} \quad \text{and} \quad X' = \begin{bmatrix} b_0 \\ b_0 \\ \vdots \\ b_0 \\ a_0 \\ b_0 \\ \vdots \\ b_0 \end{bmatrix}$$

Therefore

$$X \leq X' \qquad \text{componentwise}$$

and

$$PX \leq PX' \tag{4.2.20}$$

where

$$PX = \begin{bmatrix} P_{11} & P_{12} & \cdots & P_{1m} \\ \vdots & \vdots & & \vdots \\ P_{m1} & P_{m2} & \cdots & P_{mm} \end{bmatrix} \begin{bmatrix} x_1 \\ x_2 \\ \vdots \\ x_m \end{bmatrix}$$

and

$$PX' = \begin{bmatrix} c_1 a_0 + (1 - c_1) b_0 \\ c_2 a_0 + (1 - c_2) b_0 \\ \vdots \\ c_m a_0 + (1 - c_m) b_0 \end{bmatrix}$$

where $c_i = P_{ij}$ if a_0 occurs in the jth row of X' and hence $c_i \geq \varepsilon$, $i = 1, 2, 3, \ldots, m$.

Clearly

$$c_i a_0 + (1 - c_i) b_0 = b_0 - c_i (b_0 - a_0)$$

$$\leq b_0 - \varepsilon(b_0 - a_0) \qquad (4.2.21)$$

Using the inequality (4.2.20), we therefore have

$$b_1 \leq b_0 - \varepsilon(b_0 - a_0) \qquad (4.2.22)$$

Apply this result to the initial vector $-X$ instead of X [the maximum element of $-X$ is $-a_0$, the minimum element is $-b_0$, and the maximum element of $P(-X)$ is $-a_1$], we get

$$-a_1 \leq -a_0 - \varepsilon(-a_0 + b_0) \qquad (4.2.23)$$

Adding (4.2.22) and (4.2.23), we get the result (4.2.19). It should also be noted that (4.2.22) and (4.2.23) imply that $a_1 \geq a_0$ and $b_1 \leq b_0$. $\quad\square$

Remark 4.2.1 Lemma 4.2.1 is true even when $\varepsilon = 0$. But for the proof of Theorem 4.2.1 we need only the case $\varepsilon > 0$.

Theorem 4.2.1 *Let P be the transition probability matrix of an aperiodic, irreducible, m-state finite Markov chain. Then*

$$\lim_{n \to \infty} P^n = \Pi = \begin{bmatrix} \pi \\ \pi \\ \vdots \\ \pi \end{bmatrix} \qquad (4.2.24)$$

where $\pi = [\pi_1, \pi_2, \ldots, \pi_m]$ with $0 < \pi_j < 1$ and $\sum_1^m \pi_j = 1$.

Proof Case 1. Suppose the transition probability matrix P of the Markov chain has no zero elements. Let ε be the smallest element of P. Let ρ_j be an

m-component column vector with a 1 in the jth place and 0 elsewhere. Further let a_n and b_n be the minimum and maximum components of the vector $P^n \rho_j$. Clearly $a_0 = 0$ and $b_0 = 1$. We have

$$P^n \rho_j = P \cdot P^{n-1} \rho_j \qquad n = 1, 2, \ldots \qquad (4.2.25)$$

Writing $P^{n-1} \rho_j = X$ and applying Lemma 4.2.1 repeatedly, we get

$$b_0 \geq b_1 \geq b_2 \geq b_3 \geq \cdots \quad \text{and} \quad a_0 \leq a_1 \leq a_2 \leq a_3 \leq \cdots$$

and

$$b_n - a_n \leq (1 - 2\varepsilon)(b_{n-1} - a_{n-1}) \qquad n \geq 1 \qquad (4.2.26)$$

Let $d_n = b_n - a_n$; we then have

$$d_1 \leq (1 - 2\varepsilon)(b_0 - a_0) = (1 - 2\varepsilon)$$

$$d_2 \leq (1 - 2\varepsilon)d_1 \leq (1 - 2\varepsilon)^2$$

$$\vdots \qquad \vdots \qquad\qquad\qquad (4.2.27)$$

$$d_n \leq (1 - 2\varepsilon)d_{n-1} \leq (1 - 2\varepsilon)^n$$

which shows that as $n \to \infty$, $d_n \to 0$, and hence b_n and a_n approach a common limit.

Also $P^n \rho_j$ is the jth column of P^n, and therefore the jth column of $P^n \to$ a constant, say π_j, as $n \to \infty$. Further

$$a_n \leq \pi_j \leq b_n \qquad \text{for all } n \geq 1$$

but

$$a_1 > 0 \quad \text{and} \quad b_1 < 1$$

and hence

$$0 < \pi_j < 1$$

Clearly $\sum_{j=1}^m P_{ij}^{(n)} = 1$ for all n, which should be true of $\lim_{n \to \infty} P^n$. Hence the theorem is proved.

Case 2.   Let P have some zeros as elements. But from Theorem 3.3.5 we know that there exists an N such that for $n \geq N$, P^n has no zero entries.

Using P^N instead of P in the proof, we get

$$d_{kN} \leq (1 - 2\varepsilon_N)^k$$

where ε_N is the smallest element of P^N. It is also clear that for $0 \leq l \leq N$,

$$d_{kN+l} \leq (1 - 2\varepsilon_N)^k \tag{4.2.28}$$

Letting $k \to \infty$, we again have the same theorem. $\square$

In some cases it might be helpful to get the rate of convergence of $P_{ij}^{(n)}$ to its limit π_j. A very conservative bound for this rate of convergence is given by Theorem 4.2.2.

Theorem 4.2.2   *Let $P_{ij}^{(n)}$ be the n-step transition probability of an aperiodic, irreducible, m-state Markov chain. Let $\lim_{n \to \infty} P_{ij}^{(n)} = \pi_j$ ($j = 1, 2, 3, \ldots, m$) be the limiting probabilities. Then there exist constants c and r with c > 0 and 0 < r < 1, such that*

$$P_{ij}^{(n)} = \pi_j + e_{ij}^{(n)} \tag{4.2.29}$$

where

$$|e_{ij}^{(n)}| \leq cr^n \tag{4.2.30}$$

When the transition probability matrix P has no zero element, $c = 1$.

Proof   From Theorem 4.2.1 we know that

$$a_n \leq P_{ij}^{(n)} \leq b_n$$

and hence

$$|e_{ij}^{(n)}| \leq b_n - a_n = d_n \tag{4.2.31}$$

From (4.2.28) we also have

$$d_{kN+l} \leq (1 - 2\varepsilon_N)^k$$

Setting $n = kN + l$, we get $k = (n - l)/N$ and

$$d_n \leq (1 - 2\varepsilon_N)^{(n/N)-(l/N)} \leq (1 - 2\varepsilon_N)^{(n/N)-1}$$

$$= (1 - 2\varepsilon_N)^{-1}\left[(1 - 2\varepsilon_N)^{1/N}\right]^n \tag{4.2.32}$$

The theorem now follows by setting $c = (1 - 2\varepsilon_N)^{-1}$ and $r = (1 - 2\varepsilon_N)^{1/N}$. When the transition probability matrix P has no zero elements, we start with (4.2.27), which leads to $c = 1$. $\square$

Remark 4.2.2   The convergence property of the transition probabilities exhibited in Theorem 4.2.2 is known as *geometric ergodicity*, which indicates the geometric nature of convergence to the limiting result. For a discussion on ergodicity see Sections 3.3 and 9.3.

Remark 4.2.3   For practical purposes, depending on the accuracy required, it is possible to find an N_α^*, defined as

$$N_\alpha^* = \min\left\{ n \,\big|\, |P_{ij}^{(n)} - \pi_j| < \alpha \right\} \qquad (4.2.33)$$

The value of N_α^* can be considered the *effective steady-state time* for the Markov chain. If it is possible to obtain an upper bound for N_α^*, it can be used in decision problems to see whether the process is really in a state of equilibrium. From Theorem 4.2.2 we can obtain such a bound as follows: clearly

$$N_\alpha^* = \min\left\{ n \,\big|\, |e_{ij}^{(n)}| < \alpha \right\} \qquad (4.2.34)$$

From (4.2.30) and (4.2.34) we know that

$$cr^n < \alpha \quad \text{implies} \quad |e_{ij}^{(n)}| < \alpha \qquad (4.2.35)$$

Hence we choose the smallest n so as to satisfy

$$cr^n < \alpha$$

When the transition probability matrix has no zero entries, the inequality to be used is $r^n < \alpha$. The value of N_α^* obtained is an upper bound for the effective steady-state time.

Example 4.2.2

1. Let

$$P = \begin{bmatrix} 0.3 & 0.3 & 0.4 \\ 0.4 & 0.3 & 0.3 \\ 0.3 & 0.4 & 0.3 \end{bmatrix}$$

Now $\varepsilon = 0.3$; $r = 1 - 2\varepsilon = 1 - 0.6 = 0.4$. Suppose $\alpha = 0.001$; then the smallest n to satisfy the inequality $r^n < \alpha$ is $n = 8$.

2. Let

$$P = \begin{bmatrix} 0 & 0.6 & 0.4 \\ 0.6 & 0.4 & 0 \\ 0 & 0.4 & 0.6 \end{bmatrix}$$

We get

$$P^2 = \begin{bmatrix} 0.36 & 0.40 & 0.24 \\ 0.24 & 0.52 & 0.24 \\ 0.24 & 0.40 & 0.36 \end{bmatrix}$$

Now $\varepsilon = 0.24$; $r = (1 - 2\varepsilon)^{1/2} = (0.52)^{1/2} = 0.72$ (approximate), and $c = (1 - 2\varepsilon)^{-1} = 1.92$ (approximate). Suppose $\alpha = 0.001$; then the smallest n to satisfy the inequality $cr^n < \alpha$ is $n = 24$.

Answer

DEFINITION   *Suppose $p = (p_1, p_2, \ldots, p_m)$ is a vector of probabilities, such that $\sum_1^m p_i = 1$. Then the probability distribution $\{ p_i \}$ is said to be a stationary distribution for an m-state Markov chain with transition probability matrix P if*

$$p = pP \qquad (4.2.36)$$

It can be shown by induction that (4.2.36) also implies

$$p = pP^n \qquad n \geq 1$$

Thus, if $\{ p_i \}$ is the probability distribution of the state of the system at some stage, then it also gives the (unconditional) probability distribution of the state of the system at any step from then onward.

When the finite Markov chain is irreducible and aperiodic, the limiting distribution is unique and stationary. This is brought out in Theorem 4.2.3.

Theorem 4.2.3   *Given the transition probability matrix P of an aperiodic, irreducible, m-state finite Markov chain, there exists a unique probability vector $\pi = (\pi_1, \pi_2, \ldots, \pi_m)$ such that $\pi e = 1$ and*

$$\Pi P = \Pi \qquad and \qquad P\Pi = \Pi \qquad (4.2.37)$$

where Π is a matrix with m identical rows, each represented by π and e is a column vector consisting only of ones. The probability vector π gives the stationary distribution of the process.

Proof   From Theorem 4.2.1 we have $\lim_{n \to \infty} P^n = \Pi$. This implies $\lim_{n \to \infty} P^{n+1} = \Pi$, that is, $\lim_{n \to \infty} P^n \cdot P = \Pi$; whence $\Pi P = \Pi$. Similarly $\lim_{n \to \infty} P \cdot P^n = \Pi$, giving $P\Pi = \Pi$. To show uniqueness, let us assume that there exists a probability vector $v = (v_1, v_2, \ldots, v_m)$, with $ve = 1$ satisfying the relations

$$VP = V \quad \text{and} \quad PV = V \tag{4.2.38}$$

where

$$V = \begin{bmatrix} v \\ v \\ \vdots \\ v \end{bmatrix}$$

Consider

$$\lim_{n \to \infty} vP^n = v\Pi = \pi$$

But from (4.2.38) we have

$$vP = v$$

$$vP^2 = vP \cdot P = vP = v$$

$$\vdots$$

$$vP^n = v \qquad \text{for all } n$$

Hence

$$\lim_{n \to \infty} vP^n = v \tag{4.2.40}$$

Therefore clearly $v = \pi$. From the foregoing discussion it is also clear that

$$\pi P^n = \pi \qquad \text{for all } n$$

which shows that π is the stationary distribution.   □

Remark 4.2.4   Theorem 4.2.3 provides a set of simultaneous linear equations which are convenient for determining the stationary distribution, if it exists. We have

$$\pi P = \pi \tag{4.2.41}$$

and

$$\pi e = 1$$

Remark 4.2.5   Indiscriminate use of equations (4.2.41) for the determination of limiting probabilities should be avoided. If the Markov chain is periodic, a stationary distribution exists, but not a limiting distribution. For example, consider the periodic Markov chain with a transition probability matrix:

$$P = \begin{bmatrix} 0 & 1 \\ 1 & 0 \end{bmatrix} \qquad (4.2.42)$$

We have

$$P^n = \begin{cases} \begin{bmatrix} 1 & 0 \\ 0 & 1 \end{bmatrix} & \text{if } n \text{ is even} \\[2ex] \begin{bmatrix} 0 & 1 \\ 1 & 0 \end{bmatrix} & \text{if } n \text{ is odd} \end{cases}$$

The alternating character of P^n persists regardless of the size of n. Hence we cannot identify a unique limiting distribution for the Markov chain. On the other hand, by solving equations (4.2.41) for P given by (4.2.42), we get

$$[\pi_1, \pi_2] = [0.5, 0.5]$$

which is the stationary distribution of the Markov chain.

When the Markov chain includes more than one equivalence class, the use of equations (4.2.41) may not even give the stationary distribution. Also when there are more than one recurrent classes, the equations do not have a unique solution. In the case of transient states the corresponding probabilities can be easily seen to be zero. Writing out equations (4.2.41), we have

$$\pi_j = \sum_i \pi_i P_{ij} = \sum_i \pi_i P_{ij}^{(n)} \qquad \text{for all } n$$

When state j is transient $\lim_{n \to \infty} P_{ij}^{(n)} = 0$.

Example 4.2.3

For his yearly vacation a business executive selects one of three places—the Bahamas, Europe, or Hawaii—using the following rule: If he has been to the Bahamas the past year, he will choose Europe with probability $\frac{2}{3}$ and Hawaii with probability $\frac{1}{3}$. If he has been to Europe the past year, he will choose the Bahamas, Europe again, and Hawaii with probabilities $\frac{3}{8}$, $\frac{1}{8}$, and $\frac{1}{2}$, respectively. If he has spent his vacation in Hawaii, the Bahamas and Europe are equally likely to be chosen this year.

How would you rate his preferences after a sufficiently long time? Clearly the Markov chain has three states: B (the Bahamas), E (Europe), and H (Hawaii). The transition probability matrix of this Markov chain is given by

$$
P = \begin{array}{c} \\ B \\ E \\ H \end{array}
\begin{array}{ccc} B & E & H \end{array}
\begin{bmatrix} 0 & \frac{2}{3} & \frac{1}{3} \\ \frac{3}{8} & \frac{1}{8} & \frac{1}{2} \\ \frac{1}{2} & \frac{1}{2} & 0 \end{bmatrix}
$$

The long-run probability distribution is given by the steady-state probabilities, say π_B, π_E, and π_H, which can be determined as solutions to the set of equations (4.2.41). Thus we have

$$
[\pi_B, \pi_E, \pi_H] = [\pi_B, \pi_E, \pi_H]
\begin{bmatrix} 0 & \frac{2}{3} & \frac{1}{3} \\ \frac{3}{8} & \frac{1}{8} & \frac{1}{2} \\ \frac{1}{2} & \frac{1}{2} & 0 \end{bmatrix}
$$

giving

$$
\pi_B = \frac{3}{8}\pi_E + \frac{1}{2}\pi_H
$$

$$
\pi_E = \frac{2}{3}\pi_B + \frac{1}{8}\pi_E + \frac{1}{2}\pi_H \tag{4.2.43}
$$

$$
\pi_H = \frac{1}{3}\pi_B + \frac{1}{2}\pi_E
$$

These three equations are not independent since we have the condition

$$
\pi_B + \pi_E + \pi_H = 1
$$

Combining the first two equations of (4.2.43), we get the three independent equations as

$$
\pi_B + \pi_E + \pi_H = 1
$$

$$
4\pi_B - 9\pi_E + 8\pi_H = 0
$$

$$
2\pi_B + 3\pi_E - 6\pi_H = 0
$$

Solving these three equations simultaneously, we get

$$
\pi_B = 0.3 \quad \pi_E = 0.4 \quad \pi_H = 0.3
$$

which show that after a sufficiently long time he would prefer Europe with probability 0.40 and the Bahamas and Hawaii with probability 0.30 each.

Let us now consider the general problem of obtaining $\lim_{n \to \infty} P^n$ when the Markov chain has more than one equivalence class. The following observations are pertinent:

1. If state j is absorbing, $\lim_{n \to \infty} P_{jj}^{(n)} = 1$ and $\lim_{n \to \infty} P_{ji}^{(n)} = 0$, for $i \neq j$.

2. If state j is transient, regardless of initial state i, $\lim_{n \to \infty} P_{ij}^{(n)} = 0$.

3. Once a Markov chain enters a state belonging to a recurrent class, it stays in that class permanently.

4. Having started from a transient state i the probability that the Markov chain enters the recurrent state j when it eventually leaves the transient class is given by f_{ij} of Theorem 4.1.6. Let state j belong to a recurrent class C_l and $\sum_{k \in C_l} f_{ik} = f_i(C_l)$. Since once in the recurrent class the transitions of the Markov chain are governed by the transition probabilities within that class, we get

$$\pi_{ij} = \lim_{n \to \infty} P_{ij}^{(n)} = f_i(C_l)\pi_j \qquad i \in T, j \in C_l$$

where π_j is the limiting probability for state j within class C_l.

Based on these observations, the procedure for the determination of $\lim_{n \to \infty} P^n$ in the general case is illustrated with the help of an example.

Example 4.2.4

Consider the Markov chain with the transition probability matrix (which has been arranged in the canonical form):

$$
P = \begin{array}{c c} & \begin{array}{ccccc} 1 & \ \ 2 & \ \ 3 & \ \ 4 & \ \ 5 \end{array} \\ \begin{array}{c} 1 \\ 2 \\ 3 \\ 4 \\ 5 \end{array} & \left[\begin{array}{ccccc} 1 & 0 & 0 & 0 & 0 \\ 0 & 0.3 & 0.7 & 0 & 0 \\ 0 & 0.6 & 0.4 & 0 & 0 \\ 0.2 & 0 & 0.1 & 0.5 & 0.2 \\ 0 & 0.4 & 0 & 0.3 & 0.3 \end{array} \right] \end{array}
$$

The Markov chain has three equivalence classes.

$\{1\}$:   an absorbing state.

$\{2, 3\}$:   a recurrent class.

$\{4, 5\}$:   a transient class.

Let $\pi_{ij} = \lim\limits_{n \to \infty} P_{ij}(n)$. We have

$$\pi_{11} = 1 \qquad \pi_{ij} = 0 \qquad \text{for } j = 2, 3, 4, 5$$

$$\pi_{2j} = 0 \qquad \text{for } j = 1, 4, 5$$

$$\pi_{3j} = 0 \qquad \text{for } j = 1, 4, 5$$

$$\pi_{4j} = 0 \qquad \text{for } j = 4, 5$$

$$\pi_{5j} = 0 \qquad \text{for } j = 4, 5$$

Also

$$\pi_{i2} = \pi_2 \qquad \text{for } i = 2, 3$$

$$\pi_{i3} = \pi_3 \qquad \text{for } i = 2, 3$$

Using equations (4.2.41), we get

$$[\pi_2, \pi_3]\begin{bmatrix} 0.3 & 0.7 \\ 0.6 & 0.4 \end{bmatrix} = [\pi_2, \pi_3]$$

$$\pi_2 + \pi_3 = 1$$

Solving these equations, we get

$$\pi_2 = 0.4615 \qquad \pi_3 = 0.5385$$

To determine the first passage probabilities f_{ij} for $i = 4, 5$ and $j = 1, 2, 3$ we use Theorem 4.1.6. We have

$$Q = \begin{bmatrix} 0.5 & 0.2 \\ 0.3 & 0.3 \end{bmatrix} \qquad R = \begin{bmatrix} 0.2 & 0 & 0.1 \\ 0 & 0.4 & 0 \end{bmatrix}$$

which gives

$$(I - Q)^{-1} = \begin{bmatrix} 2.4138 & 0.6897 \\ 1.0345 & 1.7241 \end{bmatrix}$$

and

$$F = (I - Q)^{-1}R = \begin{bmatrix} 0.4828 & 0.2759 & 0.2414 \\ 0.2069 & 0.6896 & 0.1034 \end{bmatrix}$$

Since states 2 and 3 belong to the same equivalence class, the probability that the process, having initially started from state 4, enters this class is given by the sum $0.2759 + 0.2414 = 0.5173$. The corresponding probability for state 5 is given by the sum $0.6896 + 0.1034 = 0.7930$. Thus we have

$$[\pi_{42}, \pi_{43}] = 0.5173[\pi_2, \pi_3] = [0.2387, 0.2786]$$

$$[\pi_{52}, \pi_{53}] = 0.7930[\pi_2, \pi_3] = [0.3660, 0.4271]$$

Summarizing these results, we have

$$
\lim_{n \to \infty} \boldsymbol{P}^n =
\begin{array}{c}
1 \\ 2 \\ 3 \\ 4 \\ 5
\end{array}
\begin{bmatrix}
1 & 0 & 0 & 0 & 0 \\
0 & 0.4615 & 0.5385 & 0 & 0 \\
0 & 0.4615 & 0.5385 & 0 & 0 \\
0.4828 & 0.2387 & 0.2185 & 0 & 0 \\
0.2069 & 0.3660 & 0.4271 & 0 & 0
\end{bmatrix}
\begin{array}{ccccc}
1 & 2 & 3 & 4 & 5
\end{array}
$$

Answer

As given in Theorem 4.2.3 and illustrated in Examples 4.2.3 and 4.2.4, the limiting probabilities of an irreducible aperiodic Markov chain are determined by solving the set of linear simultaneous equations

$$\pi P = \pi$$

$$\pi e = 1 \qquad\qquad (4.2.44)$$

where $\pi = (\pi_1, \pi_2, \ldots, \pi_m)$ and $e^T = (1, 1, \ldots, 1)$. In so doing, available computer routines can be used. Nevertheless, in many applied problems the transition probability matrix P may exhibit some special structures which are convenient for computer calculations. As an example consider the transition probability matrix P which can be partitioned as

$$P = \begin{bmatrix} A & B \\ C & D \end{bmatrix} \qquad\qquad (4.2.45)$$

where one of the submatrices, say D, is structurally convenient for inversion. Let the dimensions of the submatrices be $A : m_1 \times m_1$, $B : m_1 \times m_2$, $C : m_2 \times m_1$, $D : m_2 \times m_2$. Let the corresponding partition in π be $(\tilde{\pi}_1, \tilde{\pi}_2)$, where $\tilde{\pi}_1 = (\pi_1, \pi_2, \ldots, \pi_{m_1})$ and $\tilde{\pi}_2 = (\pi_{m_1+1}, \pi_{m_1+2}, \ldots, \pi_m)$. Also let e_1 and e_2 be unit column vectors of m_1 and m_2 components, respectively. Now the equations in (4.2.44) can be written out as

$$[\tilde{\pi}_1, \tilde{\pi}_2]\begin{bmatrix} A & B \\ C & D \end{bmatrix} = [\tilde{\pi}_1, \tilde{\pi}_2] \qquad (4.2.46)$$

Then we get

$$\tilde{\pi}_1 e_1 + \pi_2 e_2 = 1$$

$$\tilde{\pi}_1 A + \tilde{\pi}_2 C = \tilde{\pi}_1 \qquad (4.2.47)$$

$$\tilde{\pi}_1 B + \tilde{\pi}_2 D = \tilde{\pi}_2$$

The last equation of (4.2.47) gives

$$\tilde{\pi}_2 = \tilde{\pi}_1 B (I - D)^{-1} \qquad (4.2.48)$$

provided $(I - D)^{-1}$ exists. Substituting this expression in the other two equations of (4.2.47) we get

$$\tilde{\pi}_1 \left(A + B(I - D)^{-1} C \right) = \tilde{\pi}_1$$

$$\tilde{\pi}_1 \left(e_1 + B(I - D)^{-1} e_2 \right) = 1 \qquad (4.2.49)$$

The set of equations (4.2.49) is a reduced system of equations, and if $(I - D)$ is invertible and has a nice structure that can be exploited in the inversion process (e.g., a triangular or an almost triangular matrix), the solution of the set of equations (4.2.44) can be obtained in two steps:

1.   Solve for the set (4.2.49).
2.   Determine $\tilde{\pi}_2 = (\pi_{m_1 + 1}, \pi_{m_1 + 2}, \ldots, \pi_m)$ from (4.2.48).

There are some standard transition probability matrices that lend themselves well to recursive solution techniques on the original set (4.2.44). Both procedures will be discussed further in Chapter 11. However, in addition to the limiting distribution of a Markov chain, we are interested in information regarding first passage and occupation times.

First Passage Times

Suppose the process is in state i. The time taken by the process to go from state i to state j for the first time is called the first passage time of the transition $i \to j$. When $j = i$, we call the number of steps required for such a transition the *recurrence time* of state i.

Let T_{ij} be the first passage time of the transition $i \to j$, and $f_{ij}^{(n)}$ be its distribution defined by

$$P\left(T_{ij} = n\right) = P\left[X_n = j, X_r \neq j(1 \leq r < n)|X_0 = i\right]$$

$$= f_{ij}^{(n)} \tag{4.2.50}$$

The probabilities $f_{ij}^{(n)}$ ($n = 1, 2, \ldots$) can be determined using Theorem 4.1.6 as follows.

Here we are considering a Markov chain with a single recurrent equivalence class of states. Modify the original transition probability matrix, such that state j is now absorbing. Thus the process stops as soon as it enters state j for the first time. In relation to state j, all other states of the Markov chain are transient. Let $\tilde{P}$ be the resulting transition probability matrix in the canonical form:

$$\tilde{P} = \begin{array}{c} \\ j \\ 1 \\ 2 \\ \vdots \\ i \\ \vdots \\ m \end{array} \begin{array}{c} \begin{array}{ccccccc} j & 1 & 2 & 3 & \cdot & \cdot & \cdot & m \end{array} \\ \left[\begin{array}{c|ccccccc} 1 & 0 & 0 & 0 & \cdot & \cdot & \cdot & 0 \\ \hline P_{1j} & P_{11} & P_{12} & P_{13} & \cdot & \cdot & \cdot & P_{1m} \\ P_{2j} & P_{21} & P_{22} & P_{23} & \cdot & \cdot & \cdot & P_{2m} \\ \vdots & \vdots & \vdots & \vdots & & & & \\ P_{ij} & P_{i1} & P_{i2} & P_{i3} & \cdot & \cdot & \cdot & P_{im} \\ \vdots & \vdots & \vdots & \vdots & & & & \\ P_{mj} & P_{m1} & P_{m2} & P_{m3} & \cdot & \cdot & \cdot & P_{mm} \end{array} \right] \end{array}$$

$$= \begin{bmatrix} 1 & 0 \\ R & Q \end{bmatrix} \tag{4.2.51}$$

In the Markov chain with the transition probability matrix $\tilde{P}$, the distribution of the first passage times to state j from state $i = 1, 2, 3, \ldots, j - 1, j + 1, \ldots, m$, is given by Theorem 4.1.6 as

$$F^{(n)} = Q^{n-1}R \tag{4.2.52}$$

where Q and R are as given in (4.2.51); that is, Q is the original matrix P after deleting the jth row and jth column, and R is the jth column vector of P excluding the jth element.

Since for sufficiently large n, Q^n is a null matrix, $F^{(n)}$ is easily obtained through matrix multiplications.

Alternately, one can write

$$P(T_{ij} > n) = \sum_{r \neq j} Q_{ir}^{(n)} \qquad (4.2.53)$$

where $Q_{ir}^{(n)}$ is the (i, r)th element of Q^n. We have

$$P(T_{ij} = n) = P(T_{ij} > n - 1) - P(T_{ij} > n) \qquad (4.2.54)$$

Algebraic simplifications show that this procedure also leads to the result (4.2.52).

The mean and variance of the random variable T_{ij} are easily derived using results from Section 4.1, as illustrated in the following example.

Example 4.2.5

Frequently the reliability of a system is increased by using redundant components. Consider a maintained system that contains four components in parallel. Let us assume for now that for system success at least one component should be in working order. The system, however, is inspected once every 5 min, and if a failure is detected on any component, repair starts on it immediately. Since the inspection is only once every 5 min, for the sake of simplicity we rule out the possibility of a component getting repaired and again failing within that interval. Suppose that the conditions are such that a Markov chain model can be used for the system, with the number of working components as the state and the observation epochs as just identified. By observing the system over a long period of time, the following transition probability matrix was obtained:

$$
P = \begin{array}{c} \\ 0 \\ 1 \\ 2 \\ 3 \\ 4 \end{array}
\begin{array}{ccccc}
0 & 1 & 2 & 3 & 4 \\
\left[\begin{array}{ccccc}
0.13 & 0.34 & 0.35 & 0.15 & 0.03 \\
0.02 & 0.24 & 0.42 & 0.26 & 0.06 \\
0.00 & 0.07 & 0.38 & 0.42 & 0.13 \\
0.00 & 0.03 & 0.15 & 0.53 & 0.29 \\
0.00 & 0.00 & 0.05 & 0.29 & 0.66
\end{array}\right]
\end{array}
$$

Note that the state space includes state 0 in which no component is working. From the transition probability matrix it is clear that all states communicate and belong to a single recurrent class. Now if we are interested in obtaining information on the first passage time of the Markov chain from any state (1, 2, 3, or 4) to zero, we may consider an associated Markov chain with the transition probability matrix:

$$
\tilde{P} = \begin{array}{c} \\ 0 \\ 1 \\ 2 \\ 3 \\ 4 \end{array}
\begin{array}{ccccc}
0 & 1 & 2 & 3 & 4 \\
\left[\begin{array}{ccccc}
1 & 0 & 0 & 0 & 0 \\
0.02 & 0.24 & 0.42 & 0.26 & 0.06 \\
0.00 & 0.07 & 0.38 & 0.42 & 0.13 \\
0.00 & 0.03 & 0.15 & 0.53 & 0.29 \\
0.00 & 0.00 & 0.05 & 0.29 & 0.66
\end{array}\right]
\end{array}
$$

In this Markov chain state 0 is absorbing. Consequently the mean and variance of various first passage times can be determined as described earlier.

Suppose the system is such that a minimum of two components are necessary for its successful operation. Now the first passage time of the Markov chain to states 0 or 1 gives the time to system failure. The mean and variance of this characteristic can be determined by defining an associated Markov chain with the transition probability matrix:

$$
\tilde{P} = \begin{array}{c} \\ 0 \\ 1 \\ 2 \\ 3 \\ 4 \end{array}
\begin{array}{ccccc}
0 & 1 & 2 & 3 & 4 \\
\left[\begin{array}{cc|ccc}
1 & 0 & 0 & 0 & 0 \\
0 & 1 & 0 & 0 & 0 \\
\hline
0.00 & 0.07 & 0.38 & 0.42 & 0.13 \\
0.00 & 0.03 & 0.15 & 0.53 & 0.29 \\
0.00 & 0.00 & 0.05 & 0.29 & 0.66
\end{array}\right]
\end{array}
$$

The corresponding submatrix Q is given as

$$
Q = \begin{bmatrix}
0.38 & 0.42 & 0.13 \\
0.15 & 0.53 & 0.29 \\
0.05 & 0.29 & 0.66
\end{bmatrix}
$$

and

$$
M = (I - Q)^{-1} = \begin{bmatrix}
7.06 & 16.84 & 17.06 \\
6.11 & 19.06 & 18.59 \\
6.25 & 18.73 & 21.31
\end{bmatrix}
$$

From these results, using expressions (4.1.28) and (4.1.31), we get

$$M\rho = \begin{bmatrix} 40.96 \\ 43.76 \\ 46.29 \end{bmatrix}$$

$$\|V(N_i)\| = \begin{bmatrix} 1912.93 \\ 1931.02 \\ 1935.08 \end{bmatrix}$$

which give the mean and variance of times to failure if the system starts its operation with 2, 3, and 4 components, respectively.

The mean recurrence time of any state in a Markov chain is obtained as in Theorem 4.2.4.

Theorem 4.2.4   *Let $\pi = (\pi_1, \pi_2, \ldots, \pi_m)$ be the limiting distribution of an aperiodic, irreducible m-state Markov chain whose transition probability matrix is P. Let T_{ij} be the first passage time of the transition $i \to j$, and let τ_{ij} be its expected value. Then*

$$\tau_{ii} = \frac{1}{\pi_i} \tag{4.2.55}$$

Proof   The first passage $i \to j$ can occur in one step or $n(>1)$ steps. Accordingly, we can write

$$T_{ij} = \begin{cases} 1 & \text{with probability } P_{ij} \\ T_{kj} + 1 & \text{with probability } P_{ik},\ k \neq j \end{cases}$$

Taking expectations, we get

$$\tau_{ij} = P_{ij} + \sum_{k \neq j} P_{ik}(\tau_{kj} + 1)$$

$$= \sum_{k=1}^{m} P_{ik}\tau_{kj} - P_{ij}\tau_{jj} + 1 \tag{4.2.56}$$

Let $\tau = \|\tau_{ij}\|$. Then

$$\tau = P(\tau - \tau_D) + E \tag{4.2.57}$$

where

$$\tau_D = \begin{bmatrix} \tau_{11} & & & \\ & \tau_{22} & & \bigcirc \\ & & \ddots & \\ \bigcirc & & & \tau_{mm} \end{bmatrix} \quad \text{and} \quad E = ee^T = \begin{bmatrix} 1 & 1 & \cdots & 1 \\ \vdots & & & \vdots \\ 1 & 1 & \cdots & 1 \end{bmatrix}$$

Premultiplying (4.2.57) by π, we have

$$\pi\tau = \pi P(\tau - \tau_D) + \pi E$$

$$= \pi(\tau - \tau_D) + \pi E$$

which gives

$$\pi\tau_D = \pi E \tag{4.2.58}$$

Comparing elements of (4.2.58), we find

$$\pi_i \tau_{ii} = 1 \qquad\qquad\qquad \square$$

An alternative proof for the result (4.2.55) is obtained using the relationship (3.3.9). We have

$$P_{ii}(z) = \frac{1}{1 - F_{ii}(z)} \tag{4.2.59}$$

Using Theorem B.3.1 of Appendix B, we have

$$\pi_i = \lim_{n \to \infty} P_{ii}^{(n)} = \lim_{z \to 1-} (1 - z) P_{ii}(z)$$

$$= \lim_{z \to 1-} \frac{1 - z}{1 - F_{ii}(z)}$$

$$= \lim_{z \to 1-} \left(\frac{-1}{-F_{ii}'(z)} \right)$$

$$= \frac{1}{\tau_{ii}} \tag{4.2.60}$$

where we have used the relationship $\displaystyle\lim_{z \to 1-} F_{ii}'(z) = \sum_{n=1}^{\infty} n f_{ii}^{(n)} = E[T_{ii}]$.

Occupation Times

By *occupation time* we mean the number of times (steps) the process occupies a certain state in a given time period. Let $N_{ij}^{(n)}$ be the number of times the process visits state j in n steps, given that initially the process was at state i. Clearly now $N_{ij}^{(n)}/n$ is the fraction of time the process visits j in n steps. The quantity $N_{ij}^{(n)}$ can be identified as a random variable as follows: Define an indicator random variable $Y_{ij}^{(k)}$ as

$$Y_{ij}^{(k)} = \begin{cases} 1 & \text{if } X_k = j, \text{ given } X_0 = i \\ 0 & \text{otherwise} \end{cases} \qquad (4.2.61)$$

where $\{ X_n, n = 0, 1, 2, \ldots \}$ is the Markov chain. We have

$$P\left(Y_{ij}^{(k)} = 1\right) = P_{ij}^{(k)}$$

$$P\left(Y_{ij}^{(k)} = 0\right) = 1 - P_{ij}^{(k)}$$

and hence

$$E\left(Y_{ij}^{(k)}\right) = P_{ij}^{(k)} \qquad (4.2.62)$$

We also have

$$N_{ij}^{(n)} = \sum_{k=1}^{n} Y_{ij}^{(k)}$$

and therefore

$$E\left[\frac{1}{n} N_{ij}^{(n)}\right] = \frac{1}{n} \sum_{k=1}^{n} P_{ij}^{(k)} \qquad (4.2.63)$$

To determine the long-term occupation time, we have to determine

$$\lim_{n \to \infty} E\left[\frac{1}{n} N_{ij}^{(n)}\right] = \lim_{n \to \infty} \frac{1}{n} \sum_{k=1}^{n} P_{ij}^{(k)}$$

This result is obtained by appealing to the properties of generating functions.

Let

$$\sum_{n=1}^{\infty} P_{ij}^{(n)} z^n = P_{ij}(z) \qquad |z| < 1$$

From Theorem B.3.1 of Appendix B, we have

$$\lim_{z \to 1-} (1 - z) P_{ij}(z) = \lim_{n \to \infty} P_{ij}^{(n)} = \pi_j \qquad (4.2.64)$$

Using this result in Theorem B.3.3 of Appendix B, we get

$$\lim_{n \to \infty} \frac{1}{n} \sum_{k=1}^{n} P_{ij}^{(k)} = \pi_j. \qquad (4.2.65)$$

The result (4.2.65) shows that the limiting probabilities π_j ($j = 1, 2, 3, \ldots, m$) are also fractions of time the Markov chain can be expected to occupy the various states in a large number of steps. Obviously these fractions are independent of the initial state of the process.

Referring back to Example 4.2.3, we can say that in the long run 40% of the time the business executive expects to visit Europe and 30% of the time each he expects to visit Hawaii and the Bahamas.

Suppose the Markov chain is in state i at some time. Let α_i be the number of new time periods it stays in state i until it moves out of that state. The probability distribution of α_i is easily obtained by considering the transitions as repeated trails with outcomes $i \to i$ and $i \to$ not i and probabilities P_{ii} and $1 - P_{ii}$, respectively. Thus we have a geometric distribution for α_i, given by

$$P(\alpha_i = n) = P_{ii}^n (1 - P_{ii}) \qquad (4.2.66)$$

and

$$E(\alpha_i) = \frac{P_{ii}}{1 - P_{ii}}$$

$$V(\alpha_i) = \frac{P_{ii}}{(1 - P_{ii})^2}$$

These results should not be interpreted that in any Markov chain states can be lumped together in any manner. Lumpability conditions are discussed in the next chapter. All we have done here is to consider a one-way transition $i \to$ not i in order to get the distribution of the number of steps a Markov chain will stay in a state without moving out of it.

4.3 TWO-STATE MARKOV CHAINS

The role of a two-state Markov chain is similar to the coin-tossing example in probability modeling. It is simple to conceptualize, easy to analyze, and appropriate as a mathematical model when there are only two outcomes or as a starter model in more general situations. Therefore, even though results and analyses given in the last two sections are directly applicable to this case, we shall give here some simpler analysis techniques and additional results which are particularly appropriate for the two-state Markov chains.

We shall identify the two states as 0 and 1 and assume that the transition probability matrix P is given by

$$P = \begin{matrix} 0 \\ 1 \end{matrix} \begin{bmatrix} 1-a & a \\ b & 1-b \end{bmatrix} \qquad 0 \le a, b \le 1 \qquad (4.3.1)$$

For the n-step transition probabilities, explicit expressions are given in Theorem 4.3.1.

Theorem 4.3.1. *For the two-state Markov chain with the transition probability matrix* (4.3.1), *the* n-step ($n \ge 1$) *transition probabilities are given by*

$$P^n = \begin{bmatrix} P_{00}^{(n)} & P_{01}^{(n)} \\ P_{10}^{(n)} & P_{11}^{(n)} \end{bmatrix}$$

$$= \begin{bmatrix} \dfrac{b}{a+b} + a\dfrac{(1-a-b)^n}{a+b} & \dfrac{a}{a+b} - a\dfrac{(1-a-b)^n}{a+b} \\ \dfrac{b}{a+b} - b\dfrac{(1-a-b)^n}{a+b} & \dfrac{a}{a+b} + b\dfrac{(1-a-b)^n}{a+b} \end{bmatrix}$$

$$(4.3.2)$$

Proof Using the Chapman-Kolmogorov equation (2.3.4), we have

$$P_{00}^{(1)} = 1 - a$$

$$P_{00}^{(n)} = (1-a)P_{00}^{(n-1)} + bP_{01}^{(n-1)} \qquad n > 1$$

$$= b + (1-a-b)P_{00}^{(n-1)}$$

Solving these equations recursively for $n = 1, 2, 3, \ldots$ and simplifying, we get by induction, when a and b are not simultaneously zero,

$$P_{00}^{(n)} = \frac{b}{a+b} + \frac{a(1-a-b)^n}{a+b}$$

which is the first element of the matrix in (4.3.2). The expression for $P_{10}^{(n)}$ is obtained by starting with the equations

$$P_{10}^{(1)} = b$$

$$P_{10}^{(n)} = (1-a)P_{10}^{(n-1)} + bP_{11}^{(n-1)} \qquad n > 1$$

$$= b + (1-a-b)P_{10}^{(n-1)} \qquad\qquad (4.3.3)$$

and proceeding in a similar manner. In these simplifications we use the formula for the sum of the finite geometric series

$$\sum_{r=0}^{n-2} (1-a-b)^r = \frac{1-(1-a-b)^{n-1}}{1-(1-a-b)}$$

where a and b are not simultaneously zero. $\square$

The limiting probability distribution of the states follows directly from (4.3.2), by letting $n \to \infty$ when $|1-a-b| < 1$. Then the probability vector $\pi = (\pi_0, \pi_1)$ is obtained as

$$(\pi_0, \pi_1) = \left(\frac{b}{a+b}, \frac{a}{a+b} \right) \qquad\qquad (4.3.4)$$

It may be noted that this result can also be obtained by solving appropriate equations of the type (4.2.41).

In view of the availability of explicit expressions for the n-step transition probabilities as given by (4.3.2), explicit expressions can also be given to the occupation time result of equation (4.2.63) in the case of a two-state Markov chain.

Let $\mu_{ij}^{(n)}$ $(= E(N_{ij}^{(n)}))$ be the expected number of times the process visits state j in n steps, given that initially the process was at state i. As derived in (4.2.63), we have

$$\mu_{ij}^{(n)} = \sum_{k=1}^{n} P_{ij}^{(k)} \qquad i, j = 0, 1$$

Using expressions from (4.3.2) and simplifying, we get

$$
\begin{bmatrix}
\dfrac{nb}{a+b} + \dfrac{a(1-a-b)\left[1-(1-a-b)^n\right]}{(a+b)^2} & \dfrac{na}{a+b} - \dfrac{a(1-a-b)\left[1-(1-a-b)^n\right]}{(a+b)^2} \\[3ex]
\dfrac{nb}{a+b} - \dfrac{b(1-a-b)\left[1-(1-a-b)^n\right]}{(a+b)^2} & \dfrac{na}{a+b} + \dfrac{b(1-a-b)\left[1-(1-a-b)^n\right]}{(a+b)^2}
\end{bmatrix}
$$

$$(4.3.5)$$

Frequently while using Markov chain results to problems of probability modeling, it becomes necessary to consider several processes simultaneously. The basic assumption one has to use in such cases is that of independence of these processes from each other. Furthermore, if there are m (≥ 2) states, after n (≥ 1) steps, the state of the process which has occupied a given state i initially will be multinomially distributed with probabilities $P_{ij}^{(n)}$ ($j = 1, 2, \ldots, m$). The method of extending this information to a set of independent processes is illustrated in the following example with reference to a two-state Markov chain.

Example 4.3.1

There are two food stores, A and B, in a certain area, and even though some customers go to the same store every week, there are some who often change their preference. An investigation of their preferences shows that with probability 0.15 a customer of store A one week would go over to store B the next week, and with probability 0.10 a customer of store B one week would buy from store A the next week. Initially 60% of the people buy from store A and 40% from store B. What do we expect to be the percentages after four weeks? After sufficiently long time? What are the respective variances? (See Howard, 1963.)

The transition probability matrix and its powers supply information on the behavior of single individuals rather than a group of individuals. For instance, identifying the stores A and B with states 0 and 1, the n-step transition probability $P_{11}^{(n)}$ gives the probability that a customer who was buying from store B would be buying from the same store after n weeks. But, if we are interested in the number of customers of store B after n weeks, we have to include some of those who were initially buying from store A. Therefore it is essential to consider the distribution of the number of customers buying from both the stores after n transitions. This is specifically so if we are interested in its variance or higher moments. Before we solve this specific problem, we shall first develop some theoretical results.

Let the transition probability matrix P be given by

$$P = \begin{pmatrix} 1 - a & a \\ b & 1 - b \end{pmatrix}$$

and let $\|P_{ij}^{(n)}\|$ $(i, j = 0, 1)$ be the n-step transition probability matrix of the Markov chain. Further let $c_0^{(n)}$ and $c_1^{(n)}$ be the number of individuals in the states 0 and 1 after n transitions. Initially there are $c_0^{(0)}$ and $c_1^{(0)}$ individuals in these states.

One of the basic assumptions of the model is that individuals act on their own without influencing the decision of others or without being influenced by the behavior of other individuals. Consider the number $c_0^{(n)}$ of individuals in state 0 after n steps. Some, say $c_0^{(n)}(0)$, would have been initially in state 0, and the rest, say $c_0^{(n)}(1)$, in state 1. We also have the probability that an individual initially in state 0 will be in state 0 after n steps as $P_{00}^{(n)}$ and the probability that he or she will be in state 1 as $P_{01}^{(n)}$ $(= 1 - P_{00}^{(n)})$. Similarly, the probabilities for individuals who are initially in state 1 are $P_{10}^{(n)}$ and $P_{11}^{(n)}$, respectively.

Now, consider $c_0^{(0)}$ individuals as $c_0^{(0)}$ independent repeated trials with the probability of "success" in each trial $P_{00}^{(n)}$, and $c_1^{(0)}$ individuals as $c_1^{(0)}$ independent repeated trials with "success" probability $P_{10}^{(n)}$. Using the binomial distribution, the distribution of $c_0^{(n)}$ can now be given as

$$P\left[c_0^{(n)} = k\right] = \sum_{k_1=0}^{k} P\left[c_0^{(n)}(0) = k_1\right] P\left[c_0^{(n)}(1) = k - k_1\right] \quad (4.3.6)$$

where

$$P\left[c_0^{(n)}(0) = k_1\right] = \binom{c_0^{(0)}}{k_1} \left[P_{00}^{(n)}\right]^{k_1} \left[1 - P_{00}^{(n)}\right]^{c_0^{(0)} - k_1}$$

and

$$P\left[c_0^{(n)}(1) = k - k_1\right] = \binom{c_1^{(0)}}{k - k_1} \left[P_{10}^{(n)}\right]^{k - k_1} \left[1 - P_{10}^{(n)}\right]^{c_1^{(0)} - k + k_1}$$

The mean and variance of $c_0^{(n)}$ and $c_1^{(n)}$ can now be easily obtained from their distributions. We get

$$E\left[c_0^{(n)}\right] = c_0^{(0)} P_{00}^{(n)} + c_1^{(0)} P_{10}^{(n)}$$

$$E\left[c_1^{(n)}\right] = c_0^{(0)} P_{01}^{(n)} + c_1^{(0)} P_{11}^{(n)} \quad (4.3.7)$$

and

$$V\left[c_0^{(n)}\right] = c_0^{(0)}P_{00}^{(n)}\left[1 - P_{00}^{(n)}\right] + c_1^{(0)}P_{10}^{(n)}\left[1 - P_{10}^{(n)}\right]$$

$$V\left[c_1^{(n)}\right] = c_0^{(0)}P_{01}^{(n)}\left[1 - P_{01}^{(n)}\right] + c_1^{(0)}P_{11}^{(n)}\left[1 - P_{11}^{(n)}\right] \qquad (4.3.8)$$

Clearly

$$V\left[c_0^{(n)}\right] = V\left[c_1^{(n)}\right]$$

Referring to Theorem 4.3.1 and using the explicit expressions for $P_{ij}^{(n)}$ ($i, j = 0, 1$), we can write

$$E\left[c_0^{(n)}\right] = \frac{b}{a + b}\left[c_0^{(0)} + c_1^{(0)}\right] + \frac{(1 - a - b)^n}{a + b}\left[ac_0^{(0)} - bc_1^{(0)}\right]$$

$$E\left[c_1^{(n)}\right] = \frac{a}{a + b}\left[c_0^{(0)} + c_1^{(0)}\right] - \frac{(1 - a - b)^n}{a + b}\left[ac_0^{(0)} - bc_1^{(0)}\right] \qquad (4.3.9)$$

and

$$V\left[c_0^{(n)}\right] = V\left[c_1^{(n)}\right]$$

$$= \frac{ba}{(a + b)^2}\left[c_0^{(0)} + c_1^{(0)}\right] - \frac{ba(1 - a - b)^n}{(a + b)^2}\left[c_0^{(0)} + c_1^{(0)}\right]$$

$$+ \frac{\left[1 - (1 - a - b)^n\right](1 - a - b)^n}{(a + b)^2}\left[a^2 c_0^{(0)} + b^2 c_1^{(0)}\right]$$

$$(4.3.10)$$

Clearly, as $n \to \infty$, if we write $c_0^* = \lim_{n \to \infty} c_0^{(n)}$, $c_1^* = \lim_{n \to \infty} c_1^{(n)}$, and $c^* = c_0^{(n)} + c_1^{(n)} = c_0^* + c_1^*$,

$$E\left[c_0^*\right] = \frac{b}{a + b}c^* \qquad E\left[c_1^*\right] = \frac{a}{a + b}c^* \qquad (4.3.11)$$

and

$$V\left[c_0^*\right] = V\left[c_1^*\right] = \frac{ab}{(a + b)^2}c^* \qquad (4.3.12)$$

Now going back to the example, we have $a = 0.15$ and $b = 0.10$ so that

$$P = \begin{pmatrix} 0.85 & 0.15 \\ 0.10 & 0.90 \end{pmatrix}$$

and

$$c_0^{(0)} = 60 \quad \text{and} \quad c_1^{(0)} = 40$$

After four weeks

$$E\left[c_0^{(4)}\right] = \frac{0.10}{0.25} \times 100 + \frac{(0.75)^4}{0.25}[0.15 \times 60 - 0.10 \times 40]$$

$$= 40 + 6.328125 = 46.328125$$

$$E\left[c_1^{(4)}\right] = 60 - 6.328125 = 53.671875$$

and

$$V\left[c_0^{(4)}\right] = V\left[c_1^{(4)}\right] = 24 - 7.59375 + 6.05621 = 22.46$$

After a sufficiently large number of weeks we get

$$E\left[c_0^*\right] = 40 \qquad E\left[c_1^*\right] = 60$$

and

$$V\left[c_0^*\right] = V\left[c_1^*\right] = 24$$

That means that after four weeks approximately 46.33% of the people buy from store A and 53.67% buy from store B. These percentages have a variance of 22.46. After a sufficiently long time the values are 40% and 60% with variance 24.

REFERENCES

Howard, R. A. (1963). "Stochastic Process Models for Consumer Behavior". *J. Advertising Res.* **3**, 35–42.

Kemeny, J. G., and Snell, J. L. (1960). *Finite Markov Chains*. New York: Van Nostrand. Reprinted 1976: Springer-Verlag.

FOR FURTHER READING

Bartlett, M. S. (1956). *An Introduction to Stochastic Processes*. Cambridge: Cambridge University Press.

Çinlar, E. (1975). *Introduction to Stochastic Processes*. Englewood Cliffs, N.J.: Prentice-Hall.

Cox, D. R. and Miller, H. D. (1965). *The Theory of Stochastic Processes*. London: Methuen. Reprinted 1972: Chapman and Hall.

Feller, W. (1968). *An Introduction to Probability Theory and its Applications*. 3rd ed., New York: Wiley.

Howard, R. A. (1971). *Dynamic Probabilistic Systems*. Vol. 1. New York: Wiley.

Isaacson, D. L., and Madsen, R. W. (1976). *Markov Chain Theory and Applications*. New York: Wiley.

Karlin, S., and Taylor, H. M. (1981). *A Second Course in Stochastic Processes*. New York: Academic Press.

EXERCISES

1. Determine the n-step transition probability matrix P^n of a two-state Markov chain using the following steps [assume that the transition probability matrix P is given by (4.3.1)]:

   a. Using the Cayley-Hamilton theorem (Appendix A.8) which states that a square matrix satisfies its own characteristic equation, show that

$$P^2 = (2 - a - b)P - (1 - a - b)I$$

   b. Writing $P = I + Q$ in the preceding relation, show that

$$Q^n = (-1)^{n-1}(a + b)^{n-1}Q$$

   c. Determine P^n by simplifying $(I + Q)^n$.

2. Define probability-generating functions (p.g.f)

$$\phi_{ij}(z) = \sum_{n=0}^{\infty} P_{ij}^{(n)}z^n \quad |z| < 1 \quad (i, j = 1, 2, \ldots, m)$$

   where

$$P_{ij}^{(0)} = \begin{cases} 1 & i = j \\ 0 & i \neq j \end{cases}$$

and the matrix of p.g.f.'s

$$\Phi(z) = \|\phi_{ij}(z)\|$$

Show that

$$\Phi(z) = (I - zP)^{-1} = I + zP + zP^2 + \cdots + z^nP^n + \cdots$$

Hence obtain the general expression for P^n when

a.

$$P = \begin{bmatrix} 0.2 & 0.4 & 0.4 \\ 0.1 & 0.5 & 0.4 \\ 0.6 & 0.3 & 0.1 \end{bmatrix}$$

b.

$$P = \begin{bmatrix} 1 & 0 & 0 \\ 0 & 1 & 0 \\ 0.1 & 0.5 & 0.4 \end{bmatrix}$$

3. Verify the explicit results of equation (4.3.2) for the n-step transition probabilities using the symbolic expansion of $\Phi(z)$ as

$$\Phi(z) = (I - zP)^{-1} = I + zP + z^2P^2 + z^3P^3 + \cdots + z^nP^n + \cdots$$

4. Determine the stationary distribution of the number of particles in Container A in the Ehrenfest model of Exercise 3 of Chapter 3.

5. a. Determine $P^2, P^3, \ldots, P^n, \ldots, \Pi$, where P is the transition probability matrix in Exercise 4a of Chapter 3 and $\Pi = \lim_{n \to \infty} P^n$.

   b. Determine $P^2, P^3, \ldots, P^n, \ldots, \Pi$, where P is the transition probability matrix in Exercise 5 of Chapter 3 and $\Pi = \lim_{n \to \infty} P^n$.

   c. Generalize the coin-tossing problems of Exercises 4 and 5 of Chapter 3 to define $\{X_n\}$ based on the outcomes of the $(n - k + 1)$st, $(n - k + 2)$nd, $\ldots$, nth tosses, and obtain its limiting probability matrix Π.

   Hint Observe the emerging pattern of elements in Π in parts a and b.

6. Suppose the repeated trials of the success run problem of Exercise 6 of Chapter 3 are terminated when the success run is of length k. Obtain the mean time to termination for $k = 3$.

   Hint State 3 is absorbing.

7. Determine the limiting distribution of the states in the standby machine problem of Exercise 8 of Chapter 3.

8. Obtain the transition probability matrix of the discrete time queue of Exercise 9 of Chapter 3 with the restriction that the system capacity is restricted to K customers. This means that customer arrivals stop when there are K customers in the system. Determine the limiting distribution of the number of customers in this system.

9. In a queuing system such as the one in Exercise 8, the time intervals between successive visits of the queue length process (number of customers in the system) to state 0 are called busy periods. Obtain the expected length of the busy period when the capacity K of the discrete time queue is 3 and $p = 0.4$, $r = 0.5$.

10. In an information network (e.g., a network of libraries) there are N centers. If center i cannot fulfill a request, it refers the request to center j with probability P_{ij}. Let c_{ij} be the cost of such a referral and $c_i(d)$ be the cost of disposal of the request at center i without further referring it to any other center. Let C_{ij} be the total cost to the system due to a request originating at center i and getting referred to center j before it is finally disposed of (note that a request can be referred to the same center more than once, depending on the probabilities P_{ij}). Define

$$\Gamma_{ij} = E[C_{ij}] \quad \text{and} \quad \Gamma = \|\Gamma_{ij}\|$$

Give a Markov model for the information network.

Hint Define an additional state to signify disposal of request.

Using the fundamental matrix of the resulting chain, derive an expression for the expected cost matrix Γ.

11. Specialize Exercise 10 for a strictly hierarchical network in which request referral can occur only in a specified route.

Hint Invert the resulting matrix using its special structure.

12. Random walk models are useful in analyzing situations where the fluctuations of a process are either to the right or to the left (or up or down) by one unit at a time. Example 3.1.3 is a random walk with two absorbing barriers. A random walk with two reflecting barriers can be described as follows.

Let P_{ij} be the transition probability of the random walk. Then for a state space $S = \{0, 1, 2, \ldots, N\}$ and $p + q = 1$,

$$P_{00} = q \qquad P_{01} = p \qquad P_{0j} = 0 \qquad \text{for } j \neq 0, 1$$

$$P_{i,i-1} = q \qquad P_{i,i+1} = p \qquad P_{ij} = 0 \qquad \text{for } j \neq i - 1, i + 1$$

and $i = 1, 2, 3, \ldots, N - 1$

$$P_{N,N-1} = q \qquad P_{NN} = p \qquad P_{Nj} = 0 \qquad \text{for } j \neq N - 1, N$$

Determine the limiting distribution of the state of the random walk.

13. a. A generalized cyclical random walk is defined as follows.

States $\{0, 1, \ldots, m - 1\}$ are arranged on a circle. Thus states $m - 1$ and 0 are next to each other. The random walk moves around the circle based on a transition probability matrix, whose elements P_{ij} $(i, j = 0, 1, 2, \ldots, m - 1)$ are given by

$$P_{ij} = \begin{cases} p_{j-i} & j \geq i \\ p_{m+j-i} & j < i \end{cases} \qquad i, j = 0, 1, 2, \ldots, m - 1$$

Determine the limiting distribution of the state of the random walk.

b. Let S_n be the sum of n independent rolls of a fair die. Define $X_n(K) = S_n \bmod K$ $(K \geq 6)$. Show that $\{X_n(K)\}$ is a Markov chain, and determine its state space and the transition probability matrix. Using this Markov chain find $\lim_{n \to \infty} P$ $(S_n$ is a multiple of $K)$.

Hint Solve for specific values of K $(= 8, 9,$ say), and then generalize.

14. A random walk with two absorbing barriers has the following transition probabilities for a state space $S = \{0, 1, 2, \ldots, N\}$ and $p + q = 1$.

$$P_{00} = 1 \qquad P_{0j} = 0 \qquad\qquad\qquad j \neq 0$$

$$P_{i,i-1} = q \qquad P_{i,i+1} = p \qquad P_{i,j} = 0 \qquad j \neq i - 1, i + 1$$

and $i = 1, 2, \ldots, N - 1$

$$P_{NN} = 1 \qquad P_{Nj} = 0 \qquad\qquad\qquad j \neq N$$

Starting from state i, let f_i be the probability of absorption in state 0.

(Note that this is a generalization of Example 3.1.3 and f_i is the probability of ruin for player A, if this player starts out with a capital of i.) Obtain f_i.

Hint Write $f_i = (p + q)f_i = pf_{i+1} + qf_{i-1}$. Define $g_i = f_i - f_{i+1}$, $i = 0, 1, \ldots, N - 1$, and note $g_0 = 1 - f_1$, $g_{N-1} = f_{N-1}$, and $\sum_{i=1}^{N-1} g_i = f_1$. Now derive the relation $g_i = (q/p)g_{i-1}$, and solve recursively.

15. Consider a constant number $C(\geq 1)$ of irreducible finite Markov chains with the same transition probability matrix. Let $C_j^{(n)}$ be the number of chains in state j after n transitions. Assume that $C_j^{(0)}$, $j = 1, 2, \ldots, m$; the numbers of chains in various states initially are known. Derive relations that can be used to obtain the mean and variance of $C_j^{(n)}$, $j = 1, 2, \ldots, m$, for $n = 1, 2, \ldots$.

16. Answer the questions posed in Example 3.1.1.

17. Two vendors, A and B, compete for the same market. Currently customer choice of these vendors can be represented by the transition probability matrix

$$
\begin{array}{c c}
 & \begin{array}{cc} A & B \end{array} \\
\begin{array}{c} A \\ B \end{array} &
\begin{bmatrix} 0.5 & 0.5 \\ 0.6 & 0.4 \end{bmatrix}
\end{array}
$$

What is A's long-term customer share of the market?

Suppose A would like to boost her share to 75%. What should be her strategy in terms of the ratio of the number of customers lost to the number of customers gained in a single period?

18. A country has two political parties A and B. The winners of the past 25 elections are as follows: BAAABBABABABBBBBABBBAAABBBAAB. Using a Markov chain model, determine the probability that party B will retain the power in the next two elections. Suppose, the Markov chain reflects the trend for at least 10 future elections. What is the expected number of times party A will win the elections?

19. Nine white balls are distributed between two boxes, A and B, with 4 in A and 5 in B. A red ball is in either A or B. The game consists of drawing one ball at random from each box and interchanging them without looking at them. After doing this five times, the winner is the one who gives the correct location of the red ball. If the red ball is initially in box A, model this game as a Markov chain, and find the probability that the ball is in box A at the end of the game. What would be the effect on this probability if originally the red ball was in

box B? What is the expected number of times the red ball has been in box A, assuming it was originally in A?

20. A city has its water supplied from a dam. Prompted by water scarcity in summer months, an investigation was made of the data on rainfall and the water content of the dam in the beginning of summer. From these investigations it was found that if in the beginning of one summer the dam was full, the probability that it would be full in the beginning of next summer was 0.9, whereas if in the beginning of one summer the dam was not full, the probability that it would end up being not full in the beginning of next summer was found to be 0.4. Suppose that in the beginning of last summer the dam was not full. What is the probability of finding it full after three years? What is the expected length of an interval during which it remains not full? Discuss the modifications that could be made in making the model more realistic without losing the simplicity of Markov processes.

21. A service agency employs several repairpersons. If a repairperson is on the job, the agency assigns him or her at most one job to be taken up after the present job is finished. In addition to the jobs repairpersons get from the agency they do some work of their own, but on the condition that they would be available whenever the agency needs them. Suppose an individual wishes to model the job assignments as a two-state Markov chain, in which states 0 (no agency job) and 1 (one agency job) are observed just after the completion of the agency jobs. Discuss methods of obtaining information on the following factors from the data the individual maintains of the job activities: (1) the probability of returning to a nonagency job soon after the present agency job, (2) the number of agency jobs one has to perform before returning to a nonagency job, and (3) the number of times one would have taken up one's own job for every hundred of agency jobs.

22. In maintaining summary data on accidents in a city, each hour of the day is classified as having had an accident or not. From this summary information it is hoped that a pattern of accidents will emerge. Is a Markov chain model realistic for the study of such data? If not, suggest some realistic course of action. Justify your answer.

23. Because of the delays caused by production and delivery, some demands for a certain merchandise are not met in a retail store. However, a demand analysis shows that after not being able to meet a demand, 80% of the time the next demand will be satisfied, whereas after satisfying a demand, the next demand will be satisfied only 70% of the time. Construct a Markov chain model for this situation.

Assuming that the business pattern has been stable for some time, what is the probability that the retailer will be able to satisfy the next demand that occurs? What is the expected number of demands that would go unsatisfied during a fixed period of time? Suppose that at 9:00 AM a demand goes unsatisfied. Out of the next 10 demands how many do you expect to go unsatisfied?

24. In a test paper the questions are arranged in such a way that 75% of the time a "True" answer follows a "True" answer and 35% of the time a "False" answer follows a "False" answer. Suppose the first question has "False" as the answer, and the paper has 100 questions. Obtain the approximate number of questions that have "True" as the answer.

25. Specialize the service agency problem of Exercise 10 of Chapter 3 for $N = 3$ and the job arrival distribution.

Number of jobs j	0	1	2	3	4
Probability a_j	0.3	0.3	0.2	0.1	0.1

Determine the expected number of jobs an individual would do at a stretch (not necessarily the same day) before going off to nonagency work. (Assume a uniform distribution for the number of jobs at the start of the period.)

26. In the presidential primaries problem of Exercise 12 of Chapter 3 obtain the following.

   a. Suppose a candidate is currently developing support. What are the candidate's chances of becoming a winner eventually? How many primaries would be needed for a definite outcome? Give the expected value.

   b. Suppose the candidate is currently slipping—answer the same questions as in part a.

27. Obtain the limiting distribution of the inventory level in Exercise 14 of Chapter 3, and find the expected number of days, during a 14 day period, the storekeeper will start the day with exactly 3 items. What is the probability that an order will have to be placed at the end of a day?

28. Specialize Bartkey's sampling inspection scheme of Exercise 2 of Chapter 3, for $N = 10$, $p = 0.1$, and $a = 3$. Determine the expected number of samples required to make a decision.

29. In Example 3.1.4 determine the probability that A will ultimately lose the game. Also obtain the expected duration of the game.

30. Sunny Tex encounters a carnival hawk Slicky Grease who tells him that for the nominal fee of one dollar he may buy a single coupon allowing him to play the game of Dummo. The game of Dummo requires that a player collect three coupons in order to win a Surprise. On each play of the game of Dummo, Tex has an even chance of winning or losing a coupon. Tex loses the game when he has no coupons.

Model the game of Dummo as a Markov chain, and obtain its transition probability matrix. Classify the states of the system, and arrange the matrix in the canonical form.

Obtain the fundamental matrix, and hence answer the following questions:

   a. What is the expected number of plays needed for Tex to win or lose the game?

   b. What is the probability that Tex will win the Surprise?

   c. Assuming that the Surprise is worth $3.00, would you say that Dummo is a fair game?

Note A game is "fair" if in the long run no one loses any money.

31. A structure consists of 100 joints. A joint has a probability of 0.001 of being defective. If three or more consecutive joints are defective, the structure collapses. Determine the reliability of the structure using a Markov chain model.

Hint Use the number of successive defective joints as the state of the Markov chain. State 3 is absorbing. The reliability is the probability of not being in state 3 in 100 trials.

32. Data have been gathered from hospital records to study the effect of treatment of a certain disease. A Markov model with four states 0, 1, 2, and 3 is envisaged to be used for the study: 0—state of being under treatment; 1—state of being dead immediately following treatment; 2—state of recovery when the patient is under observation; and 3—state of being lost after recovery either through death not related to the disease or by difficulties of tracing the patient.

Previous state	Following state				Total
	0	1	2	3	
0	60	12	83	0	155
2	72	0	77	27	176

Assuming that only one state transition is possible between consecutive observations, determine the probability that the patient who is under treatment survives for longer than four observation periods. What is the probability that the patient will be lost after recovery eventually?

33. For Exercise 15 of Chapter 3 answer the following questions:

   a. Obtain the fundamental matrix. How can this matrix be used to answer the next three questions, assuming that at the beginning of the game the ball is equally likely to be with Dick, Harry or Sam?

   b. What is the probability that the game ends with Mark running away with the ball?

   c. What is the probability that Joe will get the ball before Tom gets it? Will Tom ever get the ball before Joe does? If so determine the probability of this event.

   e. Assuming that the game practically ends when any one of Joe, Mark, or Tom gets the ball, what are the mean and variance of the duration of the game?

34. Obtain the limiting distribution of the Markov chain representing the number of customers in the barbershop in Example 3.1.2, and interpret the results. Obtain the expected length of time during which the barber will be continuously busy.

35. Depending on the water level in a reservoir, water supply prospects can be classified as follows:

Less than 1/4 full	Critical	(C)
Between 1/4 and 1/2 full	Poor	(P)
Between 1/2 and 3/4 full	Fair	(F)
Between 3/4 and 19/20 full	Good	(G)
More than 19/20 full or overflow	Abundant	(A)

From existing data it is found that the following transition probability matrix represents the situation fairly well over a period of time:

$$
\begin{array}{c c}
 & \begin{array}{ccccc} C & P & F & G & A \end{array} \\
\begin{array}{c} C \\ P \\ F \\ G \\ A \end{array} &
\left[\begin{array}{ccccc}
0.1 & 0.1 & 0.3 & 0.5 & 0 \\
0.3 & 0.2 & 0.2 & 0.2 & 0.1 \\
0.1 & 0.2 & 0.4 & 0.2 & 0.1 \\
0 & 0.1 & 0.2 & 0.4 & 0.3 \\
0 & 0.1 & 0.1 & 0.4 & 0.4
\end{array} \right]
\end{array}
$$

How often does the water level get to the critical level?

36. The export market of a country under stable economic conditions can be modeled as a three-state Markov chain as follows:

States

+ 1	Increases by more than 5% of the preceding year
0	Fluctuation is less than 5%
− 1	Decreases by more than 5% of the preceding year

Suppose the transition probability matrix is given as follows:

$$
\text{Last year} \quad
\begin{array}{c}
\\
+1 \\
0 \\
-1
\end{array}
\begin{array}{c}
\text{This year} \\
\begin{array}{ccc}
+1 & 0 & -1
\end{array} \\
\left[
\begin{array}{ccc}
0.8 & 0.2 & 0 \\
0.35 & 0.30 & 0.35 \\
0 & 0.4 & 0.6
\end{array}
\right]
\end{array}
$$

Determine the mean recurrence time for each state. Compare the expected lengths of the increasing and decreasing trends in economic conditions.

37. A machine component is replaced, as a rule, once in five weeks. However, it has been found to wear out in less than five weeks in some cases. To see whether to shorten the replacement cycle, data were gathered over a period of time. The overall estimates of the life of the component were as follows: 10% of the components were replaced at the end of the first week; 15% of the one week old components were replaced at the end of the second week; 35% of the two week old components were replaced at the end of three weeks; and 40% of the three week old components were replaced at the end of the fourth week. Set up the transition probability matrix. Obtain the age distribution of a component after the system has been in operation for a long time. A component is i weeks old, $i = 0, 1, 2, 3, 4$; what is the expected length of time until its replacement? Can you recommend any replacement policy based on these results?

38. A state would like to estimate the number of graduates added to its work force every year. The following information is available from existing data:

a. A projected estimate of freshman population at state schools.

   b. Estimates of the numbers of out-of-state college students who enter the state colleges at freshman, sophomore, junior, and senior levels.

   c. An estimate of the percentage of native graduates who actually look for work in the state.

   d. An estimate of the percentage of out-of-state graduates who decide to stay on to work after graduation.

   e. For each year at college, estimates of percentages of students who successfully finish the year, who drop out during the year (includes those who move out of the state), and who flunk out at the end of the year. It has been found that these percentages are fairly constant over the years.

Set up a Markov model for this problem, and describe the method of analysis.

ADVANCED THEORETICAL EXERCISES
(KEMENY AND SNELL, 1960)

39. Let $p = (p_1, p_2, \ldots, p_m)$ be the initial probability vector of an m-state Markov chain whose states $\{r + 1, r + 2, \ldots, m\}$ are transient. Let p_T be the vector with only the last $m - r$ elements of p. Let E_p and V_p denote the expectation and variance based on the initial probability vector p. Then show that

   a. $\|E_p[N_{ij}]\| = p_T M$

   b. $\|V_p[N_{ij}]\| = p_T M(2M_D - I) - p_T M_2$

   c. $E_p[N_i] = p_T M_p$

   d. $V_p[N_i] = p_T(2M - I)M_p - (p_T M_p)_2$

40. Let P be the transition probability matrix of an aperiodic, irreducible m-state Markov chain. Let Π be the matrix of limiting probabilities. Then the matrix $M = [I - (P - \Pi)]^{-1}$ is called its fundamental matrix. Show that M exists and can be written as

$$M = I + \sum_{n=1}^{\infty} (P^n - \Pi)$$

41. Prove the following properties of matrix M of Exercise 40.

   a. $PM = MP$

b. $\ Me = e\ $ where $\ e = \begin{bmatrix} 1 \\ 1 \\ 1 \\ \vdots \\ 1 \end{bmatrix}$

c. $\ \pi M = \pi$

d. $\ I - M = \Pi - PM$

42. For an aperiodic, irreducible m-state Markov chain, let $N_{ij}^{(n)}$ be the number of times the process is in state j in n steps, having initially started from i. Then show that

$$\left\| E\left[N_{ij}^{(n)} \right] - n\pi_j \right\| \to M - \Pi \qquad \text{as } n \to \infty$$

Suppose $p = (p_1, p_2, \ldots, p_m)$ is an initial probability vector for the m-state Markov chain. Then show that

$$\left\| E_p\left[N_{ij}^{(n)} \right] \right\| - n\pi \to pM - \pi \qquad \text{as } n \to \infty$$

43. Let $p = (p_1, p_2, \ldots, p_m)$ and $p' = (p_1', p_2', \ldots, p_m')$ be two initial probability vectors for the Markov chain. Then with notations as in Exercise 42, show that

$$\left\| E_p\left[N_{ij}^{(n)} \right] - E_p'\left[N_{ij}^{(n)} \right] \right\| \to (p - p')M \qquad \text{as } n \to \infty$$

44. Show that the mean first passage times τ_{ij} of Theorem 4.2.4 can be obtained as

$$\left\| \tau_{ij} \right\| = \left(I - M + EM_D \right)D_i$$

where

$$E = \begin{bmatrix} 1,1,\ldots,1 \\ \vdots \\ 1,1,\ldots,1 \end{bmatrix} \quad \text{and} \quad D_i = \begin{bmatrix} \pi_1^{-1} & & & \\ & \pi_2^{-1} & & \bigcirc \\ & & \ddots & \\ \bigcirc & & & \pi_m^{-1} \end{bmatrix}$$

and M is the fundamental matrix obtained in Exercise 40.

Chapter 5

FINITE MARKOV CHAINS: SPECIAL TOPICS AND INFERENCE

In this chapter we continue the discussion on finite Markov chains with reference to some special topics and inference problems. Under special topics we introduce three topics on Markov chains that have found use in the search for better models in the real world: (1) Markov chains of order higher than one, also known as *multiple* Markov chains, (2) *lumpable* Markov chains, and (3) *reversed* Markov chains. A Markov chain of order higher than one is a direct extension of the one-step dependence property of the Markov process and is analyzed simply by expanding the state space. In lumpable Markov chains states are lumped (or pooled) together without violating the Markovian property. Lumpability conditions are quite restrictive, and hence their use in practice seems to be limited. Reversed Markov chains, however, have found good use in applications where one has to infer properties of the system in the past, knowing present conditions. Application areas such as genetics and geosciences find ample use for this class of Markov chains.

5.1 MARKOV CHAINS OF ORDER HIGHER THAN ONE

In Section 2.3 we defined a discrete parameter discrete state Markov chain $\{ X_n, \ n = 0, 1, 2, \ldots \}$ as exhibiting the property

$$P\left(X_n = j | X_{n_1} = i_1, X_{n_2} = i_2, \ldots, X_{n_k} = i_k \right)$$

$$= P\left(X_n = j | X_{n_1} = i_1 \right) \tag{5.1.1}$$

for $n > n_1 > n_2 > \cdots > n_k$. The implication is that the dependence structure is such that, given the state of the process at any time, future development of the process can be probabilistically predicted without having any additional information. This dependence structure is very restrictive and is not sufficient to express many situations in the real world. For instance, in an information network such as a library network while routing a request or a piece of information from a node, there is every likelihood that it will not be routed back to the node from where it came immediately before. In queueing systems, for purposes of convenience we assume the arrival process to be independent of the state of the system. A more realistic model could be to assume that the arrival process depends on the state of the system during the previous few departure epochs. To allow for these modifications, the following process definition seems appropriate.

DEFINITION   *A discrete parameter discrete state Markov chain* $\{ X_n, n = 0, 1, 2, \ldots \}$ *is of order l if the following property holds*:

$$P\left(X_n = j | X_{n_1} = i_1, X_{n_2} = i_2, \ldots, X_{n_l} = i_l, \ldots, X_{n_k} = i_k \right)$$

$$= P\left(X_n = j | X_{n_1} = i_1, X_{n_2} = i_2, \ldots, X_{n_l} = i_l \right) \qquad (5.1.2)$$

for $n > n_1 > n_2 > \cdots > n_l > \cdots > n_k$.

With this terminology the Markov chains considered so far (satisfying the definition represented by equation 5.1.1) can be called Markov chains of order one. As one would surmise, the analysis of a higher-order Markov chain with large values of l is quite difficult. However, for small values of l such as $l = 2$ or 3 when the state space is small, higher-order Markov chains can be converted into Markov chains of the first order and can be analyzed in the usual manner.

Consider a second-order Markov chain $\{ X_n, n = 0, 1, 2, \ldots \}$ with states $\{1, 2\}$. Let

$$P_{ijk} = P\left(X_n = k | X_{n-1} = j, X_{n-2} = i \right) \qquad (i, j, k = 1, 2) \qquad (5.1.3)$$

Possible one-step transition probabilities are (here the word "step" denotes the gap between the epoch of most recent information and the epoch of interest):

$$P_{111} \qquad P_{112} \qquad P_{121} \qquad P_{122} \qquad P_{211} \qquad P_{212} \qquad P_{221} \qquad P_{222}$$

which can be arranged in a matrix form

ij \\ k	1	2
11	P_{111}	P_{112}
12	P_{121}	P_{122}
21	P_{211}	P_{212}
22	P_{221}	P_{222}

(5.1.4)

Clearly, for a complete representation of the state of the system at any epoch, we now need information on the previous two epochs. Since the matrix representation (5.1.4) is not convenient for matrix operations, we may represent it in the form of a 4×4 matrix.

ij \\ jk	11	12	21	22
11	P_{111}	P_{112}	0	0
12	0	0	P_{121}	P_{122}
21	P_{211}	P_{212}	0	0
22	0	0	P_{221}	P_{222}

(5.1.5)

The transition probability matrix (5.1.5) is that of a first-order Markov chain whose states are the composite states $\{11, 12, 21, 22\}$, and its analysis follows in the usual manner.

Extending this technique to higher-order Markov chains is straightforward. However, it should be noted that the number of probability elements (which could be nonzero) in the transition probability matrix of an lth order Markov chain will be 2^{l+1}. When the number of states is m, the corresponding number is m^{l+1}, arranged in a matrix of size $m^l \times m^l$.

5.2   LUMPABLE MARKOV CHAINS

In modeling natural phenomena, one approach is to start with a simple model and, once the basic features of the model are determined, to make it more realistic by extending it to include different aspects of the phenomenon. An example of such a process would be to use a three-state Markov chain for the content of a reservoir with states {high, normal, low} initially and then extend the state space identifying different levels within each state. Another approach is to start out with an elaborate model and lump the states when the identification of some states are not necessary any further. Unfortunately because of the dependence structure such lumping may result in a totally different process. Markov chains in which states can be lumped together without losing the basic Markov property are called lumpable Markov chains.

First let us consider lumpability on an intuitive basis. Consider a Markov chain with five states $\{1, 2, 3, 4, 5\}$. Suppose we wish to lump the states in such a way that the two new states are $A: \{1, 2\}$ and $B: \{3, 4, 5\}$. Let P be the transition probability matrix of the original Markov chain.

$$P = \begin{bmatrix} P_{11} & P_{12} & P_{13} & P_{14} & P_{15} \\ P_{21} & P_{22} & P_{23} & P_{24} & P_{25} \\ P_{31} & P_{32} & P_{33} & P_{34} & P_{35} \\ P_{41} & P_{42} & P_{43} & P_{44} & P_{45} \\ P_{51} & P_{52} & P_{53} & P_{54} & P_{55} \end{bmatrix} \qquad (5.2.1)$$

Assume that in P of (5.2.1) the following relations hold:

$$P_{11} + P_{12} = P_{21} + P_{22}$$

$$P_{31} + P_{32} = P_{41} + P_{42} = P_{51} + P_{52} \qquad (5.2.2)$$

Clearly because of relations (5.2.2) the following also hold:

$$P_{13} + P_{14} + P_{15} = P_{23} + P_{24} + P_{25}$$

$$P_{33} + P_{34} + P_{35} = P_{43} + P_{44} + P_{45} = P_{53} + P_{54} + P_{55}$$

Now if we set the probabilities

$$P_{AA} = P_{11} + P_{12} \qquad P_{AB} = P_{13} + P_{14} + P_{15}$$

$$P_{BA} = P_{31} + P_{32} \qquad P_{BB} = P_{33} + P_{34} + P_{35}$$

and consider a process with $\{A, B\}$ as states and

$$\tilde{P} = \begin{bmatrix} P_{AA} & P_{AB} \\ P_{BA} & P_{BB} \end{bmatrix} \qquad (5.2.3)$$

as its transition probability matrix, then the process can be seen to be a Markov chain regardless of the initial state probability vector. Thus relations such as (5.2.2) are crucial for lumpability. When such relations do not hold, it is possible to get a Markov chain for a lumped process only for specific initial state probability vectors under some special circumstances (which we do not elaborate here because of the complexity of such cases).

Based on the foregoing discussion we may define lumpability as follows.

DEFINITION *A Markov chain* $\{X_n, n = 0, 1, 2, \ldots\}$ *is strongly lumpable with respect to a partition* $\tilde{S} = \{S_1, S_2, \ldots, S_l\}$ *of the state space* **S**, *if for every initial state probability vector* $\sigma : \{\sigma_1, \sigma_2, \ldots, \sigma_m\}$ *the resulting chain* $\{\tilde{X}_n, n = 0, 1, 2, \ldots\}$ *is Markovian and the transition probabilities do not depend on the choice of* σ.

If the definition is modified to say that the resulting chain should be Markovian for at least one initial state probability vector, then we say that the Markov chain is *weakly lumpable*. In the remainder of this section we shall consider only strong lumpability, and therefore in our discussion we shall drop the qualifying adjective.

In developing an intuitive base for lumpability, we have assumed that certain relations hold, as in (5.2.2) and the following equations. In fact it is possible to establish that such relations form a set of necessary and sufficient conditions for lumpability as stated in Theorem 5.2.1.

Theorem 5.2.1 *A necessary and sufficient condition for a Markov chain to be lumpable with respect to a partition* $\tilde{S} = \{S_1, S_2, \ldots, S_l\}$ *is that for every pair of sets* S_i *and* S_j, *the probabilities*

$$\tilde{P}_{ij} = \sum_{r \in S_j} P_{kr} \qquad i, j = 1, 2, \ldots, l \tag{5.2.4}$$

for all $k \in S_i$.

The probabilities $\tilde{P}_{ij}$ $(i, j = 1, 2, \ldots, l)$ are the elements of the transition probability matrix of the lumped chain.

Example 5.2.1

Check whether the Markov chains with the following transition probability matrices are lumpable:

1.

$$P = \begin{array}{c} 1 \\ 2 \\ 3 \end{array} \begin{bmatrix} \frac{1}{4} & \frac{3}{4} & 0 \\ \frac{2}{3} & \frac{1}{3} & 0 \\ 0 & \frac{1}{2} & \frac{1}{2} \end{bmatrix}$$

2.

$$P = \begin{array}{c} 1 \\ 2 \\ 3 \\ 4 \end{array} \begin{bmatrix} \frac{1}{4} & \frac{3}{4} & 0 & 0 \\ \frac{2}{3} & \frac{1}{3} & 0 & 0 \\ 0 & \frac{1}{4} & \frac{1}{2} & \frac{1}{4} \\ 0 & 0 & \frac{1}{2} & \frac{1}{2} \end{bmatrix}$$

In matrix 1, states $\{1, 2\}$ form a recurrent class, and state 3 is transient. Clearly, for the partition of states $[\{1, 2\}, 3]$, we have

$$P_{11} + P_{12} = P_{21} + P_{22} = 1$$

$$P_{13} = P_{23} = 0$$

$$P_{31} + P_{32} = \tfrac{1}{2}$$

Comparing these relations with those implied by the conditions for lumpability [e.g., equations (5.2.2)], the Markov chain is seen to be lumpable with respect to the partition $[\{1, 2\}, 3]$. A similar analysis will show that any other partitioning will not satisfy conditions of lumpability.

In matrix 2, states $\{1, 2\}$ form a recurrent class, and states $\{3, 4\}$ form a transient class. For the partition $[\{1, 2\}, \{3, 4\}]$, we have

$$P_{11} + P_{12} = P_{21} + P_{22} = 1$$

$$P_{13} + P_{14} = P_{23} + P_{24} = 0$$

$$P_{31} + P_{32} = \tfrac{1}{4} \neq P_{41} + P_{42} = 0$$

Therefore the Markov chain is not lumpable with respect to the partition $[\{1, 2\}, \{3, 4\}]$.

However, a similar analysis will show that the Markov chain is lumpable with respect to the partition $[\{1, 2\}, 3, 4]$.

Example 5.2.2

Two brands of a certain product are in the market, and a new brand of the same product is being tested for introduction. Two procedures have been proposed for testing:

1. Test consumer brand loyalty of the new brand against the old brands taken together.
2. Test consumer brand loyalty of the new brand against the old brands considered separately.

Clearly the first alternative is simpler and hence less expensive. Under what circumstances can alternative 2 be preferred without undermining the conclusions?

In order to establish lumpability conditions, consider two possible transition structures for brand loyalty:

1.

$$
\begin{array}{c}
1 \\
2 \\
3
\end{array}
\begin{bmatrix}
0.4 & 0.3 & 0.3 \\
0.5 & 0.2 & 0.3 \\
0.3 & 0.2 & 0.5
\end{bmatrix}
$$

2.

$$
\begin{array}{c}
1 \\
2 \\
3
\end{array}
\begin{bmatrix}
0.3 & 0.3 & 0.4 \\
0.5 & 0.2 & 0.3 \\
0.3 & 0.3 & 0.4
\end{bmatrix}
$$

Let 1 and 2 be the two old brands and 3 be the new brand. Under transition structure 1, $P_{11} + P_{12} = P_{21} + P_{22} = 0.7$; $P_{13} = P_{23} = 0.3$; $P_{31} + P_{32} = 0.5$ Therefore the Markov chain is lumpable with respect to the partition $[\{1,2\},3]$. Under transition structure 2, $P_{11} + P_{12} = 0.6 \neq P_{21} + P_{22} = 0.7$; $P_{13} = 0.4 \neq P_{23} = 0.3$. Therefore the Markov chain is not lumpable with respect to the partition $[\{1,2\},3]$.

Looking at the two transition structures, it is clear that the two old brands can be lumped together only if the new brand will draw consumers in equal proportions from both of them. Otherwise, conclusions drawn from the two test procedures may not be the same regarding the new brand's appeal to the consumer.

Example 5.2.3

A library network includes four libraries; two of them are in one city, and the other two are in another city. An analysis of the information transactions in these four libraries has established the following Markovian structure for the referrals (libraries in the first city are numbered 1 and 2, and those in the second city are numbered 3 and 4):

$$
\begin{array}{c}
1 \\
2 \\
3 \\
4
\end{array}
\begin{bmatrix}
0.2 & 0.5 & 0.3 & 0 \\
0.4 & 0.3 & 0.1 & 0.2 \\
0.6 & 0 & 0.2 & 0.2 \\
0.4 & 0.2 & 0.4 & 0
\end{bmatrix}
$$

For purposes of comparing library resources of the two cities, can we pool the transaction information without upsetting the Markovian structure?

Comparing with the lumpability conditions, we have

$$P_{11} + P_{12} = P_{21} + P_{22} = 0.7$$

$$P_{13} + P_{14} = P_{23} + P_{24} = 0.3$$

$$P_{31} + P_{32} = P_{41} + P_{42} = 0.6$$

$$P_{33} + P_{34} = P_{43} + P_{44} = 0.4$$

Therefore the Markov chain is lumpable with respect to the partition $\{1,2\}$, $\{3,4\}$. Thus the answer to the question is yes.

A common Markovian model used in social processes is the simple promotion model in which the only possible changes of state are of the type $i \rightarrow i + 1$. The resulting transition probability matrix has nonzero elements only on the diagonal and a super diagonal next to it. More generally, consider a transition probability matrix that is upper triangular in nature.

$$
P = \begin{array}{c} 1 \\ 2 \\ 3 \\ \vdots \\ m \end{array}
\begin{bmatrix}
P_{11} & P_{12} & P_{13} & \cdots & P_{1m} \\
 & P_{22} & P_{23} & \cdots & P_{2m} \\
 & & P_{33} & \cdots & P_{3m} \\
 & & & \ddots & \\
0 & & & & P_{mm}
\end{bmatrix}
$$

Consider a partition of the states $[1, 2, \ldots, l,\ \{l + 1, l + 2, \ldots, m\}]$ for $l = 1, 2, \ldots, m$. Since $\sum_{j=1}^{l} P_{ij} = 0$ for $i = l + 1, l + 2, \ldots, m$, clearly the Markov chain is lumpable with respect to the partition. Hence the Markov chain for the simple promotion model is lumpable in a similar manner.

Similar discussion will show the lumpability of a Markov chain with a lower triangular transition probability matrix.

As can be seen from the preceding examples and discussion, lumpability is very restrictive, and it cannot be used to simplify analyses in many cases. For a discussion of such problems and an approximate test for lumpability, the readers are referred to Thomas and Barr (1977).

If a Markov chain is lumpable with respect to a certain partition of states, the lumped transition probability matrix $\tilde{P}$ can be obtained by pre- and post-multiplying the transition probability matrix P by suitable transform matrices. A detailed discussion of this procedure and its ramifications are given in Kemeny and Snell (1960) and Barr and Thomas (1977).

We give here only a few important results:

1. Let the original Markov chain have m states and the lumped matrix have l states corresponding to the sets $\{S_1, S_2, \ldots, S_l\}$. Let U be an $l \times m$ matrix whose ith row is a probability vector with nonzero entries for states in S_i and 0 for the remaining states. Let V be an $m \times l$ matrix whose jth column has 1's corresponding to the states in S_j and 0's otherwise. Then the lumped transition probability matrix $\tilde{P}$ is given by

$$\tilde{P} = UPV \qquad (5.2.5)$$

Also it is easy to show that

$$VUPV = PV \qquad (5.2.6)$$

and

$$UV = I \qquad (5.2.7)$$

In using (5.2.5), it is convenient to use equal probabilities for the probability vectors of matrix U. The functions of the matrices U and V are the following: V combines the probabilities in the same set, giving identical rows for elements in the same set. The matrix U then simply removes the duplication of rows.

2. The n-step transition probability of the lumped chain can be easily obtained by successive multiplication and the application of the simplifying result (5.2.6). Then we get

$$\tilde{P}^n = UP^nV \qquad (5.2.8)$$

3. The eigenvalues of $\tilde{P}$ are also eigenvalues of P.
4. Let $Y = (y_1, y_2, \ldots, y_m)$ be the left eigenvector corresponding to the eigenvalue λ of P (i.e., $YP = Y\lambda$). Then

$$(YV)\tilde{P} = (YV)\lambda \qquad (5.2.9)$$

Hence, if λ is not an eigenvector of $\tilde{P}$,

$$YV = 0 \qquad (5.2.10)$$

The last property can be used to set up a computational algorithm to determine an appropriate transform matrix V with the following steps:

1. Compute eigenvalues of P and the corresponding left eigenvectors. Generate V using property (5.2.10).

2. Determine U using (5.2.7).
3. Determine $\tilde{P}$ using (5.2.5).
4. Check eigenvalues of $\tilde{P}$ against result 3, that eigenvalues of $\tilde{P}$ are also eigenvalues of P.

For a detailed discussion of this procedure readers are referred to Barr and Thomas (1977) and Thomas (1977).

Example 5.2.1a   (continued)

Obtain the lumped transition probability matrix $\tilde{P}$ using equation (5.2.5) when

$$P = \begin{bmatrix} \frac{1}{4} & \frac{3}{4} & 0 \\ \frac{2}{3} & \frac{1}{3} & 0 \\ 0 & \frac{1}{2} & \frac{1}{2} \end{bmatrix}$$

For the partition $[\{1,2\}, 3]$, we have

$$U = \begin{bmatrix} \frac{1}{2} & \frac{1}{2} & 0 \\ 0 & 0 & 1 \end{bmatrix} \qquad V = \begin{bmatrix} 1 & 0 \\ 1 & 0 \\ 0 & 1 \end{bmatrix}$$

$$PV = \begin{bmatrix} 1 & 0 \\ 1 & 0 \\ \frac{1}{2} & \frac{1}{2} \end{bmatrix}$$

$$\tilde{P} = UPV = \begin{bmatrix} 1 & 0 \\ \frac{1}{2} & \frac{1}{2} \end{bmatrix}$$

Example 5.2.3   (continued)

Obtain the lumped transition probability matrix $\tilde{P}$ using equation (5.2.5) in the library network problem.

We have

$$P = \begin{bmatrix} 0.2 & 0.5 & 0.3 & 0 \\ 0.4 & 0.3 & 0.1 & 0.2 \\ 0.6 & 0 & 0.2 & 0.2 \\ 0.4 & 0.2 & 0.4 & 0 \end{bmatrix}$$

For the partition $[\{1,2\}, \{3,4\}]$, we have

$$U = \begin{bmatrix} \frac{1}{2} & \frac{1}{2} & 0 & 0 \\ 0 & 0 & \frac{1}{2} & \frac{1}{2} \end{bmatrix} \qquad V = \begin{bmatrix} 1 & 0 \\ 1 & 0 \\ 0 & 1 \\ 0 & 1 \end{bmatrix}$$

$$PV = \begin{bmatrix} 0.7 & 0.3 \\ 0.7 & 0.3 \\ 0.6 & 0.4 \\ 0.6 & 0.4 \end{bmatrix}$$

$$\tilde{P} = UPV = \begin{bmatrix} 0.7 & 0.3 \\ 0.6 & 0.4 \end{bmatrix}$$

5.3   REVERSED MARKOV CHAINS

Let $\{ X_n, \ n = 1, 2, \ldots \}$ be a Markov chain with transition probabilities defined as

$$P_{ij}^{(n)} = P[X_n = j | X_0 = i] \tag{5.3.1}$$

with one-step transition probability matrix P whose elements are denoted as $P_{ij} = P_{ij}^{(1)}$ $(i, j = 1, 2, \ldots, m)$, and a steady-state probability vector $\pi = (\pi_1, \pi_2, \ldots, \pi_m)$. Now consider the probability

$$R_{ij}^{(n)} = P[X_0 = j | X_n = i] \tag{5.3.2}$$

Using conditional probability arguments, we may write

$$P[X_0 = j | X_n = i] P[X_n = i] = P[X_n = i | X_0 = j] P[X_0 = j]$$

Assuming the process to be in steady state, we get

$$R_{ij}^{(n)} \pi_i = P_{ji}^{(n)} \pi_j$$

$$R_{ij}^{(n)} = \frac{\pi_j}{\pi_i} P_{ji}^{(n)} \tag{5.3.3}$$

In particular, for $n = 1$

$$R_{ij} = \frac{\pi_j}{\pi_i} P_{ji}$$

Clearly $0 \leq R_{ij} \leq 1$, and $\sum_j R_{ij} = \sum_j \pi_j P_{ji} / \pi_i = \pi_i / \pi_i = 1$. Thus the matrix R, whose elements are R_{ij}, is a stochastic matrix. Therefore we can define a Markov chain whose transitions are governed by the matrix R and which represents the reverse of the original process. Now, if we can assume steady state for the original Markov chain, it is easy to define the reverse process to provide information for its past behavior, knowing its present state. In geological investigations, where the objective is to understand the past knowing the present state, reversed Markov chains can be useful models.

Significant properties of the reversed Markov chain are easily obtained from the properties of the original chain:

1. We have

$$\sum_i \pi_i R_{ij} = \sum_i \pi_i \frac{\pi_j P_{ji}}{\pi_i}$$

$$= \sum_i \pi_j P_{ji} = \pi_j \qquad (5.3.4)$$

showing that the probability vector $\pi = \{\pi_1, \pi_2, \ldots, \pi_m\}$ is also the steady-state probability vector for the reversed chain.

2. Also

$$\sum_n R_{ij}^{(n)} = \sum_n \frac{\pi_j P_{ji}^{(n)}}{\pi_i} \qquad (5.3.5)$$

The terms on either side of the equation converge or diverge together, and hence the properties of recurrence and transience are identical in both processes.

A special case of the reversed chain is obtained when

$$R_{ij} = P_{ij}$$

that is,

$$P[X_0 = j | X_1 = i] = P[X_1 = j | X_0 = i] \qquad (5.3.6)$$

Such a process is called a reversible Markov chain, and the transition probabilities are symmetric in time.

A Markov chain is reversible if its one-step transition probabilities exhibit the following property:

$$P_{ij} = \frac{\pi_j}{\pi_i} P_{ji} \qquad (5.3.7)$$

An obvious example of a reversible chain is an irreducible finite Markov chain whose transition probability matrix is symmetric. For other theoretical examples, the readers are referred to Feller (1968, Chap. 15).

5.4   STATISTICAL INFERENCE

In this section we shall consider the estimation and hypothesis-testing problems as related to finite Markov chains with stationary (time-homogeneous) transition probability matrices. In choosing mathematical models, these statistical problems are of prime importance. Here we shall give only some simple results for the maximum likelihood estimates of the elements of the transition probability matrix and tests of hypotheses based on them. Considerable work has been done on inference problems on Markov chains, and the interested readers are referred to journal articles and books that include: Bartlett (1951), Hoel (1954), Anderson (1954), Good (1955), Anderson and Goodman (1957), Goodman (1958a, b), Billingsley (1961a, b), Goodman (1968), Lee et al. (1970), Basava and Prakasa Rao (1980), and Edwards (1980). Of the articles, Billingsley (1961a) is an excellent review paper with a long list of references. Anderson and Goodman (1957) is a comprehensive paper covering most of the important tests needed in practice.

For purposes of inference a Markov chain can be observed in two ways: (1) one observation of a chain of great length, or (2) observations on a large number of realizations of the same chain. Fortunately, due to the ergodicity of the basic process, asymptotic theory of the two procedures result in the same tests. For a discussion of the finer points of the theory, the readers are referred to Anderson and Goodman (1957).

Observe a finite Markov chain with m states $(1, 2, 3, \ldots, m)$ until n transitions have taken place. Let n_{ij} be the number of transitions from i to j $(i, j = 1, 2, 3, \ldots, m)$. Let $\sum_{j=1}^{m} n_{ij} = n_i$. These numbers can be represented as

	1	2	3	$\cdots$	m	
1	n_{11}	n_{12}	n_{13}	$\cdots$	n_{1m}	n_1
2	n_{21}	n_{22}	n_{23}	$\cdots$	n_{2m}	n_2
3	n_{31}	n_{32}	n_{33}	$\cdots$	n_{3m}	n_3
$\vdots$	$\vdots$	$\vdots$	$\vdots$		$\vdots$	$\vdots$
m	n_{m1}	n_{m2}	n_{m3}	$\cdots$	n_{mm}	n_m
						n

$$(5.4.1)$$

Let the stationary transition probability matrix of the Markov chain be P, given by

$$P = \begin{bmatrix} P_{11} & P_{12} & \cdots & P_{1m} \\ P_{21} & P_{22} & \cdots & P_{2m} \\ \vdots & \vdots & & \vdots \\ P_{m1} & P_{m2} & \cdots & P_{mm} \end{bmatrix} \qquad (5.4.2)$$

We are interested in the estimates of the elements P_{ij}; we shall denote their estimates by $\hat{P}_{ij}$ $(i, j = 1, 2, \ldots, m)$.

For a given initial state i and a number of trials n_i, the sample of transition counts $(n_{i1}, n_{i2}, \ldots, n_{im})$ can be considered as a sample of size n_i from a multinomial distribution with probabilities $(P_{i1}, P_{i2}, \ldots, P_{im})$, such that $\sum_{j=1}^m P_{ij} = 1$. The probability of this outcome can therefore be given as

$$\frac{n_i!}{n_{i1}! n_{i2}! \cdots n_{im}!} P_{i1}^{n_{i1}} P_{i2}^{n_{i2}} \cdots P_{im}^{n_{im}} \tag{5.4.3}$$

such that $\sum_{j=1}^m n_{ij} = n_i$ and $\sum_{j=1}^m P_{ij} = 1$.

Extending this argument for the m initial states $(1, 2, 3, \ldots, m)$, when the breakdown of the total number of trials n into $(n_1, n_2, \ldots, n_m)$ is given, the probability of the realization of transition counts as in (5.4.1) is given by

$$\prod_{i=1}^m \frac{n_i!}{n_{i1}! n_{i2}! \cdots n_{im}!} P_{i1}^{n_{i1}} P_{i2}^{n_{i2}} \cdots P_{im}^{n_{im}} \tag{5.4.4}$$

In (5.4.1) the row sums $(n_1, n_2, \ldots, n_m)$ are also random variables, and therefore the unconditional likelihood function $f(P)$ of the sample observation consists of another factor giving the joint distribution of these random variables. It can be shown that this distribution is independent of the probability elements P_{ij} (see Whittle, 1955), and therefore, without going into its explicit form, we shall denote it by A. The likelihood function $f(P)$ and its natural logarithm $L(P)$ can therefore be expressed as

$$f(P) = A \prod_{i=1}^m \frac{n_i!}{n_{i1}! n_{i2}! \cdots n_{im}!} P_{i1}^{n_{i1}} P_{i2}^{n_{i2}} \cdots P_{im}^{n_{im}} \tag{5.4.5}$$

$$L(P) = \ln B + \sum_{i=1}^m \sum_{j=1}^m n_{ij} \ln P_{ij} \tag{5.4.6}$$

where $\ln B$ contains all terms independent of the P_{ij}'s.

From (5.4.6) it may also be noted that n_{ij} is a sufficient statistic for the estimation of transition probability P_{ij} $(i, j = 1, 2, 3, \ldots, m)$.

To derive maximum likelihood estimates, we maximize (5.4.6) under the condition $\sum_{j=1}^m P_{ij} = 1 (i = 1, 2, 3, \ldots, m)$. Incorporating this condition in (5.4.6), we can write

$$L(P) = \ln B + \sum_{i=1}^m \sum_{j=1}^{m-1} n_{ij} \ln P_{ij}$$

$$+ \sum_{i=1}^m n_{im} \ln(1 - P_{i1} - P_{i2} - \cdots - P_{i, m-1}) \tag{5.4.7}$$

From the structure of the log likelihood function $L(P)$, it is clear that the estimates can be obtained separately for the m values of $i = 1, 2, 3, \ldots, m$. For a specific value of i we have

$$L_i(P) = \ln B + \sum_{j=1}^{m-1} n_{ij} \ln P_{ij}$$

$$+ n_{im} \ln(1 - P_{i1} - P_{i2} - \cdots - P_{i,m-1}) \qquad (5.4.8)$$

Differentiating (5.4.8) with respect to P_{ij} ($j = 1, 2, 3, \ldots, m - 1$) and setting it equal to zero, we get

$$\frac{n_{i1}}{P_{i1}} - \frac{n_{im}}{1 - P_{i1} - P_{i2} - \cdots - P_{i,m-1}} = 0$$

$$\frac{n_{i2}}{P_{i2}} - \frac{n_{im}}{1 - P_{i1} - P_{i2} - \cdots P_{i,m-1}} = 0 \qquad (5.4.9)$$

$$\vdots$$

$$\frac{n_{i,m-1}}{P_{i,m-1}} - \frac{n_{im}}{1 - P_{i1} - P_{i2} - \cdots - P_{i,m-1}} = 0$$

Combining these equations, we may write

$$\frac{n_{i1}}{P_{i1}} = \frac{n_{i2}}{P_{i2}} = \cdots = \frac{n_{i,m-1}}{P_{i,m-1}} = \frac{n_{im}}{1 - P_{i1} - P_{i2} - \cdots - P_{i,m-1}}$$

$$(5.4.10)$$

This leads us to write

$$\frac{n_{i1}}{n_{i1}} \cdot P_{i1} = P_{i1}$$

$$\frac{n_{i2}}{n_{i1}} \cdot P_{i1} = P_{i2} \qquad (5.4.11)$$

$$\vdots$$

$$\frac{n_{im}}{n_{i1}} \cdot P_{i1} = 1 - P_{i1} - P_{i2} - \cdots - P_{i,m-1}$$

Adding both sides of the equations in (5.4.11), we get

$$\frac{n_{i1} + n_{i2} + \cdots + n_{im}}{n_{i1}} \cdot P_{i1} = 1 \qquad (5.4.12)$$

which yields the estimate

$$\hat{P}_{i1} = \frac{n_{i1}}{n_i} \qquad (5.4.13)$$

In a similar manner, we can derive estimates of other elements. Thus we get

$$\hat{P}_{ij} = \frac{n_{ij}}{n_i} \qquad i,j = 1,2,3,\dots,m \qquad (5.4.14)$$

There are several hypothesis-testing problems related to Markov chains; the major ones are the following:

1. Tests for a given transition probability matrix.
2. Tests for the stationarity of the transition probability matrix.
3. Tests for the order of the Markov chain.
4. Tests for the first-order Markov dependence.

In order to find out whether the observed realization could have come from a Markov chain with a given transition probability matrix, we consider the null hypothesis:

$$H_0 : \boldsymbol{P} = \boldsymbol{P}^0 \qquad (5.4.15)$$

For large n it can be shown that n_{ij} are asymptotically normally distributed and that the statistic $n_i^{1/2}(\hat{P}_{ij} - P_{ij})$ has an asymptotic normal distribution with mean 0 and variance $P_{ij}(1 - P_{ij})$. Based on this information, a test statistic identical with the goodness of fit statistic can be obtained. Thus we have

$$\sum_{j=1}^{m} \frac{n_i \left(\hat{P}_{ij} - P_{ij}^0 \right)^2}{P_{ij}^0} \qquad i = 1,2,\dots,m \qquad (5.4.16)$$

distributed as χ^2 with $m - 1$ degrees of freedom. In obtaining (5.4.16), we have assumed that all P_{ij}^0 are nonzero. If there are some zero elements in the ith row, only the nonzero elements should be considered in (5.4.16), and the degrees of freedom should be decreased by the number of zeros in it. Alternately, we can say that

$$\sum_{i=1}^{m} \sum_{j=1}^{m} \frac{n_i \left(\hat{P}_{ij} - P_{ij}^0 \right)^2}{P_{ij}^0} \qquad (5.4.17)$$

has a χ^2 distribution with $m(m-1) - d$ degrees of freedom, where d is the number of zeros in P^0, and the summation in (5.3.17) is taken only over (i, j) for which $P_{ij}^0 > 0$.

An asymptotically equivalent test statistic is obtained by the likelihood ratio criterion based on the Neyman-Pearson lemma. The likelihood ratio criterion for H_0 given by (5.4.15) can be obtained:

$$\Lambda = \frac{f(P^0)}{f(\hat{P})} \qquad (5.4.18)$$

where $f(\hat{P})$ is the maximized value of the likelihood function (5.4.5) obtained by substituting the estimates (5.4.14) in it. From statistical theory it is known that when H_0 is true, $-2 \ln \Lambda$ has a χ^2 distribution with $m(m-1)$ degrees of freedom. In this case we have

$$-2 \ln \Lambda = 2 \left[L(\hat{P}) - L(P^0) \right]$$

$$= 2 \sum_{i=1}^{m} \sum_{j=1}^{m} n_{ij} \ln \frac{n_{ij}}{n_i P_{ij}^0} \qquad (5.4.19)$$

A similar likelihood ratio statistic can be given for the test for stationarity of the transition probability matrix. Let P_{ij}^t be the one-step transition probability of a time-dependent process $X(t)$, such that

$$P_{ij}^t = P \left[X(t+1) = j | X(t) = i \right] \qquad (5.4.20)$$

Let n^0 be the number of processes observed. It should be noted that, in order to obtain a sufficient number of observations, it is not enough to observe a single process as done earlier. In the time-homogeneous case one can either observe a single process for a sufficiently long time or observe several processes for a shorter period. Let n_{ij}^t be the number of transition $i \to j$ during the tth transition of a process. For a given initial state i, the transition counts n_{ij}^t $(t = 1, 2, \ldots, T)$ can be represented as

t \ j	1	2	$\cdots$	m	
1	n_{i1}^1	n_{i2}^1	$\cdots$	$n_{i,m}^1$	
2	n_{i1}^2	n_{i2}^2	$\cdots$	$n_{i,m}^2$	(5.4.21)
$\vdots$	$\vdots$	$\vdots$		$\vdots$	
T	n_{i1}^T	n_{i2}^T	$\cdots$	$n_{i,m}^T$	

Arguing as before, the maximum likelihood estimates of P_{ij}^t can be obtained:

$$\hat{P}_{ij}^t = \frac{n_{ij}^t}{n_i^{t-1}} \tag{5.4.22}$$

where $n_i^{t-1} = \sum_{j=1}^m n_{ij}^t$. Clearly n_i^{t-1} is the number of processes in state i at time $t - 1$.

Suppose we wish to test the null hypothesis H_0: $P_{ij}^t = P_{ij}$ ($t = 1, 2, \ldots, T$). Then the maximized likelihood function is given by $f(\hat{P}^t)$, and therefore the likelihood ratio criterion Λ is given by

$$\Lambda = \frac{f(\hat{P}^t)}{f(P)} \tag{5.4.23}$$

As before, under the null hypothesis H_0, $-2 \ln \Lambda$ has a χ^2 distribution with $(T - 1)[m(m - 1)]$ degrees of freedom. In this case

$$-2 \ln \Lambda = 2\left[L(\hat{P}^t) - L(P) \right]$$

$$= 2 \sum_{t=1}^T \sum_{i=1}^m \sum_{j=1}^m n_{ij}^t \ln \frac{n_{ij}^t}{n_i^{t-1} P_{ij}} \tag{5.4.24}$$

As discussed in Section 5.1, Markov chains of order higher than one can be reduced to a first-order Markov chain by expanding the state space to indicate the higher-order dependence. For instance, if the Markov chain is of order r, then the state space will be the set of $(r + 1)$-tuples representing the present state and the immediately preceding r states of the Markov chain. Therefore the test procedures discussed earlier can be easily modified to this case with appropriate changes in the state space. We shall illustrate this procedure by using a second-order Markov chain with stationary transition probabilities.

Let

$$P_{ijk} = P(X_n = k | X_{n-1} = j, \, X_{n-2} = i) \qquad (5.4.25)$$

and n_{ijk} be the corresponding transition count. Also let $n_{ij}^* = \Sigma_k n_{ijk}$. Now, proceeding as before, the estimate of P_{ijk} is obtained as

$$\hat{P}_{ijk} = \frac{n_{ijk}}{n_{ij}^*} \qquad (5.4.26)$$

The null hypothesis that the Markov chain is a first-order Markov chain against the alternative that it is of order two can be given as

$$H_0 : P_{ijk} = P_{jk} \qquad (i, j, k = 1, 2, \ldots, m) \qquad (5.4.27)$$

The test statistic corresponding to (5.4.27) is given as

$$\frac{\displaystyle\sum_{i,j,k} n_{ij}^* \big(\hat{P}_{ijk} - \hat{P}_{jk}\big)}{\hat{P}_{jk}} \qquad (5.4.28)$$

which has a χ^2 distribution with $m(m-1)^2$ degrees of freedom. [The degrees of freedom will be reduced by the number of zeros among $\hat{P}_{jk}$, and only $\hat{P}_{ijk}$'s corresponding to nonzero $\hat{P}_{jk}$'s will be used. Also note that the degrees of freedom is $m(m-1)^2$ instead of $m^2(m-1)$ since $\hat{P}_{jk}$'s are estimated.]

The likelihood ratio test statistic corresponding to (5.4.19) now has the form

$$-2 \ln \Lambda = 2 \sum_{i,j,k} n_{ijk} \Big[\ln \hat{P}_{ijk} - \ln \hat{P}_{jk} \Big]$$

$$= 2 \sum_{i,j,k} n_{ijk} \ln \frac{n_{ijk}}{n_{ij}^*} \cdot \frac{n_i}{n_{ij}} \qquad (5.4.29)$$

which again has a χ^2 distribution with $m(m-1)^2$ degrees of freedom.

When this procedure is extended to a test concerning a Markov chain of order r (the null hypothesis is that it is of order $r - 1$ against the alternate hypothesis that it is of order r) the corresponding χ^2 statistic has $m^{r-1}(m-1)^2$ degrees of freedom.

When $r = 1$, trivially we get a test for the null hypothesis that the observations are independent against the alternate hypothesis that the process is a first-order Markov chain. The test statistics are obtained by using an appropriate P^0 in (5.4.15) and thereafter.

The appropriate P^0 in (5.4.15) should have identical rows under the hypothesis of independence. Let P^0 consist of m identical rows $\pi = (\pi_1, \pi_2, \ldots, \pi_m)$. When these probabilities are not known, their maximum likelihood estimates can be determined as follows.

Let $n_{.j} = \sum_{i=1}^{m} n_{ij}$. The log likelihood function corresponding to (5.4.6) can now be written as

$$L(\pi) = \ln B + \sum_{j=1}^{m} n_{.j} \ln \pi_j \qquad (5.4.30)$$

Following the method used in (5.4.8)–(5.4.13), the log likelihood function (5.4.30) leads to the maximum likelihood estimate

$$\hat{\pi}_j = \frac{n_{.j}}{n} \qquad (5.4.31)$$

The χ^2 statistic of (5.4.17) now takes the form

$$\sum_{i=1}^{m} \sum_{j=1}^{m} \frac{\left(n_{ij} - n_i n_{.j}/n\right)^2}{n_i n_{.n}/n} \qquad (5.4.32)$$

with degrees of freedom $(m - 1)^2$. [Of the m^2 entries, m row sums are fixed and $(m - 1)$ probabilities are estimated. Hence the degrees of freedom $= m^2 - m(m - 1) = (m - 1)^2$.]

The likelihood ratio statistic corresponding to (5.4.19) takes the form

$$2 \sum_{i=1}^{m} \sum_{j=1}^{m} n_{ij} \ln \left(\frac{n_{ij}}{n_i n_{.j}/n} \right) \qquad (5.4.33)$$

with $(m - 1)^2$ degrees of freedom.

When the transition probability matrix has missing entries adjustments are needed in the estimation and test procedures outlined. For a discussion of these methods the readers are referred to Goodman (1968).

Example 5.4.1

This example* appears in Anderson (1954), who used the data collected by the Bureau of Applied Social Research on voter attitudes in Erie County, Ohio, during 1940 (Lazarsfeld et al., 1948). A group of about 600 people were interviewed·regarding their voting preferences (D—Democrat, R—Republican, and DK—Do not know or other candidates) in the months May–October. The Republican convention was held between the June and

*Reprinted with permission of Macmillan, Inc. from *Mathematical Thinking in the Social Sciences*, P. F. Lazarsfeld (Ed.) Copyright 1954 by The Free Press, renewed 1982 by Patricia Kendall Lazarsfeld.

July interviews, and the Democratic convention was held between the July and August interviews. The following tables give the transition counts for pairs of three successive interviews, counting only people who responded to all six interviews:

	June					July			
May	R	D	DK	Total	June	R	D	DK	Total
R	125	5	16	146	R	124	3	16	143
D	7	106	15	128	D	6	109	14	129
DK	11	18	142	171	DK	22	9	142	173
				445					445

	August			
July	R	D	DK	Total
R	146	2	4	152
D	6	111	4	121
DK	40	36	96	172
				445

Using (5.4.14), the following estimates of transition probabilities are obtained:

	June					July			
May	R	D	DK		June	R	D	DK
R	0.856	0.034	0.110		R	0.867	0.021	0.112
D	0.055	0.828	0.117		D	0.047	0.845	0.108
DK	0.064	0.105	0.831		DK	0.127	0.052	0.821

	August		
July	R	D	DK
R	0.961	0.013	0.026
D	0.050	0.917	0.033
DK	0.233	0.209	0.558

In the study of attitude changes it is now pertinent to ask whether these three sets of transition probabilities reflect the same behavior on the part of the voters over the four month period. If so, the data can be pooled to give a single transition count matrix and hence a single set of estimates. The pooled transition count matrix obtained is

	R	D	DK	Total
R	395	10	36	441
D	19	326	33	378
DK	73	63	380	516
				$\overline{1335}$

The pooled estimates of elements of the transition probability matrix obtained are

	R	D	DK
R	0.896	0.023	0.081
D	0.050	0.862	0.088
DK	0.141	0.122	0.737

This is a good example for testing the stationarity of the transition probability matrix. Calculations based on (5.4.24) give the likelihood ratio statistic

$$-2\ln\Lambda = 97.644$$

which under the null hypothesis of stationarity has a χ^2 distribution with $2 \times 3 \times 2 = 12$ degrees of freedom. From χ^2 tables we find

$$P(\chi^2 \geq 97.644) < 0.001$$

This shows that we can reject the hypothesis of stationarity even with 0.1% significance level.

Going back to the problem, the inference drawn in this study can be easily justified on the grounds of voter behavior during and after political conventions.

The estimation and hypothesis-testing procedures just described require complete information on the sample paths of the process under study. Such data are sometimes identified as micro-data because of their detailed nature. However, there are many processes in the real world from which one can get only aggregated time series data which provide information on the number (or the proportion) of processes occupying different states at different epochs of time. Since these data reflect the broader behavior of the processes, they may be identified as macro-data. We shall introduce next the least squares method for the estimation of transition probabilities from macro-data. For an extensive discussion of the problems encountered in this approach, the readers are referred to Lee et al. (1970).

Let $p_j^{(n)}$ be the unconditional distribution of the state of the Markov chain after n steps defined as

$$p_j^{(n)} = P(X_n = j)$$

Using one-step transition probabilities P_{ij} $(i, j = 1, 2, \ldots, m)$, we have

$$p_j^{(n)} = \sum_i{}' p_i^{(n-1)} P_{ij} \tag{5.4.34}$$

Let $y_j^{(n)}$ $(j = 1, 2, \ldots, m)$ be the proportion of processes observed in state j after n steps. Since relations between sample observations involve errors, we may write

$$y_j^{(n)} = \sum_i y_i^{(n-1)} P_{ij} + u_j^{(n)} \tag{5.4.35}$$

Let the observations be made over N periods. Define column vectors

$$y_j = \begin{bmatrix} y_j^{(1)} \\ y_j^{(2)} \\ \vdots \\ y_j^{(N)} \end{bmatrix} \quad P_j = \begin{bmatrix} P_{1j} \\ P_{2j} \\ \vdots \\ P_{mj} \end{bmatrix} \quad \text{and} \quad u_j = \begin{bmatrix} u_j^{(1)} \\ u_j^{(2)} \\ \vdots \\ u_j^{(N)} \end{bmatrix}$$

$$Y = \begin{bmatrix} y_1 \\ y_2 \\ \vdots \\ y_m \end{bmatrix} \quad P^* = \begin{bmatrix} P_1 \\ P_2 \\ \vdots \\ P_m \end{bmatrix} \quad \text{and} \quad U = \begin{bmatrix} u_1 \\ u_2 \\ \vdots \\ u_m \end{bmatrix}$$

and matrices Z (which is an $N \times m$ matrix)

$$
Z = \begin{bmatrix}
y_1^{(0)} & y_2^{(0)} & \cdots & y_m^{(0)} \\
y_1^{(1)} & y_2^{(1)} & \cdots & y_m^{(1)} \\
\vdots & \vdots & & \vdots \\
y_1^{(N-1)} & y_2^{(N-1)} & \cdots & y_m^{(N-1)}
\end{bmatrix}
$$

and X (which is an $Nm \times Nm$ matrix)

$$
X = \begin{bmatrix}
Z & & & \\
& Z & & 0 \\
& & \ddots & \\
& 0 & & \\
& & & Z
\end{bmatrix}
$$

Now, for $j = 1, 2, \ldots, m$ and $n = 1, 2, \ldots, N$, equation (5.4.35) may be written in matrix form as

$$
Y = XP^* + U \tag{5.4.36}
$$

We assume that matrix Z has rank m and the error terms have the property

$$
E[U] = \phi
$$

$$
E[UU^T] = \Sigma \tag{5.4.37}
$$

where Σ is a $Nm \times Nm$ nondiagonal, singular matrix.

The method of least squares requires minimizing the sum of squared error

$$
U^T U = (Y - XP^*)^T (Y - XP^*)
$$

and determining the elements P_{ij} of P^* subject to the conditions $0 \leq P_{ij} \leq 1$ and $\sum_{j=1}^{m} P_{ij} = 1$ for all i. It can be shown that the second condition is automatically satisfied if one minimizes $U^T U$ without imposing the first (non-negativity) condition (see Lee et al., 1970, p. 34). Since the non-negativity condition is not satisfied automatically, the restricted minimization problem can be formulated as a mathematical programming problem and solved using available algorithms.

REFERENCES

Anderson, T. W. (1954). "Probability Models for Analyzing Time Changes in Attitudes." In P. F. Lazarsfeld, ed., *Mathematical Thinking in the Social Sciences.* Glencoe, Ill.: The Free Press, pp. 17–66.

Anderson, T. W., and Goodman, L. A. (1957). "Statistical Inference about Markov Chains." *Ann. Math. Stat.* **28**, 89–110.

Barr, D. R., and Thomas, M. U. (1977). "An Eigenvector Condition for Markov Chain Lumpability." *Oper. Res.* **25** (6), 1028–1031.

Bartlett, M. S. (1951). "The Frequency Goodness of Fit Test for Probability Chains." *Proc. Camb. Phil. Soc.* **47**, 86–95.

Basava, I. V., and Prakasa Rao, B. L. S. (1980). *Statistical Inference for Stochastic Processes.* New York: Academic Press.

Billingsley, P. (1961a). "Statistical Methods in Markov Chains." *Ann. Math. Stat.* **32**, 12–40. This paper contains an extensive bibliography on the topic.

Billingsley, P. (1961b). *Statistical Inference for Markov Processes.* Chicago: University of Chicago Press.

Edwards, D. G. (1980). "Large Sample Tests for Stationarity and Reversibility in Finite Markov Chains." *Scand. J. Stat.* **7**, 203–206.

Feller, W. (1968). *An Introduction to Probability Theory and Its Applications.* 3rd ed. New York: Wiley, Chap. 16.

Good, I. J. (1955). "The Likelihood Ratio Test for Markov Chains." *Biometrika* **42**, 531–533. "Corrigenda." *Biometrika* **44**, 301.

Goodman, L. A. (1958a). "Simplified Runs Tests and Likelihood Ratio Tests for Markov Chains." *Biometrika* **45**, 181–197.

Goodman, L. A. (1958b). "Exact Probabilities and Asymptotic Relationships for Some Statistics from *m*-th Order Markov Chains." *Ann. Math. Stat.* **29**, 476–490.

Goodman, L. A. (1968). "The Analysis of Cross-classified Data: Independence, Quasi-independence, and Interactions in Contingency Tables with or without Missing Entries." *J. Amer. Stat. Assoc.* **63**, 1091–1131.

Hoel, P. G. (1954). "A Test for Markoff Chains." *Biometrika* **41**, 430–433.

Kemeny, J. G., and Snell, J. L. (1960). *Finite Markov Chains.* New York: Van Nostrand. Reprinted 1976, Springer-Verlag.

Lazarsfeld, P. F., Berelson, B., and Gaudet, H. (1948). *The People's Choice.* New York: Columbia University Press.

Lee, T. C., Judge, G. G., and Zellner, A. (1970). *Estimating the Parameters of the Markov Probability Model from Aggregate Time Series Data.* Amsterdam: North Holland.

Thomas, M. U. (1977). "Computational Methods for Lumping Markov Chains." *ASA Proc. Stat. Comp.*, 364–367.

Thomas, M. U., and Barr, D. R. (1977). "An Approximate Test for Markov Chain Lumpability." *J. Amer. Stat. Assoc.* **72**, 175–179.

Whittle, P. (1955). "Some Distribution and Moment Formulae for the Markov Chain." *J. Roy Stat. Soc.* B **17**, 235–242.

FOR FURTHER READING

Bartholomew, D. J. (1973). *Stochastic Models for Social Processes*. 2d ed. New York: Wiley, Chaps. 2, 3.

Bartlett, M. S. (1956). *An Introduction to Stochastic Processes*. Cambridge: Cambridge University Press.

Cox, D. R., and Miller, H. D. (1965). *The Theory of Stochastic Processes*. New York: Wiley, pp. 118–132.

Karlin, S., and Taylor, H. M. (1981). *A Second Course in Stochastic Processes*. New York: Academic Press, Chap. 10.

EXERCISES

1. A game of dice involves throwing a six-faced die marked 1 through 6. Rolling a six twice in a row is considered a win. How often can a person engaged in this game expect to win in the long run?

2. Obtain the transition probability matrix of a Markov chain whose states are three consecutive outcomes of an experiment with two possible outcomes 0 and 1, and $P(0) = p$ and $P(1) = q$, $p + q = 1$.

3. Investigate the lumpability of the states of the Markov chains with the following transition probability matrices:

   a.

   $$
   \begin{array}{c}
   1 \\ 2 \\ 3 \\ 4
   \end{array}
   \begin{bmatrix}
   0.4 & 0.2 & 0.4 & 0 \\
   0.6 & 0 & 0.4 & 0 \\
   0 & 0 & 1 & 0 \\
   0.2 & 0.5 & 0 & 0.3
   \end{bmatrix}
   $$

   Partitions to be considered [{1, 2, 4}, 3] and [{1, 2}, 3, 4].

   b.

   $$
   \begin{array}{c}
   1 \\ 2 \\ 3 \\ 4 \\ 5
   \end{array}
   \begin{bmatrix}
   1 & 0 & 0 & 0 & 0 \\
   0 & 0.3 & 0.3 & 0.4 & 0 \\
   0 & 0.5 & 0.1 & 0 & 0.4 \\
   0.2 & 0 & 0 & 0.6 & 0.2 \\
   0.2 & 0 & 0 & 0.4 & 0.4
   \end{bmatrix}
   $$

   Partitions to be considered [1, {2, 3}, {4, 5}] and [1, {2, 3, 4, 5}].

4. Determine the transition probability matrices of lumped chains in Exercise 3 whenever lumpability holds.

5. Show that a Markov chain with a symmetric transition probability matrix is reversible.

6. Show that the Markov chain of the Ehrenfest model for heat exchange of Exercise 3 of Chapter 3 is reversible.

7. Show that a two-state Markov chain with the transition probability matrix given by equation (4.3.11) is a reversible chain. Also show that the two-state chain satisfies the stronger relation

$$\pi_i P_{ij}^{(n)} = \pi_j P_{ji}^{(n)} \qquad n = 1, 2, \ldots$$

8. a. For a two-state Markov chain with a transition probability matrix given by (4.3.1), let $(\alpha, 1 - \alpha)$ be the initial state probability vector. Let $\{ X_n, n = 0, 1, 2, \ldots \}$ be the Markov chain. Determine $P(X_0 = 1 | X_n = 1)$ and the ratio of this probability to the initial probability $1 - \alpha$.

   b. While transmitting binary codes, let p be the probability that the same message will be transmitted and $q = 1 - p$ be the probability that a message switch (0 to 1 or 1 to 0) occurs. Let α and $1 - \alpha$ be the probabilities that the initial messages were 0 and 1, respectively. Suppose after n transmissions the message comes out as 1. What is the probability that the initial message was also 1?

9. A telephone receptionist at a department store can process only one call at a time. The two states the receptionist can be in are "idle" and "busy." It is conjectured that a two-state Markov chain model would be reasonable for this individual's work habits, with observations being taken once every 30 sec. The following data were obtained by observing the receptionist's state of work for a period of 25 min (0—idle; 1—busy).

$$1\ 1\ 0\ 1\ 1\ 0\ 1\ 0\ 1\ 1\ 0\ 0\ 1\ 1\ 0\ 1\ 0\ 1\ 1\ 1\ 0\ 0\ 1$$

$$1\ 1\ 1\ 0\ 1\ 0\ 0\ 1\ 1\ 1\ 0\ 1\ 1\ 0\ 1\ 0\ 1\ 0\ 1\ 1\ 1\ 0\ 1\ 0$$

   a. Derive estimates for the elements of the transition probability matrix.

   b. Is it fair to assume that the observed process is a realization of a Markov chain with the transition probability matrix

$$\begin{bmatrix} 0.2 & 0.8 \\ 0.5 & 0.5 \end{bmatrix}$$

10. An office has been receiving numerous complaints that its two-line telephone system is busy too often. A study of the system congestion is to be made using Markov models. The number of busy lines is identified as the state of the system, and observations are made every 2 minutes. For a two-hour duration the following data are recorded:

$$1\ 2\ 1\ 2\ 0\ 2\ 1\ 1\ 2\ 2\ 1\ 0\ 1\ 2\ 2\ 2\ 2\ 2\ 1$$
$$2\ 0\ 0\ 1\ 2\ 2\ 0\ 1\ 1\ 1\ 2\ 2\ 2\ 2\ 1\ 2\ 2\ 2\ 1$$
$$2\ 2\ 2\ 2\ 2\ 1\ 1\ 2\ 2\ 0\ 1\ 0\ 0\ 1\ 1\ 0\ 1\ 2\ 2$$
$$2\ 2\ 2\ 2$$

Determine the expected number of busy lines. How long do you expect the lines to be busy at a stretch? How would you determine the validity of the complaints?

11. A two-state Markov chain with a transition probability matrix

$$\begin{bmatrix} p & q \\ q & p \end{bmatrix} \quad p + q = 1$$

provides a model for a two-state process with no trend toward any of the states. A trend factor θ is incorporated into the model by modifying the transition probability matrix as

$$P = \begin{bmatrix} \dfrac{\theta p}{\theta p + q} & \dfrac{q}{\theta p + q} \\ \dfrac{\theta q}{\theta q + p} & \dfrac{p}{\theta q + p} \end{bmatrix}$$

Determine the maximum likelihood estimators for θ and p in this Markov chain [Mary H. Regier (1968), *J. Amer. Stat. Assn.* **63** (323), 993–999].

12. Using the data of Exercise 9, test whether it is an independent trials process.

13. In order to determine whether a sequence of rock types was formed randomly, the transitions between 180 consecutive sandstone (SS), mudstone (MS), lignite (LG) and limestone (LS) beds were counted. [Data from P. D. Gingerich, (1969), "Markov Analysis of Cyclic Alluvial Sediments," *J. Sedimentary Petrology* **39**, 330–332.] The tran-

sition epochs were identified as the point at which the rock types changed from one to another, regardless of the thickness of the lower bed. The following frequency distribution of rock types was observed:

Rock type	Number of beds
SS	43
MS	76
LG	45
_LS	16
	Total 180

Using this distribution, obtain a transition probability matrix under the assumption of randomness. The actual transition counts were obtained as follows:

↗	SS	MS	LG	LS
SS	0	37	3	2
MS	21	0	41	14
LG	20	25	0	0
LS	1	14	1	0

Test for the assumption of randomness.

14. Three grocery stores α, β, and γ compete for the market in a neighborhood. The stores α and β have been known to use promotional items and games such as stamps and bingo. The store γ wants to see whether these promotional items help attract more customers. At one time α is already giving out stamps and β has let it be known that it is going to start a bingo game in a month. A marketing team hired by γ targets its study with the help of 300 families, 100 buying from each store at the beginning. Surveys are conducted for six weeks, three weeks before the introduction of the bingo game and three weeks after its introduction. The five tables that follow indicate the transition counts of their preferences from one week to another during these six

weeks. What can you say about the impact of the introduction of the promotional game by β?

a.

	α	β	γ	
α	93	3	4	100
β	7	90	3	100
γ	3	2	95	100
	103	95	102	300

b.

	α	β	γ	
α	96	5	2	103
β	2	90	3	95
γ	4	5	93	102
	102	100	98	300

c.

	α	β	γ	
α	85	15	2	102
β	1	98	1	100
γ	3	14	81	98
	89	127	84	300

d.

	α	β	γ	
α	85	2	2	89
β	8	108	11	127
γ	2	2	80	84
	95	112	93	300

e.

	α	β	γ	
α	91	2	2	95
β	4	103	5	112
γ	3	1	89	93
	98	106	96	300

15. The transition counts that follow provide two sets of data on transitions among six facies for deepwater clastics of the Ordovician Tourelle Formation in Quebec, Canada. The six facies are (1) thick shale, (2) graded siltstomes and shale, (3) slurry sandstones, (4) interbedded and graded sandstones and shales, (5) amalgamated and graded sandstones, and (6) thick and coarse sandstones.

a.

Facies	1	2	3	4	5	6	n_i
1	0	2	0	2	2	0	6
2	5	0	23	31	17	8	84
3	0	21	0	45	27	8	101
4	1	54	44	0	66	25	190
5	0	6	24	81	0	38	149
6	0	5	8	31	32	0	74
$n_{.j}$	6	88	99	190	144	79	606

b.

Facies	1	2	3	4	5	6	n_i
1	0	32	2	5	2	1	42
2	37	0	8	38	13	6	102
3	0	7	0	23	13	2	45
4	4	54	16	0	49	19	142
5	0	2	14	67	0	25	108
6	1	6	4	16	26	0	53
$n_{.j}$	42	101	44	149	103	53	492

For each set test whether the formations could have been the result of random deposits of different facies.

Use both the χ^2 and the likelihood ratio statistics for the tests. (See Chapter 20 for needed adjustments when $P_{ii} = 0$ in the transition probability matrix.) [R. N. Hiscott (1981), "Chi-Square Tests for Markov Chain Analysis," *Math. Geol.* **13** (1), 69–80 and I. V. Basawa and U. N. Bhat (1985) "Chi-Square Tests for Markov Chain Analysis: Some Comments on a Paper by R. N. Hiscott", *Math. Geol.* **17**, submitted for publication.]

16. a. There are N brands of a merchandise competing in a market. Brand loyalty is defined as the proportion of customers re-purchasing the brand on the next occasion without persuasion, and purchasing pressure is defined as the proportion of customers who are persuaded to purchase a specific brand on the next occasion. For brand i ($i = 1, 2, \ldots, N$) let d_i be the coefficient of brand loyalty and w_i be the coefficient of purchasing pressure. In a Markov chain model for brand switching the coefficients of brand loyalty and purchasing pressure can be incorporated into the transition probability matrix by writing

$$P_{ij} = d_i + (1 - d_i)w_j \qquad j = i, \; j = 1, 2, \ldots, N$$

$$= (1 - d_i)w_j \qquad j \neq i, \; j = 1, 2, \ldots, N$$

where

$$0 \leq d_i \leq 1 \quad 0 \leq w_j \leq 1 \quad \text{and} \quad \sum_{j=1}^{N} w_j = 1$$

Show that the limiting distribution $\pi = (\pi_1, \pi_2, \ldots, \pi_N)$ is obtained as

$$\pi_j = \frac{\dfrac{w_j}{1 - d_j}}{\displaystyle\sum_{i=1}^{N} \frac{w_i}{1 - d_i}}$$

if $0 \le d_i < 1$, $i = 1, 2, \ldots, N$.

Interpret the results where $d_i = d$, $i = 1, 2, \ldots, N$.

b. Let w_{jt} be the purchasing pressures and y_{jt} be the brandshare at time t for brand j ($j = 1, 2, \ldots, N$). Suppose w_{jt} and y_{jt} are known for $t = 1, , \ldots, M$. Obtain the least squares estimate of the brand loyalty d in the equal brand loyalty case. [D. Whitaker (1978), "The Derivation of a Measure of Brand Loyalty Using a Markov Brand Switching Model," *J. Opl. Res. Soc.* **29**, 959–970.]

17. In a single server queueing system with Poisson arrivals, arrival rate is λ. Service times have exponential distributions with mean $1/\mu_1$ for a customer arriving into an idle system and with mean $1/\mu$ for all other customers. Let $\rho = \lambda/\mu$ and $\rho_1 = \lambda/\mu_1$. The number of customers allowed into the system is M. Let X_n be the number of customers waiting soon after the nth customer leaves the system. Then $\{X_n, n = 0, 1, 2, \ldots\}$ is a finite Markov chain with the transition probability matrix

$$P = \begin{bmatrix} k_{00} & k_{01} & k_{02} & \cdots & k_{0,M-1} & K_{0M} \\ k_{10} & k_{11} & k_{12} & \cdots & k_{1,M-1} & K_{1M} \\ & k_0 & k_1 & \cdots & k_{M-2} & K_{M-1} \\ & & & \ddots & & \\ & & & & k_0 & K_1 \end{bmatrix}$$

where

$$k_{0j} = k_{1j} = \frac{\lambda^j \mu_1}{(\lambda + \mu_1)^{j+1}} = \frac{\rho_1^j}{(\rho_1 + 1)^{j+1}} \qquad j = 0, 1, 2, \ldots$$

$$K_{0M} = K_{1M} = \sum_{j=M}^{\infty} k_{0j} = \sum_{j=M}^{\infty} k_{1j}$$

$$k_j = \frac{\lambda^j \mu}{(\lambda + \mu)^{j+1}} = \frac{\rho^j}{(\rho + 1)^{j+1}} \qquad j = 0, 1, 2, \ldots$$

and

$$K_M = \sum_{j=M}^{\infty} k_j$$

The system is observed at every departure epoch. Let n_{ij} be the number of $i \rightarrow j$ transitions observed during a given period. Determine the maximum likelihood estimates for ρ and ρ_1 using n_{ij}'s. [T. L. Goyal and C. M. Harris (1972), "Maximum-Likelihood Estimates for Queues with State Dependent Service," *Sankhya* A **34**, 65–80.]

Chapter 6

MARKOV CHAINS WITH COUNTABLY INFINITE STATES

The essential difference in the behavior of finite Markov chains and chains with countably infinite states is in their limiting behavior. For instance, in a finite Markov chain it is impossible to have all the states transient, whereas when the number of states is countably infinite, it is reasonable to think of a limit in which the process does not remain in any of the finitely identifiable states.

As an example, consider a queueing system in which the arrival rate is much larger than the departure rate and the waiting room is of infinite capacity. Under these conditions external factors are brought in to control the system (e.g., turning some of the customers away or speeding up service), but left as it is, in the long run, it is not hard to visualize a situation in which the process is not in any of the finite states. Therefore, at least for analytical purposes, such limiting behavior needs to be considered.

Apart from such oddities Markov chains with countably infinite states have properties similar to finite Markov chains. In this chapter we shall give some basic relations and theorems, mostly without proofs, and point out their importance. For the sake of brevity we shall call Markov chains with countably infinite states *infinite Markov chains*.

6.1 SOME BASIC RELATIONS

Let P be the transition probability matrix of an irreducible, infinite Markov chain with ergodic (positive recurrent and aperiodic) states. The element P_{ij} $(i, j = 0, 1, 2, \ldots)$ represents the probability of transitions from state i to state j in one step. As noted earlier, using the Chapman-Kolmogorov equation,

$$P_{ij}^{(n)} = \sum_{k=0}^{\infty} P_{ik}^{(r)} P_{kj}^{(n-r)} \tag{6.1.1}$$

for a given r, $0 \le r \le n$, we can easily see that the n-step transition probabilities $P_{ij}^{(n)}$ are given by the elements of the matrix P^n. Analogy with finite Markov chains stops here, as obtaining the nth power of an infinite matrix is a formidable task.

For the sake of completeness we shall first recall some definitions given in Chapter 3. Let $\{X_n\}$ be the Markov chain whose state space is $\{0, 1, 2, \ldots\}$. Define

$$f_{ij}^{(n)} = P\left[X_n = j, \; X_r \ne j \; (r = 1, 2, \ldots, n - 1) | X_0 = i\right] \quad (6.1.2)$$

and

$$f_{ij}^* = \sum_{n=1}^{\infty} f_{ij}^{(n)} \quad \text{and} \quad \mu_{ij} = \sum_{n=1}^{\infty} n f_{ij}^{(n)} \quad (6.1.3)$$

Clearly $f_{ij}^{(n)}$ is the probability of the first passage transition $i \to j$, and μ_{ij} is the expected value of the first passage time. When $j = i$, we shall call $f_{ii}^{(n)}$ the recurrence time distribution and $\mu_{ii} \equiv \mu_i$ the mean recurrence time of state i.

Another set of probabilities that are interesting and important in the general theory of Markov chains are taboo probabilities. These give the probabilities of transitions that take place avoiding a certain state or a set of states. We may define such taboo probabilities with reference to either ordinary transitions or first passage transitions. These are

$$_k P_{ij}^{(n)} = P\left[X_n = j, \; X_r \ne k \; (r = 1, 2, \ldots, n - 1) | X_0 = i\right] \quad (6.1.4)$$

$$_k f_{ij}^{(n)} = P\left[X_n = j, \; X_r \ne k \text{ and } j \; (r = 1, 2, \ldots, n - 1) | X_0 = i\right] \quad (6.1.5)$$

The probability $_k P_{ij}^{(n)}$ refers to the transition of the Markov chain from state i to state j in n steps, avoiding state k. Similarly $_k f_{ij}^{(n)}$ is the probability that the Markov chain enters state j for the first time at the nth step, having initially started from i and avoiding state k. Obviously

$$_j P_{ij}^{(n)} = f_{ij}^{(n)} = {_j f_{ij}^{(n)}} \quad (6.1.6)$$

A physical model will clarify the notion of taboo probabilities. In a retail store there is a regular supply of a merchandise to meet the demands for this item that arise from time to time. The retailer does not order more items than normally supplied unless the retailer is unable to meet the demand at any point of time. Assuming such a policy, we can model this situation as a

Markov chain (or a discrete-valued Markov process) with the number in stock (inventory) as the state of the process. If the cost of losing a customer is high, naturally the merchant would like to avoid the state zero in the transitions of the process. Therefore, instead of considering only the ordinary transition probabilities of the process, this individual would be inclined to consider the taboo probabilities as well, state 0 being the taboo state.

Other examples of this kind are those of avoiding the wet period of a reservoir, overflow in a finite dam, bankruptcy in any business, stoppage of work in any industry, idle period of a service counter, and so on.

Other analytical uses of taboo probabilities are given in the theorems that follow. First we shall give some useful relations that exist between ordinary, first passage, and taboo transition probabilities:

1.
$$P_{ii}^{(0)} = 1$$

$$P_{ii}^{(n)} = \sum_{k=1}^{n} f_{ii}^{(k)} P_{ii}^{(n-k)} \qquad n \geq 1 \qquad (6.1.7)$$

This follows by considering the transition $i \to i$ in n steps, as $i \to i$ for the first time in k steps and an ordinary transition $i \to i$ in $n - k$ steps. Let

$$\sum_{n=0}^{\infty} P_{ij}^{(n)} z^n = P_{ij}(z) \qquad |z| < 1 \qquad (6.1.8)$$

and

$$\sum_{n=1}^{\infty} f_{ij}^{(n)} z^n = F_{ij}(z) \qquad |z| < 1 \qquad (6.1.9)$$

Then we have, from (6.1.7),

$$P_{ii}(z) = 1 + \sum_{n=1}^{\infty} z^n \sum_{k=1}^{n} f_{ii}^{(k)} P_{ii}^{(n-k)}$$

$$= 1 + \sum_{k=1}^{\infty} f_{ii}^{(k)} z^k \sum_{n=k}^{\infty} P_{ii}^{(n-k)} z^{n-k}$$

$$= 1 + F_{ii}(z) P_{ii}(z) \qquad (6.1.10)$$

from which we get

$$P_{ii}(z) = \frac{1}{1 - F_{ii}(z)} \qquad (6.1.11)$$

[*Note:* Relations (6.1.7) and (6.1.11) are the same as given in (3.3.8) and (3.3.9), respectively.]

2.
$$P_{ij}^{(n)} = \sum_{k=1}^{n} f_{ij}^{(k)} P_{jj}^{(n-k)} \qquad n \geq 1 \qquad (6.1.12)$$

This is likewise established by considering the first passage transition $i \to j$ in $k(1 \leq k \leq n)$ steps and an ordinary transition $j \to j$ in $n - k$ steps. Again multiplying both sides by z^n and summing over n, we get

$$P_{ij}(z) = \sum_{n=1}^{\infty} z^n \sum_{k=1}^{n} f_{ij}^{(k)} P_{jj}^{(n-k)}$$

$$= \sum_{k=1}^{\infty} f_{ij}^{(k)} z^k \sum_{n=k}^{\infty} P_{jj}^{(n-k)} z^{n-k}$$

$$= F_{ij}(z) P_{jj}(z) \qquad (6.1.13)$$

Now using (6.1.11), we obtain

$$P_{ij}(z) = \frac{F_{ij}(z)}{1 - F_{jj}(z)} \qquad (6.1.14)$$

The following relations may be proved by arguments similar to those just given. These proofs are left to the reader as an exercise.

3.
$$_{k}P_{ii}^{(0)} = 1$$

$$_{k}P_{ii}^{(n)} = \sum_{r=1}^{n} f_{ii}^{(r)} {}_{k}P_{ii}^{(n-r)} \qquad (6.1.15)$$

Defining $_{k}P_{ij}(z)$ and $_{k}F_{ij}(z)$ in a manner similar to (6.1.8) and (6.1.9), we get

$$_{k}P_{ii}(z) = \frac{1}{1 - {}_{k}F_{ii}(z)} \qquad (6.1.16)$$

4.
$$_{k}P_{ij}^{(n)} = \sum_{r=1}^{n} {}_{k}f_{ij}^{(r)} {}_{k}P_{jj}^{(n-r)} \qquad (6.1.17)$$

which gives

$$_{k}P_{ij}(z) = \frac{{}_{k}F_{ij}(z)}{1 - {}_{k}F_{jj}(z)} \qquad (6.1.18)$$

5.
$$f_{ij}^{(n)} = \sum_{r=0}^{n-1} {}_jP_{ii}^{(r)} {}_if_{ij}^{(n-r)} \qquad (6.1.19)$$

giving

$$F_{ij}(z) = \frac{{}_iF_{ij}(z)}{1 - {}_jF_{ii}(z)} \qquad (6.1.20)$$

6.
$$P_{ij}^{(n)} = \sum_{r=0}^{n} P_{ii}^{(r)} {}_iP_{ij}^{(n-r)} \qquad (6.1.21)$$

These and similar relations are useful in exhibiting some of the class properties of states and the limiting behavior of the Markov chain.

6.2 SOME LIMIT THEOREMS

Next, we turn our attention to the properties and behavior of the infinite Markov chains, analogous to those of the finite Markov chains discussed in Chapter 3. When the states of the process fall into more than one equivalence class, by arranging the transition probability matrix in the canonical form as shown in (3.4.1), it is easy to see that the study of the entire process can be reduced to that of the different equivalence classes and their interrelationships. We can consider the following three possibilities:

1. A process whose state space is a single recurrent class.
2. A process whose state space is a single transient class.
3. A process whose state space contains both recurrent and transient classes.

If the Markov chain contains both recurrent and transient classes, it is intuitively clear that ultimately it should occupy one of the equivalence classes that are recurrent (recall Theorem 3.4.1 on finite Markov chains). The major problem then is to find the probability of the Markov chain leaving the transient class. This is similar to the problem of finding the first passage probability from one state to another in the same equivalence class, for in relation to the state to which the first passage is considered, all other states in that class have the same characteristics of transient states.

Suppose the Markov chain contains a single transient class. By definition there is a positive probability that ultimately the chain may not occupy one of the countably infinite states. As a natural extension of Theorem 3.4.1, one might say that as $n \to \infty$, the probability of finding X_n in one of these states is zero. This is not to say that processes with only transient states do not exist. In fact the opposite is true. As every manager knows, one of the means of easy profit is to overload the system with the hope that the resulting inconvenience may not be detrimental to the system itself. Under these circumstances the system continuously readjusts itself so that time elapsed is always finite, and only its finite time characteristics are brought to bear. Therefore, instead of investigating the limiting behavior of the system, we then look into the asymptotic behavior of system characteristics.

The study of the finite time treatment of Markov chains and their asymptotic properties are by no means simple, and therefore we shall not go into them here. We shall confine ourselves to a discussion of the limiting behavior of an irreducible aperiodic Markov chain with recurrent states. The major results are stated as theorems without proofs.

In Theorem 4.2.4, we showed for a finite Markov chain that the mean recurrence time of a state is given by the inverse of the limiting probability. In infinite chains this property provides a method of establishing the existence of the limiting distribution and obtaining it.

Theorem 6.2.1 (*1*) *Let i be a state belonging to an aperiodic recurrent equivalence class. Let $P_{ii}^{(n)}$ be the probability of the n-step transition $i \to i$, and let μ_i be its mean recurrence time [defined as in (6.1.3)]. Then $\lim_{n \to \infty} P_{ii}^{(n)}$ exists and is given by*

$$\lim_{n \to \infty} P_{ii}^{(n)} = \frac{1}{\mu_i} \tag{6.2.1}$$

(*2*) *Let j be another state belonging to the same equivalence class, and let $P_{ji}^{(n)}$ be the probability of the n-step transition $j \to i$. Then*

$$\lim_{n \to \infty} P_{ji}^{(n)} = \lim_{n \to \infty} P_{ii}^{(n)} \tag{6.2.2}$$

COROLLARY (*1*) *If i is a positive recurrent state, then*

$$\pi_i = \lim_{n \to \infty} P_{ii}^{(n)} > 0$$

(*2*) *If i is a null recurrent state, then*

$$\pi_i = \lim_{n \to \infty} P_{ii}^{(n)} = 0$$

Remark 6.2.1 Relations (6.1.7) and (6.1.12) form the basis for the proof of Theorem 6.2.1. The corollary follows directly from the theorem, since for a positive recurrent state $\mu_i < \infty$ and for a null recurrent state $\mu_i = \infty$.

Remark 6.2.2   Two more limiting results can be obtained from Theorem 6.2.1:

1. Under conditions of Theorem 6.2.1,

$$\lim_{n \to \infty} \frac{1}{n} \sum_{k=1}^{n} P_{ii}^{(k)} = \frac{1}{\mu_i} \tag{6.2.3}$$

2. When $i \leftrightarrow j$,

$$\lim_{n \to \infty} \frac{1}{n} \sum_{k=1}^{n} P_{ij}^{(k)} = \frac{1}{\mu_j} \tag{6.2.4}$$

Results (6.2.3) and (6.2.4) are meaningful even on intuitive grounds. Suppose $Y_{ij}^{(k)}$ is an indicator random variable defined as

$$Y_{ij}^{(k)} = \begin{cases} 1 & \text{if } X_k = j \text{ given } X_0 = i \\ 0 & \text{if } X_k \neq j \text{ given } X_0 = i \end{cases} \tag{6.2.5}$$

Clearly

$$P\left(Y_{ij}^{(k)} = 1\right) = P_{ij}^{(k)}$$

$$P\left(Y_{ij}^{(k)} = 0\right) = 1 - P_{ij}^{(k)}$$

Therefore

$$E\left[Y_{ij}^{(k)}\right] = P_{ij}^{(k)} \tag{6.2.6}$$

It may also be noted that $\sum_{k=1}^{n} Y_{ij}^{(k)}$ gives the number of visits to j in n steps, initially having started from i. Thus $(1/n)\sum_{k=1}^{n} P_{ii}^{(k)}$ gives the mean number of times the Markov chain returns to state i in n steps, and $(1/n)\sum_{k=1}^{n} P_{ij}^{(k)}$ gives the mean number of visits to j in n steps. As $n \to \infty$, the process behaves independently of the initial states, and the results (6.2.3) and (6.2.4) follow.

Remark 6.2.3   Relation (6.1.7), which is the fundamental equation in the proof of Theorem 6.2.1, is called the *discrete renewal equation*. It will be discussed in more detail in Chapter 8.

As in finite Markov chains, we shall call $\pi_i = \lim_{n \to \infty} P_{ii}^{(n)}$ the limiting probability of finding the process in state i. The limiting probabilities have the important property given by Theorem 6.2.2.

Theorem 6.2.2   *In an aperiodic, recurrent equivalence class* **C**, *if there exists a state i such that*

$$\lim_{n \to \infty} P_{ii}^{(n)} = \pi_i > 0$$

then $\pi_j > 0$ *for all* $j \in$ **C**.

This follows from the fact that for $i, j \in$ **C**, there exist two positive integers n and m, such that

$$P_{ij}^{(n)} > 0 \quad \text{and} \quad P_{ji}^{(m)} > 0$$

We can write

$$P_{jj}^{(m+n+\nu)} \geq P_{ji}^{(m)} P_{ii}^{(\nu)} P_{ij}^{(n)} \qquad (6.2.7)$$

As $\nu \to \infty$, we get

$$\pi_j \geq P_{ji}^{(m)} P_{ij}^{(n)} \pi_i > 0 \qquad (6.2.8)$$

Remark 6.2.4   Theorem 6.2.2 shows that positive recurrence is a class property; so also are null recurrence and transience. It is enough, therefore, to classify only one of the states of an irreducible Markov chain. This can be done either by using the basic definitions and obtaining recurrence probabilities and mean recurrence times or by other classification theorems.

Even in irreducible chains the states may exhibit different properties under varying conditions. For instance, in the queueing problem mentioned at the beginning of this chapter it is possible to show that the classification to which the states belong depends on the value of the ratio of the arrival and service rates, called *traffic intensity*. If the traffic intensity is less than one, the states of the system are positive recurrent; if it is equal to one, the states are null recurrent; and if it is greater than one, they are transient.

We shall now consider some methods of obtaining the limiting distribution of an irreducible Markov chain that is aperiodic and positive recurrent. In this connection it must be noted that Theorem 6.2.1 is not very useful, because it gives the limiting probability as the inverse of the mean recurrence time which may not be easy to obtain for all states. Theorems 6.2.3 and 6.2.4 suggest two methods for determining the limiting distribution.

Theorem 6.2.3   (*1*) *In an irreducible Markov chain with ergodic states, the limiting probabilities* $\{\pi_i\}_{i=0}^{\infty}$ *satisfy the equations*

$$\pi_j = \sum_{i=0}^{\infty} \pi_i P_{ij} \qquad j = 0, 1, 2, \ldots \qquad (6.2.9)$$

$$\sum \pi_j = 1 \qquad (6.2.10)$$

The limiting distribution obtained is stationary. (2) *Any solution of the equations*

$$\sum_{i=0}^{\infty} x_i P_{ij} = x_j \qquad j = 0, 1, 2, \ldots \qquad (6.2.11)$$

is a scalar multiple of $\{\pi_i\}_{i=0}^{\infty}$, *provided* $\Sigma |x_i| < \infty$.

Theorem 6.2.3 generalizes the result given in Theorem 4.2.3 for finite Markov chains. Unfortunately (6.2.9) and (6.2.10) cannot be used directly in obtaining π_j's as was done with (4.2.44). Here we have a set of an infinite number of equations with an infinite number of unknowns. However, these relations can be used to derive probability-generating functions which in turn give the limiting distributions.

Another method of obtaining these probabilities is to use the second part of Theorem 6.2.3. If an absolutely convergent solution for (6.2.11) can be obtained by some means, then it can be properly normalized by writing

$$\pi_j = x_j \left(\sum x_i \right)^{-1} \qquad j = 0, 1, 2, \ldots \qquad (6.2.12)$$

to determine the limiting distributions. The limiting distribution so obtained is also a stationary distribution, because from (6.2.9) we can write

$$\pi_j = \sum_{i=0}^{\infty} \pi_i P_{ij}$$

$$\pi_k = \sum_j \pi_j P_{jk} = \sum_{i=0}^{\infty} \pi_i \sum_j P_{ij} P_{jk} = \sum_i \pi_i P_{ik}^{(2)}$$

$$\vdots \qquad \qquad \vdots \qquad \qquad \vdots$$

$$= \sum_j \pi_j P_{jk}^{(n-1)} = \sum_{i=0}^{\infty} \pi_i \sum_j P_{ij} P_{jk}^{(n-1)} = \sum_i \pi_i P_{ik}^{(n)} \qquad (6.2.13)$$

giving

$$\pi_j = \sum_j \pi_i P_{ij}^{(n)} \qquad n \geq 1$$

which in fact demonstrates stationarity.

Example 6.2.1

During summer months an air conditioner repair service adopts a policy of accepting new jobs on at most two air conditioners while working on one.

Suppose a repairperson had $n(n > 1)$ more jobs waiting after starting to work on an air conditioner. Then the number of waiting jobs after the start of the next job would be (1) $n - 1$ if no new jobs arrive during work on the current job, (2) n if one new job arrives in the meantime, and (3) $n + 1$ if requests for two or more arrive. Upon finishing work on any job, if no air conditioner is waiting for service, this worker has also adopted a policy of going off to do some other work for a length of time that has the same distribution as the service time on an air conditioner. Therefore, for the purpose of modeling, the case $n = 0$ can be treated exactly as $n = 1$. Further, from experience, this worker has found that during a service period the probability distribution of repair job arrivals is as follows:

Number of arrivals	0	1	≥ 2	(6.2.14)
Probability	0.3	0.5	0.2	

Let X_n be the number of air conditioners waiting for service at the end of the nth job. To be consistent with definitions we shall define X_n to be the number actually waiting soon after the repairperson takes on a new job (which we shall exclude from X_n). Modeling $\{ X_n \}$ as a Markov chain with discrete state space $\{0, 1, 2, \ldots \}$, the transition probability matrix can be obtained:

$$P_{00} = P(0 \text{ or } 1 \text{ new job arrivals}) = 0.3 + 0.5 = 0.8$$

$$P_{01} = P(2 \text{ new job arrivals}) = 0.2$$

$$P_{i,i-1} = 0.3 \qquad P_{ii} = 0.5 \qquad P_{i,i+1} = 0.2 \qquad i > 0$$

As a matrix these probabilities can be represented as

$$P = \begin{bmatrix} 0.8 & 0.2 & 0 & 0 & 0 & \cdots \\ 0.3 & 0.5 & 0.2 & 0 & 0 & \cdots \\ 0 & 0.3 & 0.5 & 0.2 & 0 & \cdots \\ 0 & 0 & 0.3 & 0.5 & 0.2 & \cdots \\ \vdots & \vdots & \vdots & \vdots & \vdots & \end{bmatrix} \qquad (6.2.15)$$

In order to solve this example in its generality consider the job arrival distribution

Number of arrivals	0	1	≥ 2	(6.2.16)
Probability	p	q	r	

in place of (6.2.14). Note that $p + q + r = 1$. The transition probability matrix can be given as

$$
P = \begin{bmatrix}
p + q & r & 0 & 0 & 0 & \cdots \\
p & q & r & 0 & 0 & \cdots \\
0 & p & q & r & 0 & \cdots \\
0 & 0 & p & q & r & \cdots \\
\vdots & \vdots & \vdots & \vdots & \vdots &
\end{bmatrix} \tag{6.2.17}
$$

From Theorem 6.2.3 we have the limiting distribution $\pi = (\pi_0, \pi_1, \pi_2, \dots)$ satisfying $\pi = \pi P$; that is,

$$
\pi_k = \sum_{i=0}^{\infty} \pi_i P_{ik}
$$

Using the transition probability matrix (6.2.17), we get

$$
(p + q)\pi_0 + p\pi_1 = \pi_0
$$
$$
r\pi_0 + q\pi_1 + p\pi_2 = \pi_1
$$
$$
r\pi_1 + q\pi_2 + p\pi_3 = \pi_2
$$
$$
\vdots \tag{6.2.18}
$$

The first equation in (6.2.18) gives

$$
p\pi_1 = (1 - p - q)\pi_0 = r\pi_0
$$

$$
\pi_1 = \frac{r}{p}\pi_0
$$

The second equation in (6.2.18), after using the preceding relation, gives

$$
p\pi_2 = (1 - p - q)\pi_1 = r\pi_1
$$

$$
\pi_2 = \frac{r}{p}\pi_1 = \left(\frac{r}{p}\right)^2 \pi_0
$$

Proceeding in this manner, it is easy to see that

$$
\pi_n = \left(\frac{r}{p}\right)^n \pi_0 \qquad n = 1, 2, \dots \tag{6.2.19}
$$

Using the normalizing condition $\Sigma_i \pi_i = 1$, we get

$$\left[1 + \frac{r}{p} + \left(\frac{r}{p}\right)^2 + \cdots\right]\pi_0 = 1$$

$$\left(1 - \frac{r}{p}\right)^{-1}\pi_0 = 1$$

$$\pi_0 = 1 - \frac{r}{p} \qquad (6.2.20)$$

Thus we get in general

$$\pi_n = \left(1 - \frac{r}{p}\right)\left(\frac{r}{p}\right)^n$$

Clearly, in using the geometric series for the simplifications in (6.2.20), we have assumed that $r/p < 1$. If $r/p \geq 1$, naturally the limiting probabilities will be zero for finite values of n.

Going back to the original problem with the job arrival probability distribution of (6.2.14), we have

$$\pi_n = \tfrac{1}{3}\left(\tfrac{2}{3}\right)^n. \qquad (6.2.21)$$

Answer

Recursive arguments of the type used in the solution of Example 6.2.1 can be employed only in simple cases. Otherwise, we have to employ probability-generating functions in their solution. We shall illustrate this technique by slightly modifying the example.

Example 6.2.2

Consider now these modifications in Example 6.2.1.

The repairperson accepts at most three new jobs while working on one. The job arrival distribution is given as follows:

Number of arrivals	0	1	2	≥ 3	(6.2.22)
Probability	p	q	r	s	

with $p + q + r + s = 1$.

The transition probability matrix therefore has the form

$$P = \begin{bmatrix} p+q & r & s & 0 & 0 & 0 & \cdots \\ p & q & r & s & 0 & 0 & \cdots \\ 0 & p & q & r & s & 0 & \cdots \\ 0 & 0 & p & q & r & s & \cdots \\ \vdots & \vdots & \vdots & \vdots & \vdots & \vdots \end{bmatrix} \qquad (6.2.23)$$

Equations corresponding to (6.2.18) are

$$(p+q)\pi_0 + p\pi_1 = \pi_0$$
$$r\pi_0 + q\pi_1 + p\pi_2 = \pi_1$$
$$s\pi_0 + r\pi_1 + q\pi_2 + p\pi_3 = \pi_2$$
$$s\pi_1 + r\pi_2 + q\pi_3 + p\pi_4 = \pi_3 \qquad (6.2.24)$$
$$\vdots \quad\; \vdots \quad\; \vdots \quad\; \vdots \quad\; \vdots$$

Multiplying both sides of (6.2.24) by the appropriate powers of z, where $|z| < 1$, we write

$$(p+q)\pi_0 + p\pi_1 = \pi_0$$
$$r\pi_0 z + q\pi_1 z + p\pi_2 z = \pi_1 z$$
$$s\pi_0 z^2 + r\pi_1 z^2 + q\pi_2 z^2 + p\pi_3 z^2 = \pi_2 z^2$$
$$s\pi_1 z^3 + r\pi_2 z^3 + q\pi_3 z^3 + p\pi_4 z^3 = \pi_3 z^3 \qquad (6.2.25)$$
$$\vdots$$

Summing these equations and writing $\sum_{i=0}^{\infty} \pi_i z^i = \Pi(z)$, we get after some simplifications

$$p\pi_0 + sz^2\Pi(z) + rz\Pi(z) + q\Pi(z) + \frac{p}{z}\left[\Pi(z) - \pi_0\right] = \Pi(z)$$

$$\left[sz^2 + rz + q + \frac{p}{z} - 1\right]\Pi(z) = \left[\frac{p}{z} - p\right]\pi_0$$

$$\left[sz^3 + rz^2 + (q-1)z + p\right]\Pi(z) = p(1-z)\pi_0$$

$$\Pi(z) = \frac{p(1-z)\pi_0}{sz^3 + rz^2 + (q-1)z + p}$$

$$(6.2.26)$$

Noting that the sum of the coefficients of the polynomial in the denominator is zero, we conclude that $(z - 1)$ is a factor. Factoring it out and canceling, we get

$$\Pi(z) = \frac{p\pi_0}{p - (r + s)z - sz^2} \qquad (6.2.27)$$

As $z \to 1$, we should have $\sum_{i=0}^{\infty} \pi_i = 1$. Let $z \to 1$ in (6.2.27), we get

$$1 = \frac{p\pi_0}{p - (r + s) - s} = \frac{p\pi_0}{p - r - 2s}$$

that is,

$$\pi_0 = 1 - \frac{r + 2s}{p}$$

Therefore for the existence of the steady-state probabilities $\pi = (\pi_0, \pi_1, \pi_2, \dots)$ it is necessary that $(r + 2s)/p < 1$. Under this condition we have

$$\Pi(z) = \frac{p - r - 2s}{p - (r + s)z - sz^2} \qquad (6.2.28)$$

In order to determine π_i $(i = 1, 2, \dots)$, we may proceed in two ways. The first approach is to develop recursive relations from (6.2.24) so that π_i $(i = 1, 2, \dots)$ can be determined in a recursive manner. We have

$$\pi_1 = \left(\frac{1 - p - q}{p}\right)\pi_0$$

$$\pi_2 = \left(\frac{1 - q}{p}\right)\pi_1 - \left(\frac{r}{p}\right)\pi_0$$

$$\pi_n = \left(\frac{1 - q}{p}\right)\pi_{n-1} - \left(\frac{r}{p}\right)\pi_{n-2} - \left(\frac{s}{p}\right)\pi_{n-3} \qquad n \geq 3 \quad (6.2.29)$$

The second approach is by expressing the probability generating function $\Pi(z)$ as an infinite series in z. Then π_i is obtained as the coefficient of z^i in the series. For convenience we may write $\Pi(z)$ as

$$\Pi(z) = \left(\frac{r + 2s - p}{s}\right)\frac{1}{z^2 + (r + s)s^{-1}z - ps^{-1}} \qquad (6.2.30)$$

We also have

$$z^2 + \left(\frac{r+s}{s}\right) z - \frac{p}{s} = (z - z_1)(z - z_2)$$

where

$$z_1, z_2 = -\left(\frac{r+s}{2s}\right) \pm \frac{1}{2}\sqrt{\left(\frac{r+s}{s}\right)^2 + \frac{4p}{s}} \qquad (6.2.31)$$

Using z_1 and z_2, (6.2.30) may be written as

$$\Pi(z) = \left(\frac{r + 2s - p}{s}\right) \frac{1}{(z - z_1)(z - z_2)} \qquad (6.2.32)$$

Putting (6.2.32) in partial fractions (see Appendix B) and simplifying, we get

$$\Pi(z) = \frac{r + 2s - p}{s(z_1 - z_2)} \left(\frac{1}{z_2 - z} - \frac{1}{z_1 - z}\right)$$

$$= \frac{r + 2s - p}{s(z_1 - z_2)} \left[\frac{1}{z_2}\left(1 - \frac{z}{z_2}\right)^{-1} - \frac{1}{z_1}\left(1 - \frac{z}{z_1}\right)^{-1}\right]$$

$$= \frac{r + 2s - p}{s(z_1 - z_2)} \left[\frac{1}{z_2}\sum_{i=0}^{\infty}\left(\frac{z}{z_2}\right)^i - \frac{1}{z_1}\sum_{i=0}^{\infty}\left(\frac{z}{z_1}\right)^i\right] \qquad (6.2.33)$$

Now π_i $(i = 0, 1, 2, \dots)$ are obtained as coefficients of z^i on the right-hand side of (6.2.33):

$$\pi_i = \frac{r + 2s - p}{s(z_1 - z_2)} \left[\left(\frac{1}{z_2}\right)^{i+1} - \left(\frac{1}{z_1}\right)^{i+1}\right] \qquad (6.2.34)$$

Suppose we have the following numerical values for the parameters: $p = 0.4$, $q = 0.4$, $r = 0.1$, and $s = 0.1$. We have

$$\pi_0 = 1 - \frac{r + 2s}{p} = 0.250$$

From (6.2.31) we get

$$z_1 = 1.236 \qquad z_2 = -3.236$$

Substituting these values in (6.2.34), we get

$$\pi_0 = 0.250 \quad \pi_1 = 0.125 \quad \pi_2 = 0.125 \quad \pi_3 = 0.094 \quad \pi_4 = 0.078$$
$$\pi_5 = 0.063 \quad \pi_6 = 0.051 \quad \pi_7 = 0.041 \quad \pi_8 = 0.033 \quad \pi_9 = 0.027$$
$$\pi_{10} = 0.022 \quad \pi_{11} = 0.018 \quad \pi_{12} = 0.014 \quad \pi_{13} = 0.012 \quad \pi_{14} = 0.009$$
$$\pi_{15} = 0.008 \quad \pi_{16} = 0.006 \quad \pi_{17} = 0.005 \quad \pi_{18} = 0.004 \quad \pi_{19} = 0.003$$
$$\pi_{20} = 0.003 \quad \pi_{21} = 0.002 \quad \pi_{22} = 0.002 \quad \pi_{23} = 0.001 \quad \pi_{24} = 0.001$$

and so forth.   *Answer*

As an alternate method of solution, equations (6.2.24) may be treated as difference equations which can be solved using standard techniques. For instance, the equations after the second in the set (6.2.24) have the general form

$$p\pi_{n+3} + (q-1)\pi_{n+2} + r\pi_{n+1} + s\pi_n = 0 \qquad n \geq 0 \qquad (6.2.35)$$

This is a finite-homogeneous difference equation of order 3, and its solution can be determined in terms of the roots of its characteristic equation obtained as follows:

Define a forward shift operator F, such that

$$F\pi_n = \pi_{n+1} \qquad (6.2.36)$$

Using this operator F, the equation (6.2.35) can be written as

$$\left[pF^3 + (q-1)F^2 + rF + s \right]\pi_n = 0 \qquad (6.2.37)$$

The characteristic equation of this difference equation is therefore

$$pF^3 + (q-1)F^2 + rF + s = 0 \qquad (6.2.38)$$

Let F_1, F_2, and F_3 be the roots of this equation. Clearly one is a root (sum of coefficients = 0). Let $F_3 = 1$. Then F_1 and F_2 are the roots of the quadratic equation

$$pF^2 - (r+s)F - s = 0 \qquad (6.2.39)$$

It should be noted that the polynomial on the left of (6.2.39) is the same one in the denominator of (6.2.30) if we write $F = 1/z$. We have the roots

$$F_1, F_2 = \frac{r + s \pm \sqrt{(r+s)^2 + 4ps}}{2p} \qquad (6.2.40)$$

From the theory of finite difference equations (Levy and Lessman, 1961, Jordan, 1965), the homogeneous equation (6.2.37) has the solution

$$\pi_n = C_1 \cdot F_1^n + C_2 \cdot F_2^n + C_3 \cdot F_3^n \qquad (6.2.41)$$

Clearly $C_3 = 0$ since $\sum_{n=0}^{\infty} \pi_n = 1$. Also, using (6.2.27), we have

$$C_1 + C_2 = 1 - \frac{r + 2s}{p} \qquad (6.2.42)$$

and again using the condition $\sum_{n=0}^{\infty} \pi_n = 1$,

$$\frac{C_1}{1 - F_1} + \frac{C_2}{1 - F_2} = 1 \qquad (6.2.43)$$

Solving for C_1 and C_2 from (6.2.42) and (6.2.43), we get

$$C_1 = \frac{1 - F_1}{F_1 - F_2} \left(\frac{r + 2s}{p} - F_2 \right)$$

$$C_2 = \frac{1 - F_2}{F_1 - F_2} \left(F_1 - \frac{r + 2s}{p} \right) \qquad (6.2.44)$$

Substituting these values in (6.2.41), we get a complete solution for π_n, $n = 0, 1, 2, \ldots$.

As illustrated earlier, when the state space is countably infinite, the most convenient method of determining limiting probabilities depends on the structure of the transition probability matrix, and hence the structure of the steady-state equations (6.2.9). This is further illustrated by another example which is also of independent interest.

Example 6.2.3

Develop a Markov model for the number of consecutive occurrences of an event in a series of independent repeated trials, and determine its limiting distribution.

Let p be the probability of occurrence of the event (e.g., a "success" in Bernoulli trials); $q = 1 - p$. Let X_n be the number of successive occurrences of the event at the nth trial (i.e., length of the success run at the nth trial). X_n is a Markov chain. We note that $X_n = r$, $(r = 1, 2, 3, \ldots)$ if $X_{n-1} = r - 1$ and the event occurs at the nth trial; $X_n = 0$ if the event does not occur at the nth trial. The transition probability matrix is given by

$$
\mathbf{P} = \begin{array}{c}
\begin{array}{ccccccc} 0 & 1 & 2 & 3 & 4 & \cdots \end{array} \\
\begin{array}{c} 0 \\ 1 \\ 2 \\ 3 \\ \vdots \end{array}
\left[\begin{array}{cccccc}
q & p & & & & \\
q & 0 & p & & & \\
q & 0 & 0 & p & & \\
q & 0 & 0 & 0 & p & \\
\vdots & \vdots & \vdots & \vdots & \vdots & \vdots
\end{array} \right]
\end{array}
\qquad (6.2.45)
$$

The limiting probability distribution can be determined from the equation

$$
\pi_k = \sum_i \pi_i P_{ik}
$$

Using the matrix (6.2.45), we get

$$
\pi_0 = q(\pi_0 + \pi_1 + \pi_2 + \cdots)
$$
$$
\pi_1 = p\pi_0
$$
$$
\pi_2 = p\pi_1
$$
$$
\vdots
$$

Noting that $\sum_i \pi_i = 1$, we have

$$
\pi_n = p^n q \qquad n = 0, 1, 2, \ldots \qquad (6.2.46)
$$

which is a geometric probability distribution.

Answer

Remark   Example 6.2.3 provides an alternate method of establishing that the distribution of the waiting time for success in a series of repeated trials is a geometric distribution.

In some situations obtaining taboo probabilities may be simpler. From these probabilities the limiting distribution can be derived using Theorem 6.2.4, which relates taboo probabilities to the limiting distribution. Let

$$
\sum_{n=1}^{\infty} {}_0P_{0i}^{(n)} = {}_0P_{0i}^*
$$

Then we have Theorem 6.2.4.

Theorem 6.2.4 *For a recurrent irreducible Markov chain the positive sequence given by*

$$v_0 = 1 \qquad v_i = {}_0P_{0i}^* \qquad i = 1, 2, \ldots \tag{6.2.47}$$

is the unique solution (under scalar multiplicity) of the system of equations

$$v_j = \sum_{i=0}^{\infty} v_i P_{ij} \qquad j = 0, 1, 2, \ldots \tag{6.2.48}$$

Now let us turn our attention to occupation times of the Markov chain with regard to the various states. Let $N_{ij}^{(n)}$ be the occupation time of the process in state j in a total of n transitions, having started from state i initially. Dividing $N_{ij}^{(n)}$ by n, we get the mean occupation time of state j in n transitions. The limiting probabilities give the probability distribution of the state of the process after a sufficiently large number of steps. A frequency interpretation of the distribution would suggest that we should be able to interpret these limiting probabilities in terms of the mean number of times the process has occupied these different states. This is in fact so.

Theorem 6.2.5

$$\lim_{n \to \infty} \frac{1}{n} E\left(N_{ij}^{(n)} \right) = \lim_{n \to \infty} \frac{1}{n} \sum_{k=1}^{n} P_{ij}^{(k)} = \frac{1}{\mu_j} = \pi_j \tag{6.2.49}$$

The proof of this theorem follows exactly the same arguments as those used in deriving equation (4.2.63). Thus we get

$$E\left(N_{ij}^{(n)} \right) = \sum_{k=1}^{n} P_{ij}^{(k)} \tag{6.2.50}$$

and

$$\lim_{n \to \infty} \frac{1}{n} E\left(N_{ij}^{(n)} \right) = \lim_{n \to \infty} \frac{1}{n} \sum_{k=1}^{n} P_{ij}^{(k)}$$

Recalling (6.2.4) of Remark 6.2.2, we can write

$$\lim_{n \to \infty} \frac{1}{n} E\left(N_{ij}^{(n)} \right) = \frac{1}{\mu_j} = \pi_j \tag{6.2.51}$$

Finally, a word about the asymptotic normality of the occupation times. Let σ_j^2 be the variance of the recurrence times of state j, and let μ_j and σ_j^2

both be finite. Then it is possible to show that

$$P\left(a < \frac{N_{ij}^{(n)} - n/\mu_j}{\sqrt{n\sigma_j^2/\mu_j^3}} < b\right) \rightarrow \frac{1}{\sqrt{2\pi}} \int_a^b e^{-(1/2)x^2} dx \qquad \text{as } n \rightarrow \infty$$

(6.2.52)

In other words, (6.2.52) asserts that $N_{ij}^{(n)}$ is asymptotically normal with mean n/μ_j and variance $n\sigma_j^2/\mu_j^3$. An intuitive justification for this normality can be provided through the central limit theorem, noting that $N_{ij}^{(n)}$ can be represented as the sum of n indicator random variables.

6.3 BRANCHING PROCESSES

A special class of Markov chains is a Markov branching process. Using biological terminology, consider a situation in which each organism of one generation produces a random number of offspring to form the next generation. Given the probability distribution of the number of offspring produced by an organism, one is interested in characteristics such as the distribution (mean and variance in particular) of the size of the population for different generations and the probability of extinction of the population.

Examples of natural phenomena that can be modeled as branching processes are common. A classic example relates to the survival of family names. Verbal flow of information can also be modeled as a branching process. It will be illustrated later that the game of chain letters has the characteristics of a branching process. Another common example is that of electron multipliers which amplify a weak current using a series of plates that generate new electrons when hit by an electron. For the purposes of our discussion, however, we shall use the biological terminology of organism and offspring.

Suppose at the end of its lifetime an organism produces a random number S of offspring with a probability distribution

$$P(S = j) = p_j \qquad j = 0, 1, 2, \ldots \tag{6.3.1}$$

such that $p_j \geq 0$ and $\sum_j p_j = 1$. Let $S_1, S_2, \ldots, S_i, \ldots$ be independent random variables distributed identically as (6.3.1). Let X_n be the population size at the nth generation. We have

$$X_{n+1} = \sum_{k=1}^{X_n} S_k \tag{6.3.2}$$

and

$$P_{ij} = P(X_{n+1} = j | X_n = i)$$

$$= P\left(\sum_{k=1}^{i} S_k = j\right) \tag{6.3.3}$$

where S_k is the number of offspring produced by the kth member of the nth generation. It should be noted that X_{n+1} is a random sum of random variables, and since S_k, $k = 1, 2, \ldots$ are independent and identically distributed, we get

$$E[X_{n+1}] = E[X_n]E[S_k]$$

$$V[X_{n+1}] = E[X_n]V[S_k] + V[X_n]\{E[S_k]\}^2 \tag{6.3.4}$$

(Recall Review Exercise 27 of Chapter 2.)

Suppose the population starts with one organism (i.e., $X_0 = 1$). Let

$$E[S_k] = \mu \quad \text{and} \quad V[S_k] = \sigma^2 \quad k = 1, 2, 3, \ldots$$

Since $X_1 = S_1$

$$E[X_1] = \mu \quad \text{and} \quad V[X_1] = \sigma^2$$

From (6.3.4) we also have

$$E[X_n] = E[X_{n-1}]\mu \quad n > 1$$

Therefore

$$E[X_n] = \mu^n \quad n \geq 1 \tag{6.3.5}$$

Applying the variance formula from (6.3.4) recursively, we get

$$V[X_n] = \mu^{n-1}\sigma^2 + V[X_{n-1}]\mu^2$$

$$= \mu^{n-1}\sigma^2 + \mu^2\left[\mu^{n-2}\sigma^2 + V[X_{n-2}]\mu^2\right]$$

$$= \mu^{n-1}\sigma^2 + \mu^n\sigma^2 + \mu^4\left[\mu^{n-3}\sigma^2 + V[X_{n-3}]\mu^2\right]$$

$$\vdots$$

$$= \mu^{n-1}\sigma^2 + \mu^n\sigma^2 + \mu^{n+1}\sigma^2 + \cdots$$

$$+ \mu^{2n-4}\left[\mu\sigma^2 + V[X_1]\mu^2\right]$$

$$= \sigma^2\mu^{n-1}\left[1 + \mu + \mu^2 + \cdots + \mu^{n-2} + \mu^{n-1}\right]$$

giving

$$V[X_n] = \begin{cases} \dfrac{\sigma^2 \mu^{n-1}[1 - \mu^n]}{1 - \mu} & \text{if } \mu \neq 1 \\ n\sigma^2 & \text{if } \mu = 1, \quad n \geq 1 \end{cases} \qquad (6.3.6)$$

From (6.3.5) and (6.3.6) it is clear that the mean and variance of X_n increases or decreases geometrically according as $\mu > 1$ or $\mu < 1$. Also using Chebyshev's inequality, which states that

$$P[|X_n - E(X_n)| > \varepsilon] \leq \frac{V[X_n]}{\varepsilon^2} \qquad (6.3.7)$$

as $n \to \infty$, when $\mu < 1$, we also have

$$P[|X_n - 0| > \varepsilon] \to 0$$

$$P(X_n = 0) \to 1 \qquad (6.3.8)$$

When $\mu \geq 1$, the probability of extinction of the population is not easily obtained. In this case we may use a generating function approach which is also useful in the determination of the n-step transition probabilities of the branching process. Define the probability-generating function (p.g.f)

$$\phi(z) = \sum_{j=0}^{\infty} p_j z^j \qquad |z| \leq 1$$

Noting that the p.g.f of the sum of independent random variables is the product of the p.g.f.'s of individual random variables, we have

$$\sum_{j=0}^{\infty} P_{ij} z^j = [\phi(z)]^i \qquad (6.3.9)$$

Consider the n-step transition probabilities of the branching process $\{ X_n, n = 0, 1, 2, \ldots \}$, and write

$$P_{ij}^{(n)} = P[X_n = j | X_0 = i]$$

and

$$\phi_n(z) = \sum_{j=0}^{\infty} P_{ij}^{(n)} z^n \qquad |z| \leq 1$$

When $X_0 = i$, the branching processes generated by the i organisms are independent of each other, and therefore the total number of offspring at the nth generation is the sum of offspring from each organism. Again using the aforementioned p.g.f. property, we have

$$\sum_{j=0}^{\infty} P_{ij}^{(n)} z^j = \left[\phi_n(z) \right]^i \tag{6.3.10}$$

Thus, without loss of generality, we shall assume $i = 1$. From (6.3.2) we have

$$P(X_n = j) = \sum_{k=0}^{j} P(S_1 + S_2 + \cdots + S_k = j) P(X_{n-1} = k)$$

$$\sum_{j=0}^{\infty} P(X_n = j) z^j = \sum_{j=0}^{\infty} \sum_{k=0}^{j} P(S_1 + S_2 + \cdots + S_k = j) P(X_{n-1} = k) z^j$$

$$\phi_n(z) = \sum_{k=0}^{\infty} P(X_{n-1} = k) \sum_{j=k}^{\infty} P(S_1 + S_2 + \cdots + S_k = j) z^j$$

$$= \sum_{k=0}^{\infty} P(X_{n-1} = k) \left[\phi(z) \right]^k$$

that is,

$$\phi_n(z) = \phi_{n-1} \left[\phi(z) \right] \qquad n > 1 \tag{6.3.11}$$

In deriving (6.3.11), we started by conditioning on the population size at the $(n-1)$th generation. Alternately, one could start by conditioning on the size of the first generation. In this case let X_i' $(i = 1, 2, \ldots)$ be the population size of the nth generation who have descended from the ith offspring at the first generation. Then we have

$$P(X_n = j) = \sum_{k=0}^{j} P(X_1 = k) P\left(X_1' + X_2' + \cdots + X_k' = j \right)$$

Taking generating functions as before, we get

$$\phi_n(z) = \sum_{k=0}^{\infty} P(X_1 = k) \left[\phi_{n-1}(z) \right]^k$$

that is,

$$\phi_n(z) = \phi[\phi_{n-1}(z)] \qquad n > 1 \qquad (6.3.12)$$

Extending these arguments, one can easily derive the general result

$$\phi_n(z) = \phi_{n-r}[\phi_r(z)] \qquad 1 \le r < n \qquad (6.3.13)$$

The generating function relations (6.3.11)–(6.3.13) are useful in obtaining iterative solutions. The relation (6.3.11) can also be used to derive $V[X_n]$ through the implicit differentiation technique.

In order to determine the probability of ultimate extinction of the population, set $z = 0$ in $\phi_n(z)$. We have

$$\phi_n(0) = P[X_n = 0 | X_0 = 1] \qquad (6.3.14)$$

which is in fact the probability that the extinction occurs at or before the nth generation. From (6.3.12) we also have

$$\phi_n(0) = \phi(\phi_{n-1}(0)) \qquad (6.3.15)$$

As $n \to \infty$, let $\zeta = \lim_{n \to \infty} \phi_n(0)$. It can be shown that this limit exists. Taking limits in (6.3.15), it is clear that ζ satisfies the functional relation

$$x = \phi(x) \qquad (6.3.16)$$

This functional equation is a well-known relation in the theory of stochastic processes, and an analytical discussion of its roots leads us to the following result (e.g., see Karlin and Taylor, 1975, or Çinlar, 1975):

$$\zeta = \begin{cases} 1 & \text{if } \mu \le 1 \\ < 1 & \text{if } \mu > 1 \end{cases} \qquad (6.3.17)$$

where ζ is the least positive root of equation (6.3.16). Along with the information on the mean and variance of X_n, we may therefore conclude that when $\mu > 1$, there is a positive probability of indefinite growth in the population.

Example 6.3.1

An often played game is the game of a chain letter. In this game person A approaches person B with a list of a number, say n, of names and a rewarding proposition. Person A is the last name on the list. Person A would like person B to (1) send a specific gift (e.g., an amount of money, or a bottle of liquor) to the name on top of the list, (2) drop that name from the list and add his or her own name (person B) at the bottom of the list and find two other persons who would continue the process in a similar fashion. When the chain extends in this manner, person B is in a position to gain n gifts in place of the one sent, which is the main attraction of the game. (Readers are warned that this game is illegal in many states.)

The chain letter game is a good example of a branching process. If there are n names on the list, it takes n generations for person B to get his or her share of the gifts, and because of the likelihood that some branches of the expanding tree may not grow due to the inability of some individuals in the chain to find two participants, the number of gifts this person will receive is X_n.

Let p be the probability that a participant is able to recruit two new participants to continue the game. Let $q = 1 - p$. We have

$$\phi(z) = q + pz^2 \tag{6.3.18}$$

$$\mu = E(X_1) = \phi'(1) = 2p \tag{6.3.19}$$

$$\sigma^2 = \text{Var}(X_1) = \phi''(1) + \phi'(1) - [\phi'(1)]^2$$

$$= 4pq \tag{6.3.20}$$

In the nth generation, when $p \neq \frac{1}{2}$, we get

$$E(X_n) = (2p)^n$$

$$\text{Var}(X_n) = \frac{(4pq)(2p)^{n-1}[1 - (2p)^n]}{1 - 2p}$$

$$= \frac{2q(2p)^n[1 - (2p)^n]}{1 - 2p} \tag{6.3.21}$$

When $p = \frac{1}{2}$,

$$E(X_n) = 1 \qquad \text{for all } n$$

$$\text{Var}(X_n) = n \tag{6.3.22}$$

Thus when $p = \frac{1}{2}$, since $E(X_n) = 1$, the game can be considered to be fair.

For other values of p we shall illustrate the nature of the game when $n = 10$.

p	$E(X_{10})$	$\text{var}(X_{10})$
0.3	0.006	0.021
0.4	0.107	0.573
0.5	1.000	10.000
0.6	6.192	128.595
0.7	28.925	1211.596

As shown earlier, when $\mu \le 1$ (i.e., $p \le \frac{1}{2}$), the probability of extinction of the process is 1. For $p > \frac{1}{2}$ the probability is determined as the least positive root of the equation

$$x = \phi(x)$$

$$x = q + px^2$$

that is,

$$px^2 - x + q = 0$$

The roots of this quadratic equation are

$$x_1, x_2 = \frac{1 \pm (1 - 4pq)^{1/2}}{2p}$$

Noting that pq is maximum when $p = q = \frac{1}{2}$, the least positive root is

$$\zeta = \frac{1 - (1 - 4pq)^{1/2}}{2p}$$

p	Probability of extinction ζ
0.3	1.000
0.4	1.000
0.5	1.000
0.6	0.667
0.7	0.429

It should be noted that the probability of extinction does not provide any definite indication whether the extinction occurs before the person gets his or her share of the gifts. What is clear is that the longer the game has lasted, the probability of finding new participants for the game is likely to be smaller, and hence the expected return to the participant will continue to decrease.

Answer

REFERENCES

Çinlar, E. (1975). *Introduction to Stochastic Processes*. Englewood Cliffs, N.J.: Prentice-Hall.

Jordan, C. (1965). *Calculus of Finite Differences*. 3rd ed. New York: Chelsea.

Karlin, S., and Taylor, H. M. (1975). *A First Course in Stochastic Processes*. New York: Academic Press.

Levy, H., and Lessman, F. (1961). *Finite Difference Equations*. New York: Macmillan.

FOR FURTHER READING

Chung, K. L. (1967). *Markov Chains*. New York: Springer-Verlag, Part I.

Cox, D. R., and Miller, H. D. (1965). *The Theory of Stochastic Processes*. New York: Wiley, Chap. 3.

Feller, W. (1968). *An Introduction to Probability Theory and Its Applications*. 3rd ed. New York: Wiley, Chap. 15.

Gihman, I. I., and Skorohod, A. V. (1974). *The Theory of Stochastic Processes I*. Berlin: Springer-Verlag (English trans.; Russian ed., 1971).

Kannan, D. (1979). *An Introduction to Stochastic Processes*. Amsterdam: North Holland. Chaps. 2, 3, 9.

Karlin, S., and Taylor H. M. (1981). *A Second Course in Stochastic Processes*. New York: Academic Press. Chaps. 11, 12, 17, 18.

Kemeny, J. G., Snell, J. L., and Knapp, A. W. (1966). *Denumerable Markov Chains*. New York: Van Nostrand, Chaps. 4, 5, 6.

Medhi, J. (1982). *Stochastic Processes*, New Delhi: Wiley Eastern, Chaps. 3, 9.

Parzen, E. (1962). *Stochastic Processes*. San Francisco: Holden-Day, Chap. 6.

Prabhu, N. U. (1965). *Stochastic Processes*. New York: Macmillan, Chap. 2.

Rosenblatt, M. (1962). *Random Processes*. New York: Oxford University Press, Chap. 3.

Takács, L. (1960). *Stochastic Processes*. London: Methuen, Chap. 1.

EXERCISES

1. Derive the results of equations (6.1.16), (6.1.18), and (6.1.20).

2. The first passage probability of the transition $i \to j$ is f_{ij}^*; see equations (6.1.2) and (6.1.3). Show that for states i, j, k in a Markov chain

$$f_{ik}^* \geq f_{ij}^* f_{jk}^*$$

   Hint Using probabilities of the type $_k f_{ij}$ defined in (6.1.5), we have the relations

$$f_{ij}^{(n)} = {}_k f_{ij}^{(n)} + \sum_r {}_j f_{ik}^{(r)} f_{kj}^{(n-r)}$$

$$f_{ik}^{(n)} = {}_j f_{ik}^{(n)} + \sum_r {}_k f_{ij}^{(r)} f_{jk}^{(n-r)}$$

3. Solve Example 6.2.1 using probability-generating functions.

4. Let q_n ($n = 0, 1, 2, \ldots$) be the probability that a component fails when its age is n units of time. Let $p_n = 1 - q_n$. Develop a Markov model for the age of the component, and obtain its transition probability matrix. Determine the age distribution of the component from this model.

   Hint Observe that this is a generalization of the success run problem of Exercise 6 of Chapter 4.

5. Obtain the limiting distribution of the number of customers in the system in the discrete time queue of Exercise 9 of Chapter 4.

   Hint Set up equations and solve them recursively.

6. Suppose, in the service agency problem of Exercise 10 of Chapter 4, $N \to \infty$ and a full-time employee of the agency waits for new jobs to arrive when none is waiting at a service completion time. The resulting Markov chain has the transition probability matrix

$$P = \begin{bmatrix} a_0 & a_1 & a_2 & a_3 & \cdots \\ a_0 & a_1 & a_2 & a_3 & \cdots \\ & a_0 & a_1 & a_2 & \cdots \\ & & a_0 & a_1 & \cdots \\ & & \bigcirc & & \cdots \\ & & & & \cdots \end{bmatrix}$$

Let $\pi_j = \lim_{n \to \infty} P_{ij}^{(n)}$, and define

$$A(z) = \sum_{j=0}^{\infty} a_j z^j \qquad \Pi(z) = \sum_{j=0}^{\infty} \pi_j z^j \qquad |z| < 1$$

and

$$\rho = \sum_{j=0}^{\infty} j a_j$$

Show that

$$\Pi(z) = \frac{(1 - \rho)(z - 1)A(z)}{z - A(z)}$$

7. Specialize the result of Exercise 6 when

$$a_j = qp^j \qquad p, q > 0, p + q = 1$$

Determine π_j $(j = 0, 1, 2, \ldots)$ by inverting $\Pi(z)$. Also determine π_j $(j = 0, 1, 2, \ldots)$ directly without using generating functions.

Hint   Set up equations, and solve them recursively.

8. a.   A Markov chain $\{ X_n, \ n = 0, 1, 2, \ldots \}$ has a countably infinite state space $\{ 0, 1, 2, \ldots \}$ and a transition probability matrix

$$P = \begin{bmatrix} a_{00} & a_{01} & & & & \\ a_{10} & a_{11} & a_{12} & & & \bigcirc \\ & a_{21} & a_{22} & a_{23} & & \\ & & a_{32} & a_{33} & a_{34} & \\ & \bigcirc & & \cdot & \cdot & \cdot \\ & & & & \cdot & \cdot & \cdot \\ & & & & & \cdot & \cdot \end{bmatrix}$$

Determine the limiting probabilities $\pi_j = \lim_{n \to \infty} P_{ij}^{(n)}$ by solving the appropriate equations recursively.

b.   Specialize the results when $a_{00} = p + q$, $a_{01} = r$, $a_{0j} = 0$, $j \neq 0, 1$; $a_{i, i-1} = p$, $a_{ii} = q$, $a_{i, i+1} = r$, and $a_{ij} = 0$ otherwise ($i = 1, 2, \ldots$), $p, q, r > 0$, and $p + q + r = 1$.

9. A particle performs a random walk in the countably infinite space $\{0, 1, 2 \ldots\}$ based on the following transition probabilities:

$$P_{00} = q \qquad P_{01} = p \qquad P_{0j} = 0 \qquad j = 2, 3, \ldots$$

$$P_{i,i-1} = q \qquad P_{i,i+1} = p \qquad P_{ij} = 0 \qquad j \neq i - 1, i + 1, i = 1, 2, \ldots$$

$$p, q > 0 \qquad p + q = 1.$$

Classify the states of the Markov chain, and determine the limiting probabilities $\pi_j = \lim_{n \to \infty} P_{ij}^{(n)}$, $j = 1, 2, \ldots$ by solving the appropriate equations recursively.

10. Consider the random walk problem of Exercise 9 with the following modification:

$$P_{00} = 0 \qquad P_{01} = 1 \qquad P_{0j} = 0 \qquad j = 2, 3, \ldots$$

Now classify the states of the Markov chain. What can you conclude regarding the limiting probabilities in this case?

Recursively solve the set of equations

$$\sum_{i=0}^{\infty} x_i P_{ij} = x_j \qquad j = 0, 1, 2, \ldots$$

$$\sum_{i=0} x_i = 1$$

Interpret the results.

11. A gambler enters a gambling house with \$100 intending to try a hand at a popular game. The game requires the players to bet one dollar in every game. The players that win get back two dollars. Those that lose forfeit their bets. As an additional incentive the gambling house lets the players continue the game without penalty even when a player's capital is zero. (i.e., a player whose capital is zero can play at no cost, but if he or she wins, the return is only one dollar. Consequently, the gambler is allowed to play as long as he or she wants.)

Let p be the probability of win, and q $(= 1 - p)$ the probability of loss in a game. Set up a Markov model for a player's capital, and give the transition probability matrix.

Determine the limiting distribution of the capital and its expected value. Comment on the influence of the magnitude of the player's initial capital.

Specialize the results for $p = 0.45$ and $q = 0.55$.

12. A particle performs a random walk in the countably infinite space $\ldots -2, -1, 0, 1, 2, \ldots$ based on the following transition probabilities:

$$P_{i,i-1} = P_{i,i+1} = \tfrac{1}{2} \qquad P_{ij} = 0$$

$$j \neq i - 1, i + 1, \; i = \ldots -2, -1, 0, 1, 2, \ldots$$

Classify the states of the Markov chain.
Show that

$$P_{00}^{(2n)} = \binom{2n}{n} \frac{1}{2^{2n}}$$

Stirling's approximation formula is given as

$$n! \sim \sqrt{2\pi}\, n^{n+1/2} e^{-n}$$

where the sign $\sim$ is used to indicate that the ratio of the two sides tends to unity as $n \to \infty$. Using Stirling's formula, show that

$$\lim_{n \to \infty} P_{00}^{2n} = 0$$

13. Suppose the random walk problem of Exercise 12 has the following transition probabilities:

$$P_{i,i-1} = q \qquad P_{i,i+1} = p \qquad P_{ij} = 0$$

$$j \neq i - 1, i + 1, \; i = \ldots -2, -1, 0, 1, 2, \ldots$$

Classify the states of the Markov chain. Show that

$$P_{ij}^{(n)} = \binom{n}{\dfrac{n+j-i}{2}} p^{(n+j-i)/2} q^{(n-j+i)/2}$$

whenever $(n + j - i)/2$ is an integer and $0 \leq (n + j - i)/2 \leq n$. Noting that the maximum value of pq is $\tfrac{1}{4}$, show that

$$P_{ij}^{(2n)} \leq \binom{2n}{n} \frac{1}{2^{2n}} \left(\frac{p}{q}\right)^{(j-i)/2}$$

Hence conclude that

$$P_{ij}^{(2n)} \to 0 \qquad \text{as } n \to \infty$$

14. Determine the mean and the variance of the population size at the nth generation, and the probability of extinction for branching processes with the offspring distribution given as follows:

   a. $p_0 = q$, $p_1 = p$, $p + q = 1$.
   b. $p_0 = r$, $p_1 = q$, $p_2 = p$, $p + q + r = 1$.
   c. $p_0 = \frac{1}{4}$, $p_1 = \frac{1}{2}$, $p_2 = \frac{1}{4}$.
   d. $p_0 = \frac{1}{2}$, $p_2 = \frac{1}{2}$.
   e. $p_0 = \frac{1}{8}$, $p_1 = \frac{1}{2}$, $p_2 = \frac{1}{4}$, $p_3 = \frac{1}{8}$.

15. Obtain the distribution of the population size at the third generation for branching processes with the following offspring distribution, assuming the initial population size to be one:

   a. $p_0 = q$, $p_1 = p$, $p + q = 1$.
   b. $p_0 = q$, $p_2 = p$, $p + q = 1$.
   c. $p_0 = 0.2$, $p_1 = 0.6$, $p_2 = 0.2$.

16. In a single server queuing system, let a_j be the probability that j ($j = 0, 1, 2, \ldots$) customers arrive during a service period. We assume that the number of customers arriving during different service periods are independent and identically distributed random variables. A busy period is initiated when a customer arrives into an idle system, and the service starts immediately. The busy period ends when a service ends with no one waiting for service. During the intervening period the server has been kept continuously busy.

   Let

$$\phi(z) = \sum_{j=0}^{\infty} a_j z^j \qquad |z| < 1$$

   Interpret the customer arrival process during a busy period as a branching process, and show that the probability that the busy period is of finite length is given by the least positive root of the equation $x = \phi(x)$.

17. A manufacturing firm produces a single product for use of many agencies. The problem that concerns the firm is how much inventory it has on hand at the end of each day.

   Let Y_t be the amount of inventory on hand at the end of day t. Let P_t and D_t be the number of items produced and the number demanded during day t. Assume that P_t and D_t are independent random variables with distributions

$$P(P_t = k) = a_k \qquad k = 0, 1, 2, \ldots$$

$$P(D_t = r) = b_r \qquad r = 0, 1, 2, \ldots$$

Further assume that when a demand occurs, it is satisfied without delay if there is enough inventory at hand. If there are no units present to satisfy the demand, backlogging is done up to a number B to be satisfied as soon as the units are produced. (Backlogging is the process of taking orders to be satisfied later.) Excess demands will be turned away.

Model $\{ Y_t, \ t = 0, 1, 2, \ldots \}$ as a Markov chain, and give its transition probability matrix.

18. In a single server queueing system the number of customers arriving during different service periods are independent and identically distributed random variables with probability a_j for j ($j = 0, 1, 2, \ldots$) customers arrivals. Also $\sum_{j=0}^{\infty} a_j = 1$.

    Let Q_n be the number of customers in the system soon after the nth departure. Establish the relationship between Q_n and Q_{n+1}, and show that $\{ Q_n \}$ is a Markov chain. Obtain its transition probability matrix.

19. In a queueing system the time intervals between successive epochs of arrival are known as interarrival times. When there are enough customers to keep the server busy in a single server queueing system, let b_j be the probability that j ($j = 0, 1, 2, \ldots$) customers are served during an interarrival time. Also $\sum_{j=0}^{\infty} b_j = 1$. We assume that the number of customers arriving during different interarrival times are independent and identically distributed random variables. Let Q_n be the number of customers in the system just before the nth arrival. Establish the relationship between Q_n and Q_{n+1}, and show that $\{ Q_n \}$ is a Markov chain. Obtain its transition probability matrix.

20. A multiserver queueing system has s servers. The service times are constant and equal to σ units of time. The number of customers arriving during intervals of length σ are independent and identically distributed random variables. Let a_j be the probability that j ($j = 0, 1, 2, \ldots$) customers arrive during such an interval. Also $\sum_{j=0}^{\infty} a_j = 1$. Let Q_n be the number of customers in the system at time $n\sigma$. (The system starts at time zero.) Establish the relationship between Q_n and Q_{n+1}, and show that $\{ Q_n \}$ is a Markov chain. Obtain its transition probability matrix.

Chapter 7
SIMPLE MARKOV PROCESSES

7.1 EXAMPLES

In Chapter 2 the general definition of a Markov process was given as follows: Let $\{ X(t), t \in T \}$ be a stochastic process, with T as the parameter space and S as the state space. Define the conditional distribution functions of $X(t)$ as

$$F(x_0, x_1; t_0, t_1) = P[X(t_1) \le x_1 | X(t_0) = x_0]$$

$$t_0, t_1 \in T, t_0 < t_1, x_0, x_1 \in S \qquad (7.1.1)$$

Consider an arbitrary finite (or countably infinite) set of points $(t_0, t_1, t_2, \ldots, t_n, t)$, $t_0 < t_1 < t_2 < \cdots < t_n < t$ and $t, t_r \in T$ ($r = 0, 1, \ldots, n$). The process $X(t)$ is then a Markov process if we can write

$$P[X(t) \le x | X(t_n) = x_n, X(t_{n-1}) = x_{n-1}, \ldots, X(t_0) = x_0]$$

$$= P[X(t) \le x | X(t_n) = x_n]$$

$$= F(x_n, x; t_n, t) \qquad (7.1.2)$$

When T and S are discrete, we have called such processes Markov chains, and we have investigated their properties in Chapters 3–6. Here an independent treatment of some basic Markov processes with continuous parameter space and discrete state space will be given. For processes in which both S and T are the continuum of real numbers, the readers are referred to the books listed at the end of the chapter.

An important property of the Markov processes is given by the Chapman-Kolmogorov equations (2.3.2), and we shall use this in studying their behavior. Here are some of the physical phenomena for which simple Markov processes are reasonable models.

Example 7.1.1

In Example 3.1.1 the state of the computer was observed every hour. Instead, suppose we decide to observe it continuously so that, starting from some time in the morning, say 8 AM, we would like to make probability statements regarding its state at a time t ($t \geq 0$) after 8 AM. The resulting stochastic process has a continuous index parameter with the same states 0 (working) and 1 (not working).

Example 7.1.2

A sheet of metal is produced by a manufacturing process. Let $X(A)$ be the number of defects in an area A of that metal. Assuming that such defects occur at "random" (definition of this "randomness" will be given later), a good mathematical model for $X(A)$ is the Poisson process.

Example 7.1.3

There are several problems of population growth for which a certain class of simple Markov processes called "birth and death processes" has been found to provide useful mathematical models. A birth indicates an increase in the population size, whereas a death indicates a decrease. In a birth and death process model, assumptions are made of the nature of increase and decrease in the population, such that the resulting process is a Markov process. By varying such assumptions without disturbing the Markov property, different processes are obtained for different population problems. For instance, for the spread of an epidemic in a community, a pure birth process is a reasonable model if we are interested only in the number of people infected. If we are also interested in the current population size, a birth and death process would be more natural. The influence of extraneous factors can then be taken into account by introducing secondary birth or death parameters. Birth, death, and immigration process is an example.

Example 7.1.4

Birth and death process models are useful also in problems other than those related to populations. For instance, in an inventory system, suppose replenishment of stock is accomplished only by placing orders. If the demands for items occur in a Poisson process, the inventory in between replenishments can be modeled as a pure death process.

Example 7.1.5

Some common congestion processes are also good examples of birth and death process models. In a single server queueing system suppose arrivals occur as a Poisson process and the service times have an exponential distribution; then the number of customers in the system at time t is a Markov process and can be modeled as a simple birth and death process.

In view of the relevance and importance of these models we shall investigate the properties of the Poisson process and some simpler birth and death processes in this chapter. For convenience we shall use the terminology related to population problems in the discussion of these processes. Their generality and usefulness in applications will be illustrated in Chapters 11–15 while discussing a wide variety of congestion, service, and reliability problems.

7.2   THE POISSON PROCESS

Let the process $X(t)$ represent the number of times an event occurs in time $(0, t]$. Define, for $s < t$,

$$P_{ij}(s, t) = P[X(t) = j | X(s) = i] \qquad (7.2.1)$$

Further let the events occur under the following postulates:

1.   Events occurring in nonoverlapping intervals of time are independent of each other.

2.   For a sufficiently small Δt, there is a constant λ such that the probabilities of occurrence of events in the interval $(t, t + \Delta t]$ are given as follows:

   a. $P_{ii}(t, t + \Delta t) = 1 - \lambda \Delta t + o(\Delta t)$
   b. $P_{i,i+1}(t, t + \Delta t) = \lambda \Delta t + o(\Delta t)$
   c. $\displaystyle\sum_{j=i+2}^{\infty} P_{ij}(t, t + \Delta t) = o(\Delta t)$
   d. $P_{ij}(t, t + \Delta t) = 0 \qquad j < i$ $\qquad\qquad (7.2.2)$

where $o(\Delta t)$ contains all terms that tend to zero much faster than Δt; that is, $o(\Delta t)/\Delta t \to 0$ as $\Delta t \to 0$. It should be noted that $o(\Delta t)$ is a notation to represent terms of smaller order than Δt and satisfies identities such as $o(\Delta t) + o(\Delta t) = o(\Delta t)$ and $c \cdot o(\Delta t) = o(\Delta t)$ for a constant c. The need for including the $o(\Delta t)$ terms in assumptions a, b, and c is discussed in Section 7.6.

Theorem 7.2.1   *Let $X(t)$ be a process as defined at the beginning of this section. The transition distribution of the stochastic process $X(t)$ has a Poisson distribution given by*

$$P_n(t) = P[X(t) = n | X(0) = 0] = e^{-\lambda t}\frac{(\lambda t)^n}{n!} \qquad n = 0, 1, \ldots$$

$$(7.2.3)$$

Proof The independence assumption of Postulate 1 allows us to use Chapman-Kolmogorov equations, which can be given as

$$P_{in}(t, v) = \sum_k P_{ik}(t, u) P_{kn}(u, v) \qquad t < u < v \qquad (7.2.4)$$

Let $X(0) = 0$, and write $(0, t, t + \Delta t)$ instead of (t, u, v) in (7.2.4). Then we have

$$P_{0n}(0, t + \Delta t) = \sum_k P_{0k}(0, t) P_{kn}(t, t + \Delta t) \qquad (7.2.5)$$

Writing out (7.2.5) for $n = 0$, we get

$$P_{00}(0, t + \Delta t) = P_{00}(0, t) P_{00}(t, t + \Delta t) + \sum_{k=1}^{\infty} P_{0k}(0, t) P_{k0}(t, t + \Delta t)$$

and

$$P_{0n}(0, t + \Delta t) = P_{0n}(0, t) P_{nn}(t, t + \Delta t)$$

$$+ P_{0, n-1}(0, t) P_{n-1, n}(t, t + \Delta t)$$

$$+ \sum_{k \neq n, n-1}^{\infty} P_{0k}(0, t) P_{kn}(t, t + \Delta t) \qquad n > 0 \qquad (7.2.6)$$

Note that in (7.2.6) for $n = 0$, the last two terms on the right-hand side are zero, and for $n > 0$, the last term on the right-hand side is equal to $o(\Delta t)$ due to assumption c. For convenience we shall write

$$P_{0n}(0, t) \equiv P_n(t) \qquad (7.2.7)$$

Whence, using (7.2.2), we get

$$P_0(t + \Delta t) = \left[1 - \lambda \Delta t + o(\Delta t) \right] P_0(t)$$

$$P_n(t + \Delta t) = \left[1 - \lambda \Delta t + o(\Delta t) \right] P_n(t) + \left[\lambda \Delta t + o(\Delta t) \right] P_{n-1}(t) + o(\Delta t)$$

which can be rewritten as

$$P_0(t + \Delta t) = (1 - \lambda \Delta t) P_0(t) + o(\Delta t)$$

$$P_n(t + \Delta t) = (1 - \lambda \Delta t) P_n(t) + \lambda \Delta t P_{n-1}(t) + o(\Delta t) \qquad n > 0$$

$$(7.2.8)$$

Transposing $P_0(t)$ and dividing by Δt, we get

$$\frac{P_0(t + \Delta t) - P_0(t)}{\Delta t} = -\lambda P_0(t) + \frac{o(\Delta t)}{\Delta t}$$

$$\frac{P_n(t + \Delta t) - P_n(t)}{\Delta t} = -\lambda P_n(t) + \lambda P_{n-1}(t) + \frac{o(\Delta t)}{\Delta t}$$

Letting $\Delta t \to 0$, on the left-hand side we have

$$\lim_{\Delta t \to 0} \frac{P_n(t + \Delta t) - P_n(t)}{\Delta t} = \frac{d}{dt} P_n(t) = P_n'(t)$$

which we shall assume to exist (see Section 7.6). Thus we get

$$P_0'(t) = -\lambda P_0(t) \tag{7.2.9}$$

$$P_n'(t) = -\lambda P_n(t) + \lambda P_{n-1}(t) \qquad n > 0 \tag{7.2.10}$$

In this limiting operation we have used the fact that

$$\frac{o(\Delta t)}{\Delta t} \to 0 \qquad \text{as } \Delta t \to 0$$

In addition to (7.2.9) and (7.2.10) the probabilities $P_n(t)$ must satisfy the condition

$$P_n(0) = \begin{cases} 1 & \text{if } n = 0 \\ 0 & \text{if } n > 0 \end{cases} \tag{7.2.11}$$

Equations (7.2.9) and (7.2.10) form a system of difference differential equations that can be solved by a recursive method as applied to the solution of a differential equation.

Multiply both sides of (7.2.9) and (7.2.10) by $e^{\lambda t}$, and write

$$Q_n(t) = e^{\lambda t} P_n(t) \tag{7.2.12}$$

so that

$$Q_n'(t) = \lambda e^{\lambda t} P_n(t) + e^{\lambda t} P_n'(t) \tag{7.2.13}$$

Then we get

$$e^{\lambda t} P_0'(t) = -\lambda e^{\lambda t} P_0(t)$$

$$e^{\lambda t} P_n'(t) = -\lambda e^{\lambda t} P_n(t) + \lambda e^{\lambda t} P_{n-1}(t) \tag{7.2.14}$$

Therefore

$$Q_0'(t) = 0$$

$$Q_n'(t) = \lambda Q_{n-1}(t) \tag{7.2.15}$$

with the boundary condition

$$Q_0(0) = P_0(0) = 1$$

$$Q_n(0) = P_n(0) = 0 \quad n > 0 \tag{7.2.16}$$

Solving (7.2.15) recursively, we can write (with c a constant)

$Q_0'(t) = 0$ gives $Q_0(t) = c$ (7.2.16)   gives $c = 1$

$Q_1'(t) = \lambda$ gives $Q_1(t) = \lambda t + c$ (7.2.16)   gives $c = 0$

$Q_2'(t) = \lambda^2 t$ gives $Q_2(t) = \dfrac{\lambda^2 t^2}{2} + c$ (7.2.16)   gives $c = 0$

Now suppose

$$Q_{n-1}(t) = \frac{(\lambda t)^{n-1}}{(n-1)!}$$

then using (7.2.15), we have

$$Q_n'(t) = \frac{\lambda^n t^{n-1}}{(n-1)!}$$

giving

$$Q_n(t) = \frac{(\lambda t)^n}{n!} + c$$

Again using initial conditions (7.2.16), we get $c = 0$. Thus we have in general

$$Q_n(t) = e^{\lambda t} P_n(t) = \frac{(\lambda t)^n}{n!} \tag{7.2.17}$$

and hence

$$P_n(t) = e^{-\lambda t} \frac{(\lambda t)^n}{n!} \qquad \square$$

For a given time interval $(0, t]$, therefore, the number of events occurring during that interval has a Poisson distribution with mean λt. Instead of starting at $t = 0$ and $X(0) = 0$, suppose the initial observation of the process is made at s, $s > 0$, at which time $X(s) = i$. Then by the same arguments as before, by translating the time origin, it is easy to show

$$P_{in}(s, t) = e^{-\lambda(t-s)} \frac{[\lambda(t - s)]^{n-i}}{(n - i)!} \qquad (7.2.18)$$

This is true as long as λ, the rate of occurrence is not time- or state-dependent (see Exercise 9 for the corresponding result in a nonhomogeneous Poisson process).

The time intervals between successive occurrences of Poisson events are independent and have identical exponential distributions. A not too rigorous derivation of this distribution is as follows: Let $t_0 = 0, t_1, t_2, t_3, \ldots$ be the epochs at which the events occur in a Poisson process:

Let $Z_n = t_n - t_{n-1}$ $(n = 1, 2, \ldots)$. Clearly $Z_1, Z_2, \ldots$ are the random variables representing the time intervals between two successive occurrences (interoccurrence time) of Poisson events. For these random variables we have Theorem 7.2.2.

Theorem 7.2.2 *The random variables* $Z_1, Z_2, \ldots$, *representing the interoccurrence times of Poisson events, are independent and identically distributed. They have an exponential distribution given by*

$$P(Z_n \le x) = 1 - e^{-\lambda x} \qquad x \ge 0; n = 1, 2, \ldots \qquad (7.2.19)$$

Proof Considering equivalent events, we have

$$e^{-\lambda t} = P[X(t) = 0] = P(Z_1 > t) \qquad (7.2.20)$$

Therefore Z_1 has an exponential distribution with mean $1/\lambda$. Let $f_1(x)$ be the probability density of Z_1. We have

$$P(X_2 > t) = \int_u P(Z_2 > t | Z_1 = u) f_1(u) \, du$$

$$= \int_u P[X(t + u) - X(u) = 0] f_1(u) \, du$$

$$= e^{-\lambda t} \int_u f_1(u) \, du$$

$$= e^{-\lambda t} \qquad (7.2.21)$$

showing that Z_2 is also exponential with mean $1/\lambda$ and is independent of Z_1. In these simplifications we have made the crucial assumption of time-homogeneity of the Poisson process. Now let $f_2(x)$ be the probability density of $Z_1 + Z_2$. By similar arguments we can show Z_3 is also exponential with mean $1/\lambda$ and is independent of Z_1 and Z_2. The theorem now follows by induction.   □

The interoccurrence times of Poisson events therefore have a probability density function $\lambda e^{-\lambda x}$ with mean $1/\lambda$ and variance $1/\lambda^2$.

Suppose t is not a point at which an event occurs. Let $R(t)$ be the time until the next occurrence, and let $S(t)$ be the time since the last occurrence. The distributions of $R(t)$ and $S(t)$ must be known for the complete specification of the process.

Theorem 7.2.3   *The distribution of $R(t)$ is independent of t and is given by*

$$P[R(t) \leq x] = 1 - e^{-\lambda x} \qquad x \geq 0 \qquad (7.2.22)$$

Proof   Let τ $(0 \leq \tau < t)$ be the point at which the last event [say $(n-1)$th] occurred. The sequence of points can be shown diagrammatically as

The event $\{ R(t) > x \}$ implies the equivalence of two events:

$$\{ R(t) > x \} \quad \text{and} \quad \{ Z_n > t - \tau + x | Z_n > t - \tau \} \qquad (7.2.23)$$

giving

$$P[R(t) > x] = P(Z_n > t - \tau + x | Z_n > t - \tau)$$

$$= \frac{P[Z_n > t - \tau + x]}{P(Z_n > t - \tau)}$$

$$= \frac{e^{-\lambda(t - \tau + x)}}{e^{-\lambda(t - \tau)}}$$

$$= e^{-\lambda x} \qquad (7.2.24)$$

□

Theorem 7.2.4   *The distribution $S(t)$ has a probability concentration at t and is given by*

$$P[S(t) = t] = e^{-\lambda t}$$

$$P[S(t) \le x] = 1 - e^{-\lambda x} \qquad 0 \le x < t \qquad (7.2.25)$$

Proof   Suppose no event has occurred in $(0, t]$. Then we have

$$P[S(t) = t] = P(Z_1 > t) = e^{-\lambda t} \qquad (7.2.26)$$

Suppose there is at least one event occurring in $(0, t]$. Then x is such that there is at least one event occurring in the interval $(t - x, t]$. Therefore we can write

$$P[S(t) \le x] = P\{\text{There is at least one event occurring in } (t - x, t)\}$$

$$= 1 - P\{\text{No event occurs in the interval } (t - x, t)\}$$

$$= 1 - e^{-\lambda x} \qquad (7.2.27)$$

$\square$

Note that we have used Theorem 7.2.3 in the proof of this theorem.

Remark 7.2.1   In the terminology of renewal theory (Chapter 8) the random variables $R(t)$ and $S(t)$ whose distributions have been derived in Theorems 7.2.3 and 7.2.4 are known as *forward recurrence time* and *backward recurrence time*, respectively.

Remark 7.2.2   Theorem 7.2.3 gives a special property of the exponential distribution. Let Z be a random variable with an exponential distribution and

$$F(x) = P(Z \le x) \text{ and } G(x) = 1 - F(x).$$

Then we have

$$P(Z > x + y | Z > x) = P(Z > y) \qquad (7.2.28)$$

We may rewrite this relationship as

$$\frac{P(Z > x + y)}{P(Z > x)} = P(Z > y)$$

and hence

$$G(x + y) = G(x)G(y) \qquad (7.2.29)$$

It can be shown that the exponential is the only continuous distribution with this property.

In the literature the property represented by (7.2.28) has been referred to as the "memoryless," the "forgetful," or the "Markov" property of the exponential distribution. Because of this property, while dealing with Poisson events, it becomes immaterial when the last event occurred. From any point in time the length of the interval until the occurrence of the next event has an exponential distribution with the same parameter value.

Remark 7.2.3 When an exponential distribution is used as a model for life distributions, it is identified with a constant failure rate. As will be given in Section 15.2, failure rate (also known as hazard rate and instantaneous failure rate) is defined as

$$h(x) = \frac{f(x)}{1 - F(x)}$$

where $f(x)$ is the density function and $F(x)$ is the cumulative distribution function. For the exponential distribution with parameter λ we have

$$h(x) = \frac{\lambda e^{-\lambda x}}{e^{-\lambda x}} = \lambda \qquad (7.2.30)$$

which is a constant. Thus when we say that an item has a constant failure rate, we imply an exponential density for the life distribution.

Also, as will be shown in Chapter 15, the reliability $R(x)$ of an item with a constant failure rate λ (i.e., exponential lifetime) is obtained as

$$R(x) = e^{-\lambda x}$$

Remark 7.2.4 Consider two independent exponential random variables X_1 and X_2 with parameters λ_1 and λ_2, respectively. Then we have

$$P(X_1 < X_2) = \int_{x=0}^{\infty} P(X_1 < X_2 | X_2 = x) f_2(x)\, dx$$

where we have written $f_2(x)$ for the density function of the random variable

X_2. We get

$$P(X_1 < X_2) = \int_{x=0}^{\infty} P(X_1 < x) f_2(x)\, dx$$

$$= \int_{x=0}^{\infty} (1 - e^{-\lambda_1 x}) \lambda_2 e^{-\lambda_2 x}\, dx$$

$$= \frac{\lambda_1}{\lambda_1 + \lambda_2} \qquad\qquad (7.2.31)$$

The distribution of time required for the occurrence of a given number of events in the process may be needed in some situations. Let Y_n be this waiting time until the nth occurrence of a Poisson event. Clearly

$$Y_n = Z_1 + Z_2 + \cdots + Z_n$$

where $\{Z_r\}$ $(r = 1, 2, \ldots, n)$ are independent and identically distributed random variables, each having an exponential distribution given by (7.2.19). The probability density function $f(y)$ of Y_n can then be given as

$$f(y) = e^{-\lambda y} \frac{\lambda^n y^{n-1}}{(n-1)!} \qquad y > 0 \qquad\qquad (7.2.32)$$

which is a gamma probability density function with parameters n and λ. This can be derived either by noting that the distribution of Y_n is the n-fold convolution of exponential random variables (use methods of transforms and inversion) or by noting that

$$P(Y_n > y) = P(\text{Number of occurrences in}(0, y] \text{ is } \leq n - 1)$$

$$= \sum_{r=0}^{n-1} e^{-\lambda y} \frac{(\lambda y)^r}{r!} \qquad\qquad (7.2.33)$$

and using the identity

$$\int_y^{\infty} e^{-\lambda x} \frac{(\lambda x)^{n-1}}{(n-1)!} \lambda\, dx = \sum_{r=0}^{n-1} e^{-\lambda y} \frac{(\lambda y)^r}{r!} \qquad\qquad (7.2.34)$$

Also we have

$$E(Y_n) = \frac{n}{\lambda}$$

$$V(Y_n) = \frac{n}{\lambda^2}$$

One of the significant properties of the Poisson process is its additive property. Let $X_1(t)$ and $X_2(t)$ be two Poisson processes with parameters λ_1 and λ_2, respectively. Let $X(t) = X_1(t) + X_2(t)$. We have for $t \geq 0$

$$P[X_1(t) = n_1] = e^{-\lambda_1 t}\frac{(\lambda_1 t)^{n_1}}{n_1!}$$

$$P[X_2(t) = n_2] = e^{-\lambda_2 t}\frac{(\lambda_2 t)^{n_2}}{n_2!}$$

Using these results, we can show that $X(t)$ is also Poisson for $t \geq 0$:

$$P[X(t) = n] = \sum_{n_2=0}^{n} P[X_1(t) = n - n_2]P[X_2(t) = n_2]$$

$$= \sum_{n_2=0}^{n} e^{-\lambda_1 t}\frac{(\lambda_1 t)^{n-n_2}}{(n - n_2)!}e^{-\lambda_2 t}\frac{(\lambda_2 t)^{n_2}}{n_2!}$$

$$= e^{-(\lambda_1+\lambda_2)t}t^n \sum_{n_2=0}^{n} \frac{\lambda_1^{n-n_2}\lambda_2^{n_2}}{(n - n_2)!n_2!}$$

$$= e^{-(\lambda_1+\lambda_2)t}\frac{[(\lambda_1+\lambda_2)t]^n}{n!} \sum_{n_2=0}^{n} \binom{n}{n_2}\left(\frac{\lambda_1}{\lambda_1+\lambda_2}\right)^{n-n_2}\left(\frac{\lambda_2}{\lambda_1+\lambda_2}\right)^{n_2}$$

$$= e^{-(\lambda_1+\lambda_2)t}\frac{[(\lambda_1+\lambda_2)t]^n}{n!} \qquad (7.2.35)$$

Clearly this property can be extended to any number of Poisson processes.

Another useful property of the Poisson process is its relationship to the uniform distribution. Let n Poisson events occur at epochs $t_1 < t_2 < t_3 < \cdots < t_n$ in the interval $[0, T]$. Then, the random variables $t_1, t_2, \ldots, t_n$ have the same distribution as the n order statistics corresponding to the independent random variables $U_1, U_2, \ldots, U_n$, uniformly distributed in the interval

[0, T]. If $f_{t_1, t_2, \ldots, t_n}(x_1, \ldots, x_n)$ is the joint probability density function of $t_1, t_2, \ldots, t_n$, this property shows that

$$f_{t_1, t_2, \ldots, t_n}(x_1, x_2, \ldots, x_n) = \frac{n!}{T^n} \qquad 0 \leq x_1 \leq x_2 \leq \cdots \leq x_n \leq T$$

$$(7.2.36)$$

Remark 7.2.5 Events occurring in a Poisson process are often referred to as events occurring *at random*. As in statistics where the word "random" is used to denote independence of observations, the word is used here to indicate that the Poisson process represents a random distribution of an infinite number of points over the interval [0, ∞) in the sense just described. It should also be noted that Postulate 1 for the Poisson process given earlier assumes independence of occurrence of events. Thus, when we refer to events occurring at random without mentioning the underlying process, we imply that such a process is Poisson (see Exercises 3 and 4 for additional properties).

Some Discrete Analogs

Instead of regarding time as a continuous variable, consider a discrete set of time points $(0, \sigma, 2\sigma, 3\sigma, \ldots)$. When events occur, they do so at these epochs of time (in practical situations, events can be allowed to register only at the end of time intervals). Let the probability that an event occurs at a time point be p $(0 \leq p \leq 1)$, and let the probability of no occurrence be q $(= 1 - p)$. Also assume the independence of events. Under these assumptions, X_n, the number of occurrences in the time interval $(0, n\sigma]$, has a binomial distribution given by

$$P(X_n = k) = \binom{n}{k} p^k q^{n-k} \qquad k = 0, 1, 2, \ldots, n \qquad (7.2.37)$$

and therefore $\{X_n\}$ is a Bernoulli process.

With the assumption of independence of occurrence of events at successive time points, the distribution of the interoccurrence time can be easily derived. Let Z_n be the time interval between the $(n - 1)$th and the nth occurrences. Clearly

$$P(Z_n = k) = q^{k-1}p \qquad k = 1, 2, 3, \ldots, \quad n = 1, 2, 3, \ldots \qquad (7.2.38)$$

which is a *geometric* probability distribution with parameter p. It can also be shown that geometric is the only discrete distribution that exhibits the memoryless property defined in (7.2.28).

The distribution of Y_n, the waiting time until the nth occurrence, can be obtained by similar arguments. Suppose the nth event occurs at the kth time point. This is possible if $(n - 1)$ events occur in $(k - 1)$ time points and an event occurs at the kth. Thus we have

$$P(Y_n = k) = \binom{k-1}{n-1} p^{n-1} \cdot q^{k-n} \cdot p$$

$$= \binom{k-1}{n-1} p^n q^{k-n}$$

$$= \binom{-n}{k-n} p^n (-q)^{k-n} \qquad (7.2.39)$$

which is a *negative binomial* probability distribution with parameters p and n. In (7.2.39) we define

$$\binom{-n}{x} = \frac{(-n)(-n-1) \cdots (-n-x+1)}{x!}$$

Incidentally, it might be pointed out that the negative binomial distribution results from the convolution of independent and identical geometric distributions. The analogy in some of these distributions is clearly brought out in Table 7.2.1.

Table 7.2.1

Random Variable	Continuous Time	Discrete Time
Number of events	Poisson process (λ) Mean λt Variance λt	Bernoulli process (n, p) Mean np Variance $np(1 - p)$
Interoccurrence time	Exponential distribution (λ) Mean $1/\lambda$ Variance $1/\lambda^2$	Geometric distribution Mean $1/p$ Variance q/p^2
Waiting time to nth occurrence	Gamma distribution (λ) Mean n/λ Variance n/λ^2	Negative binomial distribution Mean n/p Variance nq/p^2

7.3 PURE BIRTH PROCESS

In the Poisson process the parameter λ remains constant irrespective of the number of events that have occurred before. This assumption may not be realistic in physical phenomena such as population growth. A more general process can be obtained by making the parameter λ dependent on the state of the process. Consider a process of events occurring under the following postulates: Suppose the event has occurred n times in time $(0, t]$. Then the occurrence or nonoccurrence of the event during $(t, t + \Delta t]$, for a sufficiently small Δt, is independent of the time since the last occurrence. Further, the probabilities of occurrence of events are given as follows:

1. Probability that the event occurs once is $\lambda_n \Delta t + o(\Delta t)$.
2. Probability that the event does not occur is $1 - \lambda_n \Delta t + o(\Delta t)$.
3. Probability that the event occurs more than once is $o(\Delta t)$.

Let $P_n(t)$ be the probability that n events occur in time $(0, t]$. Using Chapman-Kolmogorov equations for transitions in the intervals of time $(0, t]$ and $(t, t + \Delta t]$, we can write

$$P_0(t + \Delta t) = P_0(t)[1 - \lambda_0 \Delta t] + o(\Delta t)$$

$$P_n(t + \Delta t) = P_n(t)[1 - \lambda_n \Delta t] + P_{n-1}(t)\lambda_{n-1}\Delta t + o(\Delta t) \qquad n > 0$$

$$(7.3.1)$$

which give

$$P_0'(t) = -\lambda_0 P_0(t)$$

$$P_n'(t) = -\lambda_n P_n(t) + \lambda_{n-1}P_{n-1}(t) \qquad n > 0 \qquad (7.3.2)$$

with $P_0(0) = 1$ and $P_n(0) = 0$ for $n > 0$. These difference differential equations may be solved to give

$$P_n(t) = \sum_{\nu=0}^{n} A_n^{(\nu)}e^{-\lambda_\nu t} \qquad n \geq 0 \qquad (7.3.3)$$

where

$$A_n^{(\nu)} = \frac{\lambda_0 \lambda_1 \cdots \lambda_{n-1}}{(\lambda_0 - \lambda_\nu)(\lambda_1 - \lambda_\nu) \cdots (\lambda_{\nu-1} - \lambda_\nu)(\lambda_{\nu+1} - \lambda_\nu) \cdots (\lambda_n - \lambda_\nu)}$$

$$(7.3.4)$$

It can also be shown that $\{P_n(t)\}$ is a proper distribution $[\sum_{n=0}^{\infty} P_n(t) = 1]$ if and only if $\sum_{\nu=0}^{\infty} \lambda_\nu^{-1} = \infty$.

A Special Case: $\lambda_n = n\lambda$ (Yule Process)

When $\lambda_n = n\lambda$, (7.3.2) takes the form

$$P_n'(t) = -n\lambda P_n(t) + (n-1)\lambda P_{n-1}(t) \qquad n > 0 \qquad (7.3.5)$$

Assuming that $P_1(0) = 1$ and $P_i(0) = 0$ for $i > 1$, the solution can be obtained:

$$P_n(t) = \sum_{\nu=1}^{n} A_n^{(\nu)} e^{-\nu\lambda t} \qquad (7.3.6)$$

where

$$A_n^{(\nu)} = \frac{1 \cdot 2 \cdots (n-1)}{(1-\nu)(2-\nu) \cdots (\nu-1-\nu)(\nu+1-\nu) \cdots (n-\nu)}$$

$$= \frac{(n-1)!(-1)^{\nu-1}}{(\nu-1)!(n-\nu)!}$$

$$= (-1)^{\nu-1}\binom{n-1}{\nu-1} \qquad (7.3.7)$$

Thus we find

$$P_n(t) = e^{-\lambda t} \sum_{\nu=1}^{n} \binom{n-1}{\nu-1}(-e^{-\lambda t})^{\nu-1} \qquad (7.3.8)$$

$$= e^{-\lambda t}(1 - e^{-\lambda t})^{n-1}$$

Note that $P_n(t)$ is geometric with $p = e^{-\lambda t}$ and $q = 1 - e^{-\lambda t}$ (compare with 7.2.38). Further $1 - e^{-\lambda t}$ is the probability that a birth occurs during $(0, t]$. Now recalling the memoryless property of the exponential distribution, the expression (7.3.8) can be clearly identified as being the probability of $n - 1$ new births during that time period.

If initially we start with a population size i (number of occurrences before $t = 0$), by considering the population size at time t as having descended from these i originators, we can say that the population size at time t is the sum of i geometric random variables. This is the negative binomial distribution given by

$$P_n(t) = \binom{n-1}{n-i} e^{-i\lambda t}(1 - e^{-\lambda t})^{n-i} \qquad (7.3.9)$$

Let $X(t)$ be the population size in this case. Then we also have

$$E[X(t)] = ie^{\lambda t} \qquad (7.3.10)$$

and

$$V[X(t)] = ie^{\lambda t}[e^{\lambda t} - 1] \qquad (7.3.11)$$

The last two results follow by noting that from the mean and variance of the negative binomial distribution we have

$$E[X(t) - i] = \frac{i(1 - e^{-\lambda t})}{e^{-\lambda t}} \qquad (7.3.12)$$

and

$$V[X(t) - i] = \frac{i(1 - e^{-\lambda t})}{e^{-2\lambda t}} \qquad (7.3.13)$$

Deterministic Law of Population Growth

Instead of assuming the occurrence of events as a random phenomenon, suppose that the events occur in a deterministic manner. Let $N(t)$ be the number of events occurring during an interval $(0, t]$. Further let events occur according to the deterministic law:

$$\frac{dN(t)}{dt} = \lambda N(t) \qquad (7.3.14)$$

Solving this differential equation, one gets

$$\ln N(t) = \lambda t + c$$

$$N(t) = e^{\lambda t + c}$$

with $N(0) = i$, and therefore $e^c = i$. Thus we get

$$N(t) = ie^{\lambda t} \qquad (7.3.15)$$

It should be noted that the population size in time t obtained by assuming a deterministic law of growth is the expected value of the corresponding population size under stochastic assumptions.

7.4 PURE DEATH PROCESS

There are many physical phenomena in which there is no increase in the population size, once the process starts. For instance, consider a group of customers who enter the sidewalk from a single store; they leave the sidewalk after some time. Or consider a retailer, who starts with a certain inventory. The inventory gets depleted in the course of time and needs restocking. The mathematical model suited for such situations is the pure death process.

Given an initial population size, say $i > 0$, individuals die at a certain rate, eventually reducing the size to zero. When the population size is n, let μ_n be the death rate, defined as follows: In an interval $(t, t + \Delta t]$ the probability that one death occurs is $\mu_n \Delta t + o(\Delta t)$, that no death occurs is $1 - \mu_n \Delta t + o(\Delta t)$, and all other possibilities have a probability of magnitude $o(\Delta t)$. Also assume that the occurrence of death in $(t, t + \Delta t]$ is independent of the time since the last death. For such a process we can derive the difference differential equations as follows: Considering the transitions occurring in the intervals $(0, t]$ and $(t, t + \Delta t]$, we have

$$P_i(t + \Delta t) = P_i(t)\big[1 - \mu_i \Delta t + o(\Delta t)\big]$$

$$P_n(t + \Delta t) = P_n(t)\big[1 - \mu_n \Delta t + o(\Delta t)\big]$$

$$+ \mu_{n+1} \Delta t P_{n+1}(t) + o(\Delta t) \qquad n < i \qquad (7.4.1)$$

giving

$$P_i'(t) = -\mu_i P_i(t)$$

$$P_n'(t) = \mu_n P_n(t) + \mu_{n+1} P_{n+1}(t) \qquad (7.4.2)$$

In the special case $\mu_n = n\mu$, the solution to these equations takes the form

$$P_n(t) = \binom{i}{n} e^{-n\mu t}(1 - e^{-\mu t})^{i-n} \qquad n \leq i \qquad (7.4.3)$$

which is clearly a binomial distribution with $p = e^{-\mu t}$ and $q = 1 - e^{-\mu t}$.

The distribution of the population size at time t can also be obtained by treating the original members of the population as the initiators of a sequence of Bernoulli trials. Suppose each original member has a lifetime that is exponentially distributed with mean $1/\mu$. Then for every living member at time t the probability that it will die during $(t, t + \Delta t]$ is $\mu \Delta t + o(\Delta t)$. Assuming that the occurrence of death of the member of the population is an independent event, for n members alive at time t, the probability distribution of the number of deaths during $(t, t + \Delta t]$ can be given as follows:

$$P\{\text{No one will die during } (t, t + \Delta t]\} = \left[1 - \mu \Delta t + o(\Delta t) \right]^n$$

$$= 1 - n\mu \Delta t + o(\Delta t)$$

$$(7.4.4)$$

$$P\{\text{Someone will die during } (t, t + \Delta t]\}$$

$$= n\left(\mu \Delta t + o(\Delta t) \right)\left[1 - \mu \Delta t + o(\Delta t) \right]^{n-1}$$

$$= n\mu \Delta t + o(\Delta t) \qquad (7.4.5)$$

$$P\{ k(> 1) \text{ will die during } (t, t + \Delta t]\}$$

$$= \binom{n}{k}\left[\mu \Delta t + o(\Delta t) \right]^k \left[1 - \mu \Delta t + o(\Delta t) \right]^{n-k}$$

$$= o(\Delta t) \qquad (7.4.6)$$

These probabilities therefore imply the postulates of the pure death process. For the Bernoulli process the probability of a death before time t is the probability that the life of the individual does not exceed t and is given by $1 - e^{-\mu t}$. The probability that the individual does not die in time t is likewise given by $e^{-\mu t}$. Thus we have

$$P_n(t) = \binom{i}{n}\left(e^{-\mu t} \right)^n \left(1 - e^{-\mu t} \right)^{i-n} \qquad n \le i$$

and

$$E[X(t)] = ie^{-\mu t} \qquad (7.4.7)$$

$$V[X(t)] = ie^{-\mu t}\left[1 - e^{-\mu t} \right] \qquad (7.4.8)$$

7.5 TWO-STATE MARKOV PROCESSES

In the last three sections we considered Markov processes in which only one type of event occurred. In birth processes the occurrence of events resulted in an increase in the size of the population, and in death processes the result was a decrease in the population size. Nevertheless, natural phenomena abound in which both increase and decrease occur in the same setting. Combining the features of both birth and death processes, a "birth and death process" can be considered. These processes have proved to be realistic and useful in many problems related to the spread of epidemics, congestion in queueing systems (e.g., telephone trunking), traffic and maintenance problems, and so forth.

The simplest of birth and death processes is a two-state Markov process with states, say $\{0, 1\}$, in which a birth can occur only when the process is in state 0 and a death can occur only when the process is in state 1. Thus the only allowed state changes are $0 \to 1$ and $1 \to 0$, and therefore the process alternates between these two states. In view of this feature the process can be considered as the continuous time analog of a two-state Markov chain discussed in Section 4.3.

Let $\{X(t), t \geq 0\}$ be a two-state Markov process, with states $\{0, 1\}$. Suppose the transitions of the process are governed by the following postulates:

1. Suppose the process $X(t)$ is in state 0 at time t; then (a) there exists a $\lambda > 0$, such that during a small interval of time $(t, t + \Delta t]$ $X(t)$ changes its state from 0 to 1 with probability $\lambda \Delta t + o(\Delta t)$ and remains in state 0 with probability $1 - \lambda \Delta t + o(\Delta t)$, and (b) the occurrences of changes of state from 0 to 1 in nonoverlapping intervals of time are independent of each other.

2. Suppose $X(t) = 1$; then (a) there exists a $\mu > 0$, such that during a small interval of time $(t, t + \Delta t]$ $X(t)$ changes its state from 1 to 0 with probability $\mu \Delta t + o(\Delta t)$ and remains in state 1 with probability $1 - \mu \Delta t + o(\Delta t)$, and (b) the occurrences of changes of state from 1 to 0 in nonoverlapping intervals of time are independent of each other.

3. The probability that during $(t, t + \Delta t]$ more than one change of state occurs is $o(\Delta t)$.

The process $\{X(t), t \geq 0\}$ is time-homogeneous, and we define

$$P_{ij}(t) = P[X(t) = j | X(0) = i] \qquad i, j = 0, 1$$

Using Chapman-Kolmogorov equations for transitions in the intervals of time $(0, t]$ and $(t, t + \Delta t]$, we get

$$P_{i0}(t + \Delta t) = P_{i0}(t)[1 - \lambda \Delta t + o(\Delta t)] + P_{i1}(t)[\mu \Delta t + o(\Delta t)]$$

$$P_{i1}(t + \Delta t) = P_{i0}(t)[\lambda \Delta t + o(\Delta t)] + P_{i1}(t)[1 - \mu \Delta t + o(\Delta t)]$$

$$\text{(7.5.1)}$$

which give

$$P'_{i0} = -\lambda P_{i0}(t) + \mu P_{i1}(t)$$

$$P'_{i1}(t) = -\mu P_{i1}(t) + \lambda P_{i0}(t) \qquad \text{(7.5.2)}$$

If we note that $P_{i0}(t) + P_{i1}(t) = 1$, we need to solve only one equation in (7.5.2). Also, using the same relation, we may convert equations in (7.5.2) as linear first-order differential equations. Thus we have

$$P'_{i0}(t) = \mu - (\lambda + \mu)P_{i0}(t) \qquad \text{(7.5.3)}$$

with the initial condition

$$P_{i0}(0) = \begin{cases} 1 & \text{if } i = 0 \\ 0 & \text{if } i = 1 \end{cases}$$

Equation (7.5.3) can be solved through standard techniques (similar to the method used in deriving the Poisson process in Section 7.2) to give

$$P_{00}(t) = \frac{\mu}{\lambda + \mu} + \frac{\lambda}{\lambda + \mu} e^{-(\lambda + \mu)t}$$

$$P_{10}(t) = \frac{\mu}{\lambda + \mu} - \frac{\mu}{\lambda + \mu} e^{-(\lambda + \mu)t} \qquad \text{(7.5.4)}$$

Consequently we also get

$$P_{01} = \frac{\lambda}{\lambda + \mu} - \frac{\lambda}{\lambda + \mu} e^{-(\lambda + \mu)t}$$

$$P_{11}(t) = \frac{\lambda}{\lambda + \mu} + \frac{\mu}{\lambda + \mu} e^{-(\lambda + \mu)t} \qquad \text{(7.5.5)}$$

As $t \to \infty$, (7.5.4) and (7.5.5) give

$$\lim_{t \to \infty} P_{00}(t) = \lim_{t \to \infty} P_{10}(t) = p_0 = \frac{\mu}{\lambda + \mu}$$

$$\lim_{t \to \infty} P_{01}(t) = \lim_{t \to \infty} P_{11}(t) = p_1 = \frac{\lambda}{\lambda + \mu} \tag{7.5.6}$$

Comparing (7.5.4) and (7.5.5) with (7.5.6), we may consider the second terms of (7.5.4) and (7.5.5) as transient parts of the corresponding transition probabilities. The rate at which the transient part decreases to zero is given by the term $e^{-(\lambda+\mu)t}$, and when $t = 1/\lambda + \mu$, this term attains the value of e^{-1}. The time when the transient part reduces to e^{-1} is called the *relaxation time* in the literature, and it provides a reference point in comparing the convergence rates of transition probabilities of different processes.

Because of the alternating nature of the stochastic process, it is of interest to determine the expected amount of time the process spends in each of the states. (In the computer problem of Example 7.1.1 total uptime and down-time would correspond to these characteristics.)

Let $\mu_{ij}(t)$ be the expected length of time the Markov process $X(t)$ spends in state j during $(0, t]$, after having an initial value $X(0) = i$ $(i, j = 0, 1)$. In order to determine $\mu_{ij}(t)$, first define an indicator process

$$I_{ij}(t) = \begin{cases} 1 & \text{if } \{ X(t) = j | X(0) = i \} \\ 0 & \text{otherwise} \end{cases}$$

Clearly we have

$$P\big(I_{ij}(t) = 1\big) = P_{ij}(t)$$

and

$$E\big[I_{ij}(t)\big] = P_{ij}(t) \tag{7.5.7}$$

Thus

$$\mu_{ij}(t) = \int_0^t E\big[I_{ij}(\tau)\big] \, d\tau$$

$$= \int_0^t P_{ij}(\tau) \, d\tau \tag{7.5.8}$$

Applying (7.5.8) to (7.5.4) and (7.5.5), we get

$$\mu_{00}(t) = \frac{\mu t}{\lambda + \mu} + \frac{\lambda}{\lambda + \mu}[1 - e^{-(\lambda + \mu)t}]$$

$$\mu_{10}(t) = \frac{\mu t}{\lambda + \mu} - \frac{\mu}{\lambda + \mu}[1 - e^{-(\lambda + \mu)t}] \qquad (7.5.9)$$

and

$$\mu_{01}(t) = \frac{\lambda t}{\lambda + \mu} - \frac{\lambda}{\lambda + \mu}[1 - e^{-(\lambda + \mu)t}]$$

$$\mu_{11}(t) = \frac{\lambda t}{\lambda + \mu} + \frac{\mu}{\lambda + \mu}[1 - e^{-(\lambda + \mu)t}] \qquad (7.5.10)$$

As $t \to \infty$, we get

$$\lim_{t \to \infty} \frac{1}{t}\mu_{00}(t) = \lim_{t \to \infty} \frac{1}{t}\mu_{10}(t) = \frac{\mu}{\lambda + \mu} = p_0$$

$$\lim_{t \to \infty} \frac{1}{t}\mu_{01}(t) = \lim_{t \to \infty} \frac{1}{t}\mu_{11}(t) = \frac{\lambda}{\lambda + \mu} = p_1 \qquad (7.5.11)$$

These results justify the rather intuitive notion that the limiting probabilities also give the fraction of time the process spends in various states in the long run (see Theorem 6.2.5 for the discrete version of this property).

Another characteristic of interest in this process is the distribution of individual periods that alternate in the process. These are governed by the three postulates identified earlier, and clearly the periods during which the process occupies the states 0 and 1 are independent of each other. Let X_0 be one such period, where $X(t) = 0$, and define

$$F_0(x) = P(X_0 > x)$$

Based on Postulate 1, we have

$$F_0(x + dx) = F_0(x)[1 - \lambda \Delta x + o(\Delta x)] \qquad (7.5.12)$$

Rearranging terms and taking limits, we get

$$F_0'(x) = -\lambda F_0(x) \qquad (7.5.13)$$

The differential equation (7.5.13) has the solution (after using the initial condition $F_0(0) = 1$)

$$F_0(x) = e^{-\lambda x} \tag{7.5.14}$$

which is an exponential distribution with mean $1/\lambda$. In a similar manner we can show that the individual time period during which $X(t) = 1$ has an exponential distribution with mean $1/\mu$. This is a significant property common to all Markov processes, and it will be established more rigorously in Section 7.6.

Finally, instead of a single two-state Markov process $\{ X(t), t \geq 0 \}$ consider c^* identical and independent Markov processes in operation at the same time (e.g., consider c^* machines going through uptimes and downtimes). Further, at time t $(t \geq 0)$, let $c_0(t)$ and $c_1(t)$ be the number of Markov processes occupying states 0 and 1, respectively. Clearly $c_0(t) + c_1(t) = c^*$. For the mean and variance of $c_0(t)$ and $c_1(t)$ we have

$$E[c_0(t)] = \frac{\mu}{\lambda + \mu}c^* + \frac{\lambda c_0(0) - \mu c_1(0)}{\lambda + \mu}e^{-(\lambda+\mu)t} \tag{7.5.15}$$

$$E[c_1(t)] = \frac{\lambda}{\lambda + \mu}c^* - \frac{c_0(0) - \mu c_1(0)}{\lambda + \mu}e^{-(\lambda+\mu)t} \tag{7.5.16}$$

$$V[c_0(t)] = V[c_1(t)]$$

$$= \frac{\lambda \mu c^*}{(\lambda + \mu)^2}[1 - e^{-(\lambda+\mu)t}]$$

$$+ \frac{[\lambda^2 c_0(0) + \mu^2 c_1(0)][1 - e^{-(\lambda+\mu)t}]e^{-(\lambda+\mu)t}}{(\lambda + \mu)^2} \tag{7.5.17}$$

Let $\lim_{t \to \infty}c_0(t) = c_0^*$ and $\lim_{t \to \infty}c_1(t) = c_1^*$. Then we get

$$E[c_0^*] = \frac{\mu}{\lambda + \mu}c^* \tag{7.5.18}$$

$$E[c_1^*] = \frac{\lambda}{\lambda + \mu}c^* \tag{7.5.19}$$

$$V[c_0^*] = V[c_1^*] = \frac{\lambda \mu}{(\lambda + \mu)^2}c^* \tag{7.5.20}$$

These results can be derived the same way as in the case of two-state Markov chains of Section 4.3.

Example 7.5.1

Data collected on a computer indicate that the periods during which it is working can be modeled as independent and identically distributed negative exponential random variables with mean 30 min. Also the repair times can be modeled as independent and identically distributed negative exponential random variables with mean 10 min. A student brings a job to the center at time T and finds the computer under repair. If the student does not want to wait but wants to come back in 30 min, what is the probability that the student will find the computer under repair again upon return? During that 30 min period how long can we expect the computer to be under repair? Suppose, the student wants to come back after 2 hr. What are the student's chances then?

Recalling our remarks on equation (7.5.14), we conclude that the state of the computer can be modeled as a Markov process $X(t)$ with states 0 (working) and 1 (under repair) and parameters $\lambda = 1/30$ and $\mu = 1/10$ (the unit of time is minute) and we make the following observations:

1. The probability that the computer will be under repair (state 1) at time $T + 30$, given that it is under repair at time T, is given by the second equation in (7.5.5), with $t = 30$ and λ and μ as $1/30$ and $1/10$, respectively:

$$P_{11}(30) = \frac{1/30}{1/30 + 1/10} + \frac{1/10}{1/30 + 1/10} e^{-(1/30 + 1/10)30}$$

$$= \tfrac{1}{4} + \tfrac{3}{4}e^{-4}$$

$$= 0.2637 \qquad\qquad Answer$$

2. The expected length of time the computer would be under repair between T and $T + 30$ is given by the second equation in (7.5.10), with $t = 30$ and λ and μ as given before:

$$\mu_{11}(30) = 30\left(\tfrac{1}{4}\right) + \tfrac{90}{16}\left(1 - e^{-4}\right)$$

$$= 13.0220 \text{ min} \qquad\qquad Answer$$

3. For $t = 120$ min one finds that $e^{-(1/30 + 1/10)120} = 0.0000001125$. Thus the contribution of the second term in any of the equations of

(7.5.4) and (7.5.5) is negligible. Therefore after 2 hr.

$$P(\text{Computer working}) = 0.75$$

$$P(\text{Computer under repair}) = 0.25 \qquad Answer$$

7.6 BIRTH AND DEATH PROCESS

Let a "birth" be an event signifying an increase in population and a "death" be an event signifying a decrease in population. Suppose these two types of events occur under the following postulates:

1. *Birth:* If the population size is n (≥ 0) at time t, during the following infinitesimal interval $(t, t + \Delta t]$ the probability that a birth will occur is $\lambda_n \Delta t + o(\Delta t)$, the probability that no birth will occur is $1 - \lambda_n \Delta t + o(\Delta t)$, and the probability that more than one birth will occur is $o(\Delta t)$, where $o(\Delta t)$ is such that $\lim_{\Delta t \to 0} o(\Delta t)/\Delta t \to 0$. Births occurring in $(t, t + \Delta t]$ are independent of the time since the last occurrence.
2. *Death:* If the population size is n (> 0) at time t, during the following infinitesimal interval $(t, t + \Delta t]$ the probability that a death will occur is $\mu_n \Delta t + o(\Delta t)$, the probability that no death will occur is $1 - \mu_n \Delta t + o(\Delta t)$, and the probability that more than one death will occur is $o(\Delta t)$. Deaths occurring in $(t, t + \Delta t]$ are independent of the time since the last occurrence.
3. When the population size is 0 at time t, the probability is 0 that a death will occur during $(t, t + \Delta t]$.
4. For the same population size, births and deaths occur independent of each other.

Let $X(t)$ be the population size at time t. Define

$$P_{in}(s, t) = P[X(t) = n | X(s) = i] \qquad t > s \qquad (7.6.1)$$

This process is time-homogeneous, and therefore we shall use the definition given in (7.6.1) only when it helps understanding; otherwise, we shall write

$$P_n(t) = P[X(t) = n | X(0) = i] \qquad (7.6.2)$$

As a consequence of Postulates 1–4 we may write

$$P_{n,n-1}(t, t + \Delta t) = [\mu_n \Delta t + o(\Delta t)][1 - \lambda_n \Delta t + o(\Delta t)]$$

$$= \mu_n \Delta t + o(\Delta t)$$

$$P_{n,n}(t, t + \Delta t) = [1 - \lambda_n \Delta t + o(\Delta t)][1 - \mu_n \Delta t + o(\Delta t)]$$

$$= 1 - \lambda_n \Delta t - \mu_n \Delta t + o(\Delta t)$$

$$P_{n,n+1}(t, t + \Delta t) = [\lambda_n \Delta t + o(\Delta t)][1 - \mu_n \Delta t + o(\Delta t)]$$

$$= \lambda_n \Delta t + o(\Delta t)$$

$$\sum_{\substack{j \neq n-1, n, n+1}}^{\infty} P_{nj}(t, t + \Delta t) = o(\Delta t) \tag{7.6.3}$$

In the postulates and resulting transition probabilities given by (7.6.3) we have used λ_n and μ_n as the birth and death parameters when the population size is n, without mentioning the exact nature of their dependence to the population size. When this is so, we shall call the process a generalized birth and death process. The simplest example of the birth and death processes is obtained by setting $\lambda_n = \lambda$ $(n \geq 0)$ and $\mu_n = \mu$ $(n > 0)$. This process may therefore be called the simple birth and death process. Under the generalized assumptions only the framework of analysis can be set up, whereas by specializing the parameters further, more information can be obtained.

A general method for deriving the distribution of $X(t)$ makes use of the Chapman-Kolmogorov equation (2.3.2) for the process. For transitions occurring in nonoverlapping intervals $(0, t]$ and $(t, t + \Delta t]$, based on probabilities given in (7.6.3), the Chapman-Kolmogorov equations take the form

$$P_n(t + \Delta t) = [1 - \lambda_n \Delta t - \mu_n \Delta t + o(\Delta t)]P_n(t)$$

$$+ [\lambda_{n-1} \Delta t + o(\Delta t)]P_{n-1}(t)$$

$$+ [\mu_{n+1} \Delta t + o(\Delta t)]P_{n+1}(t) + o(\Delta t) \tag{7.6.4}$$

which, on rearranging and letting $\Delta t \to 0$, gives the difference differential equation

$$P_n'(t) = -(\lambda_n + \mu_n)P_n(t) + \lambda_{n-1}P_{n-1}(t) + \mu_{n+1}P_{n+1}(t) \tag{7.6.5}$$

with the initial conditions

$$P_n(0) = \begin{cases} 1 & \text{if } n = i \\ 0 & \text{if } n \neq i \end{cases} \tag{7.6.6}$$

The solution to this difference differential equation takes a complicated form. However, when $\lambda_n = n\lambda$ and $\mu_n = n\mu$ (when both the increase and decrease in population is linearly dependent on the population size) for $i = 1$, we get

$$P_0(t) = \xi_t = \mu \frac{1 - e^{-(\lambda-\mu)t}}{\lambda - \mu e^{-(\lambda-\mu)t}} \tag{7.6.7}$$

$$P_n(t) = (1 - \xi_t)(1 - \eta_t)\eta_t^{n-1} \tag{7.6.8}$$

where we have written $\eta_t = (\lambda/\mu)\xi_t = \lambda[1 - e^{-(\lambda-\mu)t}]/[\lambda - \mu e^{-(\lambda-\mu)t}]$. Clearly $P_n(t)$ has a geometric distribution with a modified initial term.

When the increase in population is linearly dependent on the size of the population, it is pertinent to investigate whether the population would ever become extinct. The behavior of the probability $P_0(t)$ gives a clue to this possibility, for

$$\lim_{t \to \infty} P_0(t) = \text{Probability of ultimate extinction of the population}$$

Let $\rho = \lambda/\mu$; from (7.6.7) we have

$$\lim_{t \to \infty} \xi_t = \lim_{t \to \infty} \frac{\mu[1 - e^{-(\lambda-\mu)t}]}{\lambda - \mu e^{-(\lambda-\mu)t}} \qquad \rho > 1$$

$$= \frac{\mu}{\lambda} \tag{7.6.9}$$

and

$$\lim_{t \to \infty} \xi_t = \lim_{t \to \infty} \frac{e^{(\mu-\lambda)t}[\mu e^{-(\mu-\lambda)t} - \mu]}{e^{(\mu-\lambda)t}[\lambda e^{-(\mu-\lambda)t} - \mu]} \qquad \rho \leq 1$$

$$= 1 \tag{7.6.10}$$

Thus we have

$$\lim_{t \to \infty} P_0(t) = \begin{cases} 1 & \text{if } \rho \leq 1 \\ \rho^{-1} & \text{if } \rho > 1 \end{cases} \tag{7.6.11}$$

which is in accordance with the intuitive result that when the death rate is larger than the birth rate, ultimate extinction is certain.

7.7   SOME GENERAL PROPERTIES OF MARKOV PROCESSES WITH DISCRETE STATES

In the last few sections we studied some simple Markov processes without mentioning the general properties underlying such processes. Even though much of the general treatment is beyond the scope of this book, we would like to point out a few of these properties so as to open up possible avenues of further study in this area. We shall start with the countably infinite-state Markov processes and then specialize the results to finite-state Markov processes.

Let S be the state space, and let $P(t) = \|P_{ij}(t)\|$ be the transition probability matrix of the Markov process $X(t)$, where

$$P_{ij}(t) = P[X(t) = j | X(0) = i] \qquad i, j \in S \qquad (7.7.1)$$

The probabilities $P_{ij}(t)$ can be seen to have the following properties:

1. $P_{ij}(t) \geq 0$ $t > 0$ (7.7.2)

2. $\displaystyle\sum_{j \in S} P_{ij}(t) = 1$ $t > 0$ (7.7.3)

3. $\displaystyle P_{ij}(t + s) = \sum_{k \in S} P_{ik}(t) P_{kj}(s)$ $t, s > 0$ (7.7.4)

4. $P_{ij}(t)$ continuous $t > 0$ (7.7.5)

5. $\displaystyle\lim_{t \to o} P_{ij}(t) = \begin{cases} 1 & \text{if } i = j \\ 0 & \text{if } i \neq j \end{cases}$ (7.7.6)

These properties also imply that $P_{ij}(t)$ have right-hand derivatives for every $t \geq 0$. In particular, it can be shown that

$$\lim_{t \to \infty} \frac{1 - P_{ii}(t)}{t} = -P_{ii}'(0) \qquad \text{for all } i \qquad (7.7.7)$$

exists but may be infinite, and

$$\lim_{t \to 0} \frac{P_{ij}(t)}{t} = P_{ij}'(0) \qquad \text{for all } i \text{ and } j, i \neq j \qquad (7.7.8)$$

exists and is finite. For convenience we shall denote

$$-P_{ii}'(0) = \lambda_{ii} \quad \text{and} \quad P_{ij}'(0) = \lambda_{ij} \qquad (7.7.9)$$

It should be noted that for the transition rates we have

$$\sum_{j \neq i} P'_{ij}(0) = -P'_{ii}(0)$$

that is,

$$\sum_{j \neq i} \lambda_{ij} = \lambda_{ii} \qquad (7.7.10)$$

Using transition rates λ_{ii} and λ_{ij}, we may write down the transition probabilities for the period $(t, t + \Delta t]$ as

$$P_{ij}(t, t + \Delta t) = \lambda_{ij} \Delta t + o(\Delta t) \qquad j \neq i$$

$$P_{ii}(t, t + \Delta t) = 1 - \lambda_{ii} \Delta t + o(\Delta t) \qquad (7.7.11)$$

As an example consider the Poisson process. We have assumed that

$$P_{i,i+1}(t, t + \Delta t) = \lambda \Delta t + o(\Delta t)$$

$$P_{ii}(t, t + \Delta t) = 1 - \lambda \Delta t + o(\Delta t) \qquad (7.7.12)$$

Writing $P_{ij}(t, t + \Delta t) = P_{ij}(\Delta t)$, these relations give

$$\lim_{\Delta t \to 0} \frac{P_{i,i+1}(\Delta t)}{\Delta t} = \lambda$$

and

$$\lim_{\Delta t \to 0} \frac{1 - P_{ii}(\Delta t)}{\Delta t} = \lambda \qquad (7.7.13)$$

thus showing that the Poisson process is the case when the λ's of (7.7.9) are a constant. In a similar manner other processes discussed earlier can be characterized.

The assumptions (7.7.12) made for Poisson process are consistent with the resulting exponential distribution for interevent times if we write

$$P_{i,i+1}(t, t + \Delta t) = \int_t^{t + \Delta t} \lambda e^{-\lambda x} \, dx$$

$$= 1 - e^{-\lambda \Delta t}$$

$$= \lambda \Delta t + o(\Delta t) \qquad (7.7.14)$$

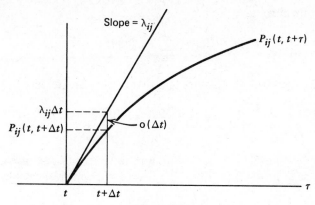

Figure 7.7.1

The need for the inclusion of $o(\Delta t)$ terms in equation (7.7.11) will be clear if we note that $P_{ij}(t)$ is a continuous function [see (7.7.5)], and the equations (7.7.11) represent linear approximations for $P_{ij}(t, t + \Delta t)$ as $\Delta t \to 0$. This explanation is illustrated in Figures 7.7.1 and 7.7.2.

The property (7.7.4) of the Markov processes is the Chapman-Kolmogorov equation. Using this identity, it is possible to derive equations that help analyze the process. We have

$$P_{ij}(t + s) = \sum_{k \in S} P_{ik}(t) P_{kj}(s) \qquad t, s > 0 \qquad (7.7.15)$$

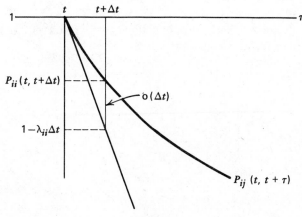

Figure 7.7.2

Set $s = \Delta t$; then we have

$$P_{ij}(t + \Delta t) = \sum_{k \in S} P_{ik}(t) P_{kj}(\Delta t) \tag{7.7.16}$$

Subtracting $P_{ij}(t)$ from both sides and dividing by Δt, we get

$$\frac{P_{ij}(t + \Delta t) - P_{ij}(t)}{\Delta t} = \frac{\sum_{k \neq j} P_{ik}(t) P_{kj}(\Delta t)}{\Delta t} + P_{ij}(t) \frac{P_{jj}(\Delta t) - 1}{\Delta t}$$

Now let $\Delta t \to 0$ in (7.7.17); substituting from (7.7.7)–(7.7.9) and assuming that the passage to the limit in (7.7.7) is uniform with respect to i, we get

$$P'_{ij}(t) = -\lambda_{jj} P_{ij}(t) + \sum_{k \neq j} \lambda_{kj} P_{ik}(t) \tag{7.7.17}$$

These are difference differential equations for a fixed i and are known as *forward Kolmogorov equations*.

Similarly, set $t = \Delta s$ in (7.7.15), and write

$$P_{ij}(\Delta s + s) = \sum_{k \in S} P_{ik}(\Delta s) P_{kj}(s) \tag{7.7.18}$$

Subtracting $P_{ij}(s)$, dividing both sides by Δs, and letting $\Delta s \to 0$, one now gets (using t instead of s)

$$P'_{ij}(t) = -\lambda_{ii} P_{ij}(t) + \sum_{k \neq i} \lambda_{ik} P_{kj}(t) \tag{7.7.19}$$

These are difference differential equations for a fixed j and are known as *backward Kolmogorov equations*. Note that in deriving (7.7.19), it is not necessary to assume that passage to the limit in (7.7.8) is uniform with respect to i. Therefore backward equations are considered to be more fundamental than forward equations.

Now define a matrix

$$A = \begin{bmatrix} -\lambda_{00} & \lambda_{01} & \lambda_{02} & \cdots \\ \lambda_{10} & -\lambda_{11} & \lambda_{12} & \cdots \\ \lambda_{20} & \lambda_{21} & -\lambda_{22} & \cdots \\ \vdots & \vdots & \vdots & \end{bmatrix} \tag{7.7.20}$$

and let

$$P'(t) = \|P'_{ij}(t)\| \tag{7.7.21}$$

For the sake of convenience we shall call matrix A a *transition rate matrix*. (This matrix is also called the *generator*). In matrix notation, the forward and backward Kolmogorov equations can be represented as

$$P'(t) = P(t)A \tag{7.7.22}$$

and

$$P'(t) = AP(t) \tag{7.7.23}$$

respectively. The initial conditions for both sets of equations are

$$P(0) = I \tag{7.7.24}$$

Formally, the solution to the set of equations (7.7.22) and (7.7.23) can be given as

$$P(t) = e^{At} = I + \sum_{n=1}^{\infty} A^n \frac{t^n}{n!} \tag{7.7.25}$$

where the last term is regarded as the definition of e^{At}. When A is a finite matrix, the series in (7.7.25) is convergent, and it is the unique solution for both forward and backward equations. When A is of infinite dimensions however, we cannot say much at this stage, as the series may not converge at all. When A is a finite matrix, and is diagonalizable, we can put A in the form

$$A = QDQ^{-1} \tag{7.7.26}$$

where D is a diagonal matrix with the distinct eigenvalues of A as elements. Let

$$D = \begin{bmatrix} d_0 & & & \\ & d_1 & & \bigcirc \\ & & \ddots & \\ \bigcirc & & & d_N \end{bmatrix} \tag{7.7.27}$$

Thus we have

$$P(t)e^{QDQ^{-1}t} = I + \sum_{n=1}^{\infty} \frac{Q(Dt)^n Q^{-1}}{n!}$$

$$= Q\left[I + \sum_{n=1}^{\infty} \frac{(Dt)^n}{n!} \right] Q^{-1}$$

$$= Qe^{Dt}Q^{-1} \tag{7.7.28}$$

where

$$e^{Dt} = \begin{bmatrix} e^{d_0 t} & & & & \\ & e^{d_1 t} & & \bigcirc & \\ & & \ddots & & \\ \bigcirc & & & e^{d_N t} \end{bmatrix} \qquad (7.7.29)$$

When A is finite, the solution exhibited in equation (7.7.25) can be extended to provide mean and variance of residence times of the Markov process in the various states. Let $\mu_{ij}(t)$ be the expected length of time the Markov process spends in state j during $(0, t]$, after having an initial value $X(0) = i(i, j = 1, 2, \ldots, m)$. Then following the arguments as in (7.5.7), we have

$$E\left[I_{ij}(t) \right] = P_{ij}(t)$$

and

$$\mu_{ij}(t) = \int_0^t E\left[I_{ij}(\tau) \right] d\tau$$

$$= \int_0^t P_{ij}(\tau) d\tau \qquad (7.7.30)$$

In matrix form, writing $\mu_{ij} = \lim_{t \to \infty} \mu_{ij}(t)$, we get

$$M = \| \mu_{ij} \| = \int_0^\infty P(t) \, dt$$

$$= \int_0^\infty e^{At} \, dt$$

$$= -A^{-1} \qquad (7.7.31)$$

(see Çinlar, 1975, Tavaré, 1979).

Let M_D be the diagonal matrix with elements $\mu_{11}, \mu_{22}, \ldots, \mu_{mm}$ and M_2, the matrix with elements μ_{ij}^2. Then, by extending the indicator variable argument, it can also be shown that the variance σ_{ij}^2 of the corresponding residence times is given by

$$\| \sigma_{ij}^2 \| = 2M M_D - M_2 \qquad (7.7.32)$$

(Matis et al., 1983).

The general Markov process as described here has the property that the amount of time the process spends in state i during a visit, say $Z(i)$, has an exponential distribution with parameter λ_{ii} (mean $1/\lambda_{ii}$). We can establish this property in the following manner (see Karlin and Taylor, 1981, p. 146).

Consider a fixed time t divided into an arbitrary number n parts. Let $X(0) = i$. Now

$$P\left[X(s) = i \text{ for } s = 0, \frac{t}{n}, \frac{2t}{n}, \dots, t \right] = \left[P_{ii}\left(\frac{t}{n}\right)^n \right] \qquad (7.7.33)$$

As $n \to \infty$, the left-hand side can be written as

$$P\left[X(s) = i, 0 \le s \le t \right] = P(Z(i) > t) \qquad (7.7.34)$$

But from (7.7.9) we have

$$\frac{1 - P_{ii}(t)}{t} = \lambda_{ii} + \frac{o(t)}{t}$$

$$P_{ii}(t) = 1 - \lambda_{ii}t + o(t)$$

and hence

$$P_{ii}\left(\frac{t}{n}\right) = 1 - \lambda_{ii}\left(\frac{t}{n}\right) + o\left(\frac{t}{n}\right)$$

Therefore

$$\left[P_{ii}\left(\frac{t}{n}\right)^n \right] = \left[1 - \lambda_{ii}\left(\frac{t}{n}\right) + o\left(\frac{t}{n}\right) \right]^n$$

$$= e^{n \ln[1 - \lambda_{ii}(t/n) + o(t/n)]} \qquad (7.7.35)$$

Using the expansion of the logarithmic series, and letting $n \to \infty$, we get

$$\lim_{n \to \infty} \left[P_{ii}\left(\frac{t}{n}\right) \right]^n = e^{-\lambda_{ii}t} \qquad (7.7.36)$$

Combining this with (7.7.34), we get

$$P(Z(i) \le t) = 1 - e^{-\lambda_{ii}t} \qquad (7.7.37)$$

7.8 LIMITING DISTRIBUTIONS

Similar to the theorems given for the limiting probabilities of Markov chains, we can also give theorems for the limiting distributions of Markov processes. We give some of the important theorems here without proofs.

Theorem 7.8.1 (1) *If the Markov process is irreducible (all states communicate), then the limiting distribution* $\lim_{t \to \infty} P_n(t) = p_n$ *exists and is independent of the initial conditions of the process. The limits* $\{ p_n, n \in S \}$ *are such that they either vanish identically (i.e.,* $p_n = 0$ *for all* $n \in S$ *) or are all positive and form a probability distribution (i.e.,* $p_n > 0$ *for all* $n \in S$, $\sum_{n \in S} p_n = 1$). (2) *The limiting distribution* $\{ p_n, n \in S \}$ *of an irreducible positive recurrent Markov process is given by the unique solution of the equation* $pA = 0$ *and* $\sum_{j \in S} p_j = 1$, *where* $p = (p_0, p_1, p_2, \ldots)$.

As $t \to \infty$, when the limit exists, it is independent of the time parameter t, and hence

$$P_n'(t) \to 0 \tag{7.8.1}$$

From the forward Kolmogorov equations (7.7.17), Condition 1 of Theorem 7.8.1, and equation (7.8.1), we therefore obtain

$$0 = -\lambda_{jj} p_j + \sum_{k \neq j} \lambda_{kj} p_k \qquad j = 0, 1, 2, \ldots \tag{7.8.2}$$

which in matrix notations can be written as

$$pA = 0 \tag{7.8.3}$$

where $p = (p_0, p_1, p_2, \ldots)$, as stated in Condition 2 of Theorem 7.8.1.

If we are interested only in the limiting distribution of the Markov process, there is no need to first write the forward equations (7.7.17) or the backward equations (7.7.19). Only the coefficients λ_{ij} $(i, j \in S)$ are needed. Recalling the definitions of these coefficients, we have

$$\lim_{\Delta t \to 0} \frac{1 - P_{jj}(\Delta t)}{\Delta t} = \lambda_{jj} \tag{7.8.4}$$

$$\lim_{\Delta t \to 0} \frac{P_{ij}(\Delta t)}{\Delta t} = \lambda_{ij} \qquad j \neq i \tag{7.8.5}$$

Among these instantaneous transition rates, λ_{jj} refers to all possible transitions out of state j, whereas λ_{ij} refers to all possible transitions from state i to state j. Rewriting (7.8.2) as

$$\lambda_{jj} p_i = \sum_{k \neq j} \lambda_{kj} p_k \qquad j = 0, 1, 2, \ldots \qquad (7.8.6)$$

we may consider them as equating all possible transitions out of state j, if the system is at j and all possible transitions into state j from wherever the system may be just the previous moment. This equation is also intuitively appealing since if there is any imbalance in these transitions, the probability may be zero that the system would be found in some of the states, which is possibly contrary to assumptions.

For the homogeneous system of equations (7.8.3), the zero vector $(0, 0, \ldots)$ is a solution, which incidentally implies that the probability is zero that the process can be found in one of the finite states as $t \to \infty$. To determine a nonzero unique solution, we therefore impose the additional condition

$$\sum_{j=0}^{\infty} p_j = 1 \qquad (7.8.7)$$

Irreducible Markov processes which yield positive limiting probabilities $\{p_j\}$ with $\sum_0^{\infty} p_j = 1$ are called *positive recurrent*. In this case the limiting distribution obtained is unique.

If the irreducible Markov process has a finite number of states, it is necessarily positive recurrent. Two other properties of limiting distributions are given in Theorems 7.8.2 and 7.8.3.

Theorem 7.8.2   *The limiting distribution of a positive recurrent irreducible Markov process is also stationary. This means that if initially we have*

$$P_n(0) = p_n \qquad n = 0, 1, 2, \ldots$$

then

$$P_n(t) = p_n \qquad \text{for all } t$$

Theorem 7.8.3   *Let $N_{ij}(t)$ be the time spent by the Markov process in state j during an interval of length t, having originally started from state i. Then*

$$\lim_{t \to \infty} P\left[\left| \frac{N_{ij}(t)}{t} - p_j \right| > \varepsilon \right] = 0 \qquad (7.8.8)$$

Comparisons may be made with the properties of the irreducible Markov chains discussed in Chapters 4 and 6.

We can now use some of these theorems to derive the limiting distributions of the Markov processes discussed earlier. A simple two-state process has been studied in detail in Section 7.5. The Poisson, pure birth, and pure death processes are not irreducible, for in the first two processes the changes of state are always of the type $j \to j + 1$ ($j = 0, 1, 2, \ldots$) and in the pure death process they are of the type $j \to j - 1$ ($j \le i$, the initial population size). The limiting distribution of the generalized birth and death process can be obtained as follows: Writing $\lim_{t \to \infty} P_n(t) = p_n$, and setting $\lim_{t \to \infty} P_n'(t) = 0$, from (7.6.5), we have

$$0 = -(\lambda_n + \mu_n)p_n + \lambda_{n-1}p_{n-1} + \mu_{n+1}p_{n+1} \qquad n \ge 1 \qquad (7.8.9)$$

$$0 = -\lambda_0 p_0 + \mu_1 p_1 \qquad (7.8.10)$$

Rearranging (7.8.10), we get

$$p_1 = \frac{\lambda_0}{\mu_1} p_0 \qquad (7.8.11)$$

For $n = 1$, (7.8.9) gives

$$\mu_2 p_2 = (\lambda_1 + \mu_1)p_1 - \lambda_0 p_0$$

$$= \lambda_1 p_1 \qquad (7.8.12)$$

using (7.8.10). This gives

$$p_2 = \frac{\lambda_0 \lambda_1}{\mu_1 \mu_2} p_0$$

Proceeding in this manner, by inductive arguments we can easily show that

$$\mu_n p_n = \lambda_{n-1} p_{n-1} \qquad n = 1, 2, \ldots \qquad (7.8.13)$$

$$p_n = \frac{\lambda_0 \lambda_1 \cdots \lambda_{n-1}}{\mu_1 \mu_2 \cdots \mu_n} p_0 \qquad (7.8.14)$$

To obtain p_0, we now use the additional equation

$$\sum_{j=0}^{\infty} p_j = 1$$

Then we get

$$p_0 = \left[1 + \sum_{n=1}^{\infty} \frac{\lambda_0 \lambda_1 \cdots \lambda_{n-1}}{\mu_1 \mu_2 \cdots \mu_n} \right]^{-1} \tag{7.8.15}$$

Thus the limiting distribution $\{ p_n \}$ is now completely determined. It should be noted, however, that $\{ p_n \}$ is a nonzero solution only if

$$1 + \sum_{n=1}^{\infty} \frac{\lambda_0 \lambda_1 \cdots \lambda_{n-1}}{\mu_1 \mu_2 \cdots \mu_n} < \infty \tag{7.8.16}$$

In particular, when $\lambda_n = \lambda$ $(n \geq 0)$ and $\mu_n = \mu$ $(n > 0)$, we have the limiting distribution of a simple birth and death process. Specializing (7.8.14) and (7.8.15), we get

$$p_0 = \left[\sum_{n=0}^{\infty} \left(\frac{\lambda}{\mu} \right)^n \right]^{-1} = 1 - \frac{\lambda}{\mu} \tag{7.8.17}$$

$$p_n = \left(1 - \frac{\lambda}{\mu} \right) \left(\frac{\lambda}{\mu} \right)^n \qquad n = 0, 1, 2, \ldots \tag{7.8.18}$$

provided $\lambda/\mu < 1$.

Remark 7.8.1 When $\lambda_n = n\lambda$ and $\mu_n = n\mu$ $(n \geq 1)$, we have a population in which the birth and death parameters are linearly dependent on the population size. If we think of the population becoming extinct whenever $X(t) = 0$, we have a Markov process in which state 0 is absorbing. Based on (7.6.7)–(7.6.11), the limiting distribution of the population size can be given as follows: When $\rho = \lambda/\mu < 1$, $p_0 = 1$, and hence

$$p_n = 0 \qquad n = 1, 2, \ldots \tag{7.8.19}$$

When $\rho = \lambda/\mu > 1$,

$$\lim_{t \to \infty} \xi_t = \rho^{-1}$$

and therefore

$$\lim_{t \to \infty} \eta_t = \lim_{t \to \infty} \frac{\lambda}{\mu} \xi_t = 1 \tag{7.8.20}$$

This gives

$$\lim_{t \to \infty} P_n(t) = \lim_{t \to \infty} (1 - \xi_t)(1 - \eta_t)\eta_t^{n-1}$$

$$= 0 \qquad\qquad (7.8.21)$$

Summarizing these results, we have

$$p_0 = \begin{cases} 1 & \text{if } \rho \le 1 \\ \rho^{-1} & \text{if } \rho > 1 \end{cases} \qquad\qquad (7.8.22)$$

and

$$p_n = 0 \qquad n > 0 \qquad\qquad (7.8.23)$$

This indicates that when $\rho > 1$, there is a positive probability that the population will grow in unlimited size.

7.9 MARKOV POPULATION PROCESSES

So far we have considered univariate processes in which the state of the process at any time is represented by a single variable. Situations in which state representation requires more than one variable are quite common. Population colonies that interact with each other, customers arriving at a service facility with different service needs, and a network of service facilities in which customers can be serviced simultaneously at more than one node are some examples where multivariate processes are needed. Spurred by the specialized results derived by J. Jackson (1957) for networks of queues and Whittle (1967, 1968) for migration processes, a systematic procedure for the analysis of such processes (with particular reference to the calculation of limiting distributions) has been given by Kingman (1969). We give here some key results derived in that paper which are becoming increasingly useful in the analysis of stochastic processes occurring in computer and communication networks.

Let $i = (i_1, i_2, \ldots, i_k)$ be a k-vector whose components i_r $(r = 1, 2, \ldots, k)$ are nonnegative integers. Let N^k be the set of k-vectors i and consider a continuous-time Markov process with state space S contained in N^k. For such a process, define transition rates $\lambda(i, j)$ with the property that $\lambda(i, i) = \sum_{j \neq i} \lambda(i, j)$ which are similar to the infinitesimal transition rates defined in Section 7.6. Write $\lambda(i, i) = \lambda(i)$. Clearly $\lambda(i)$ is the total rate of transitions out of i. A Markov process with these characteristics will be called a Markov population process if, for any i, the only values of j for which $\lambda(i, j)$ is nonzero are those with

$$j_r = i_r + 1 \qquad j_h = i_h \qquad h \neq r$$

$$j_r = i_r - 1 \qquad j_h = i_h \qquad h \neq r$$

$$j_r = i_r - 1 \qquad j_s = i_s + 1 \qquad j_h = i_h \qquad h \neq r, s \qquad (7.9.1)$$

In birth-death terminology a Markov population process is characterized by a birth, a death, and a transfer of one unit from one population colony to another. For notational simplification define a vector e that has all of its components zero except the one at the rth place, and write

$$\lambda(i, i + e_r) = \alpha_r(i)$$

$$\lambda(i, i - e_r) = \beta_r(i)$$

$$\lambda(i, i - e_r + e_s) = \gamma_{rs}(i) \qquad i \neq j \qquad (7.9.2)$$

with $\beta_r(i) = \gamma_{rs}(i) = 0$ if $i_r = 0$.

Two classes of Markov population processes of interest can be identified as open and closed systems. An open system is characterized by a state space given by

$$S = \left\{ i \in N^k, i_1 \leq N_1, i_2 \leq N_2, \ldots, i_k \leq N_k \right\}$$

where $N = (N_1, N_2, \ldots, N_k)$ and some or all the N_i may be infinite. In a closed system the total number of units in the population remains the same, and hence α_r and β_r of (7.9.2) are zero. Furthermore the state space in this case is irreducible.

Let C be an irreducible class of the state space, and let $p(j)$ be the limiting distribution of the Markov population process. (In this case it is also the stationary distribution.) The distribution $p(j)$ satisfies the equations

$$\lambda(j)p(j) = \sum_{\substack{i \neq j \\ i \in C}} \lambda(i, j)p(i) \qquad (7.9.3)$$

and

$$\sum_{j \in C} p(j) = 1 \qquad (7.9.4)$$

Note that these equations are the same as (7.8.6) and (7.8.7) in the univariate case. Also note that when C is finite, equations (7.9.3) and (7.9.4) have a unique positive solution. When C is infinite, they have at most one solution; also there is a possibility of no solution at all. Using (7.9.2), equation (7.9.3) may be rewritten as

$$\left\{ \sum_{r=1}^{k} \alpha_r(j) + \sum_{r=1}^{k} \beta_r(j) + \sum_{r,s=1}^{k} \gamma_{rs}(j) \right\} p(j)$$

$$= \sum_{r=1}^{k} \alpha_r(j - e_r) p(j - e_r) + \sum_{r=1}^{k} \beta_r(j + e_r) p(j + e_r)$$

$$+ \sum_{r,s=1}^{k} \gamma_{rs}(j + e_r - e_s) p(j + e_r - e_s) \tag{7.9.5}$$

where terms involving $p(i)$ with $i \notin C$ are to be ignored.

Without attempting a general solution for equation (7.9.5), Kingman establishes conditions under which the equation has simple product form solutions.

One class of processes for which equation (7.9.5) possesses product form solution includes reversible processes that satisfy the reversibility equation:

$$p(i)\lambda(i,j) = p(j)\lambda(j,i) \tag{7.9.6}$$

For equation (7.9.6) to hold it is necessary that

$$\{\lambda(i,j) > 0\} \leftrightarrow \{\lambda(j,i) > 0\} \tag{7.9.7}$$

where we have used the symbol $\leftrightarrow$ to denote mutual implication. Also using definitions (7.9.2) in (7.9.6), we get

$$\alpha_r(i)p(i) = \beta_r(i + e_r)p(i + e_r) \tag{7.9.8}$$

and

$$\gamma_{rs}(i + e_r)p(i + e_r) = \gamma_{sr}(i + e_s)p(i + e_s) \tag{7.9.9}$$

A necessary and sufficient condition for reversibility is the Kolmogorov cycle condition which, for any distinct states $i_1, i_2, \ldots, i_r$ in C, is given as

$$\lambda(i_1,i_2)\lambda(i_2,i_3) \cdots \lambda(i_r,i_1) = \lambda(i_1,i_r)\lambda(i_r,i_{r-1}) \cdots \lambda(i_2,i_1)$$

$$\tag{7.9.10}$$

It can be easily shown that (7.9.10) leads to the solution

$$p(i) = p(i_0) \frac{\lambda(i_0, i_1)\lambda(i_1, i_2) \cdots \lambda(i_r, i)}{\lambda(i_1, i_0)\lambda(i_2, i_1) \cdots \lambda(i, i_r)} \tag{7.9.11}$$

where $(i_0, i_1, \ldots, i_r, i)$ is any path from i_0 to i.

Under special structures for α's, β's, and γ's further simplifications result. In a closed system (for which α's and β's are zero) suppose γ_{rs} can be written as

$$\gamma_{rs}(i) = g_{rs}\phi_r(i_r)\psi_s(i_s) \tag{7.9.12}$$

where g_{rs} are positive and ϕ_r and ψ_s are positive functions except that $\phi_r(0) = 0$. Then one can show that

$$p(i) = K \prod_{r=1}^{k} f_r^{i_r} \prod_{s=0}^{i_r} \left[\frac{\psi_r(s-1)}{\phi_r(s)} \right] \tag{7.9.13}$$

where f_r are defined by the recursive relation $f_1 = 1, f_r = g_{1r}/g_{r1}$. The constant K is determined using the normalizing condition (7.9.4).

A product form solution establishes that the sizes of the population in different colonies are independent random variables. This limiting property is particularly significant if we note that in a closed system colony sizes initially are not independent.

The product form is also characteristic of some open systems. Suppose α_r and β_r of (7.9.2) are positive functions of i_r alone. Then from equation (7.9.8) we get

$$\frac{p(i + e_r)}{p(i)} = \frac{\alpha_r(i_r)}{\beta_r(i_r + 1)} \tag{7.9.14}$$

which leads to the result

$$\frac{p(i)}{p(0)} = \prod_{r=1}^{k} \left\{ \prod_{s_r=0}^{i_r-1} \frac{\alpha_r(s_r)}{\beta_r(s_r + 1)} \right\} \tag{7.9.15}$$

The reversibility condition (7.9.6) is very restrictive in many practical situations since it implies that transfers $i \to j$ and $j \to i$ should both be possible. Nevertheless, there are other situations where simple product form solutions of equations (7.9.5) are possible. Such processes are characterized by what may be called partial balance (or local balance). The processes considered by Jackson (1957), Gordon and Newell (1967), and Whittle (1967, 1968) are good examples of this phenomenon.

Suppose in (7.9.12)

$$\psi_s(i_s) \equiv 1$$

so that

$$\gamma_{rs}(i) = g_{rs}\phi_r(i_r) \tag{7.9.16}$$

Then if $p(i)$ should possess a product form, we need to have

$$p(i) = p_1(i_1)p_2(i_2) \cdots p_k(i_k) \tag{7.9.17}$$

Using this form in (7.9.5), for a closed system (for which α's and β's are zero), Kingman (1969, equation 39) derives

$$p_r(i) = p_r(0)\rho_r^i \prod_{s=0}^{i-1} \left(\frac{1}{\phi_r(s+1)} \right) \tag{7.9.18}$$

where $\rho_r \equiv \rho_r(i)$ defined in general as

$$\rho_r(i) = \frac{\phi_r(i+1)p_r(i+1)}{p_r(i)} \tag{7.9.19}$$

satisfies the relation

$$\rho_r g_{rs} = \rho_s g_{sr} \tag{7.9.20}$$

The condition derived by Whittle (1967) is obtained by summing (7.9.20) over all values of s. In either case equation (7.9.20) represents a local balance as opposed to the global or detailed balance expressed in (7.9.5). A similar equation for local balance for open systems that results in a product form solution has been derived by Whittle (1968) as

$$\left(b_r + \sum_s g_{rs} \right)\rho_r = a_r + \sum_s g_{rs}\rho_s \tag{7.9.21}$$

where $\alpha_r(i) = a_r$, $\beta_r(i) = b_r\phi_r(i_r)$, and $\gamma_{rs}(i) = g_{rs}\phi_r(i_r)$. The corresponding solution has the form

$$p(i) = C\prod_{r=1}^{k} \left\{ \frac{\rho_r^{i_r}}{\phi_r(1)\phi_r(2)\ldots\phi_r(i_r)} \right\} \tag{7.9.22}$$

Local balance illustrated in (7.9.22) is not a new phenomenon. For instance, equation (7.8.9) represents the detailed or the global balance among states in a generalized birth and death process. Equation (7.8.13), on the other hand, relates states n and $n - 1$ and thus provides a local or a partial balance in transition between those states. In the case of the closed and open systems of Markov population processes, it is interesting to note that, as established by Whittle (1967, 1968) and Kingman (1969), the existence of local balance between states is a necessary condition for product form solutions. We shall illustrate this property further in the discussion of queueing networks in Chapter 12.

7.10 STATISTICAL INFERENCE

For an applied scientist elegant mathematical models are of no use unless they come with instructions for employing them in real-life situations. Parameter estimation and hypothesis testing are two key elements necessary for the application of mathematical models. In this section we shall consider these problems as related to some simple Markov processes. For more advanced techniques the readers are referred to Basava and Prakasa Rao (1980) and journal articles referenced or recommended for further reading at the end of the chapter.

The Poisson process is the single most important stochastic process from the point of view of the practitioner. Because of the simplicity of the results and analysis techniques the first question asked about a set of events is whether they are Poisson. Therefore we shall start with inference problems of a Poisson process.

Observations on events occurring in a stochastic process can be made either over a fixed time interval or until a fixed number of events occur. In the former case the number of events is random, and in the latter case the length of the time interval is random.

Suppose a Poisson process is observed over a fixed time interval $(0, T]$. Let n events be observed during this period at time epochs $0 < t_1 < t_2 < \cdots < t_n < T$. The interoccurrence times $T_r = t_r - t_{r-1}$ $(r = 1, 2, \ldots, n)$ are independent and identically distributed random variables with an exponential distribution. Let the rate of occurrence of events be λ. The likelihood function of this realization can now be given as

$$f(\lambda) = \prod_{i=1}^{n} (\lambda e^{-\lambda T_i}) e^{-\lambda(T - t_n)}$$

$$= \lambda^n e^{-\lambda T} \tag{7.10.1}$$

Let $L(\lambda)$ be the natural logarithm of $f(\lambda)$:

$$L(\lambda) = n \ln \lambda - \lambda T$$

Equating the derivative of $L(\lambda)$ to zero and solving for λ, we get its maximum likelihood estimate $\hat{\lambda}$:

$$\frac{\partial L(\lambda)}{\partial \lambda} = \frac{n}{\lambda} - T$$

giving

$$\hat{\lambda} = \frac{n}{T} \tag{7.10.2}$$

It should be noted that n has a Poisson distribution with mean and variance λT. Thus we get

$$E(\hat{\lambda}) = \lambda; V(\hat{\lambda}) = \frac{\lambda}{T} \tag{7.10.3}$$

When a Poisson process is observed until a fixed number n of events occur, the random variable T is the sum of n exponential random variables, and therefore it has the gamma distribution (7.2.32). Nevertheless, the likelihood function remains the same as (7.10.1) and so also the maximum likelihood estimate given by (7.10.2). But, since T is a random variable, $E(\hat{\lambda})$ and $V(\hat{\lambda})$ are not the same any more.

From (7.2.32), for the distribution of T, we have

$$f_T(t) = e^{-\lambda t} \frac{\lambda^n t^{n-1}}{(n-1)!} \qquad 0 < t < \infty$$

and hence

$$E\left(\frac{1}{T}\right) = \int_0^\infty \frac{1}{t} e^{-\lambda t} \frac{\lambda^n t^{n-1}}{(n-1)!} \, dt$$

$$= \frac{\lambda}{n-1} \qquad n > 1 \tag{7.10.4}$$

giving

$$E(\hat{\lambda}) = E\left(\frac{n}{T}\right) = \left(\frac{n}{n-1}\right) \lambda \tag{7.10.5}$$

which shows that $\hat{\lambda}$ is a biased estimate. An unbiased estimate would be $\tilde{\lambda} = (n - 1)/n$ whose variance can be determined as follows:

$$V(\tilde{\lambda}) = \left(\frac{n - 1}{n}\right)^2 \cdot n^2 V\left(\frac{1}{T}\right)$$

$$= (n - 1)^2 V\left(\frac{1}{T}\right)$$

$$V\left(\frac{1}{T}\right) = E\left(\frac{1}{T^2}\right) - \left[E\left(\frac{1}{T}\right)\right]^2$$

$$= \frac{\lambda^2}{(n - 1)(n - 2)} - \frac{\lambda^2}{(n - 1)^2}$$

$$V(\tilde{\lambda}) = (n - 1)^2 \left[\frac{\lambda^2}{(n - 1)(n - 2)} - \frac{\lambda^2}{(n - 1)^2}\right]$$

$$= \frac{\lambda^2}{n - 2} \qquad n > 2 \tag{7.10.6}$$

To determine whether a process is Poisson, hypothesis testing is needed. An obvious approach would be to keep the interval of observation constant and use a goodness of fit test for the observed data against a Poisson distribution. There are several tests based on the distribution (which is exponential) of the time intervals between successive occurrences of events. This topic is covered extensively in the literature on reliability theory for which an appropriate reference is Gnedenko et al. (1969). We mention here another simple approach based on the relationship of the occurrence of Poisson events with the uniform distribution illustrated in equation (7.2.36).

Let $0 < t_1 < t_2 \cdots < t_n < T$ be the epochs of occurrence of events in a Poisson process during $(0, T]$ as mentioned earlier. Recalling the property exhibited in equation (7.2.36), let

$$S_n = \sum_{i=1}^{n} t_i$$

where S_n is the sum of n independent random variables uniformly distributed in the interval $[0, T]$. Noting that the mean and variance of a uniform random variable in $[0, T]$ are given by $T/2$ and $T^2/12$, respectively, we have

$$E[S_n] = \frac{nT}{2} \quad \text{and} \quad V[S_n] = \frac{nT^2}{12}$$

Therefore for large n, using the central limit theorem, we may conclude that

$$Z = \frac{S_n - nT/2}{\sqrt{nT^2/12}} \tag{7.10.7}$$

is a standardized normal variate.

It should be noted that the rejection of the hypothesis that the process is Poisson [i.e., when $|Z|$ of (7.10.7) is large for a given level of significance] only indicates that the process is not Poisson with a constant rate parameter. A nonhomogeneous Poisson process should be considered as a possible alternative as well. For further discussion of these inference problems, readers are referred to Cox and Lewis (1966) and Basawa and Prakasa Rao (1980).

A Markov process conveniently lends itself for the application of the maximum likelihood procedure for the estimation of its parameters. In general terms, suppose a Markov process is observed for a length of time T. Let T_i be the amount of time the process is observed in state i. With infinitesimal transition rates λ_{ii} and λ_{ij}, a constructive development of the process can be accomplished by using the property that the process resides in state i for an exponential length of time with mean $1/\lambda_{ii}$ [see result (7.7.34)] and changes its state to j with probability $\lambda_{ij}/\lambda_{ii}$. (This probability results by extending remark 7.2.4 to more than two exponential random variables and noting that $\lambda_{ii} = \sum_{j \neq i} \lambda_{ij}$.) Let n_{ij} be the number of state changes $i \rightarrow j$. Now the general form of the likelihood function $f(\lambda)$ can be given as

$$f(\lambda) = C \prod_{i,j} \left(\frac{\lambda_{ij}}{\lambda_{ii}} \right)^{n_{ij}} \left(\prod_i \lambda_{ii} e^{-\lambda_{ii} T_i} \right)$$

$$= C \left(\prod_{i,j} \lambda_{ij}^{n_{ij}} \right) \left(\prod_i e^{-\lambda_{ii} T_i} \right) \tag{7.10.8}$$

where C is a combinatorial term involving n_{ij}'s only. We illustrate this general approach in the case of the simple birth and death process.

Consider the simple birth and death process with parameters $\lambda_n = \lambda$ ($n \geq 0$) and $\mu_n = \mu$ ($n > 0$). We shall denote the estimates of λ and μ as $\hat{\lambda}$ and $\hat{\mu}$, respectively. Observe a simple birth and death process for a length of time T. During this interval let the population size be zero for a length of time T_e and nonzero for a length of time T_b, such that $T_e + T_b = T$. Further during this interval T let there be m_e changes of state $0 \rightarrow 1$, m_u changes of the type $n \rightarrow n + 1$ ($n > 0$), and m_d changes of the type $n \rightarrow n - 1$

($n > 0$). Our intention here is to derive estimates of λ and μ in terms of T_e, T_b, m_e, m_u, and m_d. A realization of the simple birth and death process has the following components:

1. A random sample of size m_e from an exponential distribution with mean λ^{-1}, such that the sum of the m_e observations is T_e—these refer to the m_e intervals preceding the transitions $0 \to 1$.

2. A random sample of size $m_u + m_d$ from an exponential distribution with mean $(\lambda + \mu)^{-1}$, such that the sum of the $m_u + m_d$ observations is T_b—these refer to the transition times when the population size is nonzero.

3. A number m_u of successes in a sequence of $m_u + m_d$ Bernoulli trials with success probability $\lambda/(\lambda + \mu)$—these refer to the m_u transitions of the type $n \to n + 1$ ($n > 0$) and m_d transitions of the type $n \to n - 1$ ($n > 0$).

4. The dependence between these observations is such that a $0 \to 1$ transition can occur only after a $1 \to 0$ transition.

5. A partial realization of a random variable whose distribution is exponential with mean λ^{-1} if the population size is zero at the end of the observation period and with mean $(\lambda + \mu)^{-1}$ otherwise.

To estimate λ and μ by the maximum likelihood method, the likelihood function giving the probability function of the kind of realization just described must be obtained. The four components of this realization make the following contributions to the corresponding likelihood function:

1. Suppose $(\tau_1, \tau_2, \ldots, \tau_{m_e})$ is a random sample of size m_e from an exponential distribution with mean λ^{-1}, such that $\tau_1 + \tau_2 + \cdots + \tau_{m_e} = T_e$. The contribution to the likelihood function from this sample is then given by

$$\prod_{i=1}^{m_e} \lambda e^{-\lambda \tau_i} = \lambda^{m_e} e^{-\lambda T_e} \qquad (7.10.9)$$

If the population size is zero at the end of the observation period, then T_e would be such that

$$\tau_1 + \tau_2 + \cdots + \tau_{m_e} + \tau = T_e$$

where τ is the time since the last transition ($1 \to 0$). Then the contribution to the likelihood function would be

$$\left(\prod_{i=1}^{m_e} \lambda e^{-\lambda \tau_i}\right) e^{-\lambda \tau} = \lambda^{m_e} e^{-\lambda T_e}$$

which is the same as given in (7.10.9).

2. By arguments similar to those given in Contribution 1, the contribution to the likelihood function from a random sample of size $m_u + m_d$ from an exponential distribution with mean $(\lambda + \mu)^{-1}$ is given by

$$(\lambda + \mu)^{m_u + m_d} e^{-(\lambda + \mu)T_b} \tag{7.10.10}$$

3. The contribution to the likelihood function arising from a sequence of $m_u + m_d$ Bernoulli trials with success probability $\lambda/(\lambda + \mu)$ resulting in m_u successes and m_d failures is given by

$$\binom{m_u + m_d}{m_u} \left(\frac{\lambda}{\lambda + \mu}\right)^{m_u} \left(\frac{\mu}{\lambda + \mu}\right)^{m_d} \tag{7.10.11}$$

4. The contribution to the likelihood function arising from the dependence between the nature of transitions is a function of m_e, m_u, and m_d, and is independent of the parameters λ and μ. We shall denote it by C_1.

Let the likelihood function be denoted by $f(\lambda, \mu)$, and let its natural logarithm be denoted by $L(\lambda, \mu)$. From (7.10.9)–(7.10.11) we have

$$f(\lambda, \mu) = \lambda^{m_e} e^{-\lambda T_e} (\lambda + \mu)^{m_u + m_d} e^{-(\lambda + \mu)T_b}$$

$$\times \binom{m_u + m_d}{m_u} \left(\frac{\lambda}{\lambda + \mu}\right)^{m_u} \left(\frac{\mu}{\lambda + \mu}\right)^{m_d} C_1 \tag{7.10.12}$$

$$= C_1 \binom{m_u + m_d}{m_u} \lambda^{m_e + m_u} \mu^{m_d} e^{-\lambda T - \mu T_b}$$

$$L(\lambda, \mu) = (m_e + m_u)\ln \lambda + m_e \ln \mu - \lambda T - \mu T_b + C \tag{7.10.13}$$

where we have written

$$\ln C_1 \binom{m_u + m_d}{m_u} = C$$

Differentiating $L(\lambda, \mu)$ with respect to λ and equating to zero, we get

$$\frac{m_e + m_u}{\lambda} - T = 0 \qquad (7.10.14)$$

giving

$$\hat{\lambda} = \frac{m_e + m_u}{T} \qquad (7.10.15)$$

Similarly, differentiating (7.10.13) with respect to μ and equating to zero, we get the estimate of μ as

$$\hat{\mu} = \frac{m_d}{T_b} \qquad (7.10.16)$$

The estimates given by (7.10.15) and (7.10.16) are the maximum likelihood estimates for the birth and death parameters λ and μ. They are also intuitively appealing because the process can be considered to be composed of two Poisson processes, the birth process and the death process, the latter being active only when the population size is nonzero. The equations (7.10.15) and (7.10.16) then give the estimates of the Poisson parameters λ and μ from $m_e + m_u$ occurrences of birth in time T and m_d occurrences of deaths in time T_b.

We shall now turn to the hypothesis-testing problem. Suppose a birth and death process has been observed for a length of time T as described, and we have the observations T_e, T_b, m_e, m_u, and m_d. Further suppose we would like to find out whether the observed process could have been the realization of a birth and death process with parameters λ^0 and μ^0. This is a hypothesis-testing problem in which the null hypothesis is H_0: $\lambda = \lambda^0$; $\mu = \mu^0$. The general problem of hypothesis testing in Markov processes using the classical Neyman-Pearson theory has been studied by Billingsley (1961), and for simple birth and death process his results specialize as follows: The maximum value of the log likelihood function $L(\lambda, \mu)$ of (7.10.13) is obtained by substituting the estimates obtained in (7.10.15) and (7.10.16). Thus we get

$$\max L(\lambda, \mu) = (m_e + m_u)\ln\left(\frac{m_e + m_u}{T}\right) + m_d\ln\left(\frac{m_d}{T_b}\right)$$

$$-(m_e + m_u) - m_d + C \qquad (7.10.17)$$

Under the null hypothesis, $L(\lambda, \mu)$ has the value

$$L(\lambda^0, \mu^0) = (m_e + m_u)\ln \lambda^0 + m_d\ln \mu^0 - \lambda^0 T - \mu^0 T_b + C \qquad (7.10.18)$$

The Neyman-Pearson statistic for testing the hypothesis H_0 is then given by

$$2\left[\max L(\lambda,\mu) - L(\lambda^0,\mu^0)\right] = 2\left[(m_e + m_u)\ln\frac{m_e + m_u}{\lambda^0 T} + m_d\ln\frac{m_d}{\mu^0 T_b}\right.$$

$$\left. -\left(m_e + m_u - \lambda^0 T\right) - \left(m_d - \mu^0 T_b\right)\right]$$

(7.10.19)

If λ^0 and μ^0 are the true parameter values, then asymptotically this statistic has a χ^2 distribution with two degrees of freedom. This information can therefore be used for testing the hypothesis H_0.

Other estimation and hypothesis-testing problems are beyond the scope of this book. In the generalized birth and death processes, however, it should be noted that when the dependence of λ_n and μ_n on n is not known, rough estimates of these parameters can be obtained by observing the process long enough (assuming the process to be of finite state space) to obtain estimates similar to (7.10.15) and (7.10.16) for every value of n. For a given value of n such an estimate would be

$$\hat{\lambda}_n = \frac{\text{Number of transitions } n \to n + 1}{\text{Sum of transition times for population size } n}$$

$$\hat{\mu}_n = \frac{\text{Number of transitions } n \to n - 1}{\text{Sum of transition times for population size } n} \qquad (7.10.20)$$

Example 7.10.1

The manager of a supermarket anticipates that, on the average, a customer will need 2 min service at the checkout counter, and he adjusts the number of counters so as to keep the arrival rate at 0.5/min. He also feels that it is reasonable to assume exponential distributions for time intervals between consecutive arrival epochs of customers at the counter as well as their service times. To check on the operation of the system, he takes observations for a 1 hr period, during which the number of customers arriving at the counter is found to be 34 and the number who leave the counter after service is found to be 23. It is also found that the counter was free for 3 min due to lack of customers.

Because of the assumptions of exponential distributions, the number of customers waiting at the counter can be modeled as a birth and death process. From the observations of 1 hr of operation of the system, in the notations given earlier, we have

$$T = 60 \text{ min} \qquad T_b = 57 \text{ min}$$

$$m_e + m_u = 34 \quad \text{and} \quad m_d = 23$$

Using (7.10.15) and (7.10.16), we have the following estimates of the arrival rate λ and service rate μ:

$$\hat{\lambda} = \frac{34}{60} = 0.57 \qquad \hat{\mu} = \frac{23}{57} = 0.40$$

But the manager wishes to find out whether these values are really different from the values he had assumed: $\lambda = 0.5$ and $\mu = 0.5$. To test this null hypothesis, we may use the likelihood ratio statistic given by (7.10.19). We have $2[\max L(\lambda, \mu) - L(\lambda^0, \mu^0)] = 2.368$. Under the null hypothesis this statistic has a χ^2 distribution with two degrees of freedom. From χ^2 tables we find

$$P(\chi^2 \geq 2.368) \cong 0.30$$

This probability is large enough to conclude that as per observations made of the operation of the system, we cannot reject the hypothesis that the parameters are still $\lambda = 0.5$ and $\mu = 0.5$.

7.11   ADDITIONAL EXAMPLES

In this section we shall illustrate the use of analysis techniques developed in this chapter with the help of examples. Additional examples are treated in application area chapters later: Queueing and service systems (Chapters 11–14); reliability systems (Chapter 15); combat as a Markov process (Chapter 16); recovery, relapse, and death due to disease (Chapter 18).

Example 7.11.1

In an immigration, birth and death process the population can increase either by birth within the population or by the arrival of immigrants. When the population size is n, let $n\lambda$ be the birth rate and $n\mu$ be the death rate. Let the arrival of immigrants be at a rate α. Assuming that assumptions necessary for the use of a Markov process model hold, determine the limiting distribution of the population size.

Instead of rederiving the equilibrium distribution in the manner of equations (7.8.9)–(7.8.16), we shall specialize the results for our case.

Combining the factors of immigration and birth together, we can identify the birth parameter λ_n as

$$\lambda_n = n\lambda + \alpha \qquad n = 0, 1, 2, \ldots$$

The death parameter μ_n in this case is equal to $n\mu$. Now specializing equations (7.8.14)–(7.8.16), we have

$$p_n = \frac{p_0}{n!\mu^n} \prod_{k=0}^{n-1} (\alpha + k\lambda)$$

$$p_0 = \left[1 + \sum_{n=1}^{\infty} \left(\frac{1}{n!\mu^n} \prod_{k=0}^{n-1} (\alpha + k\lambda) \right) \right]^{-1} \qquad (7.11.1)$$

provided the term in brackets is $< \infty$.

Example 7.11.2

Airplanes arrive at an airforce base at an average rate of λ. The probability that an airplane will require repair is p. The repair times have been found to follow an exponential distribution with parameter μ. Only one aircraft is serviced at one time. Assuming that the airplanes arrive in a Poisson process, obtain the limiting distribution of the number of airplanes either being repaired or waiting for repair at the base.

Let $X(t)$ be the number of airplanes arriving at the base during $(0, t]$. Using the Poisson assumption, we have

$$p[X(t) = n] = e^{-\lambda t} \frac{(\lambda t)^n}{n!} \qquad n = 0, 1, 2, \ldots \qquad (7.11.2)$$

Let $Y(t)$ be the number of planes that require repair during $(0, t]$. Since we can assume that the planes call for repair independently of each other, we can consider the number needing repair as following a binomial distribution for a specific number of arrivals. Using conditional probability arguments, we get

$$P[Y(t) = k] = \sum_{n=k}^{\infty} P[Y(t) = k | X(t) = n] P[X(t) = n]$$

$$= \sum_{n=k}^{\infty} \binom{n}{k} p^k (1 - p)^{n-k} e^{-\lambda t} \frac{(\lambda t)^n}{n!}$$

$$= e^{-\lambda t} \frac{p^k}{k!} \sum_{n=k}^{\infty} \frac{(\lambda t)^n (1 - p)^{n-k}}{(n - k)!}$$

$$= e^{-\lambda t} \frac{(\lambda t p)^k}{k!} \sum_{m=0}^{\infty} \frac{[\lambda t (1 - p)]^m}{m!}$$

$$= e^{-\lambda t} \frac{(\lambda p t)^k}{k!} e^{\lambda t (1 - p)}$$

$$= e^{-\lambda p t} \frac{(\lambda p t)^k}{k!} \qquad\qquad (7.11.3)$$

which indicates that $\{Y(t)\}$ is a Poisson process with parameter λp.

The limiting distribution of the number of airplanes undergoing or waiting for repair is obtained from results (7.8.17) and (7.8.18) by replacing λ by λp. Thus we get

$$p_0 = 1 - \frac{\lambda p}{\mu}$$

$$p_n = \left(1 - \frac{\lambda p}{\mu}\right)\left(\frac{\lambda p}{\mu}\right)^n \qquad n = 0, 1, 2, \ldots \qquad (7.11.4)$$

Example 7.11.3

In a department store customers first pick up merchandise on a self-serve basis and then wait for the sales clerks to ring up the sales. Suppose the customers arrive in a Poisson process with rate λ and that the amount of time they spend in picking up merchandise has an exponential distribution with mean $1/\alpha$. Let s be the number of sales clerks on duty, and let the amount of time spent in ringing up sales for each customer have an exponential distribution with mean $1/\mu$. Determine the limiting distribution of the number of customers who are still in the merchandise pickup phase and the limiting distribution of the number of customers who are in the sales ring-up phase of the system.

It is clear that this service system has two distinct components. With the Poisson assumption for the arrival process and exponential assumption for the merchandise pickup time and the sales ring-up time, we can identify two different models for the two components as follows:

Merchandise pick up: Birth and death process with $\lambda_n = \lambda$ and

$$\mu_n = n\alpha \ (n = 0, 1, 2, \ldots).$$

Sales ring up: Birth and death process with $\lambda_n = \lambda$ and

$$\mu_n = \begin{cases} n\alpha & \text{for } n = 0, 1, \ldots, s \\ s\alpha & \text{for } n = s, s+1 \ldots \end{cases}$$

For the merchandise pickup phase specializing results in (7.8.1)–(7.8.16), we get

$$p_0 = \left[1 + \sum_{n=1}^{\infty} \frac{1}{n!} \left(\frac{\lambda}{\mu} \right)^n \right]^{-1}$$

$$= e^{-\lambda/\mu}$$

$$p_n = e^{-\lambda/\mu} \left(\frac{\lambda}{\mu} \right)^n \frac{1}{n!} \tag{7.11.5}$$

which is clearly a Poisson distribution with mean λ/μ.

The birth and death process model for the sales ring-up phase is a queueing model with s servers, Poisson arrivals, and exponential service. This model is treated in more detail in Section 11.4.

An implicit assumption made in the preceding analysis is that the second phase is independent of the first phase and the arrival process to the second phase is also Poisson with rate λ. This assumption can be justified by an analysis of the departure process from the first phase of the system, which we do not go into here.

Example 7.11.4

Suppose it has been found that the repair time of a machine can be represented better by using different distribution models for different phases of repair such as finding the source of a problem and then the actual repair. The following Markov model is proposed for such cases.

The time interval the machine is in working condition has an exponential distribution with mean $1/\lambda$. The repair is in k phases, and the ith phase

($i = 1, 2, \ldots, k$) has an exponential distribution with mean $1/\mu_i$. Determine the long-run fraction of time the machine undergoes the several phases of repair.

Consider a stochastic process $X(t)$ with states $\{0, 1, 2, \ldots, k\}$, where 0 represents the state when the machine is working and $1, 2, \ldots, k$ represents the states at which it is undergoing different phases of repair, respectively. With the assumption that all time intervals are exponentially distributed, we now have a Markov process that changes its state in a cyclic manner. Let $p_0, p_1, \ldots, p_k$ be the limiting probabilities of finding the process in respective states. Balancing the transitions into and out of a state, we have

$$\lambda p_0 = \mu_k p_k$$

$$\mu_1 p_1 = \lambda p_0$$

$$\mu_2 p_2 = \mu_1 p_1$$

$$\mu_k p_k = \mu_{k-1} p_{k-1} \qquad (7.11.6)$$

These equations can be solved recursively, to give

$$p_1 = \frac{\lambda}{\mu_1} p_0$$

$$p_2 = \frac{\lambda}{\mu_2} p_0$$

$$\vdots$$

$$p_k = \frac{\lambda}{\mu_k} p_0$$

Imposing the condition $\sum_{i=0}^{k} p_i = 1$, we get

$$p_0 = \left[1 + \sum_{i=1}^{k} \frac{\lambda}{\mu_i} \right]^{-1}$$

$$p_n = \frac{\lambda}{\mu_n} \left[1 + \sum_{i=1}^{k} \frac{\lambda}{\mu_i} \right]^{-1} \qquad n = 1, 2, \ldots, k \qquad (7.11.7)$$

Remark In the preceding example the overall repair time has a distribution that is obtained by convolving k exponential distributions with different parameters. When all of them have the same parameter, the resulting distribution is gamma with the form identified in equation (7.2.32).

Example 7.11.5

A service system provides three stages of service of which the last stage is optional and requested only by a fraction α of the customers. It can be assumed that the times spent in the three stages of service are exponential with rates λ_1, λ_2, and λ_3, respectively. Assuming that there are enough customers to keep the server busy, determine the fraction of time the server spends in the three stages.

With the assumption that the server is never idle, and the services in stages are exponential, the service system can be modeled as a closed Markovian network with three nodes (when a customer leaves the network, another one takes its place). Schematically the transitions can be identified as in Figure 7.11.1. The Markov process has 3 states and the following transition rate structure:

$$p_{12}(\Delta t) = \lambda_1 \Delta t + o(\Delta t)$$

$$p_{21}(\Delta t) = \lambda_2 (1 - \alpha) \Delta t + o(\Delta t)$$

$$p_{23}(\Delta t) = \lambda_2 \alpha \Delta t + o(\Delta t)$$

$$p_{31}(\Delta t) = \lambda_3 \Delta t + o(\Delta t) \tag{7.11.8}$$

Let p_1, p_2, and p_3 be the limiting probabilities for the states of the process. Balancing transitions into a state with transitions out of it [see (7.8.6)], we get

$$\lambda_1 p_1 = \lambda_2 (1 - \alpha) p_2 + \lambda_3 p_3$$

$$\lambda_2 p_2 = \lambda_1 p_1$$

$$\lambda_3 p_3 = \lambda_2 \alpha p_2 \tag{7.11.9}$$

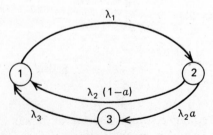

Figure 7.11.1

These equations may also be obtained by noting that the transition rate matrix is given by

$$A = \begin{pmatrix} -\lambda_1 & \lambda_1 & 0 \\ \lambda_2(1 - \alpha) & -\lambda_2 & \lambda_2\alpha \\ \lambda_3 & o & -\lambda_3 \end{pmatrix}$$

To solve equations (7.11.9), we note that the second and third equations give

$$p_2 = \left(\frac{\lambda_1}{\lambda_2} \right) p_1$$

$$p_3 = \frac{\lambda_2\alpha}{\lambda_3} p_2 = \frac{\lambda_1\alpha}{\lambda_3} p_1$$

Now using the normalizing condition $p_1 + p_2 + p_3 = 1$, we get

$$\left[1 + \frac{\lambda_1}{\lambda_2} + \frac{\lambda_1\alpha}{\lambda_3} \right] p_1 = 1$$

$$p_1 = \frac{\lambda_2\lambda_3}{\lambda_2\lambda_3 + \lambda_1\lambda_3 + \lambda_1\lambda_2\alpha}$$

which on back substitution gives

$$p_2 = \frac{\lambda_1\lambda_3}{\lambda_2\lambda_3 + \lambda_1\lambda_3 + \lambda_1\lambda_2\alpha}$$

$$p_3 = \frac{\lambda_1\lambda_2\alpha}{\lambda_2\lambda_3 + \lambda_1\lambda_3 + \lambda_1\lambda_2\alpha} \tag{7.11.10}$$

An introduction to this type of network problems is given in Chapter 12.

Remark 7.11.1   The last two examples can be solved in a simple manner by invoking properties of cyclically alternating renewal processes, which will be treated in the next chapter. They have been presented here as examples to illustrate the inherent Markovian structure.

Remark 7.11.2   By appealing to the theory of renewal processes, it can be shown that the results of the last two examples are true even for more general constituent distributions.

REFERENCES

Basawa, I. V., and Prakasa Rao, B. L. S. (1980). *Statistical Inference for Stochastic Processes*. New York: Academic Press.

Çinlar, E. (1975). *Introduction to Stochastic Processes*. Englewood Cliffs, N.J.: Prentice-Hall.

Cox, D. R., and Lewis, P. A. W. (1966). *The Statistical Analysis of Series of Events*. London: Methuen.

Gnedenko, B. V., Belyayev, Yu. K., and Solovyov, A. D. (1969). *Mathematical Methods in Reliability Theory*. New York: Academic Press (English Trans.)

Gordon, W. J., and Newell, G. F. (1967). "Closed Queueing Systems with Exponential Servers." *Oper. Res.* 15, 254–265.

Jackson, J. R. (1957). "Networks of Waiting Lines." *Oper. Res.* 5, 518–521.

Karlin, S., and Taylor, H. M. (1981). *A Second Course in Stochastic Processes*. New York: Academic Press, Chap. 14.

Kingman, J. F. C. (1969). "Markov Population Processes." *J. Appl. Prob.* 6, 1–18.

Matis, J. H., Wehrly T. E., and Metzler C. M. (1983). "On Some Stochastic Formulations and Related Statistical Moments of Pharmacokinetic Models." *J. Pharmacokinetics and Biopharmaceutics.* 11, 77–79.

Parzen, E. (1962). *Stochastic Processes*. San Francisco: Holden-Day, Chap. 7.

Tavaré, S. (1979). "A Note on Finite Homogeneous Continuous Time Markov Chains." *Biometrics* 35, 831–834.

Whittle, P. (1967). "Nonlinear Migration Processes." *Bull. Int. Stat. Inst.* 42, 642–646.

Whittle, P. (1968). "Equilibrium Distributions for an Open Migration Process." *J. Appl. Prob.* 5, 567–571.

FOR FURTHER READING

Bailey, N. T. J. (1964). *The Elements of Stochastic Processes with Applications to the Natural Sciences*. New York: Wiley, Chaps. 7, 8.

Bartlett, M. S. (1960). *Stochastic Population Models*. London: Methuen.

Billingsley, P. (1961). *Statistical Inference for Markov Processes*. Chicago: University of Chicago Press.

Çinlar, E. (1975). *Introduction to Stochastic Processes*. Englewood Cliffs, N.J.: Prentice-Hall, Chaps. 4, 8.

Cox, D. R., and Miller, H. D. (1965). *The Theory of Stochastic Processes*. New York: Wiley, Chap. 4.

Feller, W. (1968). *An Introduction to Probability Theory and Its Applications*. 3rd ed. New York: Wiley, Chap. 17.

Gihman, I. I., and Skorohod, A. V. (1969). *Introduction to the Theory of Random Processes*. Philadelphia: Saunders (English trans., Russian ed., 1965).

Kannan, D. (1979). *An Introduction to Stochastic Processes*. Amsterdam: North-Holland, Chaps. 4, 5.

Karlin, S., and Taylor, H. M. (1975). *A First Course in Stochastic Processes*. 2nd ed. New York: Academic Press, Chap. 4.

Medhi, J. (1982). *Stochastic Processes*. New Delhi: Wiley Eastern, Chap. 4.

Prabhu, N. U. (1965). *Stochastic Processes*. New York: Macmillan, Chap. 4.

Takács, L. (1960). *Stochastic Processes*. London: Methuen, Chap. 2.

EXERCISES

1. Traffic accidents in a city have been found to occur in a Poisson process at the following rates: Rush hours (7 AM to 9 AM and 4 PM to 6 PM)—four per hour; other times of the day (9 AM to 4 PM)—two per hour.

   a. What is the probability that there will be 20 or fewer accidents on one day between 7 AM and 6 PM?

   b. What is the probability that there will be 35 or more accidents on a day between 7 AM and 6 PM?

2. Customers demand two types of merchandise A and B. Demand for A is Poisson with rate λ_1, and the demand for B is Poisson with rate λ_2. During a time interval of length T, the total demand for both items is n. Show that the probability distribution of demands for each item is binomial, with success probability $\lambda_1/(\lambda_1 + \lambda_2)$.

3. In a Poisson process n events have occurred during $(0, t]$. Show that the probability distribution of the number of events that have occurred during $(0, s]$, $s < t$ is binomial with success probability s/t.

4. a. In a Poisson process n events have occurred during $(0, T]$. Let T_r $(r \leq n)$ be the time of occurrence of the rth event. Show that (T_r/T) is distributed as a beta distribution with parameters $r - 1$ and $n - r$.

      Obtain the mean and variance of T_r.

      Hint Note that T_r is the rth order statistic from a uniform distribution in $(0, T]$.

   b. Buses arrive at a bus stop once every T units of time. Customers arrive in a Poisson process. Using results from part a, show that the waiting time of a customer arriving at the bus stop (assuming that the bus has enough seats to hold all of those waiting) has a uniform distribution in $(0, T]$.

      Hint If n customers arrive during $(0, T]$, our customer could be any one of them.

5. a. A data-processing job involves three separate phases. Each phase can be assumed to have an exponential distribution. Assume that the mean times for these phases are $1/\mu_1$, $1/\mu_2$, and $1/\mu_3$, respectively.

    Determine the expected value of the total time needed for processing. Assume that the three phases have independent distributions.

    b. Suppose in part a the second phase is needed for only a fraction α of the jobs. Determine the expected value of the total time needed for processing a job arriving at the system.

    Also determine the percentages of time the job will spend in each phase.

6. A telephone switchboard has s lines. Telephone calls can be assumed to arrive in a Poisson process with parameter λ, and the call holding times can be assumed to have an exponential distribution with parameter μ. If the switchboard is busy, the caller gets a busy signal. All lines are busy at this time. What is the probability that a call will arrive when the system is still busy?

7. A Poisson process has parameter λ. Show that its probability generating function is obtained as

$$P(z) = e^{-(\lambda - \lambda z)t} \qquad |z| \leq 1$$

Suppose every Poisson event triggers X secondary events with a distribution $P(X = j) = a_j$ ($j = 1, 2, \ldots$) and a p.g.f.

$$A(z) = \sum_{j=1}^{\infty} a_j z^j \qquad |z| \leq 1$$

Show that the total number of secondary events have a p.g.f.

$$Q(z) = e^{-(\lambda - \lambda A(z))t} \qquad |z| \leq 1$$

Hence derive the distribution of the total number of secondary events when $a_1 = \frac{1}{2}$, $a_2 = \frac{1}{2}$, $a_j = 0$, $j \neq 1, 2$.

8. X and Y are independent exponential random variables with means $1/\lambda_1$ and $1/\lambda_2$, respectively. Let $Z_1 = \min(X, Y)$ and $Z_2 = \max(X, Y)$. Obtain the distributions of Z_1 and Z_2.

9. A nonhomogeneous Poisson process is defined as in Section 7.2 except that the transitions of the process during $(t, t + \Delta t)$ are governed by $\lambda(t)$, which is a function of t, instead of the constant λ. Show that for $n = 0, 1, 2, \ldots$

$$P[X(t + s) - X(t) = n] = e^{-[m(t+s) - m(t)]} \frac{[m(t + s) - m(t)]^n}{n!}$$

where $m(t) = \int_0^t \lambda(x)\, dx$.

10. At a gasoline station customer arrivals are in a nonhomogeneous Poisson process with the following rates:

7 AM–8 AM: Increases steadily from 10/hr to 20/hr.

8 AM–10 AM: Decreases steadily from 20/hr to 10/hr.

10 AM–12 noon: Remains steady at 10/hr.

12 noon–12:30 PM: Increases steadily from 10/hr to 20/hr.

12:30 PM–1:00 PM: Decreases steadily from 20/hr to 12/hr.

1 PM–5 PM: Remains steady at 12/hr.

5 PM–6 PM: Increases steadily from 12/hr to 20/hr.

6 PM–7 PM: Decreases steadily from 20/hr to 6/hr.

7 PM–10 PM: Remains steady at 6/hr.

Obtain the distribution of the number of customers coming to the gasoline station in a day.

11. When the population growth is deterministic, discuss the consequences of assuming the model as

$$\frac{dN(t)}{dt} = \lambda [N(t)]^2$$

12. In the birth and death process, with $\lambda_n = n\lambda$ and $\mu_n = n\mu$, show that

a. $\sum_n P_n(t) = 1$
b. $E[X(t)] = e^{(\lambda - \mu)t}$
c. $V[X(t)] = \dfrac{\lambda + \mu}{\lambda - \mu} e^{(\lambda - \mu)t} [e^{(\lambda - \mu)t} - 1]$

13. In the simple birth and death process, when $\lambda = \mu$, show that

$$P_0(t) = \frac{\lambda t}{1 + \lambda t} \qquad P_n(t) = \frac{(\lambda t)^{n-1}}{(1 + \lambda t)^{n+1}}$$

14. Obtain backward Kolmogorov equations for the following processes:

    a. Poisson process

    b. Pure birth process

    c. Pure death process

    d. Birth and death process

Solve the equations for the Poisson process, and show that the solution obtained is the same as the one obtained in the case of the forward Kolmogorov equations.

15. Obtain the transition rate matrices for the following processes:

    a. Poisson process

    b. Pure birth process

    c. Pure death process

    d. Birth and death process

    e. The two-state Markov process

16. Obtain the limiting distributions for the birth and death process with the following:

    a. $\lambda_n = \lambda$; $\mu_n = n\mu$

    b. $\lambda_n = \lambda$; $\mu_n = n^c\mu$, where $c > 0$

    c. $\lambda_n = \dfrac{\lambda}{n+1}$, $n \geq 0$; $\mu_n = \mu$, $n > 0$

17. Differential equations (7.5.2) can be solved using Laplace transforms. Define

$$\phi_{ij}(\theta) = \int_0^\infty e^{-\theta t} P_{ij}(t)\, dt \qquad i,j = 0,1, \qquad \mathrm{Re}(\theta) > 0$$

With the initial condition

$$P_{ij}(0) = \begin{cases} 1 & i = j \\ 0 & i \neq j \end{cases}$$

we get

$$\int_0^\infty e^{-\theta t} P_{ij}'(t)\, dt = -1 + \theta\phi_{ij}(\theta) \qquad i = j$$

$$= \theta\phi_{ij}(\theta) \qquad\qquad i \neq j$$

Using these transforms show that

$$\phi_{00}(\theta) = \frac{1}{\theta + \lambda + \mu} + \frac{\mu}{\theta(\theta + \lambda + \mu)}$$

$$= \frac{\mu}{\lambda + \mu} \cdot \frac{1}{\theta} + \frac{\lambda}{\lambda + \mu} \cdot \frac{1}{\theta + \lambda + \mu}$$

and

$$\phi_{10}(\theta) = \frac{\mu}{\theta(\theta + \lambda + \mu)}$$

$$= \frac{\mu}{\lambda + \mu} \cdot \frac{1}{\theta} - \frac{\mu}{\lambda + \mu} \cdot \frac{1}{\theta + \lambda + \mu}$$

Hence obtain the results (7.5.4) by inversion.

Hint   Use partial fraction expansion in obtaining the final form for $\phi_{00}(\theta)$ and $\phi_{10}(\theta)$; see Appendix B.

18. The two-state Markov process of Section 7.5 can be considered as the result of a limiting process on the two-state Markov chain of Section 4.3 by writing $t = n\Delta$ and the transition probability matrix

$$P = \begin{bmatrix} 1 - a & a \\ b & 1 - b \end{bmatrix} = \begin{bmatrix} 1 - \lambda\Delta & \lambda\Delta \\ \mu\Delta & 1 - \mu\Delta \end{bmatrix}$$

and letting $n \to \infty$ and $\Delta \to 0$.

a. Derive the transition probabilities (7.5.4) and (7.5.5) by applying this limiting procedure on (4.3.2).

b. Derive the results (7.5.9) and (7.5.10) for the expected length of time the process spends in the two states by applying the limiting procedure on (4.3.5).

19. Derive the expected values and variances given by (7.5.15)–(7.5.20).

20. A machine shop has two lathes. Each lathe alternates between uptime when it is in a working condition and downtime when it is being serviced. The uptimes can be assumed to have an exponential distribution with mean 2 hr, and down times can be assumed to have an exponential distribution with mean 30 min. At the beginning of the day both lathes are working. Assuming that they are used and serviced independently of each other, what is the probability that only one

lathe will be in a working condition at the end of an 8 hr workday? What is the expected length of total downtime for each lathe during that period?

21. A computing laboratory has an old computer which needs, on the average, 0.5 hr service for every 12 hr of work. The service and work periods can be considered to have exponential distributions. Buying new equipment would involve an investment of about \$100,000/year; but then the service requirements of the new equipment would only be, on the average, 1 hr for every 74 hr of work. Again, assumptions of exponential distributions are reasonable for service and work periods. The laboratory sells computer time for \$650/hr, and the service charges are \$50/hr. Do you think that buying new equipment is justified under these circumstances?

22. In an airport it has been found reasonable to assume that the arrival of airplanes follows a Poisson process at the rate of 20/hr.

    a. What is the probability that in a 3 min period no airplane will arrive?

    b. What is the probability that more than 2 airplanes will arrive during a 3 min period?

    c. It has been found that each airplane takes on the average 2 min for landing and that the landing times can be assumed to have an exponential distribution. What is the probability that at any given time the runway will not be in the process of being used? Assume that the airport has been in operation with the preceding rates continuously for a long time prior to the observation.

    d. Under conditions of part c, what do you expect to be the length of time the runway will be used in a given hour?

    e. Under steady state, how long does an airplane have to wait before getting clearance for landing?

23. a. For the study of an individual's health record a Markov model with the following states is proposed: diagnosis (0), treatment (1), temporary recovery (2), and permanent recovery (3). Death is not included as a state, with the assumption that the individual will live long enough to make the inclusion unnecessary. The amount of time spent per visit in these states can be assumed to have exponential distributions with means $1/\mu_0$, $1/\mu_1$, $1/\mu_2$, and $1/\mu_3$, respectively. Possible transitions are: $0 \rightarrow 1$, $1 \rightarrow 2$ with probability α_2, and $1 \rightarrow 3$ with probability α_3 ($\alpha_2 + \alpha_3 = 1$), $2 \rightarrow 1$, and $3 \rightarrow 0$.

Determine the fractions of time the individual spends in each of the four states in the long run.

b. In order to make the model in part a more realistic, suppose the state of death (4) is introduced. Access to this state is only from state 1 (treatment) and state 3 (permanent recovery). Now the probabilities of transition from these states are $1 \rightarrow 2$ with probability α_2, $1 \rightarrow 3$ with probability α_3, and $1 \rightarrow 4$ with probability α_4 ($\alpha_2 + \alpha_3 + \alpha_4 = 1$); $3 \rightarrow 0$ with probability β_0 and $3 \rightarrow 4$ with probability β_4 ($\beta_0 + \beta_4 = 1$). Write down the differential equations that can be used to determine the distribution of the length of the individual's life.

24. Cars arrive at the rate of λ per hour to a shopping center parking lot. The parking time for each car can be modeled as a random variable with an exponential distribution with mean $1/\mu$. Determine the limiting distribution of the number of cars parked in the lot. (Assume that the lot is large enough to accommodate all cars coming to the shopping center.)

25. In a taxi stand there is space for k taxicabs. Taxicabs arrive in a Poisson process with rate μ and pick up passengers one at a time if there are some already waiting. If there is no room to wait, they leave right away without waiting for customers. Customers arrive in a Poisson process with rate λ and wait as long as necessary to get a taxicab. Model the number of customers waiting as a Markov process and obtain its limiting distribution.

Hint If there are taxis waiting, the number of customers waiting is negative.

26. Obtain the limiting distributions for the three birth and death processes with

a. $\lambda_n = \lambda \qquad n = 0, 1, 2, \ldots, s - 1$
$\quad\ = 0 \qquad n = s, s + 1, \ldots$
$\mu_n = n\mu \qquad n = 1, 2, \ldots, s$
$\quad\ = 0 \qquad n = s + 1, s + 2, \ldots$

b. $\lambda_n = \lambda \qquad n = 0, 1, 2, \ldots$
$\mu_n = n\mu \qquad n = 1, 2, \ldots, s$
$\quad\ = s\mu \qquad n = s + 1, s + 2, \ldots$

c. $\lambda_n = \lambda \qquad n = 0, 1, 2, \ldots$
$\mu_n = n\mu \qquad n = 1, 2, 3, \ldots$

Let $p_0^{(a)}$, $p_0^{(b)}$, and $p_0^{(c)}$ be the limiting probability that the population size is zero in the three processes, respectively. Show that

$$p_0^{(a)} > p_0^{(b)} \quad \text{and} \quad p_0^{(c)} > p_0^{(b)}$$

27. a. A queueing system has one server and a waiting room of infinite capacity. Customer arrivals are in a Poisson process with parameter λ, and service times of customers can be assumed to have an exponential distribution with mean $1/\mu$. After finishing service, a customer joins the queue with probability α for a repeat service, regardless of the number of services the customer might have received before. Write down the steady-state equations for the process, and obtain the limiting distribution of the number of customers in the system.

  b. Show that the effective service time of the customer (after taking into account the customer's repeated demands for service) has an exponential distribution with mean $1/\mu$ $(1 - \alpha)$, and verify the limiting distribution derived under part a.

28. Suppose a machine has only two working parts, say A and B. Let the failure rate of the parts be λ_a and λ_b, and let the repair rates be μ_a and μ_b, respectively. If the breakdown and repair of these parts are independent, model this situation as a two-dimensional birth and death process, and derive its limiting distribution. What happens if the independence assumption is changed to the following? The parts fail independently of each other, but part A always has to be repaired before part B if they both have failed.

29. Suppose you have an investment plan composed of two subplans, say A and B. Under plan A, if you have n units in your account, the plan will increase (decrease) at a rate of $n\lambda$ $(n\mu)$. Under plan B, your money has a growth rate of α and a loss rate of β. Assuming that your capital can be modeled as a birth and death process, with the parameters given, investigate its limiting behavior. Assume that your initial capital is equally divided between the two plans.

30. In Problem 29, suppose that you have $\rho_1\%$ of your money in plan A and $\rho_2\%$ of your money in plan B $(\rho_1 + \rho_2 = 100)$. Investigate the process described under these assumptions.

31. Suppose airplanes arrive at an Air Force Base at an average rate of one every 15 min. The probability that an airplane will require repair is 0.5. The base repair shop can repair an airplane at the following rates: One every 25 min or one every 40 min. Assume that conditions necessary to

model this situation as a birth and death process hold. If the shop's officer in charge decides to run it so that the rate is one every 40 min, will this individual be considered a good supervisor? Explain your answer.

32. In Problem 31, suppose the arrival rate is λ, the probability of required repair is p, and the repair rate is μ. Find the forward equation for this process (the number of aircraft in the shop at a given time), and investigate the conditions under which the limiting distribution exists. Also obtain its limiting distribution.

33. A service station which restricts the number of jobs in the shop to two gives priority to some types of jobs. If a priority job arrives while the mechanic (there is only one mechanic) is working on a nonpriority job, the mechanic immediately starts work on the priority job. Jobs arrive in a Poisson process with rate λ. Of the arriving jobs a fraction α are priority jobs. The service times for the priority and nonpriority jobs have exponential distributions with means $1/\mu_1$ and $1/\mu_2$, respectively.

Obtain the transition rate matrix for the Markov process, and hence derive the limiting distribution of the number of the two types of jobs in the system. Find the percentage of time the system would be occupied by service to a priority item.

34. A telephone switchboard has s lines. Two classes of telephone calls arrive at the switchboard. The arrivals are in Poisson processes, and their holding times have exponential distributions. The arrival and services rates for the two classes are (λ_1, μ_1) and (λ_2, μ_2), respectively. Calls arriving when the switchboard is busy leave the system without service.

a. Obtain the limiting distribution of the number of busy lines when $s = 2$ and $s = 3$.

b. Generalize the results of part a for an arbitrary s.

c. Suggest appropriate results for the case when k (≥ 1) classes of telephone calls arrive at a switchboard with s (≥ 1) lines.

Hint Use local balance equations to simplify the solution technique.

35. In a warehouse items arrive in a Poisson process with parameter λ. Arriving items are first stacked as they come in and classified and stored by one worker; the time the worker spends in classifying and storing an item has an exponential distribution with mean $1/\mu$. Since the worker picks up the item from the top of the stack, the system can be considered to be a single-server queueing system with a "last-come, first-served" queue discipline. Write down the differential equations that lead to the determination of the time interval between the epoch of the arrival of an item and the epoch of its storing.

Hint   Observe the similarity between the waiting time in a queue with last-come, first-served queue discipline and the busy period in the regular system.

36. a. In a single-server queueing system customers arrive in a Poisson process at rate λ, and their service times have an exponential distribution with mean $1/\mu$. Suppose the system has no customers at time 0. Let T_k be the departure epoch of the kth customer. Write down differential equations leading to the determination of the distribution of T_k.

  b. Let $g_k^{(x)}$ be the probability density of T_k, and define its Laplace transform

$$\gamma_k(\theta) = \int_0^\infty e^{-\theta x} g_k(x)\, dx \qquad \mathrm{Re}(\theta) > 0$$

Obtain $\gamma_k(\theta)$ for $k = 2, 3, 4$.

37. In a single-server queueing system customers arrive in groups of size 2 in a Poisson process with parameter λ. They get served one at a time, and their service times have an exponential distribution with mean $1/\mu$. Let p_n be the probability that there are n customers in the system as $t \to \infty$. Write down the steady-state equations for the process.
Let

$$P(z) = \sum_{n=0}^\infty p_n z^n \qquad |z| \le 1$$

Using the equations show that

$$P(z) = \frac{\mu p_0(z - 1)}{-\lambda z^3 + (\lambda + \mu)z - \mu}$$

Noting that $\lim_{z \to 1} P(z) = 1$, obtain

$$P(z) = \frac{\mu(1 - \rho)(z - 1)}{-\lambda z^3 + (\lambda + \mu)z - \mu}$$

where $\rho = 2\lambda/\mu$.

Hence determine the expected number of customers in the system.

38. a. Reliability of a component is defined as the probability that the unit will perform its mission. Suppose a system has two components arranged in parallel. At least one component must be working for the system to be operative. Let the life length of each component have an exponential distribution with mean $1/\lambda$. When a component fails, repair is undertaken immediately. Let the repair times have an exponential distribution with mean $1/\mu$. Determine the reliability of the system at t and the mean time to system failure.

b. Suppose when the system fails, repair is undertaken on both components, and system becomes operative when there is at least one working component. Let the repair rate be the same on both components. Determine the long-term availability of the system. (Availability of the system is defined as the probability that it is operative.)

39. The two-component system of Exercise 38 is attached in series to a one-component system. Let the life length of this third component be exponential with mean $1/\lambda_2$ and an exponential repair rate of μ_2. Now determine the reliability at time t, mean time to failure, and the long-term availability of this composite system, assuming that the subsystems operate independently of each other.

Hint A system with subsystems arranged in series is operative only if all subsystems are working.

40. Determine the reliability at time t, mean time to failure, and the long-term availability of the triple modular redundant system of Review Exercise 16 of Chapter 2 under the following assumptions:

a. The components have an exponential life with mean $1/\lambda_1$.

b. The repair time of the components is exponential with rate μ_1.

c. The voter has an exponential life with mean $1/\lambda_2$.

d. The repair time of the voter is exponential with rate μ_2.

e. The components and the voter operate independently of each other.

41. In Exercise 38 suppose that there is a probability $1 - \alpha$ that the system fails when one of the two working units fail. Thus α is the probability that the system recovers from a component failure, and hence is sometimes known as the coverage probability in computer engineering terminology. The system does not recover from a component failure when only one component is working.

Determine the reliability at time t and mean time to failure under this coverage assumption.

42. Following the procedure outlined in Section 7.9 list the set of observations needed to estimate the parameters λ and μ in a birth and death process, with

$$\lambda_n = \lambda \qquad n = 0, 1, 2, \ldots, s - 1$$

$$= 0 \qquad n = s, s + 1, \ldots$$

$$\mu_n = n\mu \qquad n = 1, 2, \ldots, s$$

$$= 0 \qquad n > s$$

Give the maximum likelihood estimates for λ and μ in terms of the observations identified here. (*Note:* This is a telephone switchboard problem with s lines and no calls waiting.)

43. In a service facility customer arrivals can be assumed to be Poisson, and their service times can be represented by an exponential distribution. In order to estimate the arrival and service rates, the facility was observed 8 AM to 12 noon one day and the following data were recorded.

  a. The facility opens with three customers waiting at 8:00 AM.

  b. 8:00–8:42: Continuous service given to 12 customers.

  c. 8:45–9:31: Continuous service given to 13 customers.

  d. 9:38–10:07: Continuous service given to 8 customers.

  e. 10:09–10:43: Continuous service given to 10 customers.

  f. 10:49–11:33: Continuous service given to 15 customers.

  g. 11:37–12:00: Seven customers served before closing the facility at 12:00 noon; 2 customers were waiting at that time.

  h. The server was idle in between these time periods.

Obtain the maximum likelihood estimates for the arrival and service rates.

44.   a. Derive the maximum likelihood estimates of the parameters λ and μ in the two-state Markov process of Section 7.5.

  b. Give the likelihood ratio criterion for testing the hypothesis $\lambda = \lambda^\circ$ and $\mu = \mu^\circ$.

c. A telephone receptionist in a department store can process only one call at a time. The two states the receptionist can be in are "idle" and "busy." It is conjectured that a two-state Markov process model would be reasonable to represent this individual's work habits. To estimate the parameters of the process, after having settled down in the job for the day, the receptionist was observed for a period of 30 min. At the start of the 30 min period the receptionist was found to be busy and at the end idle. The total length of time the receptionist was idle was 8 min 35 sec and, including the last time, idle 17 times during that period.

Data from similar stores have indicated that, on the average, telephone receptionists spend about 40 sec of idle time between calls and the calls last for 90 sec. On this basis can we conclude that the observed process comes from a normal workday for the receptionist?

45. An automatic machine subject to two modes of failure can be modeled as a three-stage Markov process [C. L. Proctor and B. Singh (1976), *IEEE Trans. on Rel.* **R-25** (3), 210–211]. The states are identified as 0—machine is working; 1—failed in mode 1 where some unwanted operations are being carried out; 2—failed in mode 2 when the machine is not working. Let λ_n be the constant failure rate of the machine into failure mode n $(n = 1, 2)$ and μ_n $(n = 1, 2)$ be the constant repair rates in these failure modes. The following transitions are possible: $0 \to 1$, $0 \to 2$, $1 \to 0$, and $2 \to 0$. Let $P_n(t)$ be the probability that the device is in state n at time t, with $P_0(0) = 1$ and $P_n(0) = 0$, $n = 1, 2$. Determine $P_n(t)$ for $t < \infty$ as well as $t \to \infty$.

Suppose it is necessary to estimate the failure and repair rates λ_n and μ_n $(n = 1, 2)$ from data gathered from the operation of the system. Determine the minimal set of observations necessary for the estimation of these parameters.

Chapter 8

RENEWAL PROCESSES

8.1 INTRODUCTION

Consider the following examples:

1. Customers arrive at a store one at a time. Let $N(t)$ be the number of arrivals during $(0, t]$.
2. Accidents occurring in a city are reported to the central police station. Let $N(t)$ be the number of accidents reported during $(0, t]$.
3. Light bulbs are replaced as soon as they burn out. Let $N(t)$ be the number of light bulbs replaced at a single location in time $(0, t]$.
4. Starting from a reference point, the number of cars on the road are counted at a particular time (the counting is done instantaneously by aerial photography or some such device). Let $N(t)$ be the number of cars on a length t of the road.
5. A machine component is replaced as soon as it fails. Let $N(t)$ be the number of replacements made in time $(0, t]$.

In all these examples we are interested in observing the occurrence of a certain event, and $N(t)$ is the number of such events occurring in time $(0, t]$. It is a nondecreasing discrete-state process that counts the number of events occurring, and hence it may be called a counting process. Let the time intervals between consecutive epochs of occurrence of the event be independent and identically distributed random variables. Then $N(t)$ is called a renewal counting process, and we shall call the event of interest a renewal event. This is also called a recurrent event in the literature (see Feller, 1968), but we shall use the former terminology in order to avoid confusion with the classification categories of recurrence, transience, and so on. The process composed of the interoccurrence times of renewal events is called the renewal process.

The study of the renewal process is directed toward gaining information on the following major characteristics:

1. Distribution of $N(t)$.
2. Expected number of renewals in time t, $E[N(t)]$; when the time parameter is continuous, $E[N(t)]$ is also known as the renewal function.
3. The probability mass or the probability density function related to a renewal at a given time point.
4. Time needed for a specific number of events to occur.
5. Distributions of time since the last epoch of occurrence (backward recurrence time) and time until the next epoch of occurrence of the renewal event (forward recurrence time).

As in all other processes the time parameter can be considered to be either discrete or continuous. We treat these two cases separately. For a further discussion, the reader may be referred to Feller (1968) for the discrete case and Cox (1962) and Karlin and Taylor (1975) for the continuous case.

8.2   RENEWAL PROCESS WHEN TIME IS DISCRETE

Consider a sequence of repeated trials with possible outcomes E_i, $i = 1, 2, \ldots$. Suppose a certain outcome E_* is of particular interest. With regard to this outcome we say a renewal occurs at the nth trial if and only if E_* occurs at the nth trial, and the outcomes after the nth trial occur independently of the trials prior to the nth. The time interval (number of trials) between any two consecutive epochs of occurrence of E_* is known as the renewal period for the process. It may also be called the waiting time for the first occurrence of a renewal event.

Let $p^{(n)}$ be the probability that E_* occurs at the nth trial, and let $f^{(n)}$ be the probability that E_* occurs for the first time at the nth trial. Let

$$p^{(0)} = 1 \qquad f^{(0)} = 0 \tag{8.2.1}$$

and

$$f^* = \sum_{n=1}^{\infty} f^{(n)} \tag{8.2.2}$$

Clearly f^* is the probability of eventual occurrence of E_*.

A simple example of a renewal counting process is the Bernoulli process. Let a be the probability of "success" at any Bernoulli trial, and let b ($a + b = 1$) be the probability of "failure." Suppose that we are interested in the occurrence of success, the event E_*. Then, because of the independence of the trials, the probability that E_* occurs at any trial is a, and hence it is independent of n. Further the distribution $f^{(n)}$ of the renewal period is given by

$$f^{(n)} = ab^{n-1} \qquad n = 1, 2, \ldots \qquad (8.2.3)$$

Clearly $f^* = 1$ in this case. The probability distribution of the number of renewals in n trials is binomial, with probability of success a, and the expected number of renewals in n trials is na.

As related to Markov chains, if E_* refers to a certain state of the system, say i, the probabilities corresponding to $p^{(n)}$ and $f^{(n)}$ are $P_{ii}^{(n)}$ and $f_{ii}^{(n)}$. In Markov processes we study the behavior of the entire process for all values of the time parameter; in contrast, the study of renewal processes is restricted to renewal points. One of the advantages of this apparent restriction is that we do not make any assumptions regarding the behavior of the process during a renewal period. Consequently a large number of processes which do not exhibit Markovian properties can be considered in this group, and their important properties can be derived.

Similar to the classification of states done in Markov chains (Chapter 3), the renewal event E_* can be classified as

$$
\begin{array}{ll}
\text{Recurrent if} & f^* = 1 \\
\text{Transient if} & f^* < 1
\end{array}
\qquad (8.2.4)
$$

If E_* is recurrent, E_* can be further classified as null or positive recurrent, depending on whether $\sum_{n=1}^{\infty} nf^{(n)} = \infty$ or $< \infty$, respectively.

An alternate criterion for classification would be based on the probability $p^{(n)}$; we have

$$
\begin{array}{ll}
E_* \text{ recurrent if} & \sum_{n=0}^{\infty} p^{(n)} = \infty \\[2mm]
E_* \text{ transient if} & \sum_{n=0}^{\infty} p^{(n)} < \infty
\end{array}
\qquad (8.2.5)
$$

When $f^* = 1$, we may consider a proper random variable Z_r representing the length of the rth renewal period. Then $\{f^{(n)}\}$ is its probability distribution, and

$$E(Z_r) = \sum_{n=1}^{\infty} nf^{(n)} = \mu \qquad r = 1, 2, \ldots \tag{8.2.6}$$

Let $f_{(r)}^{(n)}$ be the probability that the E_* occurs at the nth trial for the rth time. Clearly for $r = 2$

$$f_{(2)}^{(n)} = f^{(1)}f^{(n-1)} + f^{(2)}f^{(n-2)} + \cdots + f^{(n-1)}f^{(1)} \tag{8.2.7}$$

Thus we get the probability distribution of the waiting time until the second occurrence of E_* as the twofold convolution of $\{f^{(n)}\}$ with itself. Generalizing this result for the waiting time until the rth occurrence, we have Theorem 8.2.1.

Theorem 8.2.1   *Let S_r be the waiting time until the rth occurrence of a renewal event E_*, and let Z_i be its ith renewal period. Further let*

$$f^{(n)} = P(Z_i = n) \qquad i = 1, 2, \ldots \tag{8.2.8}$$

$$f_{(r)}^{(n)} = P(S_r = n) \tag{8.2.9}$$

Then we have

$$S_r = Z_1 + Z_2 + \cdots + Z_r \tag{8.2.10}$$

and

$$f_{(r)}^{(n)} = \left\{ f^{(n)} \right\}^{r^*} \tag{8.2.11}$$

where r^* represents the r-fold convolution. Let

$$F(z) = \sum_{n=1}^{\infty} f^{(n)}z^n \qquad |z| < 1 \tag{8.2.12}$$

Using properties of probability generating functions of sums of independent random variables, we also get

$$F_{(r)}(z) = \sum_{n=1}^{\infty} f_{(r)}^{(n)}z^n = [F(z)]^r \tag{8.2.13}$$

In (8.2.1) we assumed $p^{(0)} = 1$; that is, the initial time point is an epoch at which E_* occurs. This may not be a realistic assumption in many practical situations. To remove this restriction, consider the initial renewal period as having a distribution

$$b^{(n)}(k) = P(Z_1 = n | \text{Last renewal occurred } k \text{ trials before } 0)$$

$$n = 1, 2, \ldots \tag{8.2.14}$$

Using the unconditional distribution of the renewal periods, we can also write

$$b^{(n)}(k) = \frac{f^{(n+k)}}{f^{(k+1)} + f^{(k+2)} + \cdots} = \frac{f^{(n+k)}}{1 - F_k} \tag{8.2.15}$$

where $F_k = 1 - \sum_{r=1}^{k} f^{(r)}$. For convenience we shall denote $b^{(n)}(k) \equiv b^{(n)}$. The probabilities $\{b^{(n)}\}$, $\{p^{(n)}\}$, and $\{f^{(n)}\}$ can be seen to be related as

$$p^{(n)} = b^{(n)} + \sum_{r=1}^{n} p^{(n-r)}f^{(r)} \qquad n \geq 1 \tag{8.2.16}$$

This is called the discrete renewal equation for probabilities of renewal.

Defining the probability generating functions as in (8.2.12) and

$$P(z) = \sum_{n=0}^{\infty} p^{(n)}z^n \qquad |z| < 1$$

$$B(z) = \sum_{n=1}^{\infty} b^{(n)}z^n \qquad |z| < 1 \tag{8.2.17}$$

from (8.2.16) we get

$$P(z) = B(z) + F(z)P(z)$$

$$P(z) = \frac{B(z)}{1 - F(z)} \tag{8.2.18}$$

If it is known that E_* has occurred at $n = 0$, we have a simpler equation:

$$p^{(0)} = 1$$

$$p^{(n)} = \sum_{r=1}^{n} p^{(n-r)}f^{(r)} \tag{8.2.19}$$

leading to the generating function

$$P(z) = \frac{1}{1 - F(z)} \tag{8.2.20}$$

Example 8.1.1

In Example 2.1.1 we considered a coin-tossing problem. It belongs to a class of problems known as "random walks," which are convenient mathematical models in many situations. A simple model is as follows: A particle is initially at zero. At every step (trial) it takes an upward or downward jump of one unit with probabilities a and b $(a + b = 1)$, respectively. Three sample functions of such processes in the symmetric case $(a = b = \frac{1}{2})$ have been represented in Figure 2.1.1.

Let E_* be the event that the particle returns to zero. This can only occur at an even number of steps. Further E_* occurs at the $2n$th trial only if the particle has taken n upward and n downward jumps. Thus we get

$$p^{(n)} = 0 \qquad \text{for } n \text{ odd}$$

$$p^{(2n)} = \binom{2n}{n} a^n b^n \qquad n = 0, 1, 2, \ldots \tag{8.2.21}$$

But we have

$$\binom{2n}{n} = \frac{2n(2n - 1) \cdots 1}{n! n!}$$

$$= 2^{2n} \frac{(-1)^n \{[-1/2][-3/2] \cdots [-(1/2)n + 1]\}}{n!}$$

$$= 2^{2n}(-1)^n \binom{-\frac{1}{2}}{n} \tag{8.2.22}$$

Therefore we can write

$$p^{(2n)} = \binom{-\frac{1}{2}}{n}(-4ab)^n \tag{8.2.23}$$

Multiplying (8.2.23) by z^{2n} and summing over n, we get

$$P(z) = \sum_{n=0}^{\infty} \binom{-\frac{1}{2}}{n}(-4abz^2)^n$$

$$= (1 - 4abz^2)^{-1/2} \tag{8.2.24}$$

Using this expression in (8.2.20), we get

$$F(z) = 1 - \frac{1}{P(z)}$$

$$= 1 - (1 - 4abz^2)^{1/2} \tag{8.2.25}$$

The probability distribution of the renewal period can be determined from its generating function (8.2.25) by inversion. Clearly $f^{(n)} = 0$ for n odd. Therefore

$$F(z) = \sum_{n=0}^{\infty} f^{(2n)} z^{2n} \tag{8.2.26}$$

Equating the right-hand sides of (8.2.25) and (8.2.26), we get

$$\sum_{n=0}^{\infty} f^{(2n)} z^{2n} = 1 - (1 - 4abz^2)^{1/2}$$

$$= 1 - \sum_{n=0}^{\infty} \binom{\frac{1}{2}}{n} (-4abz^n)^n$$

$$= 1 - \sum_{n=0}^{\infty} \binom{\frac{1}{2}}{n} (-1)^n 2^{2n} a^n b^n z^{2n} \tag{8.2.27}$$

Comparing coefficients of z^{2n}, we then get

$$f^{(0)} = 0$$

$$f^{(2n)} = \binom{\frac{1}{2}}{n} (-1)^{n+1} 2^{2n} a^n b^n$$

$$= \frac{2^n (2n - 3)(2n - 5) \cdots 5 \cdot 3 \cdot 1 \cdot a^n b^n}{n!} \qquad n \geq 1$$

$$= \frac{2n(2n - 2)(2n - 3) \cdots 3 \cdot 2 \cdot 1 \cdot a^n b^n}{n! n!}$$

$$= \frac{1}{2n - 1} \cdot \binom{2n}{n} a^n b^n \tag{8.2.28}$$

Also we find

$$f^{(2n)} = \frac{1}{2n - 1} p^{(2n)} \qquad n \geq 1 \tag{8.2.29}$$

The probability generating function $F(z)$ can also be used for classifying the renewal event E_* representing the random walk's return to zero. We have

$$f^* = \sum_{n=1}^{\infty} f^{(n)} \leq 1$$

and therefore from Abel's theorem (Appendix B)

$$f^* = \lim_{z \to 1^-} \sum_{n=1}^{\infty} f^{(n)} z^n$$

$$= 1 - (1 - 4ab)^{1/2}$$

$$= 1 - \left[(a + b)^2 - 4ab \right]^{1/2}$$

$$= 1 - |a - b| \tag{8.2.30}$$

Thus we obtain

$$\begin{aligned} f^* &= 1 &&\text{if } a = b = \tfrac{1}{2} \\ f^* &< 1 &&\text{if } a \neq b \end{aligned} \tag{8.2.31}$$

When $f^* = 1$, we can also show that

$$\mu = F'(z)\big|_{z=1} = \infty \tag{8.2.32}$$

We therefore conclude that the event E_* is transient if $a \neq b$ and null recurrent if $a = b = \tfrac{1}{2}$. This confirms our intuition about the case $a \neq b$.

When $a = b = \tfrac{1}{2}$, even though the return to zero is certain, our result shows that it may take an indefinitely long time.

Again, using an Abelian theorem on generating function transforms, we can show that

$$\lim_{n \to \infty} p^{(n)} = \lim_{z \to 1^-} (1 - z) P(z) = 0 \qquad a = b \tag{8.2.33}$$

This result can also be obtained from Theorem 8.2.2.

When the time parameter is continuous, it is convenient to have the renewal equation in terms of expected number of renewals. The corresponding discrete renewal equation for the expected number of renewals can be derived from (8.2.16) as follows: Let the indicator variable Y_r be defined as

$$Y_r = \begin{cases} 1 & \text{if } E_* \text{ occurs at the } r\text{th trial} \\ 0 & \text{otherwise} \end{cases}$$

Clearly $P(Y_r = 1) = p^{(r)}$ and

$$E(Y_r) = p^{(r)}$$

Let u_n be the expected number of renewals in n trials. Then we have

$$u_n = E(Y_1 + Y_2 + \cdots + Y_n)$$

$$= \sum_1^n p^{(r)} \tag{8.2.34}$$

From (8.2.16) we get [introducing $p^{(0)}$ and $b^{(0)}$, which are both zero when it is known that the last renewal occurred k (> 0) trials before the initial trial]

$$\sum_{r=0}^n p^{(r)} = \sum_{r=0}^n b^{(r)} + \sum_{r=0}^n \sum_{s=0}^r f^{(s)} p^{(r-s)}$$

$$= \sum_{r=0}^n b^{(r)} + \sum_{s=0}^n f^{(s)} \sum_{r=0}^{n-s} p^{(r)}$$

Writing $\sum_{r=0}^n b^{(r)} = b_n$, we therefore have

$$u_n = b_n + \sum_{s=0}^n f^{(s)} u_{n-s} \tag{8.2.35}$$

The function u_n may be called the discrete renewal function. It should be noted that (8.2.16) and (8.2.35) are identical in structure, and therefore the behavior of the probability $p^{(n)}$ and the expected number u_n should be similar. Theorem 8.2.2 gives the limiting behavior of u_n with reference to the expected renewal period μ. The corresponding behavior of $p^{(n)}$ is indicated in the corollary. The proof of the theorem is beyond the scope of this book.

Theorem 8.2.2 Let $\{f^{(n)}, n = 0, 1, 2, \ldots\}$ be such that $f^{(n)} \geq 0, \sum_0^\infty f^{(n)} = 1$, $\sum_{n=1}^\infty n f^{(n)} = \mu < \infty$, and let the greatest common divisor of the integers n for which $f^{(n)} > 0$ be 1. Let $\{b_n, n = 0, 1, 2, \ldots\}$ be such that $\sum_0^\infty |b_n| < \infty$.

Suppose the renewal equation

$$u_n = b_n + \sum_{r=0}^{n} f^{(r)} u_{n-r} \qquad n = 0, 1, 2, \ldots$$

is satisfied by a bounded sequence $\{u_n\}$ of real numbers. Then $\lim_{n \to \infty} u_n$ exists and is given by

$$\lim_{n \to \infty} u_n = \frac{\sum_{r=0}^{\infty} b_r}{\mu} \qquad (8.2.36)$$

Further, if $\mu = \infty$, the limit relation (8.2.36) is still valid if we interpret

$$\frac{\sum_{r=0}^{\infty} b_r}{\mu} = 0 \qquad (8.2.37)$$

COROLLARY   *Using results of Theorem 8.2.2 in (8.2.16), we have*

$$\lim_{n \to \infty} p^{(n)} = \frac{\sum_{r=0}^{\infty} b^{(r)}}{\mu} \qquad (8.2.38)$$

In (8.2.38), $\{ b^{(r)}, \; r = 0, 1, 2, \ldots \}$ is the probability distribution of the initial renewal period, and when this is a proper distribution, we have

$$\sum_{r=0}^{\infty} b^{(r)} = 1$$

thus giving

$$\lim_{n \to \infty} p^{(n)} = \frac{1}{\mu} \qquad (8.2.39)$$

Depending on the nature of the event E_*, we therefore have

$$\lim_{n \to \infty} p^{(n)} = \begin{cases} \dfrac{1}{\mu} > 0 & \text{if } E_* \text{ is positive recurrent} \\ 0 & \text{if } E_* \text{ is null recurrent} \end{cases} \qquad (8.2.40)$$

Example 8.1.2

Suppose there are N objects (light bulbs, machine components, household items, etc.) that need replacement as they fail. Let the lifetimes of these objects be identically distributed random variables. We can consider the replacement with respect to every item a renewal event. Suppose we are interested in the age distribution of these N objects after time n. Let $f^{(n)}$ be the probability that an item has a lifetime of length n time periods, and let $V_k^{(n)}$ be the expected number of items of age k (≥ 0) at time n with $V_k^{(0)}$ given and $\sum_{k=0}^{\infty} V_k^{(n)} = N$ ($n \geq 0$). Further let W_n be the expected number of replacements made at time n. From simple probability arguments, it can be seen that $V_k^{(n)}$ ($n = 0, 1, 2, \ldots$) are related by the recursive relation

$$V_0^{(n)} = \sum_{k=0}^{\infty} V_k^{(n-1)} \left(\frac{f^{(k+1)}}{1 - F_k} \right) \tag{8.2.41}$$

$$V_k^{(n)} = V_{k-1}^{(n-1)} - V_{k-1}^{(n-1)} \left(\frac{f^{(k)}}{1 - F_{k-1}} \right)$$

$$= V_{k-1}^{(n-1)} \left(\frac{1 - F_k}{1 - F_{k-1}} \right) \tag{8.2.42}$$

When $V_k^{(0)}$ ($k = 0, 1, 2, \ldots$) are known, $V_k^{(n)}$ can be obtained recursively from these equations. As $n \to \infty$, $\lim_{n \to \infty} V_k^{(n)}$ reduces to a much simpler form. First we note that

$$W^{(n)} = Np^{(n)} \tag{8.2.43}$$

$$\lim_{n \to \infty} W^{(n)} = \frac{N}{\mu} \tag{8.2.44}$$

Now consider $V_k^{(n)}$, the expected number of items of age k (≥ 0) at time n. For any object to be of age k at epoch n ($n \geq k$), it must be installed at epoch $n - k$ and have continued to survive for at least k time periods. Thus we have

$$V_k^{(n)} = W^{(n-k)}[1 - F_k] \tag{8.2.45}$$

As $n \to \infty$,

$$\lim_{n \to \infty} V_k^{(n)} = [1 - F_k] \lim_{n \to \infty} W^{(n-k)}$$

$$= \frac{N}{\mu}[1 - F_k] \tag{8.2.46}$$

As an illustration of this procedure, consider the following hypothetical example: suppose $N = 100$; $f^{(1)} = 0.1$, $f^{(2)} = 0.3$, $f^{(3)} = 0.5$, and $f^{(4)} = 0.1$. We have $\mu = 2.6$ and $N/\mu = 38.46$. Using (8.2.41), (8.2.42), and (8.2.46), the age distribution of these 100 objects can be obtained as follows:

k \ n	0	1	2	3	4	5	6	7	8	9	∞
0	50	38.34	33.27	42.05	38.36	36.92	39.55	38.43	37.99	38.90	38.46
1	20	45.00	34.51	29.94	37.85	34.52	33.23	35.59	34.59	34.19	34.61
2	20	13.33	34.00	23.01	19.96	25.23	23.01	22.15	23.73	23.06	23.08
3	10	3.33	2.22	5.00	3.83	3.33	4.21	3.83	3.69	3.95	3.85

8.3 RENEWAL PROCESS WHEN TIME IS CONTINUOUS

Let an event E_* occur at $t_1, t_2, t_3, \ldots$, and let $Z_r = t_r - t_{r-1}$ ($r = 2, 3, \ldots$) be independent and identically distributed random variables with

$$P(Z_r \le x) = F(x) \tag{8.3.1}$$

$$E(Z_r) = \mu \tag{8.3.2}$$

and further let $t_0 = 0$ and $Z_1 = t_1 - t_0$ be distributed as

$$P(Z_1 \le x) = F_1(x) \tag{8.3.3}$$

Let $N(t)$ be the number of times E_* occurs in time $(0, t]$. The process $N(t)$ is a renewal counting process in continuous time, and Z_r ($r = 1, 2, \ldots$) are the renewal periods.

We have defined the initial renewal period Z_1 as having a distribution $F_1(x)$, possibly different from $F(x)$, in order to represent the general situation where the initial observation point may not necessarily be a renewal point. However, if it is known that E_* has occurred at $t_0 = 0$, we can replace $F_1(x)$ by $F(x)$ in the results. We shall identify such a process as an an *ordinary renewal process*. Let $S_r = Z_1 + Z_2 + \cdots + Z_r$ ($r = 1, 2, \ldots$); clearly S_r gives the time needed for r renewals to occur. Since $Z_1, Z_2, \ldots, Z_r$ are independent random variables and $Z_2, \ldots, Z_r$ are identically distributed, the distribution of S_r can be given as

$$P(S_r \leq x) = F_1(x)^*F_{r-1}(x) \tag{8.3.4}$$

Where * denotes convolution and $F_{r-1}(x)$ is the $(r - 1)$-fold convolution of $F(x)$ with itself, $F_r^{(0)} = 0$ for $r > 0$, and $F_r^{(0)} = 1$ for $r = 0$.

Define the Laplace-Stieltjes transforms

$$\phi(\theta) = \int_0^\infty e^{-\theta x} dF(x) \qquad \text{Re}(\theta) > 0 \tag{8.3.5}$$

$$\phi_1(\theta) = \int_0^\infty e^{-\theta x} dF_1(x) \qquad \text{Re}(\theta) > 0 \tag{8.3.6}$$

From (8.3.4) we then get

$$\int_0^\infty e^{-\theta x} d_x P(S_r \leq x) = \phi_1(\theta)[\phi(\theta)]^{r-1} \tag{8.3.7}$$

Suppose a renewal process is observed at time t. Three questions may be asked at that point: (1) What is the probability that E_* will occur during the infinitesimal interval $(t, t + \Delta t]$? Suppose t is not a renewal point; then (2) What is the distribution of the time that has elapsed since the last renewal has taken place? and (3) What is the distribution of the time until the next renewal point? These two time intervals are known as backward and forward recurrence times, respectively.

One of the most common renewal counting processes is the Poisson process. When the parameter of the process is λ, we have the following results from Section 7.2:

$$P[N(t) = n] = e^{-\lambda t}\frac{(\lambda t)^n}{n!} \qquad n = 0, 1, 2, \ldots \tag{8.3.8}$$

$$E[N(t)] = \lambda t \tag{8.3.9}$$

$$P[\text{A renewal occurs between } (t, t + \Delta t]\} = \lambda \Delta t + o(\Delta t) \tag{8.3.10}$$

$$F(x) = F_1(x) = 1 - e^{-\lambda x} \qquad x \geq 0 \tag{8.3.11}$$

$$P(S_r \leq x) = \int_0^x e^{-\lambda y}\frac{\lambda^r y^{r-1}}{(r-1)!} dy \tag{8.3.12}$$

Let $R(t)$ and $S(t)$ be the forward and backward recurrence times, respectively. Then

$$P[R(t) \le x] = 1 - e^{-\lambda x} \qquad (8.3.13)$$

$$P[S(t) = t] = e^{-\lambda t}$$

$$P[S(t) \le x] = 1 - e^{-\lambda x} \qquad 0 \le x < t \qquad (8.3.14)$$

The Renewal Function and the Renewal Density

In the general case, however, we may not be able to derive such explicit expressions. Nevertheless, major properties of the process can be identified through integral equations and limiting forms.

For a given value of t, $N(t)$ is a proper random variable, and its distribution is given by Theorem 8.3.1.

Theorem 8.3.1

$$P_n(t) = P[N(t) = n] = F_n(t) - F_{n+1}(t) \qquad n = 0, 1, 2, \ldots \qquad (8.3.15)$$

and

$$U(t) = E[N(t)] = \sum_{n=1}^{\infty} F_n(t) \qquad (8.3.16)$$

where we shall assume that $F_n(t) = F_1(t)^* F_{n-1}(t)$.

Proof   Consider the two events

$$\{N(t) \ge n\} \quad \text{and} \quad \{S_n \le t\}$$

Clearly one event implies the other, and hence they are equivalent. Equating their probabilities, we get

$$P[N(t) \ge n] = P(S_n \le t) = F_n(t) \qquad (8.3.17)$$

But

$$P[N(t) = n] = P[N(t) \ge n] - P[N(t) \ge n + 1]$$

which gives

$$P_n(t) = F_n(t) - F_{n+1}(t)$$

For its expected value we have

$$E[N(t)] = \sum_{n=1}^{\infty} n P_n(t)$$

$$= P_1(t) + 2P_2(t) + 3P_3(t) + 4P_4(t) + \cdots$$

$$= P_1(t) + P_2(t) + P_3(t) + P_4(t) + \cdots$$

$$+ P_2(t) + P_3(t) + P_4(t) + \cdots$$

$$+ P_3(t) + P_4(t) + \cdots \qquad (8.3.18)$$

$$+ P_4(t) + \cdots$$

$$+ \cdots$$

$$= P[N(t) \geq 1] + P[N(t) \geq 2] + \cdots$$

$$= \sum_{n=1}^{\infty} P[N(t) \geq n]$$

$$= \sum_{n=1}^{\infty} F_n(t) \qquad (8.3.19)$$

$\square$

In the literature on renewal processes the function $U(t)$, called the renewal function, is one of the major characteristics of the process. Its derivative is called the renewal density, and it gives the probability density of a renewal at time t. Let $u(t)$ be the probability density of a renewal at time t. We have

$$u(t) = \lim_{\Delta t \to 0} \frac{P\{E_* \text{ occurs in } (t, t + \Delta t]\}}{\Delta t}$$

$$= \sum_{r=1}^{\infty} \lim_{\Delta t \to 0} \frac{P\{E_* \text{ occurs in } (t, t + \Delta t] \text{ for the } r\text{th time}\}}{\Delta t}$$

$$= \sum_{r=1}^{\infty} \lim_{\Delta t \to 0} \frac{f_r(t)\Delta t + o(\Delta t)}{\Delta t}$$

$$= \sum_{r=1}^{\infty} f_r(t) = U'(t) \qquad (8.3.20)$$

where we have assumed that $F(x)$ is absolutely continuous and $F_r'(t) = f_r(t)$. For computational purposes it is more convenient to use Laplace transforms of these functions. Let

$$U^*(\theta) = \int_0^\infty e^{-\theta t} U(t) \, dt \qquad \text{Re}(\theta) > 0 \qquad (8.3.21)$$

$$u^*(\theta) = \int_0^\infty e^{-\theta t} u(t) \, dt \qquad \text{Re}(\theta) > 0 \qquad (8.3.22)$$

By the well-known property of transforms (see Appendix B), we have

$$u^*(\theta) = \theta U^*(\theta) \qquad (8.3.23)$$

and

$$\int_0^\infty e^{-\theta t} F_n(t) \, dt = \frac{[\phi(\theta)]^n}{\theta} \qquad (8.3.24)$$

Using these results and (8.3.7) in (8.3.16), we get

$$U^*(\theta) = \frac{1}{\theta} \sum_{n=1}^\infty \phi_1(\theta)[\phi(\theta)]^{n-1}$$

$$= \frac{\phi_1(\theta)}{\theta[1 - \phi(\theta)]} \qquad (8.3.25)$$

and

$$u^*(\theta) = \frac{\phi_1(\theta)}{1 - \phi(\theta)} \qquad (8.3.26)$$

Rearranging (8.3.25) and (8.3.26), we get

$$U^*(\theta) = \frac{1}{\theta}\phi_1(\theta) + U^*(\theta)\phi(\theta) \qquad (8.3.27)$$

$$u^*(\theta) = \phi_1(\theta) + u^*(\theta)\phi(\theta) \qquad (8.3.28)$$

By inversion, transform equations (8.3.27) and (8.3.28) yield integral equations for the renewal function and the renewal density:

$$U(t) = F_1(t) + \int_0^t U(t - \tau) \, dF(\tau) \qquad (8.3.29)$$

$$u(t) = f_1(t) + \int_0^t u(t - \tau) f(\tau) \, d\tau \qquad (8.3.30)$$

It can be shown that $U(t)$ and $u(t)$ are the unique and bounded solutions of these equations. Thus the Laplace transforms (8.3.25) and (8.3.26) can be used in determining the renewal function and the renewal density of the process.

The integral equation (8.3.29) is known as the renewal equation, which in its general form can be given as

$$Z(t) = h(t) + \int_0^t Z(t - \tau) \, dF(\tau) \qquad (8.3.31)$$

where $h(t)$ is a bounded function and $F(t)$ is a distribution function. (Compare with the discrete renewal equation of Theorem 8.2.2.) It can be shown that the renewal equation (8.3.31) has a unique bounded solution given by

$$Z(t) = h(t) + \int_0^t h(t - \tau) \, dU(\tau) \qquad (8.3.32)$$

Using transforms, this result can be established as follows. We shall denote the transforms of $Z(t)$ and $h(t)$ as $Z^*(\theta)$ and $h^*(\theta)$, respectively.

From (8.3.31) we have

$$Z^*(\theta) = h^*(\theta) + Z^*(\theta)\phi(\theta)$$

$$Z^*(\theta) = \frac{h^*(\theta)}{1 - \phi(\theta)} \qquad (8.3.33)$$

Using $\phi_1(\theta) = \phi(\theta)$ in (8.3.26), we also have

$$\int_0^\infty e^{-\theta t} \, dU(t) = u^*(\theta) = \frac{\phi(\theta)}{1 - \phi(\theta)} = \frac{1}{1 - \phi(\theta)} - 1 \quad (8.3.34)$$

Substituting from (8.3.34) in (8.3.33), we get

$$Z^*(\theta) = h^*(\theta)\left[1 + u^*(\theta)\right] \qquad (8.3.35)$$

which on inversion gives (8.3.32).

Limiting Behavior

The limiting behavior of the renewal density is given in Theorem 8.3.2. For the proof of this theorem the reader may refer to advanced texts. In Section 8.2 we have already seen its discrete analog.

Theorem 8.3.2 *For the renewal process in continuous time*

$$U(t + \Delta) - U(t) \to \frac{\Delta}{\mu} \qquad t \to \infty \qquad (8.3.36)$$

$$u(t) \to \frac{1}{\mu} \qquad t \to \infty \qquad (8.3.37)$$

The important result (8.3.37) follows directly from (8.3.36) by noting that

$$u(t) = \lim_{\Delta \to 0} \frac{U(t + \Delta) - U(t)}{\Delta} \qquad (8.3.38)$$

Renewal theory results can also be used as tools for the analysis of more general stochastic processes. Theorem 8.3.3 is basic to such applications.

Theorem 8.3.3 *Let $h(t)$ be a nonnegative function of $t \geq 0$, such that*

$$\int_0^\infty h(t)\, dt < \infty$$

Then

$$\int_0^t dU(\tau) h(t - \tau) \to \frac{1}{\mu} \int_0^\infty h(t)\, dt \qquad t \to \infty \qquad (8.3.39)$$

This is called the *key renewal theorem*; using this result in (8.3.32), we get

$$Z(t) \to \frac{1}{\mu} \int_0^\infty h(t)\, dt \qquad t \to \infty \qquad (8.3.40)$$

Another limiting result which may prove to be useful in practical problems concerns the random variable $N(t)$. When $E[Z_r] = \mu$ and $V[Z_r] = \sigma^2$ are finite, it can be shown that $N(t)$ has an asymptotically normal distribution with mean t/μ and variance $t\sigma^2/\mu^3$. Thus

$$\lim_{t \to \infty} P\left(\frac{N(t) - (t/\mu)}{\sqrt{t\sigma^2/\mu^3}} \leq x \right) = \frac{1}{\sqrt{2\pi}} \int_{-\infty}^x e^{-(1/2)y^2}\, dy \qquad (8.3.41)$$

Backward and Forward Recurrence Times

For time t let $S(t)$ be the time since the last epoch of occurrence of the renewal event E_*, and let $R(t)$ be the remaining time in that renewal period. These random variables are known as *backward recurrence time* (or *current life*) and *forward recurrence time* (or *excess life*), respectively. If E_* has not occurred during $(0, t]$, we have

$$P[S(t) = t] = 1 - F_1(t) \qquad (8.3.42)$$

Let $s_t(x)$ $(0 < x < t)$ be the probability density function of $S(t)$. If $x < t$, clearly the event E_* must have occurred at $t - x$, and the last renewal period should be such that it extends beyond t. These possibilities lead us to the relation

$$s_t(x) = u(t - x)[1 - F(x)] \qquad (8.3.43)$$

As $t \to \infty$, $u(t - x) \to 1/\mu$, and hence

$$\lim_{t \to \infty} s_t(x) = \frac{1 - F(x)}{\mu} \qquad (8.3.44)$$

Let $r_t(x)$ be the probability density function of the forward recurrence time $R(t)$. Let τ $(0 \le \tau < t)$ be the last renewal point. If $\tau = 0$, then $r_t(x) = f_1(t + x)$. Otherwise, the last renewal period is such that the probability density is given by $f(t - \tau + x)$. Using these arguments, we have

$$r_t(x) = f_1(t + x) + \int_{\tau=0}^{t} u(\tau)f(t - \tau + x) \, d\tau \qquad (8.3.45)$$

As $t \to \infty$, using Theorem 8.3.3, it can be shown that the right-hand side of (8.3.45) reduces to

$$\frac{1}{\mu} \int_0^{\infty} f(\tau + x) \, d\tau = \frac{1 - F(x)}{\mu} \qquad (8.3.46)$$

Thus we have

$$\lim_{t \to \infty} r_t(x) = \frac{1 - F(x)}{\mu} \qquad (8.3.47)$$

Equations (8.3.44) and (8.3.47) show that as $t \to \infty$, the backward and forward recurrence times have the same probability density.

Now consider the renewal period $L(t)$ containing the epoch t; that is, $L(t) = S(t) + R(t)$. By arguments similar to those employed earlier, we write

$$P[L(t) \le x] = \int_t^x dF(y) + \int_{\tau=t-x}^{t} u(\tau)[F(x) - F(t - \tau)] \, d\tau$$

$$(8.3.48)$$

The first term on the right-hand side of (8.3.48) requires $x > t$; that is, the term appears only if the renewal epoch t is in the initial renewal period and therefore drops out as $t \to \infty$. In the range $0 \le x \le t$,

$$P[L(t) \le x] = \int_{\tau = t - x}^{t} u(\tau)[F(x) - F(t - \tau)]\, d\tau \qquad (8.3.49)$$

Define a function $h(t)$ as

$$h(t) = \begin{cases} F(x) - F(t) & \text{if } 0 \le t \le x \\ 0 & \text{if } t > x \end{cases}$$

The function $h(t)$ is nonnegative and nonincreasing in t. Also

$$\int_0^\infty h(t)\, dt = \int_0^x [F(x) - F(t)]\, dt$$

Using $h(t)$, (8.3.49) can be written as

$$P[L(t) \le x] = \int_0^\infty u(\tau) h(t - \tau)\, d\tau$$

Applying theorem (8.3.3), as $t \to \infty$, we get

$$\lim_{t \to \infty} P[L(t) \le x] = \frac{1}{\mu} \int_0^\infty h(y)\, dy$$

$$= \frac{1}{\mu} \int_0^x [F(x) - F(y)]\, dy$$

$$= \frac{1}{\mu} \left[xF(x) - \int_{y=0}^x \int_{z=0}^y f(z)\, dz\, dy \right]$$

$$= \frac{1}{\mu} \left[xF(x) - \int_{z=0}^x f(z) \int_{y=z}^x dy\, dz \right]$$

$$= \frac{1}{\mu} \left[xF(x) - \int_{z=0}^x (x - z) f(z)\, dz \right]$$

$$= \frac{1}{\mu} zf(z)\, dz \qquad (8.3.50)$$

Thus the probability density $h(z)$ of the recurrence time L containing a renewal epoch in the limit is obtained as

$$h(z) = \frac{1}{\mu} z f(z)$$

We also get

$$E[L] = \frac{1}{\mu} \int_0^\infty z^2 f(z) \, dz \qquad (8.3.51)$$

If X and Y are two random variables, Schwarz's inequality gives

$$\{ E[XY] \}^2 \leq E[X^2] E[Y^2]$$

Setting $Y =$ a constant gives

$$\{ E[X] \}^2 \leq E[X^2] \qquad (8.3.52)$$

Applying (8.3.52) to (8.3.51), we obtain

$$E[L] = \frac{1}{\mu} \int_0^\infty z^2 f(z) \, dz \geq \frac{\mu^2}{\mu} = \mu \qquad (8.3.53)$$

showing that the mean length of a renewal period containing a renewal epoch is longer than the mean length of an arbitrary renewal period. In particular, when the renewal periods are exponential with mean μ,

$$f(x) = \frac{1}{\mu} e^{-x/\mu}$$

$$h(x) = \left(\frac{1}{\mu} \right)^2 x e^{-x/\mu}$$

$$E[L] = \frac{1}{\mu} \int_0^\infty x^2 \frac{1}{\mu} e^{-x/\mu} \, dx$$

$$= \frac{2\mu^2}{\mu} = 2\mu \qquad (8.3.54)$$

Implications of the result (8.3.53) are explained in Feller (1966, p. 185) as related to the so-called inspection paradox:

Consider items that are replaced as soon as they fail. The replacement process now forms a renewal process. In order to get information on the lifetimes of such items, suppose the items are sampled at some epoch $t > 0$ and their total lifetimes are observed. Does the life distribution obtained in this manner represent the life distribution of the items in the population? Clearly from what we have just seen the lifetimes of sampled items have a different distribution than those in the population. One way of avoiding this situation is to observe the total lifetime of the first item installed after the sampling epoch.

Another implication of property (8.3.53) is the need to write $E[S_{N(t)+1}] = E[Z_1][U(t) + 1]$ rather than $E[S_{N(t)}] = E[Z_1]U(t)$.

The Stationary Process

Suppose the renewal process has been in operation for a sufficiently long time. Then we can use the limiting form of the forward recurrence time as given by (8.3.47) in lieu of $F_1(x)$. We then have

$$\phi_1(\theta) = \int_0^\infty e^{-\theta x} \frac{[1 - F(x)]}{\mu} \, dx$$

$$= \frac{1 - \phi(\theta)}{\mu\theta} \qquad (8.3.55)$$

Substituting this result in (8.3.26), we get

$$u^*(\theta) = \frac{1}{\mu\theta}$$

which can be inverted easily to give

$$u(t) = \frac{1}{\mu} \qquad (8.3.56)$$

Thus when the initial renewal period is assumed to have the form given by (8.3.47), the renewal density is independent of the time parameter. Further we get

$$E[N(t)] = U(t) = \frac{t}{\mu} \qquad (8.3.57)$$

and therefore

$$E[N(s + t) - N(s)] = \frac{t}{\mu} \qquad (8.3.58)$$

It can be shown that the distribution of the number of renewals during

$(s, s + t]$, that is, $N(s + t) - N(s)$, is also independent of s, thus making it independent of the origin of time. Therefore the renewal process in this case is called *stationary*. It turns out that $f_1(x) = [1 - F(x)]/\mu$ is also the necessary condition for the stationarity of the renewal process.

8.4 SPECIAL TOPICS

In this section se shall introduce some variants of renewal processes that have found significant use in applications.

Alternating Renewal Processes

In Example 3.2.1 we described the state of a computer going through two periods alternately: work and repair. In our discussion at that time we assumed that those periods could be modeled as having exponential distributions. Suppose the distributions needed for the model are not exponential. Then the resulting process is not Markovian. If we retain the assumption of independence between neighboring periods, we can use a renewal process model to analyze the system. Because of the alternating nature of the renewal periods, we call the resulting process an alternating renewal process.

In an alternating renewal process let the alternating periods be denoted by X's and Y's such that $(X_1, Y_1, X_2, Y_2, \ldots)$ be the sequence of renewal periods. Clearly, if we consider $Z_i = X_i + Y_i$, the sequence of independent and identically distributed random variables $(Z_1, Z_2, \ldots)$ form a regular renewal process which alternates between an X-state and a Y-state within a renewal period.

Let $\phi_1(\theta)$ and $\phi_2(\theta)$ be the Laplace transforms of the probability distribution of $\{X_i\}$ and $\{Y_i\}$, respectively. Then the Laplace transform of the probability distribution of $\{Z_i\}$ is given by their product $\phi_1(\theta)\phi_2(\theta)$. Let $U_1^*(\theta)$ be the renewal function for the renewal process $\{Z_i\}$ when the renewal is marked by a transition from a Y-state to an X-state. Similarly let $U_2^*(\theta)$ be the renewal function for the renewal process when the renewal is marked by a transition from an X-state to a Y-state. Using arguments similar to those leading to (8.3.25), we have

$$U_1^*(\theta) = \frac{\phi_1(\theta)\phi_2(\theta)}{\theta[1 - \phi_1(\theta)\phi_2(\theta)]} \tag{8.4.1}$$

$$U_2^*(\theta) = \frac{\phi_1(\theta)}{\theta[1 - \phi_1(\theta)\phi_2(\theta)]} \tag{8.4.2}$$

A significant result for use in practice is the probability of occurrence of any state in the long run. (In the computer problem of Example 3.2.1 this is the probability of finding the computer working or under repair in the long run.) Let $E[X_i] = \mu_1$ and $E[Y_i] = \mu_2$ and $\mu_1 + \mu_2 = \mu$. Intuitively we should expect:

$$p_X = P[\text{Process is in } X\text{-state as } t \rightarrow \infty]$$

$$= \frac{\mu_1}{\mu_1 + \mu_2} \tag{8.4.3a}$$

$$p_Y = P[\text{Process is in } Y\text{-state as } t \rightarrow \infty]$$

$$= \frac{\mu_2}{\mu_1 + \mu_2} \tag{8.4.3b}$$

The validity of these results is easily established using the renewal equation as follows (Karlin and Taylor, 1975, p. 201).

Let $p_X(t)$ be the probability that the process is found in X-state at time t. We have

$$p_X(t) = \int_{z=0}^{t} p_X(t|Z_1 = z) \, dF(z)$$

$$p_X(t|Z_1 = z) = \begin{cases} 1 & t < X_1 \\ 0 & X_1 < t \le z \\ p_X(t-z) & t > z \end{cases} \tag{8.4.4}$$

Let

$$I_{X_1}(t) = \begin{cases} 1 & \text{if } X_1 \text{ covers } t \\ 0 & \text{if } X_1 \text{ does not cover } t \end{cases}$$

Hence

$$P[t \text{ is covered by } X_1] = E\left[I_{X_1}(t)\right] \tag{8.4.5}$$

For $p_X(t)$ we have

$$p_X(t) = P[t \text{ is covered by } X_1] + \int_{z=0}^{t} p_X(t-z) \, dF(z) \tag{8.4.6}$$

which is a renewal equation of the form (8.3.31) with a solution

$$p_X(t) = E\left[I_{X_1}(t)\right] + \int_{\tau=0}^{t} E[I_X(t-\tau)] \, dU(\tau) \tag{8.4.7}$$

Using Theorem 8.3.3, we get

$$p_X = \lim_{t \to \infty} p_X(t) = \frac{1}{\mu} \int_0^\infty E\left[I_{X_1}(t) \right] dt$$

$$= \frac{1}{\mu} E\left[\int_0^\infty I_{X_1}(t)\, dt \right]$$

$$= \frac{1}{\mu} E[X_1] = \frac{\mu_1}{\mu} \qquad\qquad (8.4.8)$$

Example 8.4.1

In a queueing system busy periods during which service is rendered alternates with idle periods during which the server is idle. A common practice is to start serving as soon as a customer arrives at an empty system and not to stop serving as long as there are customers in the system. With this structure, time periods between successive epochs when busy periods start are independent and identically distributed random variables, thus defining a renewal process. If in addition we assume that there is only one server and the arrivals are in a Poisson process, the idle period has an exponential distribution which is independent of the preceding busy period. Now we have an alternating renewal process with the busy and idle periods as the component renewal periods. Let $E[B]$ and $E[I]$ be the corresponding mean values, respectively. As $t \to \infty$, the probability that the system will be found busy or idle is given by,

$$P[\text{System busy as } t \to \infty] = \frac{E[B]}{E[B] + E[I]}$$

$$P[\text{System idle as } t \to \infty] = \frac{E[I]}{E[B] + E[I]}$$

which can be simplified further by writing $E[I] = 1/\lambda$.

Example 8.4.2

Counter models are classical examples for the application of renewal theory. A counter is a device used to detect the registration of impulses due to radioactive substances. The problem with this device is that once it registers an impulse, it is unable to register any other impulse for a length of time called the resolving time, dead time, or locked time. Generally, two types of counters are considered. The type I counter gets locked only with a registration. But in a type II counter, the dead time gets extended with every impulse hitting the counter. We shall explain this in terms of notations.

Let the arrival of impulses occur in a Poisson process. Let t_1 ($= T_1$) be the arrival epoch for the first impulse. Also let Y_1 be the dead time following the registration of impulse at time t_1. A type I counter is ready to register another signal after a time $t_1 + Y_1$. Let the second registration be at T_2. Then $T_2 - (t_1 + Y_1)$ is the forward recurrence time (or excess life) of the arrival process. Because of the memoryless property of the Poisson process this forward recurrence time also has the same exponential distribution as the distribution of interarrival times of impulses in the original process. Thus the impulse registration process occurring at epochs $(T_1, T_2, \ldots)$ is a renewal process, and the process of epochs $(T_1, T_1 + Y_1, T_2, T_2 + Y_2, T_3, T_3 + Y_3, \ldots)$ is an alternating renewal process. Clearly the probability that the counter will be found locked as $t \to \infty$ is given by $E(Y)/[E(Y) + E(X)]$, where we have used Y and X to represent dead time and the impulse interarrival time in the limit.

In a type II counter every arriving impulse locks the counter for a certain length of time, say Y, whether registered or not. Thus an impulse arriving at t at a locked counter extends the dead time until $t + Y$; similarly impulses arriving after t lock the counter for a length of time Y from the epoch of their arrival. The impulse registration epochs $(T_1, T_2, \ldots)$ again constitute a renewal process with dead times (possibly extended in this case) and the forward recurrence time of a Poisson process as component alternating renewal periods. In this case the derivation of dead time distribution is quite complicated, and we shall not go into it here. For a detailed discussion of counter models readers are referred to Bharucha-Reid (1960).

Cyclically Alternating Renewal Process

The alternating renewal process with two components in a renewal period can be generalized to include k components. A real-world example is a service that is provided in k stages (or phases) where the amount of time required for each stage of service has a specific distribution. For this process the preceding results can be easily generalized.

Consider random variables $X_1, X_2, \ldots$, such that X_j ($j = i, k + i, 2k + i, \ldots$) are independent and identically distributed random variables with distribution function $F_i(x)$ $i = 1, 2, \ldots, k$. Let

$$Z_1 = X_1 + X_2 + \cdots + X_k$$

$$Z_j = X_{(j-1)k+1} + X_{(j-1)k+2} + \cdots + X_{jk} \quad j = 2, 3, \ldots$$

Then $(X_1, X_2, \ldots, X_k, X_{k+1}, \ldots)$ is a cyclically alternating renewal process with $Z_1, Z_2, \ldots$ as renewal periods. Identifying state i corresponding to X_j ($j = i, k + i, 2k + i, \ldots$), we may consider the generalized renewal process as occupying states $(1, 2, \ldots, k, 1, 2, \ldots, k, \ldots)$ in a cyclical manner. Let $p_i(t)$ be the probability that the process is in state i at time t. Also let $E[X_j] = \mu_i < \infty$ ($j = i, k + i, 2k + i, \ldots$). Generalizing the results (8.4.3), we then get

$$\lim_{t \to \infty} p_i(t) = \frac{\mu_i}{\mu_1 + \mu_2 + \cdots + \mu_k} \qquad (8.4.9)$$

Markov Renewal Processes (Semi-Markov Processes)

A cyclically alternating renewal process passes through states $(1, 2, \ldots, k, 1, 2, \ldots)$ in a cyclic fashion. Consider a modification to this scheme by which the process passes through different states as determined by the Markovian transitions of a k-state transition probability matrix. The resulting process is called a Markov renewal process or a semi-Markov process. The underlying Markov chain may be considered to be a discrete time process embedded in a continuous time process.

Let $\{X_n, n = 0, 1, 2, \ldots\}$ be a stochastic process assuming values in the countable set $S = \{0, 1, 2, \ldots\}$. Let $T_0, T_1, T_2, \ldots$ be the transition epochs on the nonnegative half of the real line, such that $0 = T_0 \leq T_1 \leq T_2 \ldots$. The two-dimensional process $(X, T) = \{X_n, T_n; n = 0, 1, 2, \ldots\}$ is called a *Markov renewal process* (MRP) if it has the property

$$P[X_{n+1} = j, T_{n+1} - T_n \leq t | X_0, X_1, \ldots, X_n; T_0, \ldots, T_n]$$

$$= P[X_{n+1} = j, T_{n+1} - T_n \leq t | X_n] \qquad j \in S, t \geq 0 \quad (8.4.10)$$

We assume that it is time-homogeneous. Define the probability

$$Q_{ij}(t) = P[X_{n+1} = j, T_{n+1} - T_n \leq t | X_n = i] \qquad (8.4.11)$$

Then $Q = \{Q_{ij}(t), i, j \in S, t \geq 0\}$ is called the *semi-Markov kernel* for the process. Let

$$P_{ij} = \lim_{t \to \infty} Q_{ij}(t) \quad \text{and} \quad F_{ij}(t) = \frac{Q_{ij}(t)}{P_{ij}}$$

Now P_{ij} is the transition probability of the embedded Markov chain $\{X_n, n = 0, 1, 2, \ldots\}$ and hence $\sum_{j \in S} P_{ij} = 1$ and $F_{ij}(t)$ (which is defined as 1 when $P_{ij} = 0$, since $Q_{ij}(t) = 0$ for all t then) is a cumulative distribution function of the sojourn time of the process in state i defined as

$$F_{ij}(t) = P[T_{n+1} - T_n \le t | X_n = i, X_{n+1} = j] \qquad (8.4.12)$$

Let $N_j(t)$ be the number of visits of the process to state j during $(0, t]$, and define $N(t) = \sum_{j \in S} N_j(t)$. Let $Y(t) = Y_{N(t)}$ denote the state of the Markov renewal process (X, T) as defined in (8.4.10). The stochastic process $\{Y(t), t \ge 0\}$ is called a *semi-Markov process* (SMP). The vector process $N(t) = \{N_0(t), N_1(t), \dots\}$ may be identified as the *Markov renewal counting process*.

Thus a Markov renewal process represents the transition epoch and the state of the process at that epoch. A semi-Markov process represents the state of the Markov renewal process at an arbitrary time point, and the Markov renewal counting process records the number of times the process has visited each of the states up to time t. A Markov renewal process becomes a Markov process if the sojourn times are exponentially distributed independent of the next state; it becomes a Markov chain if the sojourn times are all equal to one; and it becomes a renewal process if there is only one state.

The sojourn time $T_{n+1} - T_n$ ($n = 0, 1, 2, \dots$) is dependent only on X_n, and not any of the X_i's or the time periods $T_{i+1} - T_i$ ($i = 0, 1, 2, \dots, n - 1$). By itself $\{X_n\}$ is Markovian and hence the time epochs $T_n^{(j)}$'s at which the process enters state j successively form a renewal process (if $X_0 = j$, it is an ordinary renewal process). Clearly $N_j(t)$ defined earlier is the corresponding renewal counting process. Let

$$R_{ij}(t) = E[N_j(t) | X_0 = i] \qquad (8.4.13)$$

Extending renewal theory terminology, we may call $R_{ij}(t)$ the *Markov renewal function*.

Let $G_{ij}(t)$ be the distribution of the first passage time for the transition from state i to state j. Now $G_{jj}(t)$ is the renewal period distribution for state j. From (8.3.16) we have

$$R_{jj}(t) = \sum_{n=0}^{\infty} G_{jj}^{(n)}(t) \qquad (8.4.14)$$

where $G_{jj}^{(n)}(t)$ is the n-fold convolution of $G_{jj}(t)$ with itself. Extending this result and conditioning on the first entrance time to state j,

$$R_{ij}(t) = \int_0^t dG_{ij}(s) R_{jj}(t - s) \qquad i \ne j \qquad (8.4.15)$$

Other results from renewal theory follow in a similar manner.

To capitalize on the Markovian properties of the process, define

$$Q_{ij}^{(n)}(t) = P[X_n = j, T_n \le t | X_0 = i] \qquad (8.4.16)$$

which is the extension of the n-step transition probability to the MRP. We have

$$R_{ij}(t) = \sum_{n=0}^{\infty} Q_{ij}^{(n)}(t) \qquad (8.4.17)$$

where

$$Q_{ij}^{(0)}(t) = \begin{cases} 1 & \text{if } i = j \\ 0 & \text{if } i \neq j \end{cases}$$

and

$$Q_{ij}^{(n+1)}(t) = \sum_{k \in S} \int_0^t d Q_{ik}(s) Q_{kj}^{(n)}(t - s) \qquad n \ge 1 \qquad (8.4.18)$$

Define Laplace-Stieltjes transforms

$$\phi_{ij}^{(n)}(\theta) = \int_0^{\infty} e^{-\theta t} d Q_{ij}^{(n)}(t) \qquad \text{Re}(\theta) > 0, \, n = 0, 1, 2, \ldots$$

and write $\phi_{ij}^{(1)}(\theta) = \phi_{ij}(\theta)$ and $\boldsymbol{\phi}^{(n)}(\theta)$ as the matrix of transforms $\phi_{ij}^{(n)}(\theta)$. Also let $\rho_{ij}(\theta)$ be the Laplace-Stieltjes transform of $R_{ij}(t)$ and $\rho(\theta)$ be the matrix of these transforms. Then we have the following results:

$$\boldsymbol{\phi}^{(n)}(\theta) = [\boldsymbol{\phi}(\theta)]^n$$

$$\rho(\theta) = I + \boldsymbol{\phi}(\theta) + \boldsymbol{\phi}^2(\theta) + \cdots \qquad (8.4.19)$$

and hence $\rho(\theta)$ is given by

$$\rho(\theta) = [I - \boldsymbol{\phi}(\theta)]^{-1} \qquad (8.4.20)$$

provided S is finite.

As can be seen from the preceding results, most of the important properties of renewal processes and Markov chains carry through to Markov renewal processes with necessary modifications. A few key results are listed as follows:

1. The classification of states is based on the classification of states in the embedded Markov chain or the corresponding renewal processes.

2. The well-known renewal equation of renewal theory extends to MRP, and corresponding properties hold with necessary modifications.

3. If state j is transient and $i \neq j$, then

$$\lim_{t \to \infty} R_{ij}(t) = G_{ij}^* R_{jj}^* \qquad (8.4.21)$$

where we have written $G_{ij}(\infty) = G_{ij}^*$ and $R_{ij}(\infty) = R_{ij}^*$.

If state j is recurrent and aperiodic, then

$$\lim_{t \to \infty} \left[R_{ij}(t) - R_{ij}(t - \Delta) \right] = \frac{\Delta}{\mu_j} G_{ij}^* \qquad (8.4.22)$$

where μ_j is the mean renewal period of state j [recall (8.3.36)].

4. Suppose the embedded Markov chain is irreducible and recurrent. Let $\pi = (\pi_0, \pi_1, \dots)$ be a solution of the equations

$$\sum_{i \in S} \pi_i P_{ij} = \pi_j \qquad j \in S$$

and let m_j be the mean sojourn time for state j. Then

$$\frac{1}{\mu_j} = \frac{\pi_j}{\displaystyle\sum_{k \in S} \pi_k m_k} \qquad j \in S \qquad (8.4.23)$$

5. Suppose the embedded Markov chain is irreducible and recurrent, and π is defined as in the preceding result. Then for a certain class of functions h (directly Riemann integrable functions when the state space is finite; see, Çinlar, 1975, for conditions in the general case)

$$\lim_{t \to \infty} \sum_{j \in S} \int_0^t dR_{ij}(s) h_j(t - s) = \frac{1}{\displaystyle\sum_{k \in S} \pi_k m_k} \int_0^\infty \sum_{r \in S} \pi_r h_r(s) \, ds$$

$$(8.4.24)$$

[recall (8.3.40)].

Consider the semi-Markov process $Y(t)$ defined earlier. Let the transition probability be defined as

$$P_{ij}(t) = P\left[Y(t) = j | X_0 = i \right]$$

Using renewal theory concepts and conditioning on the last transition epoch, we get

$$P_{ij}(t) = \int_0^t dR_{ij}(s) H_j(t-s)$$

where

$$H_j(t) = 1 - \sum_{k \in S} Q_{jk}(t) \qquad j \in S, \, t \geq 0$$

If the embedded Markov chain $\{X_n\}$ is irreducible, aperiodic, and recurrent, using (8.4.24), it can be shown that

$$\lim_{t \to \infty} P_{ij}(t) = \frac{\pi_j m_j}{\sum_{k \in S} \pi_k m_k} \qquad j \in S \qquad (8.4.25)$$

Let $M_{ij}(t)$ be the expected occupation time of the process in state j during $(0, t]$ having initially started at i. Then we get [using indicator function arguments; e.g., see (7.5.8)]

$$M_{ij}(t) = \int_0^t P_{ij}(u) \, du$$

$$= \int_0^t dR_{ij}(s) \int_0^{t-s} H_j(u) \, du \qquad (8.4.26)$$

As $t \to \infty$, writing $M_{ij}^* = M_{ij}(\infty)$, we get

$$M_{ij}^* = R_{ij}^* m_j \qquad (8.4.27)$$

showing that $M_{ij}^* = \infty$ if j is recurrent and $G_{ij}^* > 0$ or j is transient, $G_{ij}^* > 0$, and $m_j = \infty$. When $G_{ij}^* = 0$, $R_{ij}^* = 0$ from (8.4.21). The following ratio result is also worth noting:

$$\lim_{t \to \infty} \frac{1}{t} M_{ij}(t) = \frac{\pi_j m_j}{\sum_{k \in S} \pi_k m_k} \qquad (8.4.28)$$

which is a direct analog of the result for Markov processes [recall (6.2.4)].

For a detailed discussion of properties of semi-Markov and Markov renewal processes, the readers are referred to Çinlar (1975) and Medhi (1982). For semi-Markov process analogs of Theorems 4.1.2–4.1.5, reference can be made to Kao (1974).

Cumulative Processes (Renewal Reward Processes)

In applied problems often the process of interest is not the renewal process itself but an associated process that is triggered by renewal events. For instance, in dam theory, it is not enough to model the input of water through rains as a renewal process; we need information on the amount of water being added to the dam at every such epoch. In a service system, with every completed service one can associate a profit (or cost), and this secondary process is crucial in the operation of the system. To allow for these situations, we may define a cumulative process (also known as a renewal reward process) as follows.

Let $(Z_1, Z_2, \ldots)$ define the renewal process as before. Let $t_0, t_1, t_2, \ldots$ be the renewal epochs and $N(t)$ be the renewal counting process. Associate a process composed of independent and identically distributed random variables Y_i $(i = 1, 2, 3, \ldots)$ at renewal epochs t_i $(i = 1, 2, 3, \ldots)$, respectively. Let

$$W(t) = \sum_{i=1}^{N(t)+1} Y_i \tag{8.4.29}$$

For convenience write $\lim_{i \to \infty} Y_i = Y$. Let $\omega(t) = E[W(t)]$. For $\omega(t)$ we may write down the renewal equation

$$\omega(t) = E[Y_1] + \int_0^t \omega(t - \tau)\, dF(\tau) \tag{8.4.30}$$

Based on (8.3.32), the solution to this equation can be written down as

$$\omega(t) = E[Y_1] + \int_0^t E[Y_1]\, dU(\tau)$$

giving

$$\omega(t) = E[Y_1][1 + U(t)] \tag{8.4.31}$$

As $t \to \infty$, when we apply (8.3.37), equation (8.4.31) gives

$$\lim_{t \to \infty} \frac{\omega(t)}{t} = \frac{E[Y]}{\mu} \tag{8.4.32}$$

which is clearly an intuitive result. For the two examples cited earlier (8.4.32) provides the expression for the average rainfall and the average profit (or cost), respectively.

Superposed Processes

One of the common assumptions made in applied stochastic processes is that a sequence of events such as customer arrivals to a service facility and occurrence of demands for some manufactured item is a Poisson process. This assumption in most cases is justified based on empirical data. The major question is whether this assumption is justifiable based on theoretical considerations.

As shown in Chapter 7, if we superimpose two or more Poisson processes, the resulting superposed process is also Poisson. It should be noted here that this property, taken as an exact statement, is unique to Poisson processes. In the case of renewal processes (which belong to the general class of point processes and of which the Poisson process is a special case) we state the following properties without elaboration. In these statements we describe the superposed process as that resulting from combining the renewal epochs of several renewal processes:

1. In general, the superposed process is not a renewal process. The interoccurrence times of the combined renewal epochs are not independent and identically distributed.
2. If the renewal epochs of each individual renewal process are extremely rare (the corresponding renewal processes may be called infinitesimal), then the superposed process is Poisson.
3. The additive property of the Poisson process is unique among renewal processes.

For a discussion of these properties, the readers are referred to Karlin and Taylor (1975, p. 221).

8.5 RENEWAL THEORY AS AN ANALYSIS TECHNIQUE

Stochastic processes encountered in mathematical models present different degrees of difficulty. So far we have been able to present some methods of analysis if they were well-defined simple Markov or renewal processes. But if the processes we encounter do not belong to any of these categories, we should either look for more powerful methods or try to use the available techniques in modified forms. For instance, a process can be made Markovian by defining a sufficient number of supplementary variables; as a consequence the analysis may become unmanageable. Another method would be to identify a certain event E_* that specifies the state of the process when it occurs and to consider the entire process as a renewal process with regard to E_*. Then the study of the general stochastic process can be broken down into the study of the renewal process related to E_* and the transitions of the stochastic process within a renewal period. The limiting behavior of the general process follows from Theorem 8.3.3. As an illustration of this method, consider the following example.

Customers arrive in a queuing system in a renewal process and get served by a single server. The service times of customers are independent random variables identically distributed as an exponential distribution. Let $Q(t)$ be the number of customers in the system at time t. Under these conditions the stochastic process $Q(t)$ is non-Markovian, and therefore methods given in Chapter 7 cannot be used in its analysis. The process $Q(t)$ alternates between periods during which $Q(t) > 0$ and $Q(t) = 0$. These periods are called "busy" and "idle," respectively. The time interval from the epoch of commencement of one busy period to the epoch of commencement of the following busy period is known as the "busy cycle." It can be shown that the epochs of commencement of busy periods are renewal points for the process, with busy cycles as renewal periods. Deriving the distribution of the busy cycle and the transition probabilities of the $Q(t)$ process during a busy cycle is much simpler than the analysis of the general process $Q(t)$. Let

$$_0P_{ij}(t) = P[Q(t) = j, Q(\tau) > 0 \text{ for all } \tau \in (0, t] | Q(0) = i] \quad (8.5.1)$$

and

$$P_{ij}(t) = P[Q(t) = j | Q(0) = i] \quad (8.5.2)$$

Further let $F(t)$ be the distribution function of the busy cycle, and let $dU(t)$ be the corresponding renewal density. The transition probability $P_{ij}(t)$ can then be derived as follows: If

$$Q(\tau) > 0 \quad 0 < \tau \le t$$

then $P_{ij}(t)$ is given by $_0P_{ij}(t)$ of (8.5.1). Otherwise, let τ $(0 < \tau \le t)$ be the last epoch of commencement of a busy period. The probability density of this occurrence between $(\tau, \tau + d\tau]$ is $dU(\tau)$, and the transition probability of $Q(t)$ between τ and t is given by $_0P_{0j}(t - \tau)$. Combining these probabilities, we get

$$P_{ij}(t) = {}_0P_{ij}(t) + \int_0^t dU(\tau) {}_0P_{0j}(t - \tau) \quad j > 0 \quad (8.5.3)$$

As $t \to \infty$, it can be shown that when the stability conditions hold, $\lim_{t \to \infty} P_{ij}(t) \to 0$, and from Theorem 8.3.3

$$\lim_{t \to \infty} P_{ij}(t) = \frac{1}{\mu} \int_0^{\infty} {}_0P_{0j}(t)\, dt \qquad (8.5.4)$$

where μ is the expected length of the busy cycle.

In giving this example, the intention has been the illustration of the general technique of renewal theory in the study of general stochastic processes. To work out the details, sophisticated mathematical techniques are needed.

REFERENCES

Bharucha-Reid, A. T. (1960). *Elements of the Theory of Markov Processes and Their Applications*. New York: McGraw-Hill.

Çinlar, E. (1975). *Introduction to Stochastic Processes*. Englewood Cliffs, N.J.: Prentice-Hall, Chaps. 9, 10.

Cox, D. R. (1962). *Renewal Theory*. London: Methuen.

Feller, W. (1968). *An Introduction to Probability Theory and Its Applications*. Vol. 1. 3rd ed. New York: Wiley, Chap. 13.

Kao, E. P. C. (1974). "A Note on the First Two Moments of Times in Transient States in a Semi-Markov Process." *J. Appl. Prob.* **11**, 193–198.

Karlin, S., and Taylor, H. M. (1975). *A First Course in Stochastic Processes*. 2nd ed. New York: Academic Press, Chap. 5.

Medhi, J. (1982). *Stochastic Processes*. New Delhi: Wiley Eastern, Chaps. 6, 7.

FOR FURTHER READING

Cox, D. R., and Miller, A. D. (1965). *The Theory of Stochastic Processes*. New York: Wiley, Chap. 9.

Feller, W. (1966). *An Introduction to Probability Theory and Its Applications*. Vol. 2. New York: Wiley, Chap. 11.

Parzen, E. (1962). *Stochastic Processes*. San Francisco: Holden-Day, Chap. 5.

Prabhu, N. U. (1965). *Stochastic Processes*. New York: MacMillan, Chaps. 5, 6.

EXERCISES

1. In a discrete time renewal process let $\{f^{(n)}\}$ be the probability distribution of the renewal period. For

$$f^{(n)} = p(1 - p)^{n-1} \qquad n = 1, 2, \ldots, \quad 0 < p < 1$$

obtain $F(z)$ and $P(z)$ of (8.2.12) and (8.2.20). Hence show that the probability that the renewal occurs at epoch n is p.

2. Determine the probability that a renewal occurs at epoch n when the probability distribution of the renewal period is Poisson.

3. Following the method described in Example 8.2.2, determine the long term age distribution of 500 objects that need replacement when they fail according to the following life distribution $f^{(n)}$:

   a. $f^{(n)} = 0.2$ for $n = 1, 2, 3, 4, 5$.
   b. $f^{(n)}$ is truncated Poisson with $\lambda = 3$, $n = 1, 2, 3, \ldots$.
   c. $f^{(n)} = (0.3)(0.7)^{n-1}$, $n = 1, 2, \ldots$.

4. Compare the age distributions obtained in Exercise 3 for $n \to \infty$ with the results one would get using a Markov chain approach as in Exercise 4, Chapter 6.

   Hint Note that for an $i \to j$ transition in a Markov chain, we assume that the process is in state i.

5. Determine $P[N(t) = n]$, $U(t)$, and $u(t)$ when the distribution function $F(x)$ of the renewal period is given by the following:

   a. $F(x) = 1 - e^{-\lambda x} \qquad x \geq 0$
   b. $F(x) = \begin{cases} 0 & x < b \\ 1 & x \geq b \end{cases}$

6. Determine the Laplace transform $u^*(\theta)$ for the renewal density for the following renewal period distributions when the process is (a) ordinary renewal process and (b) stationary renewal process:

$$f(x) = \alpha\lambda_1 e^{-\lambda_1 x} + (1 - \alpha)\lambda_2 e^{-\lambda_2 x} \qquad x > 0$$

$$f(x) = \lambda^2 x e^{-\lambda x} \qquad x > 0$$

7. The life distribution of a machine component has a probability density

$$f(x) = \frac{x}{400} e^{-x/20} \qquad x > 0$$

   Suppose the process starts with a new component and when a

component fails, it is replaced with a new one. Show that the expected number of replacements in time t is given by

$$\frac{t}{40} - \frac{1}{4} + \frac{1}{4}e^{-t/10}$$

Hint Invert the renewal function for the process.

8. A machine component is found to have a life distribution whose probability density is given by

$$f(x) = e^{-\lambda x}\frac{\lambda^k x^{k-1}}{(k-1)!} \qquad k \geq 1; \ x > 0$$

Consider the replacement of this component as a renewal process, and show that the residual life distribution of the component as $t \to \infty$ has the probability density

$$\frac{\lambda e^{-\lambda x}}{k}\sum_{r=0}^{k-1}\frac{(\lambda x)^r}{r!}$$

9. The central processing unit of a computing laboratory has enough jobs at hand to keep it busy all the time. The processing time on each job is found to have an exponential distribution with mean 100 msec, and between any two jobs a time interval of length 10 msec is taken up in swapping jobs. Let $N(t)$ be the number of jobs completed in time t. Show that

$$P[N(t) < n] = \sum_{k=0}^{n-1}e^{-(t-10n)/100}\left(\frac{t-10n}{100}\right)^k \cdot \frac{1}{k!}$$

Hint Invert $[\phi(\theta)]^n$, and use (7.2.34).

10. Establish the "memoryless" property of the exponential distribution by considering the forward recurrence time of a renewal process in which renewal periods have an exponential distribution.

11. Obtain discrete time analogs of results (8.3.42)–(8.3.47) for the forward and backward recurrence times. Hence establish the "memoryless" property of the geometric distribution. (Define this property similar to the exponential case.)

12. In a renewal process the renewal periods have mean μ and standard deviation σ. For $\theta \to 0$, show that the Laplace transform $\phi(\theta)$ of the

renewal period distribution can be expressed as

$$\phi(\theta) = 1 - \theta\mu + \tfrac{1}{2}\theta^2(\mu^2 + \sigma^2) + o(\theta^2)$$

where $o(x)/x \to 0$ as $x \to 0$. Determine the Laplace transform $U^*(\theta)$ of the renewal function as

$$U^*(\theta) = \frac{1}{\theta^2\mu} + \frac{1}{\theta}\frac{\sigma^2 - \mu^2}{2\mu^2} + o\left(\frac{1}{s}\right)$$

Hence show that for large t we have

$$U(t) = \frac{t}{\mu} + \frac{\sigma^2 - \mu^2}{2\mu^2} + o(1)$$

where $o(1)$ is a function of t which $\to 0$ as $t \to \infty$.

Hint Relate the Taylor Series expansion of $\phi(\theta)$ with its moment-generating properties.

13. Discuss and interpret the following three cases with reference to the number of renewals using the expression for $U(t)$ in Exercise 12:

   a. $\sigma = \mu$.

   b. $\sigma < \mu$.

   c. $\sigma > \mu$.

(See Cox, 1962, p. 47.)

14. In a continuous time renewal process let $N(t)$ be the number of renewals during $(0, t]$. Let $C(t) = E[N(t)(N(t) + 1)]$ and $C_o(t)$ and $C_s(t)$ be this expectation under the assumptions that the process is an ordinary renewal process and a stationary renewal process, respectively. Let $C_o^*(\theta)$ and $C_s^*(\theta)$ be the corresponding Laplace transforms. Establish the following results:

   a. $V[N(t)] = C(t) - U(t) - [U(t)]^2$

   b. $C^*(\theta) = \dfrac{2}{\theta}\dfrac{\phi_1(\theta)}{[1 - \phi(\theta)]^2}$

   c. $C_o^*(\theta) = \dfrac{2}{\theta}\dfrac{\phi(\theta)}{[1 - \phi(\theta)]^2}$

   d. $C_s^*(\theta) = \dfrac{2}{\mu\theta^2[1 - \phi(\theta)]}$

   $\qquad = \dfrac{2}{\mu\theta}\left\{U_o^*(\theta) + \dfrac{1}{\theta}\right\}$

e. $C_s(t) = \dfrac{2}{\mu} \displaystyle\int_0^t U_o(x)\, dx + \dfrac{2t}{\mu}$

f. $V[N_s(t)] = \dfrac{2}{\mu} \displaystyle\int_0^t \left[U_o(x) - \dfrac{x}{\mu} + \dfrac{1}{2} \right] dx$

where we have written $N_s(t)$ to indicate that it is a stationary renewal process and $U_o(x)$ and $U_o^*(\theta)$ for the renewal function and its transform in an ordinary renewal process (see Cox, 1962, pp. 55–58.)

15. Let μ_k' be the kth central moment of the renewal period. Let γ_k and ν_k be the kth central moments of the limiting distributions of the forward recurrence time and the recurrence time containing a renewal epoch. Show that

$$\gamma_k = \frac{\mu_{k+1}'}{(k+1)\mu}$$

and

$$\nu_k = \frac{1}{\mu}\mu_{k+1}'$$

16. A counter is used to detect and register instantaneous pulse-type signals. After a signal is registered, the counter needs some readjustment time which is a lost time in the registration process, and we shall identify this period as dead time. The counter is ready to register signals again at the end of the dead time. Clearly signals arriving during dead times will not be registered by the counter. Suppose signals arrive in a Poisson process with rate λ. Assume that dead times are independent and identically distributed random variables with an exponential distribution with mean $1/\mu$. Assuming that the counter has just registered a signal at $t = 0$, determine the expected number of signals registered during $(0, t]$.

17. An alternate method of establishing results (8.4.3a) and (8.4.3b) for an alternating renewal process starts with the integral equation

$$p_x(t) = \left[1 - F_x(t) \right] + \int_0^t u_2(v)\left[1 - F_x(t - v) \right] dv$$

Justify this equation. Taking Laplace transforms and using the obvious notations, show that

$$p_x^*(\theta) = \frac{1 - \phi_1(\theta)}{\theta\left[1 - \phi_1(\theta)\phi_2(\theta) \right]}$$

$$= U_2^*(\theta) - U_1^*(\theta) + \frac{1}{\theta}$$

and hence that

$$p_x(t) = U_2(t) - U_1(t) + 1$$

Noting that for $\theta \to 0$ a Taylor series expansion gives

$$\phi_i(\theta) = 1 - \theta\mu_i + o(\theta) \qquad i = 1, 2$$

show that

$$p_x^*(\theta) = \frac{\mu_1}{\mu_1 + \mu_2} \cdot \frac{1}{\theta} + o\left(\frac{1}{\theta}\right)$$

which on inversion leads to (8.4.3a). [See Exercise 12 for the definition of $o(\theta)$.]

18. In the queueing system of Example 8.4.1 let service times V_n, $n = 1, 2, \ldots$ be independent and identically distributed random variables with mean $E(V)$ and the Poisson arrival rate be λ. In such a queue it can be shown that the length of a busy period that starts with n customers has the same distribution as the sum of n busy periods starting with only one customer. This leads to the relation

$$E(B) = E(V) + E(N_V)E(B)$$

when N_V is the number of customers arriving during a service period. Justify this relation.

Using this relation and the results of Example 8.4.1, show that

$$E(B) = \frac{E(V)}{1 - \lambda E(V)}$$

$$P[\text{System busy as } t \to \infty] = \lambda E(V)$$

19. a. A machine is serviced when it fails or when it has been in use for T units of time. Let $f(x)$ be the probability density of the length of life of the machine. Determine the mean time to service, given that it has been put into operation after service at $t = 0$.

b. Let $\$C_1$ be the cost of service when it fails and $\$C_2$ be the cost of service after it has been in use for T units of time. Determine the overall service cost per unit time.

c. Specialize the results in part b when the life distribution of the machine is exponential with mean $1/\lambda$.

    d. Suppose servicing times are independent and identically distributed random variables with mean $1/\mu$ and the life distribution is exponential as assumed in part c. Determine the fraction of time the machine will be working.

20. Solve Exercise 18a of Chapter 7 using a semi-Markov process model.

21. A Markov chain has three states $\{0, 1, 2\}$ and the transition probability matrix

$$\begin{bmatrix} 0 & \frac{1}{2} & \frac{1}{2} \\ \frac{1}{2} & 0 & \frac{1}{2} \\ 0 & \frac{1}{2} & \frac{1}{2} \end{bmatrix}$$

Consider a visit to state 0 as a discrete renewal event, and show that the expected value of its renewal period is 6 (transitions).

Chapter 9

STATIONARY PROCESSES

9.1 DEFINITION

Stationarity has been mentioned previously in two connections:

1. To indicate that the transition probability matrix of a Markov chain does not change with time (the Markov chain has a stationary transition probability matrix).
2. To indicate that the limiting distribution of a Markov process (or a Markov chain), once attained, does not change with time (the limiting distribution is stationary).

In both cases stationarity referred to the time-invariance property of some aspect of the stochastic process. It may now be asked whether these indicate any general properties regarding the basic stochastic processes. To answer this question, we shall first define stationarity (which is of two kinds) of the stochastic processes in general.

Let $\{X(t), t \in T\}$ be a stochastic process with finite second moments. Further let

$$\mu(t) = E[X(t)] \qquad \text{for all } t \in T \qquad (9.1.1)$$

$$\mu_2(t) = E\{[X(t)]^2\} \qquad \text{for all } t \in T \qquad (9.1.2)$$

$$\mu_{11}(s, t) = E[X(s)X(t)] \qquad \text{for all } s, t \in T \qquad (9.1.3)$$

Then $\mu(t)$ is the mean of $X(t)$, and the variance of $X(t)$ is given by

$$\sigma^2(t) = E\{[X(t)]^2\} - \{E[X(t)]\}^2$$

$$= \mu_2(t) - [\mu(t)]^2 \qquad (9.1.4)$$

Further the covariance function is given by

$$\gamma(s,t) = E[X(s)X(t)] - E[X(s)]E[X(t)]$$

$$= \mu_{11}(s,t) - \mu(s)\mu(t) \qquad (9.1.5)$$

Based on these definitions, the correlation function (also called *autocorrelation function*) can be given as

$$\rho(s,t) = \frac{\gamma(s,t)}{\sigma(s)\sigma(t)} \qquad (9.1.6)$$

These moment functions of the stochastic process $X(t)$ play essentially the same role as the first two moments in describing the behavior of a random variable (or a distribution function). When it is known that a random variable has a normal distribution, its mean and variance describe the distribution completely. A multinormal distribution is completely defined by its mean vector and the variance-covariance matrix. Similarly a Gaussian stochastic process (also known as the normal process) is completely defined by its mean value function $\mu(t)$ and the covariance function $\gamma(s,t)$. When the transition distribution function of the stochastic process $X(t)$ is not known, however, the information supplied by the mean value and covariance functions is limited. The general properties of the process must then be obtained through limit theorems and their convergence properties, in many ways similar to the study of sequences of random variables. Let

$$F(x_1, x_2, \ldots, x_n; t_1, t_2, \ldots, t_n)$$

$$= P[X(t_1) \le x_1, X(t_2) \le x_2, \ldots, X(t_n) \le x_n] \qquad (9.1.7)$$

for $t_1, t_2, \ldots, t_n \in T$ and $n = 1, 2, \ldots$ be the joint distribution function of the random variables $X(t_1), X(t_2), \ldots, X(t_n)$.

DEFINITION   *The stochastic process* $\{X(t), t \in T\}$ *is said to be stationary in the strict sense if and only if*

$$F(x_1, x_2, \ldots, x_n; t_1 + h, t_2 + h, \ldots, t_n + h)$$

$$= F(x_1, x_2, \ldots, x_n; t_1, t_2, \ldots, t_n) \qquad (9.1.8)$$

As a consequence we have

$$P[X(t_1) \le x_1] = F(x_1, \infty, \ldots, \infty; t_1, t_2, \ldots, t_n)$$

$$= F(x_1; t_1)$$

$$= F(x_1; 0) \tag{9.1.9}$$

which is independent of t, and hence

$$E[X(t)] = \mu \tag{9.1.10}$$

and

$$V\{X(t)\} = \sigma^2 \tag{9.1.11}$$

where μ and σ^2 are constants. Without loss of generality assume $\mu = 0$ and $\sigma^2 = 1$. For $t > s$ the covariance function $\gamma(s, t)$ is then given by

$$\gamma(s, t) = E[X(s)X(t)]$$

$$= \iint_{x_1, x_2 \in S} x_1 x_2 \, dF(x_1, x_2; s, t)$$

$$= \iint_{x_1, x_2 \in S} x_1 x_2 \, dF(x_1, x_2; 0, t - s)$$

$$= E[X(0)X(t - s)] \tag{9.1.12}$$

which is a direct consequence of (9.1.5) and (9.1.8). From the right-hand side of (9.1.12) we therefore have

$$\gamma(s, t) = \gamma(0, t - s) \tag{9.1.13}$$

which depends only on the difference between t and s instead of the individual values of t and s. The length $t - s$ of the time period is called the *lag*.

It should also be noted that when these results are true, we have the correlation function

$$\rho(s, t) = \gamma(0, t - s) = \rho(t - s) \tag{9.1.14}$$

when we assume $\sigma^2 = 1$. The results (9.1.10), (9.1.11) and (9.1.13) are therefore implied by the definition of strict stationarity. This does not mean that all stochastic processes for which results (9.1.10), (9.1.11), and (9.1.13) are true are stationary in the strict sense, because these results do not imply (9.1.8). When this is the case we speak of stationarity in the weak or wide sense.

DEFINITION   *The stochastic process* $\{ X(t), t \in T \}$ *is said to be stationary in the wide sense* (*also called weakly stationary or covariance stationary*) *if and only if*

$$E[X(t)] = \mu \qquad \text{for all } t \in T$$

$$V[X(t)] = \sigma^2 \qquad \text{for all } t \in T$$

and

$$\gamma(t, t + h) = \gamma(h) \qquad \text{for all } t \in T \qquad (9.1.15)$$

For stationarity in the wide sense to imply stationarity in the strict sense, it is therefore necessary that the distribution function $F(x_1, x_2, \ldots, x_n; t_1, t_2, \ldots, t_n)$ of $X(t)$ depend only on the parameters involved in the definition of the wide sense stationarity. The stationary Gaussian process is one such example.

Suppose $\{ X(t), t \in T \}$ is a stochastic process such that for $t_1, t_2, \ldots, t_n \in T$, $n = 1, 2, \ldots$ the random variables $X(t_1), X(t_2), \ldots, X(t_n)$ have an n-variate multinormal distribution with

$$E[X(t_i)] = \mu(t_i)$$

$$V[X(t_i)] = \gamma(t_i, t_i)$$

$$\text{Cov}[X(t_i), X(t_j)] = \gamma(t_i, t_j) \qquad (9.1.16)$$

$i, j = 1, 2, \ldots, n$; $i \neq j$. Then $\|\gamma(t_i, t_j)\|$ is the variance-covariance matrix of the distribution function. Under these conditions $\{ X(t), t \in T \}$ is called a *Gaussian process* (also known as a *normal process*). In addition, if

$$\mu(t_i) = \mu \qquad \text{for all } t_i \in T$$

$$\gamma(t_i, t_i) = \sigma^2 \qquad \text{for all } t_i \in T$$

$$\gamma(t_i, t_j) = \gamma(|t_i - t_j|) \qquad \text{for all } t_i, t_j \in T; \ i \neq j \qquad (9.1.17)$$

then such a Gaussian process is covariance stationary. Since the distribution of the process is completely defined by the parameters just described, it is also stationary in the strict sense.

Suppose $\{ X_n, n = 0, 1, \ldots \}$ is a covariance stationary stochastic process with a discrete parameter n and (in this chapter we shall use the same notation for mean values in discrete as well as continuous parameter space processes)

$$E[X_n] = \mu \qquad\qquad n = 0, 1, \ldots$$

$$V[X_n] = \sigma^2 = \gamma(0) \qquad n = 0, 1, \ldots$$

and

$$\mathrm{Cov}[X_n, X_{n+h}] = \gamma(h)$$

Let

$$Z_n = \alpha_0 X_n + \alpha_1 X_{n-1} + \cdots + \alpha_k X_{n-k} \qquad (9.1.18)$$

Then we have

$$E[Z_n] = \mu \left(\sum_{i=0}^{k} \alpha_i \right)$$

$$V[Z_n] = \sum_i \sum_j \alpha_i \alpha_j \gamma(|i - j|) \qquad (9.1.19)$$

Similarly it can be shown that $\mathrm{Cov}[Z_n, Z_{n+h}]$ is also independent of n, thus establishing the covariance stationarity of linear functions of covariance stationary stochastic processes. In particular, the first difference $\{ X_n - X_{n-1}, n = 1, 2, \ldots \}$ and higher differences obtained through successive differencing are all stationary in the wide sense. Therefore differencing is one of the methods used in the analysis of nonstationary time series (see Box and Jenkins, 1976).

An interesting special case of the Gaussian process is the Gaussian Markov process. As the name implies, it is a Gaussian process that is also Markovian. Using the properties of the process, it can be shown that in this case

$$\gamma(s, t) = \gamma(s - t) = \gamma(0) e^{-\lambda |s - t|} \qquad \lambda > 0 \qquad (9.1.20)$$

Let $(x_1, x_2, \ldots, x_n)$ be n observations of a covariance stationary stochastic process $\{ X_n, n = \ldots, -1, 0, 1, 2, \ldots \}$ made at n successive epochs. Let

Γ_n be the matrix of covariances between these observations,

$$\Gamma_n = \begin{bmatrix} \gamma(0) & \gamma(1) & \gamma(2) & \cdots & \gamma(n-1) \\ \gamma(1) & \gamma(0) & \gamma(1) & \cdots & \gamma(n-2) \\ \gamma(2) & \gamma(1) & \gamma(0) & \cdots & \gamma(n-3) \\ \vdots & \vdots & \vdots & & \vdots \\ \gamma(n-1) & \gamma(n-2) & \gamma(n-3) & \cdots & \gamma(0) \end{bmatrix}$$

An important property of Γ_n is that it is positive definite. This can be established by considering the variance of a linear function Z_n of random variables $X_1, X_2, \ldots, X_n$:

$$Z_n = \sum_{j=1}^{n} a_j X_j$$

From (9.1.19) we have

$$V[Z_n] = \sum_i \sum_j a_i a_j \gamma(|i - j|)$$

which is necessarily greater than zero since not all a's are zero.

The positive definiteness of the covariance matrix Γ_n carries through to the corresponding correlation matrix

$$P_n = \begin{bmatrix} 1 & \rho(1) & \rho(2) & \cdots & \rho(n-1) \\ \rho(1) & 1 & \rho(1) & \cdots & \rho(n-2) \\ \rho(2) & \rho(1) & 1 & \cdots & \rho(n-3) \\ \vdots & & & & \\ \rho(n-1) & \rho(n-2) & \rho(n-3) & \cdots & 1 \end{bmatrix}$$

Positive definiteness of a matrix implies that its determinant and all its principal minors are greater than zero. Therefore for different values of n we get the following:

$n = 2$:

$$\begin{vmatrix} 1 & \rho(1) \\ \rho(1) & 1 \end{vmatrix} > 0$$

hence $-1 < \rho(1) < +1$.

$n = 3$:

$$\begin{vmatrix} 1 & \rho(1) \\ \rho(1) & 1 \end{vmatrix} > 0 \qquad \begin{vmatrix} 1 & \rho(2) \\ \rho(2) & 1 \end{vmatrix} > 0$$

and

$$\begin{vmatrix} 1 & \rho(1) & \rho(2) \\ \rho(1) & 1 & \rho(1) \\ \rho(2) & \rho(1) & 1 \end{vmatrix} > 0$$

hence $-1 < \rho(1) < +1$, $-1 < \rho(2) < +1$, and $\rho(2) > 2\rho^2(1) - 1$.

In the remaining sections of this chapter we shall concentrate on some general properties of covariance stationary stochastic processes.

Analysis of general stochastic processes can be carried out in two ways: (1) in the time domain and (2) in the frequency domain. Analysis in the time domain uses the properties of the covariance function and the correlation function for different time lags, and the frequency analysis of the process employs a harmonic analysis on a periodic function representation of the stochastic process. Since harmonic analysis requires more sophisticated mathematical background than assumed so far, we shall only give a brief introduction to the basic concept of spectral distribution in Section 9.4.

The properties discussed in this chapter are extensively used in the analysis of time series that can be represented as stationary or nonstationary stochastic processes. Two basic models used in representing time series are the moving average and autoregressive schemes. These are introduced as examples in the next section, and their covariance functions are identified. Further discussion of the analysis of these models in the context of time series will be reserved for Chapter 21 on time series analysis.

9.2 SOME EXAMPLES

Example 9.2.1 The Two-State Markov Process

From (7.5.4) and (7.5.5) we have

$$P_{00}(t) = \frac{\mu}{\lambda + \mu} + \frac{\lambda}{\lambda + \mu} e^{-(\lambda + \mu)t}$$

$$P_{01}(t) = \frac{\lambda}{\lambda + \mu} - \frac{\lambda}{\lambda + \mu} e^{-(\lambda + \mu)t}$$

$$P_{10}(t) = \frac{\mu}{\lambda + \mu} - \frac{\mu}{\lambda + \mu} e^{-(\lambda + \mu)t}$$

$$P_{11}(t) = \frac{\lambda}{\lambda + \mu} + \frac{\mu}{\lambda + \mu} e^{-(\lambda + \mu)t} \tag{9.2.1}$$

From these transition probability distributions the unconditional probability distribution $P_0(t), P_1(t)$ of the state of the system at time t is determined by assuming an initial probability vector $\{p_0, p_1\}$ for the state of the process at $t = 0$ and writing

$$P_0(t) = p_0 P_{00}(t) + p_1 P_{10}(t)$$

$$p_1(t) = p_0 P_{01}(t) + p_1 P_{11}(t) \tag{9.2.2}$$

Substituting from (9.2.1), equations (9.2.2) yield

$$P_0(t) = \frac{\mu}{\lambda + \mu} + \left[p_0 - \frac{\mu}{\lambda + \mu} \right] e^{-(\lambda + \mu)t} \tag{9.2.3}$$

$$P_1(t) = \frac{\lambda}{\lambda + \mu} + \left[p_1 - \frac{\lambda}{\lambda + \mu} \right] e^{-(\lambda + \mu)t} \tag{9.2.4}$$

Now suppose

$$p_0 = \frac{\mu}{\lambda + \mu} \qquad p_1 = \frac{\lambda}{\lambda + \mu} \tag{9.2.5}$$

Then we get

$$P_0(t) = \frac{\mu}{\lambda + \mu}$$
$$\qquad\qquad\qquad \text{for all } t \geq 0 \tag{9.2.6}$$
$$P_1(t) = \frac{\lambda}{\lambda + \mu}$$

which gives

$$E[X(t)] = \frac{\lambda}{\lambda + \mu} \tag{9.2.7}$$

$$V[X(t)] = \frac{\lambda\mu}{(\lambda + \mu)^2} \tag{9.2.8}$$

$$\text{Cov}[X(s), X(t)] = E[X(s)X(t)] - \left(\frac{\lambda}{\lambda + \mu} \right)^2 \tag{9.2.9}$$

but for $t \geq s$,

$$E[X(s)X(t)] = P[X(s) = 1 \quad \text{and} \quad X(t) = 1]$$

$$= P[X(s) = 1] P[X(t) = 1 | X(s) = 1]$$

$$= \frac{\lambda}{\lambda + \mu} \cdot P_{11}(t - s)$$

$$= \frac{\lambda}{\lambda + \mu} \left[\frac{\lambda}{\lambda + \mu} + \frac{\mu}{\lambda + \mu} e^{-(\lambda + \mu)(t - s)} \right] \tag{9.2.10}$$

Substituting this value in (9.2.9), we get

$$\gamma(s,t) = \text{Cov}[X(s), X(t)] = \frac{\lambda\mu}{(\lambda + \mu)^2} e^{-(\lambda+\mu)(t-s)} \qquad (9.2.11)$$

and

$$\rho(s,t) = e^{-(\lambda+\mu)(t-s)} \qquad (9.2.12)$$

which depends only on the difference $t - s$. Thus the process $X(t)$ is stationary in the wide sense. In this case, however, the stationarity is stronger, for consider the joint distribution of the process at two points $t, t + h$ ($h > 0$). The possible values are as follows:

$X(t)$	$X(t + h)$
0	0
0	1
1	0
1	1

In considering one set of values, say $\{0, 1\}$, we get

$$P[X(t) = 0, X(t + h) = 1] = P[X(t) = 0] P[X(t + h) = 1 | X(t) = 0]$$

$$= \frac{\mu}{\lambda + \mu} P_{01}(h)$$

$$= \frac{\lambda\mu}{(\lambda + \mu)^2} [1 - e^{-(\lambda+\mu)h}] \qquad (9.2.13)$$

which depends only on the time increment h, not on the specific value of t. Thus it is seen that the process $X(t)$ is also stationary in the strict sense.

Similar conclusions can be drawn for the two-state Markov chain. It is suggested that the reader derive results analogous to (9.2.3)–(9.2.13) in this case.

In the two-state Markov process discussed earlier, the stationary process is generated by starting with suitable initial probability vector. This result can be generalized to show that in the case of all Markov processes and chains with limiting distributions appropriate initial conditions can be used to generate stationary processes. From these discussions it is also clear that to be able to generate such stationary processes, the transition rate matrix of the continuous parameter process or the transition probability matrix of the Markov chain should be stationary, meaning that it should be time-homogeneous.

For the purposes of modeling in physical problems the occurrence of stationary processes is more natural. If one has to observe the process at some time t, for all practical purposes it can be considered to have started in the remote past and assumed that the state distribution is already stationary.

Example 9.2.2 Purely Random Process (White Noise)

A sequence of uncorrelated random variables with mean zero and constant variance $\sigma^2 > 0$ is called a *purely random process*. Clearly the covariance function of this process is given by

$$\gamma(h) = \begin{cases} \sigma^2 & \text{if } h = 0 \\ 0 & \text{if } h \neq 0 \end{cases}$$

Thus the process is covariance stationary. If we assume that the random variables have identical distributions, then the process is also strictly stationary.

The purely random process is often called "white noise" for reasons that will be clear later. When the parent population is normally distributed, we may qualify white noise as being Gaussian. Physical phenomena that can be approximated as a white noise include the current at the anode of a vacuum tube and small random voltage fluctuations between the ends of a resistor due to the random thermal motion of the conducting electrons inside the resistor. Furthermore the purely random process is an essential component of other stochastic process models, as will be seen later.

Example 9.2.3 A Moving Average Process

Let $\{Y_n\}_{n=0}^{\infty}$ be a sequence of uncorrelated random variables with

$$E[Y_n] = 0 \qquad n = 0, 1, 2, \dots$$

and

$$V[Y_n] = \sigma_y^2 \qquad n = 0, 1, 2, \dots \qquad (9.2.14)$$

A moving average of order k is generated by defining a process $\{X_n, n \geq k\}$ as

$$X_n = \alpha_0 Y_n + \alpha_1 Y_{n-1} + \cdots + \alpha_k Y_{n-k}$$

where α_i, $i = 0, \dots, k$ are some constants. These are called the weights of the moving average. Clearly we have

$$E[X_n] = 0 \qquad\qquad n = k, k+1, \ldots$$

$$V[X_n] = \left(\sum_{i=0}^{k} \alpha_i^2 \right) \sigma_y^2 \qquad n = k, k+1, \ldots$$

and

$$\gamma(n+h, n) = \gamma(h) = \begin{cases} \left(\displaystyle\sum_{m=0}^{k-h} \alpha_m \alpha_{m+h} \right) \sigma_y^2 & \text{for } h = 0, 1, \ldots, k \\ 0 & \text{for } h = k+1, k+2, \ldots \end{cases}$$

$$(9.2.15)$$

The stochastic process $\{ X_n, n \geq k \}$ is therefore covariance stationary.
   As a special case, consider

$$\alpha_i = \frac{1}{k+1} \qquad i = 0, 1, 2, \ldots, k \qquad\qquad (9.2.16)$$

Then

$$V[X_n] = \frac{\sigma_y^2}{k+1}$$

and

$$\gamma(h) = \begin{cases} \dfrac{k-h+1}{(k+1)^2} \sigma_y^2 & \text{for } h = 0, 1, \ldots, k \\ 0 & \text{for } h = k+1, \ldots \end{cases} \qquad (9.2.17)$$

Clearly the correlation function is given by

$$p(h) = \frac{\gamma(h)}{\sqrt{V(X_n)V(X_{n+h})}}$$

$$= \begin{cases} 1 - \dfrac{h}{k+1} & \text{for } h = 0, 1, \ldots, k \\ 0 & \text{for } h = k+1, \ldots \end{cases} \qquad (9.2.18)$$

   Moving averages play an important role in smoothing and prediction problems in time series analysis. We shall pursue this problem in more detail in Chapter 21.

Example 9.2.4   Autoregressive Process of the First Order

Another class of stochastic models used in time series analysis come under autoregressive schemes. Let $\{Y_n\}_{n=-\infty}^{\infty}$ be a sequence of uncorrelated random variables with zero mean and constant variance as in the second example. Define

$$X_n = \alpha X_{n-1} + Y_n \qquad n = \ldots, -1, 0, 1, \ldots \qquad (9.2.19)$$

where α is a constant. This is a recurrence relation for $\{X_n\}_{n=-\infty}^{\infty}$, and hence we can write

$$X_n = \alpha X_{n-1} + Y_n$$

$$X_{n-1} = \alpha X_{n-2} + Y_{n-2}$$

$$\vdots$$

$$X_{n-N+1} = \alpha X_{n-N} + Y_{n-N}$$

Thus we get

$$X_n = \alpha(\alpha X_{n-2} + Y_{n-1}) + Y_n$$

$$= \alpha^2 X_{n-2} + \alpha Y_{n-1} + Y_n$$

$$\vdots$$

$$= \alpha^N X_{n-N} + \sum_{i=0}^{N-i} \alpha^i Y_{n-i} \qquad (9.2.20)$$

As it is, clearly the scheme $\{X_n\}$ is nonstationary. Now let $|\alpha| < 1$ and $N \to \infty$; (9.2.20) yields for the autoregressive process defined in (9.2.19)

$$X_n = \sum_{i=0}^{\infty} \alpha^i Y_{n-i} \qquad (9.2.21)$$

It should be noted that the right-hand side converges to a well-defined random variable. We have

$$E(X_n) = 0$$

and

$$V(X_n) = \left(\sum_{i=0}^{\infty} \alpha^{2i} \right) \sigma_y^2$$

$$= \frac{1}{1 - \alpha^2} \sigma_y^2 \qquad \text{for all } n$$

$$= \sigma_x^2 \qquad\qquad\qquad (9.2.22)$$

For the covariance function, we get

$$\gamma(n + h, n) = E\left[\left(\sum_{i=0}^{\infty} \alpha^i Y_{n+h-i} \right) \left(\sum_{j=0}^{\infty} \alpha^j Y_{n-j} \right) \right]$$

$$= E\left[\left(\sum_{r=-\infty}^{n+h} \alpha^{n+h-r} Y_r \right) \left(\sum_{s=-\infty}^{n} \alpha^{n-s} Y_s \right) \right]$$

$$= \left(\sum_{r=-\infty}^{n} \alpha^{n-r} \alpha^{n+h-r} \right) \sigma_y^2$$

$$= \left(\sum_{i=0}^{\infty} \alpha^i \alpha^{i+h} \right) \sigma_y^2$$

$$= \alpha^h \frac{\sigma_y^2}{1 - \alpha^2}$$

$$= \alpha^h \sigma_x^2 \qquad\qquad\qquad (9.2.23)$$

The covariance function is independent of n, and hence the autoregressive process is covariance stationary. The correlation function of the process is given by

$$\rho(h) = \alpha^h \qquad\qquad\qquad (9.2.24)$$

Example 9.2.5 A Periodic Process

Consider the process $\{ X_n, \ n = 0, \pm 1, \pm 2, \dots \}$ given by

$$X_n = A \cos \omega n + B \sin \omega n \qquad\qquad (9.2.25)$$

where A and B are identically distributed uncorrelated random variables with mean 0 and variance σ^2, and $\omega \in [0, \pi]$ is a given frequency. Then we have

$$E[X_n] = 0$$

and

$$\gamma(n, n + h) = E[X_n X_{n+h}]$$

$$= E\left[\{A\cos\omega n + B\sin\omega n\}\{A\cos\omega(n + h) + B\sin\omega(n + h)\}\right]$$

$$= E\left[A^2\cos\omega n\cos\omega(n + h) + B^2\sin\omega n\sin\omega(n + h)\right]$$

$$= \sigma^2\cos\omega h \tag{9.2.26}$$

In the derivation of (9.2.26) we have used the properties

$$\cos(a - b) = \cos a\cos b + \sin a\sin b$$

and $E[AB] = 0$.

Equation (9.2.26) indicates that the process is covariance stationary. Furthermore, if A and B are normally distributed random variables, the process $\{X_n,\ n = 0, \pm 1, \ldots\}$ is a Gaussian process and is also strictly stationary.

Now, we generalize the representation (9.2.25) of $\{X_n\}$ as follows: Let $A_0, A_1, \ldots, A_m$ and $B_0, B_1, \ldots, B_m$ be uncorrelated random variables with zero mean. Also assume that A_i and B_i $(i = 0, 1, 2, \ldots, m)$ have a common variance σ_i^2. Now for distinct frequencies $\omega_0, \omega_1, \ldots, \omega_m \in [0, \pi]$ for $n = 0, \pm 1, \pm 2, \ldots$ we write

$$X_n = \sum_{k=0}^{m} \{A_k\cos n\omega_k + B_k\sin n\omega_k\} \tag{9.2.27}$$

Proceeding as before, for this process we get

$$\gamma(n, n + h) = \sum_{k=0}^{m} \sigma_k^2\cos h\omega_k \tag{9.2.28}$$

Let $\sigma^2 = \sigma_0^2 + \sigma_1^2 + \cdots + \sigma_m^2$ and $p_k = \dfrac{\sigma_k^2}{\sigma^2}$. Then the covariance function $\gamma(n, n + h)$ can be written as

$$\gamma(h) = \sigma^2 \sum_{k=0}^{m} p_k\cos h\omega_k \tag{9.2.29}$$

where p_k is the contribution of the frequency ω_k to the covariance of the function. This representation is significant when we later note that covariance stationary processes can be represented as the sum of periodic functions with random coefficients.

The form of $\gamma(h)$ in (9.2.29) can be further generalized if we note that $\{p_k\}$ is a probability distribution. Now using a continuum of frequencies in $[0, \pi]$ we may write

$$\gamma(h) = \sigma^2 \int_0^\pi \cos h\omega \, dF(\omega) \qquad (9.2.30)$$

where $F(\omega)$ is the distribution function of a random variable representing these frequencies. Later we shall in fact identify $F(\omega)$ as the spectral distribution function corresponding to the covariance function $\gamma(h)$.

Example 9.2.6 Processes with Independent Increments

Consider a continuous-parameter stochastic process $\{X(t), t \geq 0\}$. If for all choice of indexes $t_0 < t_1 < t_2 < \cdots < t_n$ the n random variables

$$X(t_1) - X(t_0), X(t_2) - X(t_1), \ldots, X(t_n) - X(t_{n-1})$$

are mutually independent, the stochastic process is said to have *independent increments*. If in addition $X(t_k + h) - X(t_{k-1} + h)$ and $X(t_k) - X(t_{k-1})$ have the same distribution for all choices of indexes t_k and t_{k-1} and every $h > 0$, the process is said to have stationary independent increments.

The Poisson process is a good example of a stochastic process with stationary independent increments. Let $X(t)$ be a Poisson process with parameter λ. Then from Section 7.2

$$P[X(t) = n] = e^{-\lambda t}\frac{(\lambda t)^n}{n!} \qquad n = 0, 1, 2, \ldots \qquad (9.2.31)$$

with

$$E[X(t)] = \lambda t \qquad V[X(t)] = \lambda t$$

and

$$\begin{aligned}
\gamma(s, t) &= \operatorname{Cov}[X(s), X(t)] \\
&= \operatorname{Cov}[X(s), X(t) - X(s) + X(s)] \\
&= \operatorname{Cov}[X(s), X(t) - X(s)] + \operatorname{Cov}[X(s), X(s)] \\
&= V[X(s)] = \lambda s
\end{aligned}$$

since $X(s)$ and $X(t) - X(s)$ are independent. Further note that we have assumed $s < t$. Thus the general form of $\gamma(s, t)$ can be given as

$$\gamma(s, t) = \lambda \min(s, t) \qquad \text{for all } s, t \geq 0 \qquad (9.2.32)$$

Obviously $X(t)$ is not a stationary process, for $E[X(t)]$ and $V[X(t)] \to \infty$ as $t \to \infty$.

Now consider the increment process of $X(t)$ defined by

$$I(t) = X(t + h) - X(t)$$

for a given positive constant h. By arguments similar to the preceding ones, we have

$$E[I(t)] = \lambda h \qquad V[I(t)] = \lambda h$$

and

$$
\begin{aligned}
\gamma(s, t) &= \text{Cov}[I(s), I(t)] \\
&= \text{Cov}[X(s + h) - X(s), X(t + h) - X(t)] \\
&= \text{Cov}[X(s + h) - X(t) + X(t) - X(s), X(t + h) - X(t)] \\
&= \text{Cov}[X(s + h) - X(t), X(t + h) - X(t)] \\
&\quad + \text{Cov}[X(t) - X(s), X(t + h) - X(t)]
\end{aligned}
$$

But $X(t) - X(s)$ and $X(t + h) - X(t)$ are nonoverlapping and hence have zero covariance. Further

$$\text{Cov}[X(s + h) - X(t), X(t + h) - X(t)] = 0 \qquad \text{if } |t - s| > h$$

and if $|t - s| \leq h$, we can write it as

$$
\begin{aligned}
\text{Cov}&[X(s + h) - X(t), X(t + h) - X(s + h) + X(s + h) - X(t)] \\
&= \text{Cov}[X(s + h) - X(t), X(s + h) - X(t)] \\
&= V[X(s + h) - X(t)] \\
&= \lambda[h - |t - s|] \qquad\qquad\qquad\qquad\qquad (9.2.33)
\end{aligned}
$$

since $\text{Cov}[X(s + h) - X(t), X(t + h) - X(s + h)] = 0$. This shows that the increment process is covariance stationary.

Another example of a stochastic process with stationary independent increments is the Wiener process, defined as follows: A stochastic process $\{X(t), t \geq 0\}$ is said to be a *Wiener process* (*Brownian motion process*) if the following conditions hold:

1. $\{X(t), t \geq 0\}$ has stationary independent increments.
2. For every $t > 0$, $X(t)$ is normally distributed.
3. For all $t > 0$, $E[X(t)] = 0$.
4. $X(0) = 0$.
5. $V[X(t) - X(S)] = \sigma^2|t - s|$.   (9.2.34)

By arguments similar to those employed in deriving the covariance function of the Poisson process, we can show that the covariance function of the Wiener process is given by

$$\gamma(s, t) = \sigma^2 \min(s, t) \qquad \text{for all } s, t \geq 0 \qquad (9.2.35)$$

The Wiener process has applications in the study of the movement of particles immersed in liquid or gas (Brownian motion), in quantum mechanics, as a model for price fluctuations in stock and commodity markets, and as the asymptotic distribution of goodness of fit tests for distribution functions. Interested readers may refer to the books cited at the end of the chapter.

9.3   ERGODIC THEOREMS

In order to relate theoretical models to observed phenomena through sample observations, it is necessary to identify the properties of sample functions. When we identify expected values of process characteristics, we speak of hypothetical averages that can be determined from an ensemble of processes. However, observations are made on a stochastic process over a period of time, and thus only time averages are available. Under what conditions do these two averages correspond? The following two theorems provide ergodic properties of the process mean and the covariance function.

Theorem 9.3.1   (1) *Let* $\{X_n, \ n = 0, 1, 2, \dots\}$ *be a covariance stationary stochastic process. Consider the sequence of sample means*

$$m_N = \frac{1}{N} \sum_{n=0}^{N-1} X_n \qquad (9.3.1)$$

In order for $\lim_{N \to \infty} V(m_N) = 0$, *it is necessary and sufficient that*

$$\lim_{N \to \infty} \frac{1}{N} \sum_{h=0}^{N-1} \gamma(h) = 0$$

(2) *For* $\lim_{N \to \infty} V(m_N) = 0$, *a sufficient condition is that* $\lim_{h \to \infty} \gamma(h) = 0$.

Theorem 9.3.2 (1) *Let* $\{ X_n, \ n = 0, 1, 2, \ldots \}$ *be a covariance stationary Gaussian process with mean zero and covariance function* $\gamma(h)$. *For a fixed h let*

$$X_n^*(h) = X_n X_{n+h} \qquad n = 0, 1, 2, \ldots$$

Then clearly $E X_n^*(h) = \gamma(h)$. *Now write*

$$m_N^*(h) = \frac{1}{N} \sum_{n=0}^{N-1} X_n^*(h) \tag{9.3.2}$$

In order for $\lim_{N \to \infty} V(m_N^*(h)) = 0$, *it is necessary and sufficient that*

$$\lim_{N \to \infty} \frac{1}{N} \sum_{k=0}^{N-1} \left[\gamma^2(k) + \gamma(k-h)\gamma(k+h) \right] = 0 \tag{9.3.3}$$

(2) *For* $\lim_{N \to \infty} V(m_N^*(h)) = 0$, *a sufficient condition is that* $\lim_{h \to \infty} \gamma(h) = 0$.

We will not venture to prove these theorems. The importance of the theorems will be clear if we note that if the variance of the sample mean tends to zero, it implies that the sample mean in fact approaches the population mean. Thus, if the sample size is large enough, sample means can be considered to be good estimates of population means.

It should also be noted that the sufficient condition in both cases is

$$\lim_{h \to \infty} \gamma(h) = 0 \tag{9.3.4}$$

indicating that the correlation structure of the process should be such that as observations fall farther apart, they become essentially uncorrelated.

Similar theorems can be stated for continuous parameter stochastic processes.

9.4 COVARIANCE STATIONARY PROCESSES IN THE FREQUENCY DOMAIN

Consider a complex-valued random variable

$$X = X_1 + iX_2$$

where X_1 and X_2 are real valued random variables and $i = \sqrt{-1}$. For complex-valued random variables the following definitions are in order:

$$E[X] = E[X_1] + iE[X_2]$$

If X and Y are two such complex-valued random variables, then

$$\text{Cov}[X, Y] = E\left[(X - E(X))\overline{(Y - E(Y))}\right]$$

where $\overline{Y - E(Y)}$ is the complex conjugate of $Y - E(Y)$. (When $X = X_1 + iX_2$, its complex conjugate is $X_1 - iX_2$.)

In particular,

$$V[X] = E\left[|X - E(X)|^2\right]$$

where $|X - E(X)|$ is the modulus of $X - E(X)$. Also note that $X\overline{X} = X_1^2 + X_2^2 = |X|^2$.

Consider a stochastic process $\{X(t), t \geq 0\}$ given by

$$X(t) = \sum_{j=-n}^{n} Z(\lambda_j) e^{i\lambda_j t} \qquad (9.4.1)$$

where $Z(\lambda_j)$, $j = -n, -n + 1, \ldots, n - 1, n$ are complex-valued random variables. Using the preceding definitions and the properties of stochastic processes, it can be shown that the necessary and sufficient conditions for $X(t)$ to be real valued and covariance stationary are the following:

$$\lambda_{-j} = -\lambda_j$$

and

$$Z(\lambda_{-j}) = Z(-\lambda_j) = \overline{Z(\lambda_j)} \qquad (9.4.2)$$

(*Note*: A number α is real valued if and only if $\alpha - \bar{\alpha} = 0$.)

Under these conditions it is also seen that

$$E\big[Z(\lambda_j)\big] = 0 \qquad \text{for all } j \neq 0$$

and

$$\text{Cov}\big[Z(\lambda_j), Z(\lambda_k)\big] = 0 \qquad \text{for } j \neq k$$

Based on these properties, the following conclusions emerge regarding the covariance stationary process $X(t)$ defined in (9.4.1):

1. $X(T)$ is a sum of periodic functions with random coefficients. This follows from the fact that the function $e^{i\lambda_j t}$ is periodic since

$$e^{i\lambda_j(t + 2\pi/\lambda_j)} = e^{i\lambda_j t} e^{i2\pi}$$

$$= e^{i\lambda_j t}(\cos 2\pi + i\sin 2\pi)$$

$$= e^{i\lambda_j t} \qquad (9.4.3)$$

This function has a period $2\pi/\lambda_j$ and hence its frequency (1/period) is $\lambda_j/2\pi$, and its angular frequency $= 2\pi \times (\text{frequency}) = \lambda_j$. (Sometimes the term frequency is used to mean angular frequency.)

2. After much simplification, $X(t)$ can be put in the form

$$X(t) = \sum_{0}^{n} \big[A_j \cos\lambda_j t + B_j \sin\lambda_j t\big]$$

where

$$A_0 \equiv Z(0), \qquad\qquad B_0 \equiv 0$$

$$A_j = 2\,\text{Re}\,Z(\lambda_j) \qquad B_j = -2I_m Z(\lambda_j) \qquad (9.4.4)$$

Furthermore one can show that $\text{Cov}(A_j, B_k) = 0$ for all j and k and

$$E\big[A_j\big] = E\big[B_j\big] = 0$$

$$V(A_j) = V(B_j) = 2V\big(Z(\lambda_j)\big) \equiv \sigma_j^2$$

Recall the process defined in (9.2.35).

3. The covariance function of $X(t)$ is given by

$$\gamma(h) = \sum_{j=-n}^{n} E\big[|Z(\lambda_j)|^2\big] e^{i\lambda_j h} \qquad (9.4.5)$$

This result follows by simplifying $E[X(t + h)\overline{X(t)}]$ in terms of $Z(\lambda_j)$ and noting that $\text{Cov}[Z(\lambda_j), Z(\lambda_k)] = 0$ for $j \neq k$.

Let $A = \{\lambda_{-n}, \ldots, \lambda_n\}$, and define measures

$$f(\lambda_j) = E\left[|Z(\lambda_j)|^2\right]$$

$$F(A) = \sum_{\lambda_j \in A} f(\lambda_j)$$

In terms of these we may write

$$\gamma(h) = \sum_{-n}^{n} e^{i\lambda_j h} f(\lambda_j)$$

which, when extended to the continuum of frequencies, can be written as

$$\gamma(h) = \int_{-\infty}^{\infty} e^{i\lambda h} \, dF(\lambda) \qquad (9.4.6)$$

Equation (9.4.6) is called *spectral representation* for the covariance function, and F is the *spectral distribution function*. In particular we get

$$\gamma(0) = \int_{-\infty}^{\infty} dF(\lambda) \qquad (9.4.7)$$

Noting that $\gamma(0)$ is the variance of the process, (9.4.7) indicates that F provides the distribution of the variance corresponding to the frequencies; that is, the contribution to the variance from frequencies in the range (λ_1, λ_2), $\lambda_1 < \lambda_2$ is given by $F(\lambda_2) - F(\lambda_1)$.

We may caution here that because of (9.4.7) $F(\infty) = \gamma(0)$ instead of 1. Hence if one wants to go by the strict definition of the distribution, normalizing $F(\lambda)$ by $\gamma(0)$ is necessary.

If the distribution F admits a derivative we call the derivative as the *spectral density*. It may be noted from (9.4.6) that $\gamma(h)$ is in fact the characteristic function of the distribution function F.

If the parameter space $T = (-\infty, \infty)$ and if $\int_{-\infty}^{\infty} |\gamma(h)| \, dh < \infty$ and $\int_{-\infty}^{\infty} |\gamma(h)|^2 \, dh < \infty$, the inverse relation to (9.4.6) can be written as

$$f(\lambda) = \frac{1}{2\pi} \int_{-\infty}^{\infty} \gamma(h) e^{-i\lambda h} \, dh \qquad -\infty < \lambda < \infty \qquad (9.4.8)$$

If the parameter space $T = \{0, \pm 1, \pm 2, \ldots\}$, (9.4.6) takes the form

$$\gamma(h) = \int_{-\pi}^{\pi} e^{i\lambda h} \, dF(\lambda) \qquad (9.4.9)$$

In this case if $\sum_{-\infty}^{\infty} [\gamma(h)]^2 < \infty$, we get the inverse relation (which holds in most cases of our interest) as

$$f(\lambda) = \frac{1}{2\pi} \sum_{h=-\infty}^{\infty} \gamma(h) e^{-i\lambda h} \qquad (9.4.10)$$

An alternate form of (9.4.10) is obtained by noting that for real-valued processes, $\gamma(h)$ is an even function. Then one can write

$$f(\lambda) = \frac{1}{2\pi} \left[\gamma(0) + 2 \sum_{h=1}^{\infty} \gamma(h) \cos \lambda h \right] \qquad (9.4.11)$$

If we use the normalized form

$$f^*(\lambda) = \frac{f(\lambda)}{\gamma(0)}$$

we get

$$f^*(\lambda) = \frac{1}{2\pi} \left[1 + 2 \sum_{h=1}^{\infty} \rho(h) \cos \lambda h \right] \qquad (9.4.12)$$

which involves the correlation function $\rho(h)$ in place of the covariance function $\gamma(h)$.

Example 9.4.1

A sequence of uncorrelated random variables with constant variance $\sigma^2 > 0$ is called the white noise. Clearly the covariance function for this process is given by

$$\gamma(h) = \begin{cases} \sigma^2 & \text{if } h = 0 \\ 0 & \text{if } h \neq 0 \end{cases}$$

Noting that $\sum_{h=-\infty}^{\infty} [\gamma(h)]^2 = \sigma^4 < \infty$, using (9.4.10), we get

$$f(\lambda) = \frac{\sigma^2}{2\pi} \qquad \lambda \in [-\pi, \pi] \qquad (9.4.13)$$

which when divided by σ^2 is a uniform density in $[-\pi, \pi]$. Therefore all frequencies in $[-\pi, \pi]$ contribute equally to the variance of the process. (This property is responsible for the name of the process since the color white is made up of a mixture of all other colors in equal amounts.)

Example 9.4.2

Consider the autoregressive process of Example 9.2.3. The parameter space $T = \{0, \pm 1, \pm 2, \dots\}$, and the covariance function is given by equation (9.2.23):

$$\gamma(h) = \sigma_x^2 \alpha^{|h|} \qquad h = 0, \pm 1, \dots$$

where $|\alpha| < 1$.

Checking the condition preceding (9.4.10), we find

$$\sum_{-\infty}^{\infty} [\gamma(h)]^2 = \sigma_x^4 \sum_{h=-\infty}^{\infty} \alpha^{2|h|}$$

$$= \sigma_x^4 \left[\sum_0^\infty \alpha^{2h} + \sum_0^\infty \alpha^{2h} - 1 \right]$$

$$= \sigma_x^4 \left[(1 - \alpha)^{-2} + (1 - \alpha)^{-2} - 1 \right]$$

$$< \infty$$

Thus we write

$$f(\lambda) = \frac{1}{2\pi} \sum_{h=-\infty}^{\infty} \gamma(h) e^{-i\lambda h}$$

$$= \frac{\sigma_x^2}{2\pi} \left[\sum_0^\infty (\alpha e^{-i\lambda})^h + \sum_0^\infty (\alpha e^{i\lambda})^h - 1 \right]$$

$$= \frac{\sigma_x^2}{2\pi} \left[\frac{1}{1 - \alpha e^{-i\lambda}} + \frac{1}{1 - \alpha e^{i\lambda}} - 1 \right]$$

$$= \frac{\sigma_x^2}{2\pi} \left[\frac{1 - \alpha^2}{1 - 2\alpha \cos\lambda + \alpha^2} \right] \qquad (9.4.14)$$

While deriving (9.4.13), we have used the formula

$$\cos\lambda = \frac{e^{i\lambda} + e^{-i\lambda}}{2}$$

In order to discuss the properties of $f(\lambda)$, we consider the two cases $\alpha > 0$ and $\alpha < 0$ separately.

CASE 1: $\alpha > 0$. Note that when $\cos(-\pi) = -1$, $\cos 0 = +1$, and $\cos \pi = -1$,

$$\lambda \in [0, \pi], \qquad \cos \lambda \text{ decreases}$$
$$\lambda \in [-\pi, 0], \qquad \cos \lambda \text{ increases}$$

When $\alpha > 0$, $f(\lambda)$ changes with $\cos \lambda$. Also when $\alpha = 0$, $f(\lambda) = \sigma_x^2/2\pi$. When $\alpha \to 1$,

$$f(\lambda) \to \infty \qquad \text{if } \lambda = 0$$
$$\to 0 \qquad \text{if } \lambda \in (0, \pi] \quad \text{or } [-\pi, 0).$$

When $0 < \alpha < 1$,

$$f(\lambda) = \frac{\sigma_x^2}{2\pi} \frac{1+\alpha}{1-\alpha} \qquad \text{if } \lambda = 0$$
$$= \frac{\sigma_x^2}{2\pi} \frac{1-\alpha}{1+\alpha} \qquad \text{if } \lambda = +\pi \quad \text{or} \quad -\pi$$

Therefore frequencies near zero dominate as $\alpha \to 1$. Hence the period of the function gets larger.

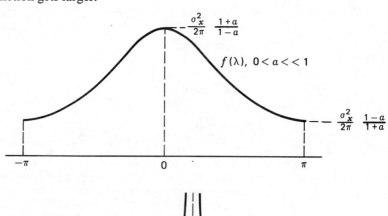

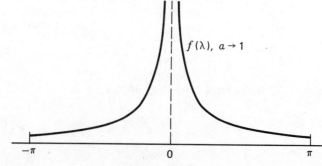

CASE 2: $\alpha < 0$.   When $\alpha < 0$, the changes in $f(\lambda)$ are in opposite direction to the changes in $\cos\lambda$. Also when $\alpha = 0$, $f(\lambda) = \sigma_x^2/2\pi$, and when $\alpha \rightarrow -1$,

$$f(\lambda) \rightarrow 0 \qquad \text{for } \lambda \in (\pi, \pi)$$
$$\rightarrow \infty \qquad \text{for } \lambda = +\pi \quad \text{or} \quad -\pi$$

Thus we have the following two graphs:

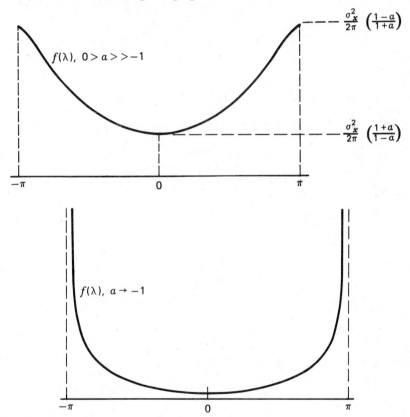

Therefore in this case the contribution to the variance is primarily from high frequencies. Hence the period of the function gets smaller.

Another approach to the analysis of stochastic processes in the frequency domain is to consider the process as the result of a transformation of some other process which we may or may not have observed and investigate the properties of the transformation relative to the process. Let $X(t)$ be the stochastic process that undergoes a transformation to give an output process

$Y(t)$. For instance if $X(t)$ is the voice signal sent over telephone lines, $Y(t)$ is the voice received at the other end. In a similar manner all physical phenomena with an input and output process pair can be identified as the result of appropriate transformations.

The terminology used in identifying the transformation is a "filter," and an important class of filters is known as time-invariant linear filters or just linear filters. If L is a linear filter, then it has the following basic properties:

1. If $\{X_1(t)\}$ and $\{X_2(t)\}$ are two processes, then for constants α and β,

$$L[\alpha X_1(t) + \beta X_2(t)] = \alpha L[X_1(t)] + \beta L[X_2(t)]$$

2. If $Y(t) = L[X(t)]$,

$$L[X(t + h)] = Y(t + h)$$

Applying the linear filter L to the representation

$$X(t) = \sum_{j=-n}^{n} Z(\lambda_j)e^{i\lambda_j t}$$

of equation (9.4.1), we get

$$L[X(t)] = \sum_{j=-n}^{n} Z(\lambda_j)L(e^{i\lambda_j t})$$

$$= \int_{-\infty}^{\infty} L(e^{i\lambda t})\, dZ(\lambda) \qquad (9.4.15)$$

Hence the effect of L on $X(t)$ is determined by the effect of L on $e^{i\lambda t}$. Write

$$\xi_\lambda(t) = L(e^{i\lambda t}) \qquad (9.4.16)$$

Consider

$$\xi_\lambda(t + h) = L(e^{i\lambda(t+h)})$$

$$= e^{i\lambda h}L(e^{i\lambda t})$$

$$= e^{i\lambda h}\xi_\lambda(t)$$

Now setting $t = 0$, we get

$$\xi_\lambda(h) = \xi_\lambda(0)e^{i\lambda h}$$

$$= B(\lambda)e^{i\lambda h} \qquad (9.4.17)$$

The function $B(\lambda) = \xi_\lambda(0)$ is called the *transfer function* of the filter. Thus we observe that the effect of L on $e^{i\lambda t}$ is to multiply it by the transfer function.

A similar analysis shows that the covariance function of $Y(t) = L[X(t)]$ has the spectral representation

$$\gamma_y(h) = \int e^{i\lambda h} |B(\lambda)|^2 \, dF(\lambda) \qquad (9.4.18)$$

which can be compared with the representation given in (9.4.6). Consequently the spectral density $f_y(\lambda)$ of the $Y(t)$ process is obtained as

$$f_y(\lambda) = |B(\lambda)|^2 f(\lambda) \qquad (9.4.19)$$

where $f(\lambda)$ is given by (9.4.8) or (9.4.10) according as the parameter space is $T = (-\infty, \infty)$ or $T = \{0, \pm 1, \pm 2, \dots\}$. Thus, if $f(\lambda)$ is known (e.g., if $Y(t)$ can be represented as the output of a linear filter acting on a white noise input), the determination of the spectral density is accomplished by the determination of $B(\lambda)$, which is done by expressing $L(e^{i\lambda t}) = B(\lambda)e^{i\lambda t}$; that is, $B(\lambda)$ is the coefficient of $e^{i\lambda t}$ in $L(e^{i\lambda t})$.

Example 9.4.3

For the parameter space $T = \{0, \pm 1, \pm 2, \dots\}$ consider the Y_n process given as

$$Y_n = \sum_{j=-\infty}^{\infty} \alpha_j X_{n-j} \qquad (9.4.20)$$

where α_j's are real and satisfy the relation $\Sigma \alpha_j^2 < \infty$, and $\{X_n, \ n = 0, \pm 1, \dots\}$ is a white noise process with variance σ^2. We have

$$L(e^{i\lambda n}) = \sum_{j=-\infty}^{\infty} \alpha_j e^{i\lambda(n-j)}$$

$$= \left(\sum_{j=-\infty}^{\infty} \alpha_j e^{-i\lambda j} \right) e^{i\lambda n}$$

that is,

$$B(\lambda) = \sum_{j=-\infty}^{\infty} \alpha_j e^{-i\lambda j} \qquad (9.4.21)$$

The spectral density of the white noise is given by

$$f(\lambda) = \frac{\sigma^2}{2\pi}$$

Thus the spectral density of the Y_n process is obtained as

$$f_y(\lambda) = \frac{\sigma^2}{2\pi} \left| \sum_{j=-\infty}^{\infty} \alpha_j e^{-i\lambda j} \right|^2 \qquad (9.4.22)$$

It should be cautioned that in the foregoing representations, for the sake of easy understanding, we have ignored some of the finer mathematical sophistications. Nevertheless, the results provide sufficient rigor for applications in time series analysis, which will be illustrated in Chapter 21.

For elaborations on the methods given here and other related topics, the readers are referred to Jenkins and Watts (1968), Koopmans (1974), Fuller (1976), and Priestley (1981). Our treatment has followed largely that of Koopmans.

Remarks As stated elsewhere, mathematical sophistication needed for the study of stationary processes is more than what we have assumed for this text. Therefore in this chapter we have had to give only a general discussion of the concepts and state the necessary results without proofs. Nevertheless, an understanding of the basic properties of stationary processes is essential in applications specifically related to time series. For further discussion, the readers are referred to the books Parzen (1962), Prabhu (1965), Cox and Miller (1965), Karlin and Taylor (1975), and Priestley (1981).

REFERENCES

Box, G. E. P., and Jenkins, G. M. (1976). *Time Series Analysis*. 2nd ed. San Francisco: Holden-Day.

Cox, D. R., and Miller, H. D. (1965). *The Theory of Stochastic Processes*. London: Chapman and Hall, Chaps. 7, 8.

Fuller, Wayne A. (1976). *Introduction to Statistical Time Series*. New York: Wiley.

Jenkins, G. M., and Watts, D. G. (1968). *Spectral Analysis*. San Francisco: Holden-Day.

Karlin, S., and Taylor, H. M. (1975). *A First Course in Stochastic Processes*. New York: Academic Press.

Koopmans, L. H. (1974). *The Spectral Analysis of Time Series*. New York: Academic Press.

Parzen, E. (1962). *Stochastic Processes*. San Francisco: Holden-Day, Chaps. 1–3.

Prabhu, N. U. (1965). *Stochastic Processes*. New York: Macmillan, Introduction, Chap. 1.

Priestley, M. B. (1981). *Spectral Analysis and Time Series*. Vol. 1. *Univariate Series*, New York: Academic Press.

FOR FURTHER READING

Chatfield, C. (1980). *The Analysis of Time Series: An Introduction.* 2nd ed. New York: Chapman and Hall.

Cramér, H., and Leadbetter, M. R. (1967). *Stationary and Related Stochastic Processes.* New York: Wiley.

Dhrymes, P. J. (1974). *Econometrics: Statistical Foundations and Applications.* New York: Springer-Verlag.

Doob, J. L. (1963). *Stochastic Processes.* New York: Wiley.

Dym, H., and McKean, H. P. (1972). *Fourier Series and Integrals.* New York: Academic Press.

Feller, W. (1970). *An Introduction to Probability Theory and Its Applications.* Vol. 2. 2nd ed. New York: Wiley.

Granander, U., and Rosenblatt, M. (1957). *Statistical Analysis of Stationary Time Series.* New York: Wiley.

Loéve, M. (1963). *Probability Theory.* 3rd ed. New York: Van Nostrand.

Yaglom, A. M. (1962). *An Introduction to the Theory of Stationary Random Functions.* Englewood, N.J.: Prentice-Hall.

EXERCISES

1. A sequence of independent and identically distributed random variables $\{\ldots, X_{-2}, X_{-1}, X_0, X_1, X_2, \ldots\}$ is usually called white noise. Determine the covariance function $\gamma(m, n)$ and correlation function $\rho(m, n)$ for the white noise process.

2. Obtain the covariance and correlation functions for the following stochastic processes and say whether they are covariance stationary:

   a. *Bernoulli process* $\{S_n,\ n = 1, 2, \ldots\}$:

   $$P(S_n = k) = \binom{n}{k} p^k (1 - p)^{n-k} \qquad k = 0, 1, 2, \ldots, n$$

   b. *Wiener process* $\{X(t), t \geq 0\}$: $X(0) = 0$, has stationary independent increments and for (t_1, t_2) with $t_2 > t_1$, $X(t_2) - X(t_1)$ is normally distributed with mean zero and variance $\sigma^2(t_2 - t_1)$.

   Hint $\begin{aligned} E[X(s)X(t)] &= E[X(s)(X(t) - X(s) + X(s))] \\ &= E[X(s)(X(t) - X(s))] + E[X^2(s)]. \end{aligned}$

3. A process used in modeling random telegraph signals is defined as follows: $\{X(t), t \geq 0\}$ is such that

   $$X(t) = X(0)(-1)^{N(t)}$$

where $X(0)$ is a random variable defined by

$$P(X(0) = -1) = P(X(0) = +1) = \tfrac{1}{2}$$

and $N(t)$ is a Poisson process, say with parameter λ. Show that the covariance function of this process can be obtained as

$$\gamma(s, t) = e^{-2\lambda|t - s|}$$

and hence it is covariance stationary.

4. A stochastic process $\{B(t), t \in [0,1]\}$ is called a Brownian bridge process if

$$B(t) = (1 - t)X\left(\frac{t}{1 - t}\right)$$

where $X(t)$ is a Wiener process with $\sigma^2 = 1$ (see Exercise 2 for definition). Show that its covariance function can be obtained as

$$\gamma(s, t) = \min(s, t) - st$$

5. In a general random walk the increments of the process occurring at transition epochs are independent and identically distributed random variables. Thus a random walk $\{S_n, n = 0, 1, 2, \dots\}$ is defined as

$$S_n = S_{n-1} + X_n$$

where $\{X_n, n = 1, 2, \dots\}$ are i.i.d. random variables with $E[X_n] = \mu$ and $V[X_n] = \sigma^2$. Obtain the covariance function $\gamma(m, n)$ for the process. Is the process stationary?

6. $\{X_n, n = 0, 1, 2, \dots\}$ is a covariance stationary stochastic process. Define processes $\{X_n^{(i)}, n = 0, 1, 2, \dots\}$ $i = 1, 2, \dots$ of differences

$$X_n^{(1)} = X_n - X_{n-1}$$

$$X_n^{(2)} = X_n^{(1)} - X_{n-1}^{(1)}$$

$$\vdots$$

and show that these processes are also covariance stationary.

7. Derive results analogous to (9.2.3)–(9.2.13) for a two-state Markov chain.

8. Starting with suitable initial probability vectors show that irreducible and positive recurrent Markov chains generate stationary processes.

9. $\{X(t), t \geq 0\}$ is an irreducible and positive recurrent Markov process with discrete states $i = 0, 1, 2, \ldots$. Let $p = (p_0, p_1, p_2, \ldots)$ be the limiting distribution of the process. Show that if one starts with $P(X(0) = i) = p_i$, $i = 0, 1, 2, \ldots$, $\{X(t), t \geq 0\}$ is a stationary process.

   Hint Let $t \to \infty$ in the Chapman-Kolmogorov relation for the process $P_{ij}(t + s) = \sum_k P_{ik}(t)P_{kj}(s)$.

10. Show that the moving average processes

$$X_n = Y_n + \alpha Y_{n-1}$$

$$X_n = Y_n + \frac{1}{\alpha} Y_{n-1}$$

have the same correlation function.

11. Determine the covariance function and the correlation function of the following moving average processes:

   a. $X_n = Y_n - 0.5Y_{n-1} - 0.5Y_{n-2}$
   b. $X_n = Y_n + 0.6Y_{n-1} - 0.2Y_{n-2} - 0.1Y_{n-3}$

12. Find the covariance and correlation functions of the following autoregressive processes:

   a. $X_n = 0.8X_{n-1} + Y_n$
   b. $X_n = 0.4X_{n-1} + Y_n$
   c. $X_n = -0.5X_{n-1} + Y_n$

   Plot the correlation functions for these autoregressive processes for lag values $h = -8, -7, \ldots 0, \ldots, +7, +8$.

13. Define a backward shift operator B as

$$B^j X_n = X_{n-j} \qquad \text{for all } j$$

Using the operator B the pth-order autoregressive process,

$$X_n = \alpha_1 X_{n-1} + \alpha_2 X_{n-2} + \cdots + \alpha_p X_{n-p} + Y_n$$

can be expressed as

$$\left(1 - \alpha_1 B - \alpha_2 B^2 \cdots - \alpha_p B^p\right) X_n = Y_n$$

Show that X_n can be expressed as an infinite moving average process

$$X_n = Y_n + \beta_1 Y_{n-1} + \beta_2 Y_{n-2} \cdots$$

where the β's are to be appropriately determined. Give the covariance function of $\{X_n\}$ using the β's.

14. $\{X_n, n = -2, -1, 0, 1, 2, \ldots\}$ is a covariance stationary stochastic process with $V(X_n)$ finite. Starting with a pth-order autoregressive scheme for X_n,

$$X_n = \alpha_1 X_{n-1} + \alpha_2 X_{n-2} + \cdots + \alpha_p X_{n-p} + Y_n$$

derive the Yule-Walker equations

$$\rho(h) = \alpha_1 \rho(h-1) + \cdots + \alpha_p \rho(h-p) \qquad \text{for all } h > 0$$

Hint   Multiply by X_{n-h}, and take expectations.

15. Show that the random walk process defined in Exercise 5 is a process with stationary independent increments.

16. Show that the stationary Gaussian Markov process is ergodic.

17. Show that under conditions (9.4.2), $X(t)$ of equation (9.4.1) is real valued.

18. Investigate the properties of the spectral density for a process whose covariance function is given by

$$\gamma(h) = \sigma^2 e^{-h^2}$$

Hint   Use the fact that $(1/\sqrt{2\pi})(e^{-\frac{1}{2}h^2})$ is a normal density.

19. Investigate the properties of the spectral density for a process whose covariance function is given by

$$\gamma(h) = \sigma^2 e^{-\alpha|h|}$$

Hint   Use the properties of a gamma distribution in simplifications.

Chapter 10
MARKOV DECISION PROCESSES

10.1 INTRODUCTION

Mathematical model building is accomplished through four distinct phases: (1) problem identification, (2) idealization, (3) model construction, and (4) implementation. The implementation phase can be further characterized by three analytic processes: behavioral (or descriptive), statistical, and operational. So far in this book we have considered only the descriptive and statistical aspects of simple stochastic processes. From an operational point of view the decision maker has to choose between alternate models or parameters in the same model, and therefore it is essential that some analytic procedures be available for this decision process. We shall illustrate some of the techniques employed in this process with regard to systems that may be modeled as simple Markov processes.

Markov decision processes use the fundamental properties of Markov processes in arriving at decisions with the help of analytical and computational techniques known as linear, nonlinear, and dynamic programming. As we have not covered and do not intend to cover these techniques in this book, here we shall stop short of giving complete solutions to the decision problems and restrict ourselves to merely formulating the problem and indicating solution methods and sources. This, we hope, will spur the reader to learn more about such techniques and their relevance to the applications of stochastic processes.

Markov decision processes bring together the study of sequential decision problems of statistics (Wald, 1947, 1950) and the dynamic programming technique of applied mathematics and operations research (Bellman, 1957a). The first explicit formulation of a Markov decision process is due to Bellman (1957b). It was followed by Howard's (1960) pioneering monograph reporting extensive work on such processes which aroused great interest in the area. For a comprehensive discussion of significant contributions that have been made since then, the readers are referred to Derman (1970), Ross (1970, 1983), Denardo (1982), and Heyman and Sobel (1983).

Consider a Markov chain with state space S. Suppose with every state we associate a decision to be chosen out of a set D (in its generality D may depend on the state as well as the transition epoch). Let $^kP_{ij}$ be the probability of the one-step transition $i \rightarrow j$ ($i, j \in S$) under decision $k \in D$. Also we associate a reward $^kR_{ij}$ with decision k and transition $i \rightarrow j$. Knowing the set of alternatives in the decision set and the corresponding transition probabilities and rewards, the objective of the process is to select the optimal decision under certain criteria. When we associate rewards with every decision, maximization of expected reward over a given time horizon is the natural criterion. If costs are associated with decisions (costs are essentially negative rewards) minimization of expected costs is called for.

Let $^kV_i^{(n)}$ be the expected total earnings in n future transitions if decision k is made when the process is in state i. For the optimal decision $k = o$ if it exists, we have

$$^oV_i^{(n)} = \max_{k \in D} \sum_{j \in S} {}^kP_{ij}\left[{}^kR_{ij} + {}^oV_j^{(n-1)}\right] \qquad n = 1, 2, \ldots, i \in S \quad (10.1.1)$$

This is a functional equation satisfied by the expected reward. An alternate formulation would be to use average reward in place of total reward. In either case, as $n \rightarrow \infty$, an optimal policy satisfying the corresponding functional equation may not exist. For a discussion of conditions under which an optimal policy exists, the readers are referred to Derman (1970), Ross (1970, 1983), and Heyman and Sobel (1983). For our discussion we assume that the Markov chain is irreducible and that the state space of the Markov chain and the decision set D are finite, in which case optimal decisions are known to exist.

The technical difficulties associated with the expected reward case do not exist when the expected rewards are discounted every period with a factor α, $0 < \alpha < 1$. Then the functional equation takes the form

$$^oV_i^{(n)} = \max_{k \in D} \sum_{j \in S} {}^kP_{ij}\left[{}^kR_{ij} = \alpha^oV_j^{(n-1)}\right] \qquad n = 1, 2, \ldots, i \in S \quad (10.1.2)$$

The optimal policy in this case is identified as being α-optimal.

In the next section we explicitly formulate and give partial results to decision problems related to a system that can be modeled as a Markov chain. These results are primarily due to Howard (1960).

10.2   MARKOVIAN REWARD PROCESSES

Consider an aperiodic irreducible Markov chain with m states ($m < \infty$) and the transition probability matrix

$$P = \begin{bmatrix} P_{11} & P_{12} & \cdots & P_{1m} \\ \vdots & \vdots & & \vdots \\ P_{m1} & P_{m2} & & P_{mm} \end{bmatrix}$$

With every transition $i \to j$ associate a reward R_{ij}. If we consider an inventory system for which the transition probability matrix P represents the probabilities of transition of the inventory levels at successive epochs of time, then R_{ij} can be the profit derived out of it; or it could be merely the cost of storing items (negative reward). Let $V_i^{(n)}$ be the expected total earnings (reward) in the next n transitions, given that the system is in state i at present. A simple recurrence relation can be given for $\{V_i^{(n)}\}_{n=1}^{\infty}$ as follows:

$$V_i^{(n)} = \sum_{j=1}^{m} P_{ij}\left[R_{ij} + V_j^{(n-1)} \right] \qquad i = 1, 2, \ldots, m; \; n = 1, 2, 3, \ldots$$

$$(10.2.1)$$

Let

$$\sum_{j=1}^{m} P_{ij} R_{ij} = Q_i \qquad (10.2.2)$$

Equation (10.2.1) can now be written as

$$V_i^{(n)} = Q_i + \sum_{j=1}^{m} P_{ij} V_j^{(n-1)} \qquad (10.2.3)$$

Setting values $n = 1, 2, \ldots$, we get

$$V_i^{(1)} = Q_i + \sum_{j=1}^{m} P_{ij} V_j^{(0)}$$

$$V_i^{(2)} = Q_i + \sum_{j=1}^{m} P_{ij}\left[Q_j + \sum_{k=1}^{m} P_{jk} V_k^{(0)} \right]$$

$$= Q_i + \sum_{j=1}^{m} P_{ij} Q_j + \sum_{k=1}^{m} \sum_{j=1}^{m} P_{ij} P_{jk} V_k^{(0)}$$

$$= Q_i + \sum_{j=1}^{m} P_{ij} Q_j + \sum_{k=1}^{m} P_{ik}^{(2)} V_k^{(0)} \qquad (10.2.4)$$

where $P_{ij}^{(n)}$ is the (i, j)th element of the matrix P^n. Let

$$
V^{(n)} = \begin{bmatrix} V_1^{(n)} \\ V_2^{(n)} \\ \vdots \\ V_m^{(n)} \end{bmatrix} \qquad Q = \begin{bmatrix} Q_1 \\ Q_2 \\ \vdots \\ Q_m \end{bmatrix} \tag{10.2.5}
$$

Equation (10.2.4) can be put in matrix notation as

$$
V^{(2)} = Q + PQ + P^2 V^{(0)} \tag{10.2.6}
$$

Extending this to a general n, we get

$$
V^{(n)} = Q + PQ + P^2 Q + \cdots + P^{n-1} Q + P^n V^{(0)}
$$

$$
= \left[I + \sum_{k=1}^{n-1} P^k \right] Q + P^n V^{(0)} \tag{10.2.7}
$$

From Theorem 4.2.2 we have

$$
P_{ij}^{(n)} = \pi_j + e_{ij}^{(n)} \tag{10.2.8}
$$

where $\{ \pi_j \}_{j=1}^m$ is the limiting distribution of the Markov chain and

$$
|e_{ij}^{(n)}| \leq c r^n \tag{10.2.9}
$$

with $c > 0$ and $0 < r < 1$. Therefore, as $n \to \infty$,

$$
e_{ij}^{(n)} \to 0
$$

geometrically. Let

$$
\| e_{ij}^{(n)} \| = \eta^{(n)} \tag{10.2.10}
$$

and

$$
\Pi = \begin{bmatrix} \pi_1 & \pi_2 & \cdots & \pi_m \\ \pi_1 & \pi_2 & \cdots & \pi_m \\ \vdots & \vdots & & \vdots \\ \pi_1 & \pi_2 & \cdots & \pi_m \end{bmatrix} \tag{10.2.11}
$$

In matrix notation, we can write P^n as

$$P^n = \Pi + \eta^{(n)} \tag{10.2.12}$$

Substituting this form in (10.2.7), we get

$$V^{(n)} = \left[I + \sum_{k=1}^{n-1} \left(\Pi + \eta^{(k)} \right) \right] Q + \left(\Pi + \eta^{(n)} \right) V^{(0)}$$

$$= \Pi V^{(0)} + \sum_{k=0}^{n-1} \eta^{(k)} Q + n \Pi Q + \eta^{(n)} V^{(0)} \tag{10.2.13}$$

In deriving (10.2.13), we have noted that $\eta^{(0)} = I - \Pi$.
Now consider the sum

$$\sum_{k=0}^{n-1} \eta^{(k)} = \sum_{k=0}^{\infty} \eta^{(k)} - \sum_{k=n}^{\infty} \eta^{(k)} \tag{10.2.14}$$

It should be noted that each term in $\eta^{(k)}$ is less than or equal to $cr^k (c > 0,$ $0 < r < 1)$ in absolute value, and hence for large n the second term on the right-hand side of (10.2.14) approaches an $m \times m$ matrix with zero elements. For the same reason the last term in (10.2.13) approaches a null matrix for large n. Thus asymptotically we have

$$V^{(n)} \doteq \Pi V^{(0)} + \sum_{k=0}^{\infty} \eta^{(k)} Q + n \Pi Q \tag{10.2.15}$$

which gives

$$V_i^{(n)} \doteq \sum_{j=1}^{m} \pi_j V_j^{(0)} + \sum_{j=1}^{m} \gamma_{ij} Q_j + n \sum_{j=1}^{m} \pi_j Q_j \tag{10.2.16}$$

where we have written $\sum_{k=0}^{\infty} \eta^{(k)} = \| \gamma_{ij} \|$. Writing

$$\sum_{j=1}^{m} \pi_j V_j^{(0)} + \sum_{j=1}^{m} \gamma_{ij} Q_j = v_i \tag{10.2.17}$$

and

$$\sum_{j=1}^{m} \pi_j Q_j = q \tag{10.2.18}$$

(10.2.16) can be put in the form

$$V_i^{(n)} = v_i + nq \tag{10.2.19}$$

which shows that for large n, $V_i^{(n)}$ is a linear function of q for every i. Further, for different values of i, $V_i^{(n)}$ are represented by parallel straight lines with slope q and intercepts v_i ($i = 1, 2, \ldots, m$).

So far we have considered the transition probability matrix P and the reward matrix R as given. Instead, suppose the decision maker has other alternatives and so is able to alter the elements of P and R. To incorporate this feature, define D as the decision set, and under a decision $k \in D$ let $^kP_{ij}$ and $^kR_{ij}$ be the probability of the transition $i \to j$ and the corresponding reward, respectively for $^kV_i^{(n)}$ the expected total earnings in n transitions under decision k, we have the recurrence relations ($k = o$ represents the optimal decision)

$$^oV_i^{(n)} = \max_{k \in D} \sum_{j=1}^{m} {}^kP_{ij}\left[{}^kR_{ij} + {}^oV_j^{(n-1)}\right] \qquad n = 1, 2, \ldots; \ i = 1, 2, \ldots, m$$

$$\tag{10.2.20}$$

giving

$$^oV_i^{(n)} = \max_{k \in D}\left[{}^kQ_i + \sum_{j=1}^{m} {}^kP_{ij}\, {}^oV_j^{(n-1)}\right] \qquad i = 1, 2, \ldots, m; \ n = 1, 2, \ldots$$

$$\tag{10.2.21}$$

where we have written $\sum_{j=1}^{m} {}^kP_{ij}\, {}^kR_{ij} = {}^kQ_i$. Recursive relation (10.2.20) gives an iterative procedure to determine the optimum decisions $d_i^{(n)} \in D$, for $i = 1, 2, \ldots, m$ and $n = 1, 2, \ldots$. This is a standard technique in dynamic programming, and it has been shown (Bellman, 1957, Chap. 11) that this iteration process will converge on the best alternative for each state as $n \to \infty$.

Since the procedure is based on the value of the policy (total earning) for any n, it is called the value iteration method. The method is based on recursively determining the optimum policy for every n, that would give the maximum value. This enumeration technique is illustrated in the following example.

Example 10.2.1

The manufacturer of a certain product has found tough competition in business and would like to use analytical techniques in making decisions for advertising and investing in the research and development of new products. The following information is available from current market surveys done on the product: using a week as the time unit, the product undergoes state changes between favorable (1) and unfavorable (2) states based on the following transition probability matrix:

$$P = \begin{bmatrix} 0.5 & 0.5 \\ 0.25 & 0.75 \end{bmatrix}$$

Sales figures give the following reward (profit) matrix corresponding to these transitions (each unit representing $1000):

$$R = \begin{bmatrix} 10 & 5 \\ 4 & -2 \end{bmatrix}$$

When the product is in state 1, let there be two alternatives open to the manufacturer:

Alternative 1:   Continue without change.
Alternative 2:   Increase advertising.

Let the corresponding transition probabilities and rewards be given as

$$[^1P_{11}, {}^1P_{12}] = [0.5, 0.5]$$

$$[^1R_{11}, {}^1R_{12}] = [10, 5]$$

and

$$[^2P_{11}, {}^2P_{12}] = [0.75, 0.25]$$

$$[^2R_{11}, {}^2R_{12}] = [6, 3]$$

When the product is in state 2, let the two alternatives open to him be as follows:

Alternative 1:   Continue without change.
Alternative 2:   Invest in research and development.

Then suppose

$$[^1P_{21}, {}^1P_{22}] = [0.25, 0.75]$$

$$[^1R_{21}, {}^1R_{22}] = [4, -2]$$

and

$$[^2P_{21}, {}^2P_{22}] = [0.8, 0.2]$$

$$[^2R_{21}, {}^2R_{22}] = [-2, -25]$$

We shall use these values in (10.2.21) to determine the best policies for every n. We have

$${}^1Q_1 = {}^1P_{11}{}^1R_{11} + {}^1P_{12}{}^1R_{12} = 7.5$$

$${}^1Q_2 = {}^1P_{21}{}^1R_{21} + {}^1P_{22}{}^1R_{22} = -0.5$$

$${}^2Q_1 = {}^2P_{11}{}^2R_{11} + {}^2P_{12}{}^2R_{12} = 5.25$$

$${}^2Q_2 = {}^2P_{21}{}^2R_{21} + {}^2P_{22}{}^2R_{22} = -6.6 \qquad (10.2.22)$$

Let ${}^oV_i^{(0)} = 0$ for $i = 1, 2$. Then for $n = 1$ in (10.2.21) we find ${}^oV_i^{(1)} = \max_{1,2} {}^kQ_i$, and hence

$$d_1^{(1)} = 1 \quad \text{and} \quad d_2^{(1)} = 1 \qquad (10.2.23)$$

Let ${}^oV_1^{(1)}$ and ${}^oV_2^{(1)}$ be the maximum earnings corresponding to $d_1^{(1)}$ and $d_2^{(1)}$, respectively. We have ${}^oV_1^{(1)} = 7.5$ and ${}^oV_2^{(1)} = -0.5$. For $n = 2$, from (10.2.21), we have

$${}^oV_i^{(2)} = \max_{1,2} \left[{}^kQ_i + \sum_{j=1}^{2} {}^kP_{ij} \, {}^oV_j^{(1)} \right]$$

which gives for

$i = 1$ and $k = 1$ ${}^1V_1^{(2)} = 7.5 + 0.5 \times 7.5 + 0.5 \times (-0.5) = 11.0$

$i = 1$ and $k = 2$ ${}^2V_1^{(2)} = 5.25 + 0.75 \times 7.5 + 0.25 \times (-0.5) = 10.75$

$i = 2$ and $k = 1$ ${}^1V_2^{(2)} = -0.5 + 0.25 \times 7.5 + 0.75 \times (-0.5) = 1.00$

$i = 2$ and $k = 2$ ${}^2V_2^{(2)} = -6.6 + 0.8 \times 7.5 + 0.2 \times (-0.5) = -0.61$

Clearly

$$d_1^{(2)} = 1 \quad \text{with } {}^oV_1^{(2)} = 11.0$$

$$d_2^{(2)} = 1 \quad \text{with } {}^oV_2^{(2)} = 1.00 \qquad (10.2.24)$$

Proceeding in this manner, we get

$$d_1^{(3)} = 2 \qquad \text{with } {}^oV_1^{(3)} = 13.75$$

$$d_2^{(3)} = 1 \qquad \text{with } {}^oV_2^{(3)} = 3.00$$

$$d_1^{(4)} = 2 \qquad \text{with } {}^oV_1^{(4)} = 16.3125$$

$$d_2^{(4)} = 1 \qquad \text{with } {}^oV_1^{(4)} = 5.1875$$

$$d_1^{(5)} = 2 \qquad \text{with } {}^oV_2^{(5)} = 18.7812$$

$$d_2^{(5)} = 2 \qquad \text{with } {}^oV_2^{(5)} = 7.4875 \qquad (10.2.25)$$

and so on. One major drawback of the method is that there is no way to say when the policy converges into a stable policy; therefore the value iteration procedure is useful only when n is fairly small. In practice, we are confronted with systems that operate over a long period of time, and in such cases it is better to make use of the asymptotic result derived in (10.2.19) to arrive at the best alternative for every initial state i. Such a procedure has also been given by Howard (1960), which is known as the policy iteration method. This method has two phases: (1) value determination operation and (2) policy improvement routine. In the value determination operation relative values of v_i ($i = 1, 2, \ldots, m - 1$) and q are obtained setting $v_m = 0$ by the following arguments: Substituting (10.2.19) in (10.2.3), we get

$$nq + v_i = Q_i + \sum_{j=1}^{m} P_{ij}\big[(n-1)q + v_j\big] \qquad n > 1, \, i = 1, 2, \ldots, m$$

$$q + v_i = Q_i + \sum_{j=1}^{m} P_{ij}v_j \qquad\qquad n = 1 \qquad (10.2.26)$$

Since there are m equations in $m + 1$ unknowns, q and v_i ($i = 1, 2, \ldots, m$), setting $v_m = 0$, we can obtain the relative values of the remaining unknowns by solving these equations simultaneously. The P_{ij} and Q_i values in (10.2.3) are based on the existing policy.

In the policy improvement routine, the best alternate policy is determined so as to maximize

$${}^kQ_i + \sum_{j=1}^{m} {}^kP_{ij}v_j \qquad (10.2.27)$$

where v_j $(j = 1, 2, \ldots, m)$ are the relative values obtained in the earlier operation. The expression (10.2.27) is obtained by substituting from (10.2.19) in the right-hand side expression of (10.2.21) and neglecting the constant term which is unnecessary for comparison purposes.

The iteration policy is continued until one gets the same optimal policy on two successive iterations. It can be shown that the iteration cycle will terminate at the policy that has the largest gain attainable within the realm of the problem. For further details and extensions of this method, see Howard (1960, pp. 32–75).

Consider the decision problem with discounting. Let α $(0 < \alpha < 1)$ be the discounting factor. In this case the result corresponding to (10.2.19) is independent of n. This can be seen as follows: Let $V_i^{(n)}$ be the expected total earnings for n transition periods during which the Markov chain operates with a transition probability matrix $\|P_{ij}\|$ and reward matrix $\|R_{ij}\|$. The sequence $\{V_i^{(n)}\}_{n=0}^{\infty}$ satisfies the recurrence relations

$$V_i^{(n)} = \sum_{j=1}^{m} P_{ij}\left[R_{ij} + \alpha V_j^{(n-1)} \right]$$

$$= Q_i + \alpha \sum_{j=1}^{m} P_{ij} V_j^{(n-1)} \qquad i = 1, 2, \ldots, m; \; n = 1, 2, \ldots$$

$$(10.2.28)$$

Solving these equations recursively, in matrix notations we have

$$V^{(1)} = Q + \alpha P V^{(0)}$$

$$V^{(2)} = Q + \alpha P Q + (\alpha P)^2 V^{(0)}$$

$$\vdots$$

$$V^{(n)} = Q + \alpha P Q + (\alpha P)^2 Q + \cdots + (\alpha P)^{n-1} Q + (\alpha P)^n V^{(0)}$$

$$= \left[I + \sum_{k=1}^{n-1} (\alpha P)^k \right] Q + (\alpha P)^n V^{(0)}$$

$$= \left[\sum_{k=0}^{n-1} \alpha^k \left(\Pi + \eta^{(k)} \right) \right] Q + \alpha^n \left[\Pi + \eta^{(n)} \right] V^{(0)}$$

$$= \frac{1 - \alpha^n}{1 - \alpha} \Pi Q + \sum_{k=0}^{\infty} \alpha^k \eta^{(k)} Q - \sum_{k=n}^{\infty} \alpha^k \eta^{(k)} Q + \alpha^n \Pi V^{(0)} + \alpha^n \eta^{(n)} V^{(0)}$$

$$(10.2.29)$$

For large n, $\alpha^n \rightarrow 0$ as well as $\eta^{(n)} \rightarrow \phi$, the null matrix. Therefore, writing

$$\sum_{k=0}^{\infty} \alpha^k \eta^{(k)} = \|\gamma_{ij}(\alpha)\|$$

from (10.2.29) we get

$$V_i^{(n)} \doteq \frac{1}{1-\alpha} \sum_{j=1}^{m} \pi_j Q_j + \sum_{j=1}^{m} \gamma_{ij}(\alpha) Q_j \tag{10.2.30}$$

giving

$$V_i^{(n)} \doteq \frac{1}{1-\alpha} q + v_i(\alpha) \tag{10.2.31}$$

and showing that the present worth of expected future earnings of the process do not grow without bounds with n, as they did in the no-discounting case.

Even though one could use the same decision process as employed in the no-discounting case, as $n \rightarrow \infty$, from (10.2.29) we get

$$\lim_{n \rightarrow \infty} V^{(n)} = (1 - \alpha P)^{-1} Q \tag{10.2.32}$$

The term $(1 - \alpha P)^{-1}$ exists since αP is a substochastic matrix. Introducing the decision set D, for the optimal decision $k = o$, we may write

$$^{o}V = (1 - \alpha - {}^{o}P)^{-1} {}^{o}Q \tag{10.2.33}$$

where the vectors ^{o}V and ^{o}Q, and the matrix ^{o}P correspond to the optimal decision. Using this result, a policy iteration algorithm can be given that consists of the two phases identified earlier, viz, value determination and policy improvement. For a description of this algorithm and the existence of the optimal policy, see Denardo (1982, p. 167).

10.3 CHOOSING THE BEST TRANSITION PROBABILITY MATRIX

The policy iteration method described in Section 10.2 accomplished essentially the following: A Markov chain (system) can be in any one of the $m(< \infty)$ states. Associated with each state i ($i = 1, 2, \ldots, m$) let P_i be the set of available alternate probability vectors $({}^{k}P_{i1}, {}^{k}P_{i2}, \ldots, {}^{k}P_{im})$, $k \in D_i$.

We wish to select one vector out of P_i in such a way that when it is used for the one-step transition probabilities out of state i in the Markov chain, the ultimate gain is maximized, assuming that the system stays in operation for a sufficiently long time. Repeating the iteration procedure for all the states $i = 1, 2, \ldots, m$, we can assemble a transition probability matrix for the Markov chain that can be considered to be the best under the given circumstances. The iteration method used in achieving this result has been shown to be a special extension of the simplex method of linear programming allowing for multiple substitutions (see de Ghellinck, 1960). In this section we shall formulate the problem of choosing the best transition probability matrix of a simple inventory system as a direct linear programming problem. The formulation of such a problem as a linear programming problem is originally due to D'Epenoux (1960), Manne (1960), and Oliver (1960), but here we shall give the elegant and more general formulation due to Wolfe and Dantzig (1962).

Let an inventory system have m attainable levels of inventory (e.g., a capacity for storing $m - 1$ units). Let c_{ij} $(i, j = 1, 2, \ldots, m)$ be the cost attached to the transition $i \to j$. In terms of rewards considered in Section 10.2, $c_{ij} = -R_{ij}$. Let the system be operating for a sufficiently long time. We are interested in minimizing the total expected cost associated with each transition. The problem is therefore to choose a transition probability vector $^kP_i = (^kP_{i1}, ^kP_{i2}, \ldots, ^kP_{im})$, $k \in D_i$ and $i = 1, 2, \ldots, m$, such that the total expected cost,

$$\sum_{i=1}^{m} \sum_{j=1}^{m} \pi_i{}^k P_{ij} c_{ij} \tag{10.3.1}$$

is minimized, where $\pi = (\pi_1, \pi_2, \ldots, \pi_m)$ is the corresponding limiting distribution. In linear programming terms we may then say that we need to determine ^kP_i, $k \in D_i$ and $i = 1, 2, \ldots, m$, such that (10.3.1) is minimized for all $(\pi_1, \pi_2, \ldots, \pi_m)$, and hence

$$\sum_{1}^{m} \pi_j = 1 \quad \text{and} \quad \sum_{i=1}^{m} \Pi_i{}^k P_i = \pi \qquad \pi_i \geq 0 \tag{10.3.2}$$

To put (10.3.1) and (10.3.2) in a convenient form, we write

$$^kQ_i = \left[^kP_{i1}, \ldots, ^kP_{ii} - 1, \ldots, ^kP_{im} \right] \tag{10.3.3}$$

and define $\bar{c}_{ij}$, such that

$$\bar{c}_{ij}{}^kQ_{ij} = c_{ij}{}^kP_{ij} \tag{10.3.4}$$

Let D_i' be the decision set corresponding to ^kQ_i. Then the transformed programming problem is to determine ^kQ_i, $k \in D_i'$ and $i = 1, 2, \ldots, m$, such that

$$\sum_i \sum_j \pi_i^k Q_{ij} \bar{c}_{ij} \tag{10.3.5}$$

is minimized for all π, such that

$$\sum_1^m \pi_i = 1 \quad \text{and} \quad \sum_{i=1}^m \pi_i^k Q_i = 0 \qquad \pi_i \geq 0 \tag{10.3.6}$$

Clearly the solutions to problems (10.3.1), (10.3.2) and (10.3.5), (10.3.6) are the same. The problem (10.3.5), (10.3.6) is solved using the decomposition principle developed earlier by Dantzig and Wolfe (1960); for details of this technique and the solution of the present problem, the reader is referred to the articles cited earlier.

For a slightly different formulation of a similar problem, see Manne (1960). In this article the author discusses the possibilities of extending the formulation to include seasonal variations in inventory systems or multilocation inventory problems.

10.4 REMARKS

Inventory and queueing theory provide two major areas where decision problems under Markovian setting are appropriate. For instance, optimization problems in queueing systems have been formulated as continuous time Markov decision processes by Miller (1968a, 1968b, 1969), Crabill (1972, 1975) and others. For extensive bibliographical listings on the control of queues that include techniques based on Markov decision processes, the reader may be referred to Sobel (1973), Stidham and Prabhu (1973), and Heyman and Sobel (1983). For some Markov decision problems in inventory theory, see Veinott (1966), Ross (1970, 1983), Denardo (1982), and Heyman and Sobel (1983).

From the investigations of Markov decision problems a common operational characteristic can be identified. Once the functional equation to be satisfied by the optimal policy is formulated, the theory of optimization is brought to bear in order to establish the optimal policy if it exists. Then the properties need to be investigated are those of the underlying controlled stochastic processes, which require considerably more mathematical sophistication than assumed as the necessary background for this text. For the interested reader in addition to the books referred to earlier (Derman, 1970,

Ross, 1970, 1983, Denardo, 1982, and Heyman and Sobel, 1983), Howard (1971), Dynkin and Yushkevich (1979), Gihman and Skorohod (1979), and Whittle (1982) may be recommended. Articles listed for further reading provide a sampling of investigations covering various application areas.

REFERENCES

Bellman, R. (1957). *Dynamic Programming*. Princeton: Princeton University Press.

Bellman, R. (1957). "A Markovian Decision Process." *J. Math. Mech.* **6**, 679–684.

Crabill, T. B. (1972). "Optimal Control of a Service Facility with Variable Exponential Service Times and Constant Arrival Rate." *Management Sci.* **18**, 560–566.

Crabill, T. B. (1975). "Optimal Control of a Maintenance System with Variable Services Rates." *Oper. Res.* **22**, 736–745.

Dantzig, G. B., and Wolfe, P. (1960). "Decomposition Principle for Linear Programs." *Oper. Res.* **8**, 101–111.

Denardo, E. V. (1982). *Dynamic Programming: Theory and Application*. Englewood Cliffs, N.J.: Prentice-Hall.

D'Epenoux, F. (1960). "Sur un problém de production et de stockage dans l'aleatoire." *Rev. Fr. Informat. Rech. Opérationnelle*, 3–16. [Eng. transl.: *Management Sci.* **10**, 98–108, (1963)]

de Ghellinck, G. (1960). "Les problemes de decisions sequentielles." *Cahiers Centre d'Etude Rech. Opérationelle* **2** (2), 161–179.

Derman, C. (1970). *Finite State Markovian Decision Processes*. New York: Academic Press.

Dynkin, E. B., and Yushkevich, A. A. (1979). *Controlled Markov Processes*. Berlin: Springer-Verlag (English trans.; Russian ed. 1975). Includes a good bibliography of papers published in probability journals.

Gihman, I. I., and Skorohod, A. V. (1979). *Controlled Stochastic Processes*. Berlin: Springer-Verlag (English trans.; Russian ed. 1977).

Heyman, D. P., and Sobel, M. J. (1983). *Stochastic Models in Operations Research*. Vol. 2. New York: McGraw-Hill.

Howard, R. A. (1960). *Dynamic Programming and Markov Processes*. Cambridge, Mass.: The MIT Press.

Howard, R. A. (1971). *Dynamic Probabilistic Systems*. Vols. 1, 2. New York: Wiley.

Manne, A. S. (1960). "Linear Programming and Sequential Decisions." *Management Sci.* **6** (3), 259–267.

Miller, B. L. (1968). "Finite State Continuous Time Markov Decision Processes with a Finite Planning Horizon." *SIAM J. Control* **6**, 266–280.

Miller, B. L. (1968). "Finite State Continuous Time Markov Decision Processes with an Infinite Planning Horizon." *J. Math. Anal. Appl.* **22**, 552–569.

Miller, B. L. (1969). "A Queueing Reward System with Several Customer Classes." *Management Sci.* **16**, 234–245.

Oliver, R. M. (1960). "A Linear Programming Formulation of Some Markov Decision Processes." Presented at the Meeting of the Institute of Management Sciences, Monterey, April 1960.

Ross, S. M. (1970). *Applied Probability Models with Optimization Applications*. San Francisco: Holden-Day.

Ross, S. M. (1983). *Introduction to Stochastic Dynamic Programming*. New York: Academic Press.

Sobel, M. J. (1973). "Optimal Operation of Queues." *Mathematical Methods in Queueing Theory*. Proceedings of a Conference, May 10–12, 1973. New York: Springer-Verlag, 231–261. Includes a good bibliography.

Stidham, S., Jr., and Prabhu, N. U. (1973). "Optimal Control of Queueing Systems." *Mathematical Methods in Queueing Theory*. Proceedings of a Conference, May 10-12, 1973. New York: Springer-Verlag, 262–294. Includes a good bibliography.

Veinott, A. F., Jr. (1966). "The Status of Mathematical Inventory Theory." *Management Sci.* **12**, 745–777.

Wald, A. (1947). *Sequential Analysis*. New York: Wiley.

Wald, A. (1950). *Statistical Decision Functions*. New York: Wiley.

Whittle, P. (1982). *Optimization Over Time: Dynamic Programming and Stochastic Control*. New York: Wiley.

Wolfe, P., and Dantzig, G. B. (1962). "Linear Programming in a Markov Chain." *Oper. Res.* **10**, 702–710.

FOR FURTHER READING

Albright, S. C. (1974). "A Markov-Decision-Chain Approach to a Stochastic Assignment Problem." *Oper. Res.* **22**, 61–64.

Albright, S. C., and Winston, W. (1979). "Markov Models of Advertising and Pricing Decisions." *Oper. Res.* **27**, 668–681.

Beebe, J. H., Beightler, C. S., and Stark, J. P. (1968). "Stochastic Optimization of Production Planning." *Oper. Res.* **16**, 799–818.

Blackwell, D. (1962). "Discrete Dynamic Programming." *Ann. Math. Stat.* **33**, 719–726.

Blackwell, D. (1965). "Discounted Dynamic Programming." *Ann. Math. Stat.* **36**, 226–235.

Crabill, T. B., Gross, D., and Magazine, M. J. (1977). "A Classified Bibliography of Research on Optimal Design and Control of Queues." *Oper. Res.* **25**, 219–232.

de Cani, J. S. (1964). "A Dynamic Programming Algorithm for Embedded Markov Chains where the Planning Horizon is at Infinity." *Management Sci.* **10**, 716–733.

Denardo, E. V. (1968). "Separable Markov Decision Problems." *Management Sci.* **14**, 451–462.

Derman, C. (1962). "On Sequential Decisions and Markov Chains." *Mangement Sci.* **9**, 16–24.

Derman, C. (1963a). "On Optimal Replacement Rules When Changes of State are Markovian." *Mathematical Optimization Techniques.* Berkeley: University of California Press, Chap. 9.

Derman, C. (1963b). "Optimal Replacement and Maintenance under Markovian Deterioration with Probability Bounds for Failure." *Management Sci.* 9(3), 478–481.

Derman, C. (1966). "Denumerable State Markovian Decision Processes—Average Cost Criterion." *Ann. Math. Stat.* 37, 1545–1553.

Derman, C., and Klein, M. (1965). "Some Remarks on Finite Horizon Markovian Decision Models." *Oper. Res.* 13, 272–278.

Dobbie, J. M. (1974). "A Two-Cell Model of Search for a Moving Target." *Oper. Res.* 22, 79–92.

Emmons, H. (1972). "The Optimal Admissions Policy to a Multi-Server Queue with Finite Horizon." *J. Appl. Prob.* 9, 103–116.

Evans, J. P. (1969). "Duality in Markov Decision Problems with Countable Action and State Spaces." *Mangement Sci.* 15, 626–638.

Evans, R. V. (1971). "Programming Problems and Changes in the Stable Behavior of a Class of Markov Chains." *J. Appl. Probl.* 8, 543–550.

Fox, B. (1966). "Markov Renewal Programming by Linear Fractional Programming." *SIAM J. Appl. Math.* 14, 1418–1432.

Grinold, R. C. (1974). "A Generalized Discrete Dynamic Programming Model." *Management Sci.* 20, 1092–1103.

Hastings, N. A. J. (1968). "Some Notes on Dynamic Programming and Replacement." *Oper. Res. Quart.* 19, 453–464.

Howard, R. A., and Matheson, J. E. (1972). "Risk-Sensitive Markov Decision Processes." *Management Sci.* 18, 356–369.

Ignall, E., and Kolesar, P. (1974). "Optimal Dispatching of an Infinite Capacity Shuttle Control at a Single Terminal." *Oper. Res.* 22, 1008–1024.

Jewell, W. S. (1963). "Markov Renewal Programming, I & II." *Oper. Res.* 11, 938–971.

Klein, M. (1962). "Inspection-Maintenance-Replacement Schedules under Markovian Deterioration." *Management Sci.* 9(1), 25–32.

Lippman, S. A. (1975). "On Dynamic Programming with Unbounded Rewards." *Management Sci.* 21, 1225–1233.

Lippman, S. A. (1976). "Countable-State, Continuous-Time, Dynamic Programming with Structure." *Oper. Res.* 24, 477–490.

MacQueen, J. (1966). "A Modified Programming Method for Markovian Decision Problems." *J. Math. Anal. Appl.* 14, 38–43.

Martin-Löf, A. (1967). "Optimal Control of a Continuous-Time Markov Chain with Periodic Transition Probabilities." *Oper. Res.* 15, 872–881.

Norman, J. M., and White, D. J. (1968). "A Method for Approximate Solutions to Stochastic Dynamic Programming Problems Using Expectations." *Oper. Res.* 16, 296–306.

Porteus, E. L. (1975). "Bounds and Transformations for Discounted Finite Markov Decision Chains." *Oper. Res.* 23, 761–784.

Riis, J. O. (1965). "Discounted Markov Programming in a Periodic Process." *Oper. Res.* **13**, 920–929.

Rosenfield, D. (1976). "Markovian Deterioration with Uncertain Information." *Oper. Res.* **24**, 141–155.

Ross, S. M. (1968). "Non-discounted Denumerable Markovian Decision Models." *Ann. Math. Stat.* **39**, 412–423.

Satia, J. K., and Lave, R. E. (1973). "Markovian Decision Processes with Probabilistic Observation of States." *Management Sci.* **20**, 1–13.

Satia, J. K., and Lave, R. E., Jr. (1973). "Markovian Decision Processes with Uncertain Transition Probabilities." *Oper. Res.* **21**, 728–740.

Shapiro, J. F. (1968). "Turnpike Planning Horizons for a Markovian Decision Model." *Management Sci.* **14**, 292–300.

Su, S. Y., and Deininger, R. A. (1972). "Generalization of White's Method of Successive Approximations to Periodic Markovian Decision Processes." *Oper. Res.* **20**, 318–326.

Veinott, A. F., Jr. (1966). "On Finding Optimal Policies in Discrete Dynamic Programming with No Discounting." *Ann. Math. Stat.* **37**, 1286–1294.

White, D. J. (1963). "Dynamic Programming, Markov Chains, and the Method of Successive Approximations." *J. Math. Anal. Appl.* **6**, 373–376.

Chapter 11
QUEUEING PROCESSES

11.1 INTRODUCTION

Congestion is a natural phenomenon in real systems. A service facility gets congested if there are more people than the server can possibly handle. A sidewalk gets congested when there are too many pedestrians using the sidewalk at the same time. There is congestion in a service station if there are too many machines awaiting service. There is congestion in a computing center when the number of jobs demanding service exceeds the number it can process. In all these situations the uncertainties related to the system characteristics such as the arrival of customers, the time needed for service, and the processes occurring in the analysis of these systems are better represented by stochastic processes. Queueing is a mechanism that is used to handle congestion. It helps to organize the various elements of the system in a manner conducive to modeling.

A system consisting of a servicing facility, a process of arrival of customers who wish to be served by the facility, and the process of service is called a queueing system. As an idealized model this can represent an actual queue situation in front of a counter, a group of pedestrians on a sidewalk, a system of machine repair, vehicles waiting to cross a major road, telephone calls waiting to be connected at a switchboard and many other situations. The four common characteristics of such systems are the following:

1. *Input process:* If the arrivals and service rates are strictly according to schedule, a queue can be avoided. But in practice this is not the case, and in most situations arrivals are controlled by factors external to the system. Therefore the best that can be done is to represent the input process in terms of random variables. Some factors needed for the complete specification of an input process are the source of arrivals, the type of arrivals, and the interarrival times.

2. *Service mechanism:* The uncertainties involved in the service mechanism are the number of servers, the number of customers getting served at

any time, and the duration of service. Again, random variable representations of these characteristics seem to be essential.

3. *Queue discipline:*   All other factors regarding the rules of conduct of the queue can be pooled under this heading. One of these is the rule followed by the server in taking the customers into service. In this context rules such as "first-come, first-served," "last-come, first-served," and "random selection for service" are self-explanatory. In many situations "priority" disciplines need to be introduced so as to make the system more realistic. Some other factors that may be included here are based on customer behavior, such as balking, reneging, and jockeying.

4. *Number of queues:*   When there is only one server in the system, specifications of the preceding three factors give a complete description. However, in problems of communication networks and job-shop scheduling, one has to deal with more than one queue (server) in series and/or in parallel.

In this chapter we shall consider only some special cases of these characteristics. Except in Section 11.7, the arrivals are assumed to be occurring one at a time from an infinite source in a Poisson process. In the machine-servicing process discussed in Section 11.7 we shall consider a finite source for arrivals. Except in Section 11.6, we shall assume that the service times have exponential distributions, and the basic queue discipline will always be assumed to be first-come, first-served, allowing the variations in customer and server behavior. Further, even when there are more than one servers, we shall assume that there is only one queue in the system.

It is convenient to simplify the description of a queueing system by a notational representation. Out of the four characteristics of a queueing system the third mentioned, queue discipline, is rather qualitative, and it is better described in the proper context. The number of queues will always be one in the systems we consider. The remaining characteristics can be summarized by specifying the input distribution, service time distribution, and the number of servers. This can be written as

Input distribution/Service time distribution/Number of servers

in that order. Some standard notations used are G (or GI, as originally used by Kendall, 1953, to represent "general independent" inputs) for an arbitrary distribution, M for Poisson (if arrivals) or exponential distributions (for interarrival or service times), D for a constant length of time (for interarrival or service times), and E_k for the Erlangian distribution (Gamma distribution with one of the parameters, k, an integer). For instance, suppose the arrivals are Poisson, service times are exponential, and there are

s servers. Then we shall denote such a system by $M/M/s$. In this notation some system variants we shall be considering are $M/M/s$, $M/M/\infty$, $M/D/1$, and $M/G/1$.

Due to the random fluctuations in the input and service process, system characteristics lend themselves to modeling after stochastic processes. Some of them are the following:

1. *Number of customers in the system:* Let $Q(t)$ be the number of customers in the system (called queue length) at time t. Clearly this has a continuous time parameter t and can take on values $\{0, 1, 2, 3, \dots\}$. Based on the uncertainties of the arrival and service process, $Q(t)$ is a stochastic process. It should be noted that we have defined $Q(t)$ so as to include customers who are in service. Except in some cases of group service or specialized disciplines, the number of customers who are being served at time t is known when $Q(t)$ is known, and hence the actual queue length can be easily obtained.

By observing the system at discrete epochs such as arrival or departure epochs, we also define a discrete parameter, discrete state stochastic process $\{Q_n, n \geq 0\}$ to represent the number of customers in the system at such time points.

One major advantage of the processes $Q(t)$ and Q_n is that they are independent of the queue discipline, provided the server does not remain idle as long as there is a customer demanding service and every customer is given complete service at one time.

2. *Waiting time:* Suppose a customer wishes to join the queue at time t. At that point, in addition to the number of customers in the system, information regarding the time the customer may have to wait until served would help the customer in deciding to join the system. Let $W(t)$ be the time the customer has to wait until entering service after deciding to join the system at time t. When the queue discipline is first-come, first-served, $W(t)$ is simply the aggregate of remaining service time of those who are already in the system. Clearly $W(t)$ is a continuous parameter and continuous state space stochastic process.

Another waiting time process can be defined with reference to every customer joining the system. Let $C_1, C_2, C_3, \dots$ be the customers arriving at the system, and let W_n be the time the customer C_n has to wait until entering service. Even though the processes $W(t)$ and W_n are related in the same system, W_n is conditional on observing the system at discrete epochs of time, whereas $W(t)$ makes no such assumption. To highlight this distinction, we may call W_n the actual waiting time of the nth customer and $W(t)$ the virtual waiting time at time t. Even though one might feel that W_n and $W(t)$ should have the same limiting distribution (as $n \to \infty$ or $t \to \infty$), in general there is no guarantee that it would be so. A similar statement can be made regarding Q_n and $Q(t)$, the number of customers in the system.

3. *Busy period and busy cycle:* The number of times the processes $Q(t)$ and $W(t)$ visit state 0 yields information regarding its operational efficiency, and under most of the assumptions made for queueing systems this is a renewal counting process. The basic renewal period for this process is the busy cycle which is the time interval between two consecutive epochs at which the server starts work after being idle for a positive length of time. The busy cycle is made up of two parts: busy period in which the server remains continuously busy and idle period during which the server does not work.

In the next few sections we shall discuss mainly the limiting distributions of the number of customers in the system and the waiting times for some simple queueing systems. Queueing theory literature is vast, and the techniques of analysis are much more sophisticated than the material covered in this text. The list of references given at the end of the chapter should prove helpful for readers who develop interest in these processes.

11.2 A GENERALIZED QUEUEING MODEL

Under certain assumptions on the arrival and service processes, a large class of queueing systems can be modeled as a generalized birth and death process described in Section 7.5:

1. *Arrivals:* If the number of customers in the system is n at time t, during the following infinitesimal interval $(t, t + \Delta t]$ let the probability that one new customer arrives be $\lambda_n \Delta t + o(\Delta t)$, the probability that no customer arrives be $1 - \lambda_n \Delta t + o(\Delta t)$, and the probability that more than one customer arrives be $o(\Delta t)$, where $o(\Delta t)$ is such that $o(\Delta t)/\Delta t \to 0$ as $\Delta t \to 0$. Also assume that the customer arrivals during $(t, t + \Delta t]$ are independent of the time since the last arrival.

2. *Service:* Let the service process be such that if the number of customers in the system at time t is n, during the following infinitesimal interval $(t, t + \Delta t]$ the probability that there is one service completion is $\mu_n \Delta t + o(\Delta t)$, the probability that there is no service completion is $1 - \mu_n \Delta t + o(\Delta t)$, and the probability that there is more than one service completion is $o(\Delta t)$. Also assume that the service completion epoch is independent of the time it started.

3. Further we shall assume that the dependence between arrivals and departures is restricted only to Assumptions 1 and 2 in which they both depend on the population size. To be realistic, we shall assume $\mu_0 = 0$ and $\lambda_0 > 0$.

Under these conditions the queueing process defined is identical to the generalized birth and death process of Section 7.6. Let $Q(t)$ be the number of customers in the system at time t, and define

$$P_n(t) = P[Q(t) = n] \tag{11.2.1}$$

From equations (7.6.5) and (7.6.6) we have

$$P_0'(t) = -\lambda_0 P_0(t) + \mu_1 P_1(t) \tag{11.2.2}$$

$$P_n'(t) = -(\lambda_n + \mu_n)P_n(t) + \lambda_{n-1}P_{n-1}(t) + \mu_{n+1}P_{n+1}(t) \tag{11.2.3}$$

with the initial conditions

$$P_n(0) = \begin{cases} 1 & \text{if } n = i \\ 0 & \text{if } n \neq i \end{cases} \tag{11.2.4}$$

As $t \to \infty$, based on equation (7.8.16), we may conclude that the limiting distribution of $Q(t)$ exists only if

$$1 + \sum_{n=1}^{\infty} \frac{\lambda_0 \lambda_1 \cdots \lambda_{n-1}}{\mu_1 \mu_2 \cdots \mu_n} < \infty \tag{11.2.5}$$

Let

$$\lim_{t \to \infty} P_n(t) = p_n \qquad n = 0, 1, 2, \ldots \tag{11.2.6}$$

The probabilities p_n ($n = 0, 1, 2, \ldots$) can be derived recursively from

$$\lambda_0 p_0 = \mu_1 p_1 \tag{11.2.7}$$

$$(\lambda_n + \mu_n)p_n = \lambda_{n-1}p_{n-1} + \mu_{n+1}p_{n+1} \qquad n > 0 \tag{11.2.8}$$

$$\sum_{n=0}^{\infty} p_n = 1 \tag{11.2.9}$$

The steady-state equations (11.2.7) and (11.2.8) can be obtained either by letting $t \to \infty$ in (11.2.2) and (11.2.3) or directly by considering the possible transitions in and out of a state as described following Theorem 7.8.1. Equations (11.2.7)–(11.2.9) yield the solution

$$p_0 = \left[1 + \sum_{n=1}^{\infty} \frac{\lambda_0 \lambda_1 \cdots \lambda_{n-1}}{\mu_1 \mu_2 \cdots \mu_n} \right]^{-1} \tag{11.2.10}$$

$$p_n = \frac{\lambda_0 \lambda_1 \cdots \lambda_{n-1}}{\mu_1 \mu_2 \cdots \mu_n} p_0 \qquad n > 0 \tag{11.2.11}$$

In Sections 11.3–11.5 we shall specialize this system by assuming different dependence structure of λ_n and μ_n on n. By this method we are able to analyze the $Q(t)$ processes of some simple queueing systems and solve realistic examples based on them. However, it is necessary to make clear at the outset that we should not view these idealized models as cure-alls for real-life problems. The merits and limitations of such models should be clearly understood, and solutions to real problems should be taken in that spirit.

While specializing the results stated in this section to give the limiting behavior of $Q(t)$ in most cases, we shall not derive them again unless some special technique needs demonstrating. However, to gain sufficient practice in the analysis of such systems, the reader would be well advised to treat each example as a separate problem and derive the results starting from fundamentals.

11.3   THE QUEUE $M/M/1$

Consider a queueing system in which arrivals occur one at a time in a Poisson process with parameter λ. These customers get served at a single counter, and their service times are independent and identically distributed random variables. Let the common distribution of these random variables be exponential with mean μ^{-1}. This implies that as long as there are enough customers for service, the service completions occur in a Poisson process with parameter μ. For the purposes of considering the waiting times of customers, we shall assume that the customers are served in the order of their arrival (first-come, first-served). We shall also assume that the service does not stop as long as there are customers to be served.

Let $Q(t)$ be the number of customers in the system at time t, and define

$$P_n(t) = P[Q(t) = n] \qquad (11.3.1)$$

Clearly $\{Q(t)\}$ is the queue length process described in Section 11.2 with $\lambda_n = \lambda$ and $\mu_n = \mu$ for all n. The difference differential equations for $P_n(t)$ take the form

$$P_0'(t) = -\lambda P_0(t) + \mu P_1(t)$$

$$P_n'(t) = -(\lambda + \mu)P_n(t) + \lambda P_{n-1}(t) + \mu P_{n+1}(t) \qquad n > 0$$

$$(11.3.2)$$

In this and the systems to be considered later, we shall not try to solve these equations for finite time. In many real situations solutions for finite time (transient solutions) are better approximations than limiting solutions, but they require better tools and techniques than we have developed so far. Let

$$\rho = \frac{\lambda}{\mu} = \frac{\text{Arrival rate}}{\text{Service rate}} = \frac{\text{Mean service time}}{\text{Mean interarrival time}} = \left\{ \begin{array}{l} \text{Traffic intensity} \\ \quad \text{of the system} \end{array} \right.$$

$$(11.3.3)$$

As noted in (11.2.5), the limiting distribution exists if

$$1 + \rho + \rho^2 + \rho^3 + \cdots < \infty$$

that is, if

$$\rho < 1 \qquad (11.3.4)$$

Then we have

$$1 + \rho + \rho^2 + \rho^3 + \cdots = \frac{1}{1 - \rho}$$

Denoting the limiting distribution of $Q(t)$ by $\{ p_n \}$ and specializing (11.2.11), we get

$$p_n = \rho^n p_0 \qquad n > 0 \qquad (11.3.5)$$

which gives

$$p_0 = \left[1 + \rho + \rho^2 + \cdots \right]^{-1} = 1 - \rho$$

$$p_n = (1 - \rho)\rho^n \qquad n \geq 0 \qquad (11.3.6)$$

It should be noted that this is a geometric distribution. Writing $Q(\infty) = Q$, we have

$$E(Q) = \frac{\rho}{1 - \rho} \qquad (11.3.7)$$

$$V(Q) = \frac{\rho}{(1 - \rho)^2} \qquad (11.3.8)$$

Let $W(t)$ be the virtual waiting time process defined in Section 11.1. Clearly it is given by the sum of the service times of those customers who are actually waiting and the remaining service time of the customer who is already in service. The service time distribution is exponential, and from Theorem 7.2.3 we find that the distribution of this remaining service time is the same as that of the original random variable. Therefore, if there are n customers in the system at time t, the waiting time $W(t)$ is given by the sum of n random variables which are independently and identically distributed with the exponential probability density

$$\mu e^{-\mu x} \qquad x > 0 \qquad\qquad (11.3.9)$$

From (7.2.32) we also know that the probability density of this sum is the gamma density

$$g(x) = e^{-\mu x}\frac{\mu^n x^{n-1}}{(n-1)!} \qquad x > 0 \qquad\qquad (11.3.10)$$

Write $W(\infty) = W$; then for

$$F(x) = P(W \le x) \qquad\qquad (11.3.11)$$

we have

$$F(0) = P(Q = 0) = 1 - \rho \qquad\qquad (11.3.12)$$

and

$$
\begin{aligned}
dF(x) &= \sum_{n=1}^{\infty} (1-\rho)\rho^n e^{-\mu x}\frac{(\mu x)^{n-1}}{(n-1)!}\,\mu\,dx \\[2mm]
&= \lambda(1-\rho)e^{-\mu x}\sum_{n=1}^{\infty}\frac{(\lambda x)^{n-1}}{(n-1)!}\,dx \\[2mm]
&= \lambda(1-\rho)e^{-(\mu-\lambda)x}\,dx \\[2mm]
&= \lambda(1-\rho)e^{-\mu(1-\rho)x}\,dx \qquad\qquad (11.3.13)
\end{aligned}
$$

showing that the limiting waiting time distribution is also exponential with a probability concentration at the origin.

The mean and variance of the waiting time W can be obtained from (11.3.13):

$$E(W) = \frac{\rho}{\mu(1-\rho)} \qquad\qquad (11.3.14)$$

$$V(W) = \frac{\rho(2-\rho)}{\mu^2(1-\rho)^2} \qquad\qquad (11.3.15)$$

In practical problems the distribution function of the waiting time W furnishes useful information regarding the system. The distribution function $F(x)$ gives the probability that the waiting time of the customer in the system is no more than x. We have

$$
\begin{aligned}
F(x) &= P(W \leq x) \\
&= P(W = 0) + P(0 < W \leq x) \\
&= 1 - \rho + \int_0^x \lambda(1 - \rho)e^{-(\mu-\lambda)y}\,dy \\
&= 1 - \rho + \frac{\lambda(1 - \rho)}{\mu - \lambda}\int_0^x (\mu - \lambda)e^{-(\mu-\lambda)y}\,dy \\
&= 1 - \rho + \rho[1 - e^{-(\mu-\lambda)x}] \\
&= 1 - \rho e^{-\mu(1-\rho)x}
\end{aligned}
\tag{11.3.16}
$$

If one wants to consider the total delay of the customer in the system, then the length of service has to be added to the waiting time. Let $G(x)$ be the probability that the total time spent by the customer in the system is no more than x. We have

$$
G(x) = P(W + S \leq x) \tag{11.3.17}
$$

where S is the random variable representing the service time which has the probability density given by (11.3.9). By standard operations of convolution we get

$$
\begin{aligned}
G(x) &= \int_0^x F(x - y)\mu e^{-\mu y}\,dy \\
&= \int_0^x \left[1 - \rho e^{-\mu(1-\rho)(x-y)}\right]\mu e^{-\mu y}\,dy \\
&= 1 - e^{-\mu(1-\rho)x}
\end{aligned}
\tag{11.3.18}
$$

after some simplification. This is again an exponential distribution. The expected total stay of the customer in the system is therefore

$$
E(W + S) = \frac{1}{\mu(1 - \rho)} \tag{11.3.19}
$$

The result (11.3.19) can also be obtained from the relationship that exists between the expected number of customers in the system and the expected stay $E(W + S)$ as follows: Consider the customers $C_1, C_2, \ldots$ getting service in the order of their arrival. Let S_n be the service time of the customer C_n.

Clearly the number of customers in the system at the nth departure epoch (at which time C_n leaves the system) is the number of customers who have arrived during $W_n + S_n$, where S_n is the service time of C_n. If Q_n is the number of customers left by the nth departing customer, λ being the arrival rate, we can write

$$E(Q_n) = \lambda E[W_n + S_n]$$

When the arrival process is Poisson, the limiting distributions of the random variables W_n and $W(t)$ have the same form. Writing $W_\infty = W$, we have

$$E(Q) = \lambda E(W + S) \tag{11.3.20}$$

This is known as Little's (1961) formula in queueing literature, and it holds good in more general situations as well.

We shall now work out a few examples illustrating the results obtained so far in this section.

Example 11.3.1

It has been found reasonable to assume that the arrivals at a telephone booth form a Poisson process with a mean of 12 per hour. An exponential distribution with mean 2 min has also been found to be a good fit for the distribution of the length of telephone calls.

1. What is the probability that an arrival will find the telephone occupied?

This is given by the probability that there is at least one person in the system at the arrival time. We have

$$P(Q > 0) = 1 - P(Q = 0) = \rho$$

But ρ is given by

$$\rho = \frac{\lambda}{\mu} = \frac{\text{Mean service time}}{\text{Mean interarrival time}} = \frac{2}{5} = 0.4 \qquad \textit{Answer}$$

2. What is the average length of the queue when it forms?

Here we need the expectation of Q conditional on Q being greater than 0. Thus we have

$$E(Q|Q > 0) = \frac{E(Q)}{P(Q > 0)} = \frac{\rho/(1 - \rho)}{\rho}$$

$$= \frac{1}{1 - \rho} = \frac{1}{0.6} = 1.67 \qquad \textit{Answer}$$

3.   It is the policy of the telephone company to install additional booths if the customers wait on the average at least 3 min for the phone. By how much must the flow of arrivals increase in order to justify the second booth?

From (11.3.14) we have

$$E(W) = \frac{\rho}{\mu(1 - \rho)}$$

$$= \frac{\lambda}{\mu(\mu - \lambda)}$$

Given $\mu = 0.5$, the telephone company will install additional booths if

$$E(W) = \frac{\lambda}{0.5(0.5 - \lambda)} \geq 3$$

Solving this equation, we find that $\lambda \geq 0.3$. With $\lambda = 0.3$, the mean number of arrivals per hour is 18. Therefore, for an additional booth to be justified, the arrival rate should increase by 6 per hour.   *Answer*

Example 11.3.2

At a service station the rate of service is μ cars per hour and the arrival of cars for service is λ per hour. It is also found that it is reasonable to assume that the arrivals are Poisson and the service times are exponential. The cost incurred by the service station due to waiting cars is $\$C_1$ per car per hour, and the operating and service costs are $\$\mu C_2$ per hour when the service rate is μ. Determine the service rate that results in the least expected cost.

From (11.3.7) we have

$$E(Q) = \frac{\rho}{1 - \rho} = \frac{\lambda}{\mu - \lambda}$$

After the system has settled down to an equilibrium behavior, we can assume that on the average there are $\lambda/(\mu - \lambda)$ cars waiting at the service station. Hence the total cost per hour to the station can be given as

$$C = \frac{\lambda C_1}{\mu - \lambda} + \mu C_2$$

Minimizing C through standard techniques, we get

$$\frac{\partial C}{\partial \mu} = \frac{-\lambda C_1}{(\mu - \lambda)^2} + C_2 = 0$$

giving

$$(\mu - \lambda)^2 = \frac{\lambda C_1}{C_2}$$

$$\mu = \lambda + \sqrt{\frac{\lambda C_1}{C_2}} \quad \text{or} \quad \lambda - \sqrt{\frac{\lambda C_1}{C_2}}$$

Since for stability we need $\mu > \lambda$, of these two values only $\mu = \lambda + \sqrt{\lambda C_1/C_2}$ is admissible. This value also results in the least expected cost; for

$$\frac{\partial^2 C}{\partial \mu^2} = \frac{2\lambda C_1}{(\mu - \lambda)^3}$$

$$= \frac{2C_2}{\sqrt{\lambda C_1/C_2}} > 0 \quad \text{if } \mu = \lambda + \sqrt{\frac{\lambda C_1}{C_2}} \qquad \textit{Answer}$$

Suppose, however, the cost structure is such that $\$C_1$ is the cost to the station incurred per hour of total delay (wait + service) for each car and $\$\mu C_2$ is the cost per car for actual service. Now the total cost per car is

$$C = \frac{C_1}{\mu - \lambda} + \mu C_2$$

where we have used equation (11.3.19) for the expected total delay. Assuming the rate of arrival λ as a constant, the objective function to be minimized is C, which yields the best value of μ as

$$\mu = \lambda + \sqrt{\frac{C_1}{C_2}} \qquad \textit{Answer}$$

11.4   THE QUEUE $M/M/s$

Let the arrivals occur in a Poisson process with parameter λ, and let service times be independent and identically distributed random variables with an exponential distribution with mean μ^{-1}. Let there be s (≥ 1) servers in the system working in parallel independently of each other. Arriving customers queue up in a single line in the order of their arrival. A server who is free takes the customer at the head of the queue for service. A customer arriving when more than one server is free gets served by any one of them.

It should noted that in this system all servers offer service at the same rate. In actual practice there can be systems where the service rates are different with different servers. To distinguish between these two cases, we shall call these homogeneous and heterogeneous servers, respectively. A computing center with two computers, one of which is fast and the other slow, and checkout counters at supermarkets where there is at least one fast checkout counter are examples of systems with heterogeneous servers.

By arguments similar to those employed in deriving (7.4.4)–(7.4.6), it can be shown that in between arrivals the number of customers in the system behaves like a pure death process with parameters

$$\mu_n = s\mu \qquad n \geq s$$

$$= n\mu \qquad n \leq s \qquad (11.4.1)$$

As before, define $Q(t)$ as the number of customers in the system at time t and $P_n(t)$ as its probability distribution (conditional on the initial value). The forward Kolmogorov equations (11.2.2) and (11.2.3) now take the form

$$P_0'(t) = -\lambda P_0(t) + \mu P_1(t)$$

$$P_n'(t) = -(\lambda + n\mu)P_n(t) + \lambda P_{n-1}(t) + (n+1)\mu P_{n+1}(t)$$

$$0 < n < s \quad (11.4.2)$$

$$P_n'(t) = -(\lambda + s\mu)P_n(t) + \lambda P_{n-1}(t) + s\mu P_{n+1}(t) \qquad n \geq s$$

As $t \to \infty$, the limiting distribution $\{p_n\}$ of the queue length Q $[= Q(\infty)]$ is obtained:

$$p_0 = \left[\sum_{r=0}^{s} \frac{(s\rho)^r}{r!} + \frac{\rho^{s+1}s^s}{s!}(1-\rho)^{-1} \right]^{-1} \qquad (11.4.3)$$

$$p_n = \frac{(s\rho)^n}{n!} p_0 \qquad n \leq s$$

$$= \frac{s^s\rho^n}{s!} p_0 \qquad n \geq s \qquad (11.4.4)$$

where we have defined $\rho = \lambda/s\mu$ = traffic intensity of the system. We also have

$$p_n = \rho^{n-s} p_s \qquad n \geq s \qquad (11.4.5)$$

showing that when $Q \geq s$, the system behaves like a single server $M/M/1$ with service rate $s\mu$ [see (11.3.5)]. It is found that the condition for existence of the limiting distribution is given by

$$\rho = \frac{\lambda}{s\mu} < 1 \tag{11.4.6}$$

From (11.4.3)–(11.4.5) the expected queue length can be obtained:

$$E(Q) = s\rho + \frac{\rho p_s}{(1 - \rho)^2} \tag{11.4.7}$$

Let Q' be the number of customers actually waiting in the long run. Its expected value is

$$E(Q') = \sum_{n=s+1}^{\infty} (n - s) p_n$$

$$= \frac{\rho p_s}{(1 - \rho)^2} \tag{11.4.8}$$

The expression for the variance of Q gets cumbersome, and therefore we shall not present it here.

As in the case of the system $M/M/1$ let us denote the long-run waiting time of a customer by W. Then for

$$F(x) = P(W \leq x)$$

we have

$$F(0) = P(Q < s)$$

$$= p_0 \sum_{n=0}^{s-1} \frac{(s\rho)^n}{n!} \tag{11.4.9}$$

and

$$dF(x) = P(x < W \leq x + dx)$$

$$= \sum_{n=s}^{\infty} p_n e^{-s\mu x} \frac{(s\mu x)^{n-s}}{(n - s)!} s\mu \, dx$$

$$= \sum_{n=s}^{\infty} \rho^{n-s} p_s e^{-s\mu x} \frac{(s\mu x)^{n-s}}{(n - s)!} s\mu \, dx$$

$$= s\mu \, p_s e^{-(s\mu - \lambda)x} \, dx \tag{11.4.10}$$

From this we get its mean:

$$E(W) = \frac{p_s}{s\mu(1-\rho)^2} \tag{11.4.11}$$

In a single-server system, looking at the limiting probabilities [see (11.3.6)], it is clear that the traffic intensity ρ is the probability that the server is busy at any time (or the fraction of time the server is busy). Is this true of the multiserver system $M/M/s$? Consider the probability that a particular server is busy at an arbitrary time point in the system $M/M/s$. It should be noted that when the number of customers in the system is greater than or equal to s all the servers are busy, and when there are only k $(<s)$ customers in the system, the probability that the server under consideration will be busy is k/s. Thus we have the required probability:

$$\left[\sum_{n=s}^{\infty} \frac{(s\rho)^n}{s^{n-s}s!} + \frac{s-1}{s} \cdot \frac{(s\rho)^{s-1}}{(s-1)!} + \frac{s-2}{s} \cdot \frac{(s\rho)^{s-2}}{(s-2)!} + \cdots + \frac{1}{s}s\rho\right]p_0$$

$$= p_0\left[\rho + s\rho^2 + \cdots + \frac{(s\rho)^{s-2}}{(s-2)!}\rho + \frac{(s\rho)^{s-1}}{(s-1)!}\rho + \sum_{n=s+1}^{\infty} \frac{(s\rho)^{n-1}\rho}{s^{n-1-s}s!}\right]$$

$$= \rho p_0\left[\sum_{n=0}^{s-1} \frac{(s\rho)^n}{n!} + \sum_{n=s}^{\infty} \frac{(s\rho)^n}{s^{n-s}s!}\right] \tag{11.4.12}$$

The term in brackets in (11.4.12) is in fact p_0^{-1}, and hence we have the result that the probability that a particular server is busy at an arbitrary time point (in the long run) is also the traffic intensity ρ.

A Comparison of the Systems $M/M/2$ and $M/M/1$ with the Same ρ

Specializing the results (11.4.3)–(11.4.5) and (11.4.7) for the two-server case, we get

$$E(Q) = \frac{2\rho}{(1+\rho)(1-\rho)} \tag{11.4.13}$$

Let Q_1 be the long-run queue length in a system $M/M/1$ with the same ρ as before. Then we have

$$E(Q_1) = \frac{\rho}{1-\rho}$$

from (11.3.7). Comparing these two means, we get

$$E(Q) - E(Q_1) = \frac{2\rho}{(1 + \rho)(1 - \rho)} - \frac{\rho}{1 - \rho}$$

$$= \frac{\rho}{1 + \rho} > 0 \qquad (11.4.14)$$

This shows that, maintaining the same traffic intensity, a single-server system results in a smaller expected queue length than a two-server system. Clearly this is a case where the quality of service is preferable to the quantity of service offered.

Example 11.4.1

A supermarket has three clerks ringing up sales at the counters. Service time for each customer is exponential with mean 5 min, and people arrive in a Poisson process at the rate of 30 per hour.

1.   Under steady-state conditions, what is the probability that all the clerks will be busy?
We have, with notations described before, $\mu = 1/5$, $s = 3$, and $\lambda = 1/2$; hence, $\rho = 5/6$. Specializing (11.4.3) and (11.4.4) for these values, we get

$$p_0 = \tfrac{4}{89} \qquad p_1 = \tfrac{10}{89}$$

$$p_2 = \tfrac{25}{178} \qquad p_3 = \tfrac{125}{1068}$$

Clearly the probability that all clerks are busy is equivalent to

$$1 - p_0 - p_1 - p_2 = \tfrac{125}{178} \qquad \qquad \textit{Answer}$$

2.   What is the expected number of people waiting to be served?
The required value is $E(Q')$ as given by (11.4.8). Substituting the known results, we get

$$E(Q') \cong 3.5 \qquad \qquad \textit{Answer}$$

3.   What is the expected length of time customers have to wait for service?
The required value is $E(W)$ as given by (11.4.11). For this we get

$$E(W) \cong 7.0 \text{ min} \qquad \qquad \textit{Answer}$$

4. If a customer has to wait, what is the expected length of his waiting time?

This is the conditional expectation of W, given that a customer has to wait. We get

$$\frac{E(W)}{P(Q \geq 3)} = 10 \text{ min} \qquad\qquad \textit{Answer}$$

Example 11.4.2

A university computing center has two computers of equivalent capacity. Computing jobs arriving at the center have been found to have an exponential service time with mean 3 min. The jobs are of two types, university jobs and external jobs. These are found to arrive in Poisson processes with rates 18 and 15 per hour, respectively. Is there any advantage in using one computer exclusively for university jobs and the other for external jobs?

We shall consider this question with reference to the waiting times of the two kinds of computing jobs. When the computers are used separately, let W_1 and W_2 be the waiting times of university and external jobs, respectively. Treating these computers as two single-server systems, with reference to W_1 we have

$$\lambda = \tfrac{18}{60} = \tfrac{3}{10} \qquad \mu = \tfrac{1}{3}$$

so that

$$\rho = \frac{\lambda}{\mu} = \frac{9}{10}$$

and hence,

$$E(W_1) = \frac{9/10}{(1/3)[1 - (9/10)]} = 27 \text{ min}$$

With reference to W_2, we have

$$\lambda = \tfrac{15}{60} = \tfrac{1}{4} \qquad \mu = \tfrac{1}{3}$$

so that

$$\rho = \tfrac{3}{4}$$

and hence

$$E(W_2) = \frac{3/4}{(1/3)(1/4)} = 9 \text{ min}$$

When two computers handle both types of jobs, we can use the two-server model with

$$\lambda = \lambda_1 + \lambda_2 = \tfrac{3}{10} + \tfrac{1}{4} = \tfrac{11}{20} \qquad \mu = \tfrac{1}{3}$$

and

$$\rho = \frac{\lambda}{2\mu} = \frac{11}{20} \cdot \frac{3}{2} = \frac{33}{40}$$

Hence from (11.4.3), (11.4.4), and (11.4.11) we get

$$p_0 = \tfrac{7}{73} \qquad p_2 = \tfrac{7623}{58400}$$

and $E(W) \cong 6.45$ min. Clearly it is much better to pool the jobs and consider the system as a two-server system. *Answer*

11.5 SOME VARIANTS OF THE QUEUEING SYSTEMS $M/M/1$ AND $M/M/s$

Specializing the general queueing model of Section 11.2 to different types of arrival and service parameters, we can arrive at realistic modifications of the two queueing systems discussed in Sections 11.3 and 11.4. Here we shall briefly discuss a few of these variants just to indicate the wide variety of systems that can be modeled as birth and death processes. For the sake of simplicity we shall discuss only the limiting distributions of the process $Q(t)$.

The Queue $M/M/1$ with Balking

If an arriving customer for some reason decides not to join the queue, we say that the customer balks. Balking can be due to an estimation of one's chances of getting an early service or due to the fact that one is forced to balk by the system itself (as when the waiting room is full in a system with a finite waiting room). To account for this behavior of the customer, let us define λ_n as the arrival rate of the customers when there are n customers in the system. Let μ be the service rate.

Let p_n be the limiting distribution of the number of customers in the system. The steady-state equations for p_n can be derived from (11.2.7) and (11.2.8), and their solutions from (11.2.10) and (11.2.11) by setting $\mu_n = \mu$, $n = 1, 2, \ldots$. The specific λ values that follow have significant practical implications.

A Queue with Discouragement: Poisson Queue (Haight, 1957)
Let

$$\lambda_n = \frac{\lambda}{n+1} \qquad n = 0, 1, 2, \ldots \tag{11.5.1}$$

We then have

$$p_0 = \frac{1}{1 + (\lambda/\mu) + (\lambda^2/2!\mu^2) + \cdots} = e^{-\rho} \tag{11.5.2}$$

where we have written $\lambda/\mu = \rho$. Then

$$p_n = e^{-\rho} \frac{\rho^n}{n!} \qquad n = 0, 1, 2, \ldots \tag{11.5.3}$$

which is a Poisson distribution with mean ρ.

A Binomial Queue (Haight, 1957)
Let

$$\lambda_n = \frac{N-n}{N(n+1)} \qquad n < N$$

$$= 0 \qquad n \geq N \tag{11.5.4}$$

We then get

$$p_0 = \left(1 + \frac{1}{N\mu}\right)^{-N}$$

and

$$p_n = \binom{N}{n}\left(\frac{1}{N\mu}\right)^n\left(1 + \frac{1}{N\mu}\right)^{-N}$$

$$= \binom{N}{n}\left(\frac{1}{1+N\mu}\right)^n\left(\frac{N\mu}{1+N\mu}\right)^{N-n} \qquad n \geq 0 \tag{11.5.5}$$

which is a binomial distribution in which the probability of success is $(1 + N\mu)^{-1}$.

A Negative Binomial Queue (Haight, 1957)
Let

$$\lambda_n = \frac{N+n}{N(n+1)} \qquad n \geq 0 \tag{11.5.6}$$

When then get

$$p_0 = \left(1 - \frac{1}{N\mu}\right)^N$$

and

$$p_n = \binom{-N}{n}\left(1 - \frac{1}{N\mu}\right)^N\left(-\frac{1}{N\mu}\right)^n \qquad n \geq 0 \qquad (11.5.7)$$

which is a negative binomial distribution.

A Finite Waiting Room Queue
Let

$$\lambda_n = \lambda \qquad 0 \leq n < N$$

$$= 0 \qquad n \geq N \qquad (11.5.8)$$

After simplifications we find

$$p_0 = \frac{1 - \rho}{1 - \rho^{N+1}}$$

$$p_n = \frac{1 - \rho}{1 - \rho^{N+1}}\rho^n \qquad n = 1, 2, \ldots, N \qquad (11.5.9)$$

A Queue with Service-Dependent Balking (Morse, 1958)
Let

$$\lambda_n = \lambda e^{-\alpha n/\mu} \qquad n \geq 0; \ \alpha > 0 \qquad (11.5.10)$$

Define $\gamma = e^{-\alpha/2\mu}$. Then we get

$$p_0 = \left(\sum_{n=0}^{\infty} \rho^n \gamma^{n(n-1)}\right)^{-1}$$

$$p_n = \rho^n \gamma^{n(n-1)} p_0 \qquad n > 0 \qquad (11.5.11)$$

The results given can be used either to obtain the characteristics of the limiting queue length when balking behavior is known or to derive balking characteristics when the limiting distribution is known. In either case, in some situations these variants may turn out to be more realistic than the standard queue $M/M/1$.

The Queue $M / M / 1$ with State-Dependent Service

Consider a system in which service rate is dependent on the number of customers in the system. This seems realistic in systems where the server reacts to the number of people waiting. Let us discuss two types of models.

Service Rate with a Pressure Coefficient (Conway and Maxwell, 1962)

In the generalized queueing model of Section 11.2, assume

$$\lambda_n = \lambda \qquad n \geq 0$$

$$\mu_n = n^c \mu \qquad n > 0 \qquad (11.5.12)$$

Simplifying (11.2.10) and (11.2.11), we find

$$p_0 = \left[\sum_{n=0}^{\infty} \frac{\rho^n}{(n!)^c} \right]^{-1}$$

$$p_n = \frac{\rho^n}{(n!)^c} p_0 \qquad n \geq 0 \qquad (11.5.13)$$

A Linearly Dependent Service Rate
Let

$$\lambda_n = \lambda \qquad n \geq 0$$

$$\mu_n = n\mu \qquad n \geq 0 \qquad (11.5.14)$$

Then we get

$$p_0 = e^{-\rho} \qquad p_n = e^{-\rho} \frac{\rho^n}{n!} \qquad n \geq 0 \qquad (11.5.15)$$

where we have written $\lambda/\mu = \rho$. In this case it should be noted that the actual service time of a customer is not exponential with mean $(n\mu)^{-1}$, as it is possible for the service rate to change with the arrival of new customers. Further the queueing model with parameters λ_n and μ_n as defined in (11.5.14) represents the system with an infinite number of servers (no waiting) or a single-server system in which service rate is always μ and customers get impatient and leave the system at a rate $(n - 1)\mu$.

The Queue $M/M/1$ with an Additional Server for Longer Queues (Singh, 1968)

In many queue situations, the number of servers is dependent on the queue length. For instance, in a bank more and more windows are opened for service when the queues in front of the already open windows get too long. This procedure is sometimes used even in the case of airlines, buses, and so forth. The basic advantage in such a system is the inherent flexibility of service. Here we shall consider a simplified system of this type.

Suppose the arrivals are in a Poisson process with parameter λ and service times of customers are exponential with mean μ^{-1}. As long as the number of customers in the system is less than or equal to N, there is only one server in the system. As soon as the number goes beyond N, an additional server is brought in who will be taken off when the number falls to N or below. This latter simplification may not be very realistic in real-life queues, but it has been included to make the model simple enough for its inclusion here. Assuming the limiting distribution of the number of customers in the system to be $\{p_n\}$, as in earlier models we obtain the steady-state equations

$$\lambda p_0 = \mu p_1$$

$$(\lambda + \mu) p_n = \lambda p_{n-1} + \mu p_{n+1} \quad 1 \le n < N$$

$$(\lambda + \mu) p_N = \lambda p_{N-1} + 2\mu p_{N+1}$$

$$(\lambda + 2\mu) p_n = \lambda p_{n-1} + 2\mu p_{n+1} \quad n > N \qquad (11.5.16)$$

Solving these recursively and writing $\lambda/\mu = \rho$, we get

$$p_0 = \frac{(1-\rho)(2-\rho)}{2 - \rho - \rho^{N+1}} \qquad (11.5.17)$$

$$p_n = \rho^n p_0 \quad 1 \le n \le N$$

$$= \frac{1}{2^{n-N}} \rho^n p_0 \quad n \ge N \qquad (11.5.18)$$

and

$$E(Q) = \frac{\rho(2-\rho)}{(1-\rho)(2 - \rho - \rho^{N+1})} - \frac{(2-\rho)(N + 1 - N\rho)\rho^{N+1}}{(1-\rho)(2 - \rho - \rho^{N+1})}$$

$$+ \frac{(1-\rho)(2N + 2 - N\rho)\rho^{N+1}}{(2-\rho)(2 - \rho - \rho^{N+1})} \qquad (11.5.19)$$

Suppose the cost to the system due to the presence of a customer is $\$C_1$ and the cost of bringing in an additional server is $\$C_2$. It is profitable to use the additional server only if its cost is less than that due to the waiting customers.

Let Q_1 be the long-run queue length of the standard queue $M/M/1$ with the same parameters. The modified system is therefore recommended only if

$$C_1[E(Q_1) - E(Q)] > C_2 P(Q > N) \tag{11.5.20}$$

We also have

$$E(Q_1) = \frac{\rho}{1 - \rho}$$

and

$$P(Q > N) = \sum_{n=N+1}^{\infty} p_n = \frac{(1 - \rho)\rho^{N+1}}{2 - \rho - \rho^{N+1}} \tag{11.5.21}$$

Inequality (11.5.20) can now be simplified to

$$\frac{C_2}{C_1} < \frac{N}{1 - \rho} + \frac{2}{(1 - \rho)(2 - \rho)} \tag{11.5.22}$$

or

$$N > \frac{C_2}{C_1}(1 - \rho) - \frac{2}{2 - \rho} \tag{11.5.23}$$

From these two inequalities, depending on the decision variable, the largest value of $C^* = C_2/C_1$ or the least value of N can be determined when ρ and any one of N or C^* are given. Table 11.5.1 has been obtained from these inequalities for the purpose of illustration.

A more sophisticated approach to this kind of problem is to cast it as a decision problem for the parameter μ in the single-server system $M/M/1$ and use optimization techniques of mathematical and dynamic programming to arrive at optimal decision rules. An excellent example of such an investigation is that of Crabill (1969, 1972). His major result is as follows: Consider an $M/M/1$ queue with arrival rate λ and two possible service rates μ_1 and μ_2, such that $0 < \lambda < \mu_1 < \mu_2$. Let C_i ($i = 1, 2$) be the service cost per unit time when the service rate μ_i is being used, and let ω_i be the customer waiting or work-in process cost per unit time when there are i customers in the system. Further let $0 \le C_1 \le C_2$ and ω_i be a positive nondecreasing function, such that $\omega_i \to \infty$ as $i \to \infty$ and $\sum_{i \in S} \omega_i \rho_1^i < \infty$ where $\rho_1 = \lambda/\mu_1$. Then there exists an optimal stationary policy (resulting in the least long-run expected cost) characterized by a single positive finite

Table 11.5.1

ρ	$C^* = C_2/C_1$								
N	0.1	0.2	0.3	0.4	0.5	0.6	0.7	0.8	0.9
1	2.28	2.64	3.11	3.75	4.67	6.07	8.46	13.33	28.18
2	3.39	3.89	4.54	5.41	6.67	8.57	11.80	18.33	38.18
3	4.50	5.14	5.96	7.08	8.67	11.07	15.13	23.33	48.18
4	5.61	6.39	7.39	8.75	10.67	13.57	18.46	28.33	58.18
5	6.72	7.64	8.82	10.41	12.67	16.07	21.08	33.33	68.18
10	12.28	13.89	15.96	18.75	22.67	28.57	38.46	58.33	118.18
15	17.84	20.14	23.10	27.08	32.67	31.07	55.13	83.33	168.18
20	23.39	26.39	30.24	35.41	42.67	53.57	71.79	108.33	218.18

number N, such that it is optimal to use the slow service rate μ_1 when the number of customers in the system is less than or equal to N and the faster service rate μ_2 when the number in the system is greater than N. Extensions of these results to k ($< \infty$) different service rates and the inclusion of fixed service rate changeover costs may also be found in Crabill (1969, 1972).

The Loss System $M/M/s$

Queueing systems discussed thus far in this chapter, with some exceptions, allow for the delay in the commencement of service. Such systems are called delay systems. As against these, we can think of systems which are known as loss systems.

Suppose there are s lines to a telephone switchboard. Calls arriving when all the lines are busy do not get registered, and when the caller tries again it is more realistic to consider it as a new call. Thus it is convenient to model this situation as a loss system in which customers arriving when all the servers are busy are lost to the system. If one needs to consider the finite size of the source of arrivals, this can be done by using a state-dependent arrival rate.

Consider a loss system with s servers, Poisson arrivals with rate λ_n, when the number of customers in the system is n, and exponential service with mean μ^{-1}. The steady-state equations for the queue length process are

$$\lambda_0 p_0 = \mu p_1$$

$$(\lambda_n + n\mu)p_n = \lambda_{n-1}p_{n-1} + (n+1)\mu p_{n+1} \qquad 0 < n < s$$

$$s\mu p_s = \lambda_{s-1}p_{s-1} \qquad\qquad\qquad (11.5.24)$$

Solving these recursively, we get

$$p_0 = \left[1 + \sum_{n=1}^{s} \frac{\lambda_0 \lambda_1 \cdots \lambda_{n-1}}{n! \mu^n} \right]^{-1}$$

and

$$p_n = \frac{\lambda_0 \lambda_1 \cdots \lambda_{n-1}}{n! \mu^n} p_0 \qquad 0 < n \le s \qquad (11.5.25)$$

When designing such a system, one of the objectives would be to minimize the probability of losing customers. The proportion of lost customers can be obtained as follows: We have

$$\text{Probability of losing a customer} = p_s$$

$$\text{Expected number of lost customers during} \atop \text{a unit service interval} = \lambda_s \cdot \frac{1}{\mu} \cdot p_s$$

$$\text{Expected number of arrivals during a unit} \atop \text{service interval} = \sum_{n=0}^{s} \lambda_n \cdot \frac{1}{\mu} \cdot p_n$$

Thus we get

Proportion of lost customers

$$= \frac{\text{Expected number of lost customers}}{\text{Total number of arrivals}}$$

$$= \left[\frac{\lambda_1 \lambda_2 \cdots \lambda_s}{s! \mu^s} \right] \left[1 + \frac{\lambda_1}{\mu} + \frac{\lambda_1 \lambda_2}{\mu^2 2!} + \cdots + \frac{\lambda_1 \lambda_2 \cdots \lambda_s}{\mu^s s!} \right]^{-1} \qquad (11.5.26)$$

This is called the Engset-O'Dell formula. When $\lambda_n = \lambda$, with $\lambda/\mu = \rho$, we get

$$p_0 = \left[1 + \rho + \frac{\rho^2}{2!} + \cdots + \frac{\rho^s}{s!} \right]^{-1}$$

$$p_n = \frac{\rho^n}{n!} p_0 \qquad 0 \le n \le s$$

The limiting probability p_s gives the fraction of time customers will be lost as well as the probability of losing a customer. In telephone systems this is known as Erlang's loss formula. It can be given as

$$p_s = \frac{\rho^s/s!}{1 + \rho + \rho^2/2! + \cdots + \rho^s/s!} \qquad (11.5.27)$$

11.6   THE QUEUES $M/G/1$ AND $G/M/1$

When the arrival process is Poisson and the service times have an exponential distribution, the number of customers in the system is a Markov process with discrete state space (see Sections 11.3–11.5 for a discussion of its limiting properties). In this section we shall consider a more general system in which either the service times or the interarrival times will be assumed to have an arbitrary distribution.

The Queue $M/G/1$

Let customers arrive in a Poisson process with parameter λ and get served by a single server. Let the service times of these customers be independent and identically distributed random variables $\{S_n\}$, with

$$P(S_n \le x) = B(x) \qquad (11.6.1)$$

and

$$E(S_n) = b$$

$$V(S_n) = \sigma_S^2 \qquad (11.6.2)$$

For the complete description of the state of the system at any time, we need information regarding the number of customers in the system as well as the remaining service time of the customer currently being served. Only then can information on future behavior be given independently of past behavior. Let $Q(t)$ be the number of customers in the system at time t, and let $R(t)$ be the remaining service time of the customer currently being served. Now the vector $[Q(t), R(t)]$ is a vector Markov process because of the property just mentioned. Therefore, for a complete investigation into the behavior of $Q(t)$, the vector process $[Q(t), R(t)]$ has to be studied by some means. Nevertheless, for understanding the general behavioral pattern, one can study this process at some discrete set of points conveniently defined so as to make the study simple.

Suppose $t_0 = 0, t_1, t_2, \ldots$ are the points of departure of customers from the system. At these points the remaining service time of the customer is zero, and hence the queue length process $Q(t)$ can be studied alone. Let Q_n be the number of customers in the system soon after the nth departure. In terms of $Q(t)$ we therefore write $Q(t_n + 0) = Q_n$. The property that $\{Q_n\}$ is a Markov chain can also be seen from the following representation: Let

S_n be the service time of the nth customer, and let X_n be the number of customers arriving during S_n. With the Poisson assumption for the arrival process we have

$$k_j = P(X_n = j) = \int_0^\infty P(X_n = j|S_n)P(t < S_n \le t + dt) \qquad (11.6.3)$$

$$= \int_0^\infty e^{-\lambda t}\frac{(\lambda t)^j}{j!}\,dB(t) \qquad j = 0,1,2,\ldots \qquad (11.6.4)$$

Let

$$\psi(\theta) = \int_0^\infty e^{-\theta t}\,dB(t) \qquad \mathrm{Re}(\theta) > 0 \qquad (11.6.5)$$

$$K(z) = \sum_{j=0}^\infty k_j z^j \qquad |z| \le 1 \qquad (11.6.6)$$

Equation (11.6.5) defines the Laplace-Stieltjes transform of the distribution $B(x)$, and (11.6.6) defines the generating function of the probability distribution $\{k_j\}$. From the properties for Laplace-Stieltjes transforms and probability-generating functions (see Appendix B), we have

$$b = E(S_n) = -\psi'(0)$$
$$E(S_n^2) = \psi''(0)$$
$$E(X_n) = K'(1)$$
$$E(X_n^2) = K''(1) + K'(1)$$
$$E(X_n^3) = K'''(1) + 3K''(1) + K'(1) \qquad (11.6.7)$$

From (11.6.4) we get

$$K(z) = \int_0^\infty e^{-\lambda t}\sum_{j=0}^\infty \frac{(\lambda tz)^j}{j!}\,dB(t)$$

$$= \psi(\lambda - \lambda z) \qquad (11.6.8)$$

and

$$K'(z) = -\lambda\psi'(\lambda - \lambda z)$$
$$K''(z) = \lambda^2\psi''(\lambda - \lambda z)$$
$$K'''(z) = -\lambda^3\psi'''(\lambda - \lambda z)$$

giving

$$E(X_n) = -\lambda\psi'(0) = \lambda b = \rho$$

$$E(X_n^2) = \lambda^2 E(S_n^2) + \rho$$

$$E(X_n^3) = \lambda^3 E(S_n^3) + 3\lambda^2 E(S_n^2) + \rho \qquad (11.6.9)$$

Consider the relationship between Q_n and Q_{n+1}. We have

$$Q_{n+1} = \begin{cases} Q_n - 1 + X_{n+1} & \text{if } Q_n > 0 \\ 1 - 1 + X_{n+1} & \text{if } Q_n = 0 \end{cases} \qquad (11.6.10)$$

For convenience define the Heaviside function $H(X)$ as follows:

$$H(X) = \begin{cases} 1 & \text{if } X > 0 \\ 0 & \text{if } X \le 0 \end{cases} \qquad (11.6.11)$$

The function has the properties

$$H^2(X) = H(X)$$

$$XH(X) = X \qquad \text{if } X \ge 0 \qquad (11.6.12)$$

Using the function $H(X)$, we write (11.6.10) in the convenient form

$$Q_{n+1} = Q_n - H(Q_n) + X_{n+1} \qquad (11.6.13)$$

As can be seen from (11.6.13), Q_{n+1} depends only on Q_n and the independent random variable X_{n+1}, and therefore $\{Q_n\}$ is a Markov chain. This is called an embedded Markov chain of the process $Q(t)$, and the departure points are the transition epochs for the Markov chain (Kendall, 1951, 1953). The transition probability matrix of this Markov chain is

$$P = \begin{array}{c} \\ 0 \\ 1 \\ 2 \\ 3 \\ 4 \\ \vdots \end{array} \begin{array}{cccccc} 0 & 1 & 2 & 3 & \cdots \\ \begin{bmatrix} k_0 & k_1 & k_2 & k_3 & \cdots \\ k_0 & k_1 & k_2 & k_3 & \cdots \\ 0 & k_0 & k_1 & k_2 & \cdots \\ 0 & 0 & k_0 & k_1 & \cdots \\ 0 & 0 & 0 & k_0 & \cdots \\ \vdots & \vdots & \vdots & \vdots & \end{bmatrix} \end{array} \qquad (11.6.14)$$

For a complete investigation of the behavior of this Markov chain, the reader is referred to the books and articles mentioned at the end of the chapter. As can be seen from its structure, when $k_0 > 0$ and $k_0 + k_1 < 1$, this chain is irreducible (all states communicate with each other) and aperiodic. It can also be shown that when $\rho < 1$ [defined in (11.6.9)], the chain is positive recurrent; when $\rho = 1$, the chain is null recurrent; and when $\rho > 1$ the chain is transient. When $\rho < 1$, using Theorem 6.2.3, the probability-generating function for the limiting distribution can be obtained.

Let π_n ($n = 0, 1, 2, \dots$) be the limiting distribution of the Markov chain whose transition probability matrix is given by (11.6.14). Using Theorem 6.2.3 we note that $\pi = (\pi_0, \pi_1, \pi_2, \dots)$ satisfies the equation

$$\pi = \pi P \qquad (11.6.15)$$

Writing out these equations, we have

$$\pi_0 = \pi_o k_0 + \pi_1 k_0$$

$$\pi_1 = \pi_0 k_1 + \pi_1 k_1 + \pi_2 k_0$$

$$\pi_2 = \pi_0 k_2 + \pi_1 k_2 + \pi_2 k_1 + \pi_3 k_0$$

$$\pi_3 = \pi_0 k_3 + \pi_1 k_3 + \pi_2 k_2 + \pi_3 k_1 + \pi_4 k_0$$

$$\vdots \qquad\qquad (11.6.16)$$

In order to use probability-generating functions, define

$$\Pi(z) = \sum_{n=0}^{\infty} \pi_n z^n \qquad |z| \le 1 \qquad (11.6.17)$$

Multiplying the equations in (11.6.16) with appropriate powers of z and summing, we get

$$\Pi(z) = \pi_0 K(z) + \pi_1 K(z) + \pi_2 z K(z) + \pi_3 z^2 K(z) + \cdots$$

$$= \pi_0 K(z) + \frac{K(z)}{z}\left(\pi_1 z + \pi_2 z^2 + \pi_3 z^3 + \cdots\right)$$

that is,

$$\Pi(z) = \pi_0 K(z) + \frac{K(z)}{z}\left[\Pi(z) - \pi_0\right]$$

which gives

$$\Pi(z)\left[1 - \frac{K(z)}{z}\right] = \pi_0 K(z)\left[1 - \frac{1}{z}\right]$$

$$\Pi(z) = \frac{\pi_0 K(z)[z-1]}{z - K(z)} \tag{11.6.18}$$

The unknown quantity π_0 in (11.6.18) is determined using the normalizing condition $\Sigma_n \pi_n = 1$. We have

$$\Pi(1) = \sum_{n=0}^{\infty} \pi_n = 1$$

Letting $z \to 1$ in (11.6.18), we get

$$1 = \frac{\lim_{z \to 1} \pi_0\left[K(z) - (z-1)K'(z)\right]}{\lim_{z \to 1}\left[1 - K'(z)\right]}$$

$$= \frac{\pi_0}{1 - \rho} \qquad \rho < 1$$

giving

$$\pi_0 = 1 - \rho \tag{11.6.19}$$

Thus we get

$$\Pi(z) = \frac{(1-\rho)(z-1)K(z)}{z - K(z)} \tag{11.6.20}$$

For specific forms of $K(z)$, the limiting distribution π can be determined by expanding (11.6.20) as an infinite series in z. The mean and variance of the limiting queue length can be obtained using standard properties of probability-generating functions. However, it is much simpler to determine them directly from the random variable equation (11.6.13). Squaring both sides of the equation, we have

$$Q_{n+1}^2 = Q_n^2 + H^2(Q_n) + X_{n+1}^2 - 2Q_n H(Q_n) - 2X_{n+1}H(Q_n) + 2Q_n X_{n+1} \tag{11.6.21}$$

When $\rho < 1$, we also have

$$\lim_{n \to \infty} E(Q_{n+1}^r) = \lim_{n \to \infty} E(Q_n^r) \qquad r = 1, 2, 3, \dots \qquad (11.6.22)$$

Taking expectations in (11.6.21), letting $n \to \infty$, and dropping the suffix n to denote long-run random variables, we get

$$0 = E[H(Q)] + E(X^2) - 2E(Q) - 2E[XH(Q)] + 2E(QX)$$

$$(11.6.23)$$

Here we have used (11.6.12) to simplify the equation.

Taking expectations in (11.6.13), letting $n \to \infty$, and using (11.6.22), we also get

$$E[H(Q)] = E(X) \qquad (11.6.24)$$

Thus (11.6.23) gives

$$2E(Q)[1 - E(X)] = E(X) + E(X^2) - 2E(X)E[H(Q)]$$

$$= E(X) + E(X^2) - 2[E(X)]^2$$

$$E(Q) = \frac{1}{2}\left\{ E(X) + \frac{V(X)}{1 - E(X)} \right\}$$

which gives after substitution from (11.6.9)

$$E(Q) = \rho + \frac{\lambda^2 E(S^2)}{2(1 - \rho)} \qquad (11.6.25)$$

The result (11.6.25) is often referred to as the Pollaczek–Khintchine formula in the literature (see Gross and Harris, 1974, p. 226).

To determine $V(Q)$, we take the third power of (11.6.13) and write

$$Q_{n+1}^3 = Q_n^3 - H^3(Q_n) + X_{n+1}^3 - 3Q_n^2 H(Q_n)$$

$$+ 3Q_n H^2(Q_n) + 3H^2(Q_n)X_{n+1} - 3H(Q_n)X_{n+1}^2$$

$$+ 3X_{n+1}^2 Q_n + 3X_{n+1}Q_n^2 - 3Q_n H(Q_n)X_{n+1} \qquad (11.6.26)$$

Extending the relations given by (11.6.12), we get

$$H^3(X) = H(X)$$

$$X^2 H(X) = X^2$$

$$XH^2(X) = X \qquad (11.6.27)$$

Using these relations in (11.6.26), taking expectations, and letting $n \to \infty$, we find

$$0 = -E[H(Q)] + E(X^3) - 3E(Q^2) + 3E(Q) + 3E[H(Q)X]$$

$$-3E[H(Q)X^2] + 3E[X^2 Q] + 3E[XQ^2] - 6E[QX]$$

Rearranging and using (11.6.24) gives

$$3E(Q^2)[1 - E(X)] = -E(X) + E(X^3)$$

$$+ 3E(Q)[1 + E(X^2) - 2E(X)]$$

$$+ 3E(X)[E(X) - E(X^2)] \qquad (11.6.28)$$

and in turn

$$E(Q^2) = \frac{\lambda^3 E(S^3)}{3(1-\rho)} + \lambda^2 E(S^2) + E(Q)\left[1 + \frac{\lambda^2 E(S^2)}{1-\rho}\right] \qquad (11.6.29)$$

Thus we get

$$V(Q) = E(Q^2) - [E(Q)]^2$$

$$= E(Q)[1 - 2\rho + E(Q)] + \lambda^2 E(S^2) + \frac{\lambda^3 E(S^3)}{3(1-\rho)} \qquad (11.6.30)$$

Alternately, substituting for $E(Q)$, we may write

$$V(Q) = \rho(1-\rho) + \frac{\lambda^2 E(S^2)}{2(1-\rho)}\left[3 - 2\rho + \frac{\lambda^2 E(S^2)}{2(1-\rho)}\right] + \frac{\lambda^3 E(S^3)}{3(1-\rho)}$$

$$(11.6.31)$$

Expression (11.6.25) for $E(Q)$ can also be written as

$$E(Q) = \rho + \frac{\rho^2}{2(1 - \rho)} + \frac{\lambda^2 \sigma_S^2}{2(1 - \rho)} \qquad (11.6.32)$$

which clearly shows that the mean queue length increases with the variance of the service time distribution. For instance, when the service time is constant $\sigma_S^2 = 0$,

$$E(Q) = \rho + \frac{\rho^2}{2(1 - \rho)} = \frac{\rho}{1 - \rho}\left(1 - \frac{\rho}{2}\right) \qquad (11.6.33)$$

If instead the service time has an exponential distribution

$$B(x) = 1 - e^{-\mu x} \qquad x > 0$$

then

$$\sigma_S^2 = \frac{1}{\mu^2} = b^2$$

and therefore

$$E(Q) = \frac{\rho}{1 - \rho}$$

[Also see (11.3.7).] Similarly, substituting appropriate moments in (11.6.31), we can moreover show that the variance of the limiting queue length in an $M/D/1$ queue is smaller than the variance in an $M/M/1$ queue.

The expected waiting time of a customer can be obtained from (11.3.20) and the result (11.6.25). We have

$$E(W + S) = \frac{E(Q)}{\lambda}$$

$$= \frac{1}{\lambda}\left[\rho + \frac{\lambda^2(b^2 + \sigma_S^2)}{2(1 - \rho)}\right]$$

$$E(W) = \frac{\lambda(b^2 + \sigma_S^2)}{2(1 - \rho)} \qquad (11.6.34)$$

Suppose the queue $M/G/1$ described before is modified to incorporate a finite waiting room of size N. Assuming that customers arriving at the system when the waiting room is full depart without service, the transition probability matrix of the embedded Markov chain $\{Q_n\}$ takes the form

$$
P = \begin{array}{c@{\,}c}
 & \begin{array}{cccccccc} 0 & 1 & 2 & 3 & \cdots & N-1 & N \end{array} \\
\begin{array}{c} 0 \\ 1 \\ 2 \\ 3 \\ \vdots \\ N \end{array} &
\left[\begin{array}{ccccccc}
k_0 & k_1 & k_2 & k_3 & \cdots & k_{N-1} & \overline{A}_{N-1} \\
k_0 & k_1 & k_2 & k_3 & \cdots & k_{N-1} & \overline{A}_{N-1} \\
0 & k_0 & k_1 & k_2 & \cdots & k_{N-2} & \overline{A}_{N-2} \\
0 & 0 & k_0 & k_1 & & k_{N-3} & \overline{A}_{N-3} \\
\vdots & \vdots & \vdots & \vdots & \vdots & \vdots & \vdots \\
0 & 0 & 0 & 0 & \cdots & k_0 & \overline{A}_0
\end{array}\right]
\end{array}
\qquad (11.6.35)
$$

where we have written $k_j + k_{j+1} + \cdots = \overline{A}_{j-1}$. Let

$$
\pi_j = \lim_{n \to \infty} P(Q_n = j)
$$

The limiting distribution $\pi = (\pi_0, \pi_1, \ldots, \pi_N)$ is determined from the set of equations

$$
\pi P = \pi
$$

$$
\sum_{i=0}^{N} \pi_i = 1 \qquad (11.6.36)
$$

Writing out the first N equations of (11.6.36), we have

$$
k_0 \pi_0 + k_0 \pi_1 = \pi_0
$$

$$
k_1 \pi_0 + k_1 \pi_1 + k_0 \pi_2 = \pi_0
$$

$$
\vdots
$$

$$
k_{N-1} \pi_0 + k_{N-1} \pi_1 + \cdots + k_0 \pi_N = \pi_{N-1} \qquad (11.6.37)
$$

A convenient way of solving the set of equations (11.6.37) is to define

$$
\nu_0 \equiv 1 \quad \text{and} \quad \nu_i = \frac{\pi_i}{\pi_0} \qquad (11.6.38)
$$

and to rewrite (11.6.37) in terms of ν_i $(i = 1, 2, \ldots, N)$ as

$$\nu_1 = \frac{1 - k_0}{k_0}$$

$$\nu_2 = \frac{1 - k_1}{k_0}\nu_1 - \frac{k_1}{k_0}$$

$$\vdots$$

$$\nu_j = \frac{1 - k_1}{k_0}\nu_{j-1} - \frac{k_2}{k_0}\nu_{j-2} - \cdots - \frac{k_{j-1}}{k_0}\nu_1 - \frac{k_{j-1}}{k_0}$$

$$\vdots$$

$$\nu_N = \frac{1 - k_1}{k_0}\nu_{N-1} - \frac{k_2}{k_0}\nu_{N-2} - \cdots - \frac{k_{N-1}}{k_0}\nu_1 - \frac{k_{N-1}}{k_0} \qquad (11.6.39)$$

The recursive relations (11.6.39) completely determine ν_i $(i = 1, 2, \ldots, N)$. Further we have

$$\sum_{i=0}^{N} \nu_i = 1 + \sum_{i=1}^{N} \frac{\pi_i}{\pi_0} = \frac{\sum_{i=0}^{N} \pi_i}{\pi_0} = \frac{1}{\pi_0}$$

Thus we get

$$\pi_0 = \left[1 + \sum_{i=1}^{N} \nu_i\right]^{-1}$$

and

$$\pi_i = \frac{\nu_i}{1 + \sum_{i=1}^{N} \nu_i} \qquad (11.6.40)$$

A significant outcome of this solution technique is that for the same set of parameters, regardless of the value of the system capacity N, the quantities $\nu_0, \nu_1, \nu_2, \ldots$ are the same. Therefore, if one wants to find the limiting distribution of the queue size for two values of N, say n and $n + 1$, one first determines the set of values $(\nu_0, \nu_1, \ldots, \nu_n)$ and extends the set to include ν_{n+1} by using the recursive relations (11.6.39).

Consider a second modification of the $M/G/1$ queue to include service in groups of size b in addition to a finite waiting room of size N. The transition probability matrix of the embedded Markov chain $\{Q_n\}$, now takes the form

$P =$

$$
\begin{array}{c}
\\
0 \\
1 \\
\vdots \\
b \\
b+1 \\
b+2 \\
\vdots \\
N
\end{array}
\begin{array}{cccccccccc}
0 & 1 & 2 & \cdot & \cdot & N-b & N-b+1 & \cdot & \cdot & N-1 & N \\
\left[\begin{array}{c} k_0 \end{array}\right. & k_1 & k_2 & \cdot & \cdot & k_{N-b} & k_{N-b+1} & \cdot & \cdot & k_{N-1} & \overline{A}_{N-1} \\
k_0 & k_1 & k_2 & \cdot & \cdot & k_{N-b} & k_{N-b+1} & \cdot & \cdot & k_{N-1} & \overline{A}_{N-1} \\
\vdots & \vdots & \vdots & \vdots & & \vdots & \vdots & & & \vdots & \vdots \\
k_0 & k_1 & k_2 & \cdot & \cdot & k_{N-b} & k_{N-b+1} & \cdot & \cdot & k_{N-1} & \overline{A}_{N-1} \\
0 & k_0 & k_1 & \cdot & \cdot & k_{N-b-1} & k_{N-b} & \cdot & \cdot & k_{N-2} & \overline{A}_{N-2} \\
0 & 0 & k_0 & \cdot & \cdot & k_{N-b-2} & k_{N-b-1} & \cdot & \cdot & k_{N-3} & \overline{A}_{N-3} \\
\vdots & \vdots & \vdots & \vdots & & \vdots & \vdots & & & \vdots & \vdots \\
0 & 0 & 0 & \cdot & \cdot & k_0 & k_1 & \cdot & \cdot & k_{b-1} & \overline{A}_{b-1} \end{array}\right]
\end{array}
$$

$$(11.6.41)$$

The recursive solution technique employed for the determination of the limiting distribution of the queue with a transition probability matrix (11.6.35) cannot be used in this case. However, one can use the reduced system method illustrated in Section 5.2 by partitioning P as indicated in (11.6.41). Thus we get

$$P = \begin{bmatrix} A & B \\ C & D \end{bmatrix} \qquad (11.6.42)$$

where C is a triangular submatrix

$$
C = \begin{bmatrix}
k_0 & k_1 & \cdot & \cdot & \cdot & k_{N-b} \\
 & k_0 & \cdot & \cdot & \cdot & k_{N-b-1} \\
 & & & \cdot & & \cdot \\
 & & & & & k_0
\end{bmatrix} \qquad (11.6.43)
$$

which is invertible as long as $k_0 > 0$. A convenient partitioning of $\pi = (\tilde{\pi}_1, \tilde{\pi}_2)$ now leads us to the reduced system of equations

$$\tilde{\pi}_2 = \tilde{\pi}_1 (I - A) C^{-1}$$

$$\tilde{\pi}_1 [B + (I - A) C^{-1} (D - I)] = 0$$

$$\tilde{\pi}_1 [e_1^T + (I - A) C^{-1} e_2^T] = 1 \qquad (11.6.44)$$

where e_1 and e_2 are unit vectors of appropriate length. The equations (11.6.44) can be solved in the usual manner.

The Queue $M / D / 1$

Consider a single-server queueing system with Poisson arrivals and constant service times, say of length b units of time. Instead of considering the departure points as in the general case $M/G/1$, consider $(0, b, 2b, 3b, \ldots)$ as the discrete set of points, and let $Q_n = Q(nb)$. The process $\{Q_n\}$ is also a Markov chain embedded in the process $Q(t)$. Its behavior is identical to the chain $\{Q_n\}$ of the queue $M/G/1$, with departure points as the transition epochs. Verification of this property is left as an exercise for the reader.

The Queue $G / M / 1$

Consider a single-server queueing system in which interarrival times have an arbitrary distribution and the service times have an exponential distribution. As it is, the $Q(t)$ process in this system is also non-Markovian. However, it is possible to identify an embedded Markov chain $\{Q_n\}$ using the arrival points as the transition epochs. Derivation of its transition probability matrix is left as an exercise for the reader.

Example 11.6.1

Customer arrivals at a department store are at the rate of λ per unit time, and it has been found that a Poisson process is a reasonably good fit for the arrival process. The amounts of time spent by the customers while ringing up their merchandise are independent and identically distributed random variables with mean b_1 and the second moment b_2. Each customer spends on the average a length of time proportional to a ringing up time in shopping. In deciding about the number of checkout counters to be installed, the management has to take into account the following costs:

1. $\$C_1$ per unit time due to a waiting customer.
2. $\$C_2$ per unit time for maintaining service at a checkout counter.

Assuming that an arriving customer ends up at any one of the checkout counters with equal probability, determine the optimum number of counters that would minimize the total expected cost to the system. Suggest a method of solution when λ varies in time.

Solution Let s be the number of checkout counters, and let L_s be the long-run queue length for a checkout counter when there are s counters in the system. It should be noted that the shopping time of the customers does not depend on s, and therefore there is no need to include it in the decision model. Consider each counter as an $M/G/1$ queue with an arrival rate λ/s and service rate $1/b_1$. In (11.6.30), writing $b_1^2 + \sigma_s^2 = b_2$, we therefore have

$$E[L_s] = \frac{\lambda b_1}{s} + \frac{(\lambda/s)^2 b_2}{2[1 - (\lambda b_1/s)]}$$

$$= \frac{\lambda b_1}{s} + \frac{\lambda^2 b_2}{2s(s - \lambda b_1)} \tag{11.6.45}$$

Let C be the total cost per unit time of being in the system. Using queue length for measuring total time in the system,

$$E(C) = sC_1 \left[\frac{\lambda b_1}{s} + \frac{\lambda^2 b_2}{2s(s - \lambda b_1)} \right] + sC_2 \tag{11.6.46}$$

Using standard methods, we find that

$$s = \lambda b_1 + \sqrt{\frac{\lambda^2 b_2 C_1}{2C_2}} \tag{11.6.47}$$

minimizes $E(C)$, as given by (11.6.46). Let s_0 be this value. The number of counters is integer valued, and therefore the optimum value of s is given by

$$\begin{array}{ll} [s_0] & \text{if } E(C)|_{s=[s_0]} < E(C)|_{s=[s_0+1]} \\ [s_0 + 1] & \text{if } E(C)|_{s=[s_0]} > E(C)|_{s=[s_0+1]} \end{array} \tag{11.6.48}$$

where $[x]$ denotes the greatest integer contained in x.

As a numerical example let the arrivals be at the rate of 2 $(= \lambda)$ per minute, and let the ringing up times be exponential with mean 3 $(= b_1)$ minutes. Then the second moment $b_2 = 9$. Further let $C_1 = 0.5$ and $C_2 = 2.5$. Clearly

$$s_0 = 2 \times 3 + \sqrt{\frac{4 \times 9 \times 0.5}{2.5}}$$

$$= 6 + \sqrt{7.2}$$

$$= 8.683$$

Also

$$E(C)|_{s=8} = 27.5$$

$$E(C)|_{s=9} = 28.5$$

Hence the optimum value of s is found to be 8.

If the arrival rate varies during a day, it should be possible to divide the working hours into periods during which the arrival rates remain fairly constant and then use the preceding decision rule for every period. Even though such a procedure is not justified because of lack of steady-state conditions, the results obtained would seem to be good approximations for the transient case.   *Answer*

11.7 THE MACHINE INTERFERENCE PROBLEM

In this world of machines and automation, employing the right number of repairpersons to look after a number of machines is a problem of some significance. The number of repairpersons should be decided so as to minimize the loss due to idle machines and repairpersons. In this section we shall consider only a simplified model of such a situation. For further discussion see Palm (1947), Benson and Cox (1951), and Cox and Smith (1961), and references cited by them. Palm's results have been quoted by Feller (1968).

Suppose there are M machines, each of which calls for service after being in use for an interval of time that has an exponential distribution with mean λ^{-1}. Under this assumption, when there are m machines working at time t, the probability that any one of these machines would call for service in the infinitesimal interval $(t, t + \Delta t]$ is $m\lambda \Delta t + o(\Delta t)$. Let the service times of these machines be exponential random variables with mean μ^{-1}. Suppose there are s repairpersons (called "operatives"). We shall call $\rho = \lambda/\mu$ the servicing factor. First we shall derive some behavioral results of this process which can then be used in decision problems.

Let $N(t)$ be the number of machines not working at time t. Define

$$P_n(t) = P[N(t) = n | N(0) = i] \tag{11.7.1}$$

The stochastic process $N(t)$ can now be modeled as a birth and death process, with parameters

$$\lambda_n = (M - n)\lambda \tag{11.7.2}$$

and

$$\mu_n = n\mu \qquad 1 \le n \le s$$
$$= s\mu \qquad n \ge s$$

Let $\lim_{t \to \infty} P_n(t) = p_n$, $n = 0, 1, 2, \ldots, M$. For these probabilities the steady-state equations can be written down as

$$M\lambda p_0 = \mu p_1$$

$$[(M - n)\lambda + n\mu] p_n = (M - n + 1)\lambda p_{n-1}$$

$$+ (n + 1)\mu p_{n+1} \qquad 1 \le n < s$$

$$[(M - n)\lambda + s\mu] p_n = (M - n + 1)\lambda p_{n-1}$$

$$+ s\mu p_{n+1} \qquad s \le n < M$$

$$s\mu p_M = \lambda p_{M-1} \tag{11.7.3}$$

Solving these equations recursively, we get

$$p_n = \binom{M}{n}\rho^n p_0 \qquad 0 \le n \le s$$

$$= \binom{M}{n}\frac{n!}{s!s^{n-s}}\rho^n p_0 \qquad s \le n \le M \tag{11.7.4}$$

Clearly p_0 has to be obtained from the condition

$$\sum_0^M p_n = 1$$

In particular, when $s = 1$, we get

$$p_n = \frac{M!}{(M - n!)}\rho^n p_0 \tag{11.7.5}$$

and

$$p_0\left[1 + \frac{M!}{(M - 1)!}\rho + \frac{M!}{(M - 2)!}\rho^2 + \cdots + M!\rho^M\right]^{-1} \tag{11.7.6}$$

Two measures of effectiveness for the system can be defined as follows:

$$\text{Machine availability} = 1 - \frac{E(N)}{M} \tag{11.7.7}$$

$$\text{Operative utilization} = \sum_{n=0}^{s} \frac{np_n}{s} + \sum_{n=s+1}^{M} p_n \tag{11.7.8}$$

Table 11.7.1

ρ	Machine Availability			Operative Utilization		
M	0.1	0.5	1.0	0.1	0.5	1.0
5	0.8721	0.3853	0.1994	0.7151	0.9633	0.9969
10	0.7854	0.2000	0.1000	0.9816	1.0000	1.0000
50	0.6423	0.1333	0.0667	0.9998	1.0000	1.0000

Clearly the machine availability represents the fraction of total running time on all machines, and the operative utilization represents the fraction of time any operative would be working [compare with the result (11.4.12)]. When $s = 1$, the operative utilization has the simple form $1 - p_0$. Table 11.7.1 gives the values of these measures when there is only one operative in the system. Table 11.7.2 gives the operative utilization for different pairs of (M, s) values with the same machine per repairperson ratio. From the results of Table 11.7.2 one can conclude that it is better to use operatives in a pool rather than assigning a certain number of machines to each operative.

Other measures of effectiveness of the system are coefficient of loss for machines, defined as

$$\frac{\text{Average number of machines actually waiting}}{\text{Total number of machines}}$$

Table 11.7.2

ρ	M	s	Operative Utilization
0.45	4	1	0.881
	8	2	0.934
	16	4	0.994
0.05	15	1	0.656
	30	2	0.682
	60	4	0.705

and the coefficient of loss for operatives, defined as

$$\frac{\text{Average number of idle operatives}}{\text{Total number of operatives}}$$

11.8 REMARKS

The study of queueing systems has seen phenomenal growth during the 1960s and the 1970s. In the last six sections we have introduced only a few basic systems as illustrations of applications of stochastic processes. Readers interested in the study and application of queueing theory are referred to the books that have appeared on the subject as well as journal articles which explore its frontiers. The following recent books may be suggested: Chaudhry and Templeton (1983), Cohen (1983), Cooper (1981), Gross and Harris (1974, 1985), Kleinrock (1975, 1976), and Neuts (1981). For an extensive bibliography of significant contributions, see Bhat (1978).

Queueing theory is an applied probability area where stochastic models are developed in order to study real-world problems. Major areas of application for queueing theory are communications and computers. These two areas are further explored in the next two chapters. Queueing models also find applications in diverse areas such as hospital systems and traffic problems, and we provide a selected bibliography of such applications under the heading "For Further Reading." Major journals reporting methodological advances in queueing theory are *Advances in Applied Probability*, *Applied Probability*, *Bell System Technical Journal*, *Management Science*, *Mathematics of Operations Research*, *Naval Research Logistics Quarterly*, *Operations Research*, and *Stochastic Processes and Applications*. Readers interested in applications are referred to appropriate area journals.

REFERENCES

Benson, F., and Cox, D. R. (1951, 1952). "The Productivity of Machines Requiring Attention at Random Intervals." *J. Roy. Stat. Soc.* **B13**, 65–82; **14**, 200–219.

Bhat, U. N. (1978). "Theory of Queues." *Handbook of Operations Research*. Vol. 1 (Ed. by J. J. Moder and S. E. Elmaghraby.) New York: Van Nostrand-Reinhold, pp. 352–397.

Chaudhry, M. L., and Templeton, J. G. C. (1983), *A First Course in Bulk Queues*. New York: Wiley. 2nd ed. (1985).

Cohen, J. W. (1983). *The Single Server Queue*. 2nd ed. Amsterdam: Elsevier.

Conway, R. W., and Maxwell, W. L. (1962). "A Queueing Model with State Dependent Service Rates." *J. Ind. Eng.* **12**, 132–136.

Cooper, R. B. (1981). *Introduction to Queueing Theory*. 2nd ed. Amsterdam: North Holland.

Cox, D. R., and Smith, W. L. (1961). *Queues*. London: Methuen (Wiley).

Crabill, T. B. (1969). "Optimal Control of a Queue with Variable Service Rates." Ph.D. dissertation, Tech. Rep. 75. Department of Operations Research, Cornell University, Ithaca, N.Y.

Crabill, T. B. (1972). "Optimal Control of a Service Facility with Variable Exponential Service Times and Constant Arrival Rate." *Management Sci.* **18**, 560–566.

Feller, W. (1968). *An Introduction to Probability Theory and Its Applications*. 3rd ed. New York: Wiley, Chaps. 13–15.

Gross, D., and Harris, C. M. (1974). *Fundamentals of Queueing Theory*. New York: Wiley. 2nd ed. (1985).

Haight, F. A. (1957). "Queueing with Balking." *Biometrika*, **44**, 360–369.

Kendall, D. G. (1951). "Some Problems in the Theory of Queues." *J. Roy. Stat. Soc.* **B13**, 151–185.

Kendall, D. G. (1953). "Stochastic Processes Occurring in the Theory of Queues and Their Analysis by the Method of Imbedded Markov Chains." *Ann. Math. Stat.* **24**, 338–354.

Kleinrock, L. (1975). *Queueing Systems*. Vol. 1: *Theory*. New York: Wiley.

Kleinrock, L. (1976). *Queueing Systems*. Vol. 2: *Computer Applications*. New York: Wiley.

Little, J. D. C. (1961). "A Proof of the Queueing Formula $L = \lambda W$." *Oper. Res.* **9**(3), 383–387.

Morse, P. M. (1958). *Queues, Inventories and Maintenance*. New York: Wiley.

Neuts, M. F. (1981). *Matrix-Geometric Solutions in Stochastic Models—An Algorithmic Approach*. Baltimore, Md.: Johns Hopkins University Press.

Palm, C. (1947). "The Distribution of Repairmen in Servicing Automatic Machines" (in Swedish). *Indust. Nord.* **75**, 75–70, 90–94, 119–123.

Singh, V. P. (1968). "Queueing Systems with Balking and Heterogeneous Servers." Ph.D. dissertation, Tech. Memo. 113. Department of Operations Research, Case Western Reserve University, Cleveland, Ohio.

FOR FURTHER READING

Ashcroft, H. (1950). "The Productivity of Several Machines under the Care of One Operator." *J. Roy. Stat. Soc.* **B12**, 145–151.

Assad, A. A. (1981). "Analytical Models in Rail Transportation: An Annotated Bibliography." *INFOR* **19**, 59–80.

Beneš, V. E. (1963). *General Stochastic Processes in the Theory of Queues*. Reading, Mass.: Addison-Wesley.

Bhat, U. N. (1968). *A Study of the Queueing Systems M/G/1 and GI/M/1*. New York: Springer-Verlag.

Bhat, U. N. (1969). "Sixty Years of Queueing Theory." *Management Sci.* **15**, 280–284.

Bookbinder, J. H., and Martell, D. L. (1979). "Time-Dependent Queueing Approach to Helicopter Allocation for Forest Fire Initial Attack." *INFOR* 17, 58–70.

Brigham, G. (1955). "On a congestion Problem in an Aircraft Factory." *J. Oper. Res. Soc. Amer.* 3, 412–428.

Brockmeyer, E., Halstrøm, H. L., and Jensen, A. (1948). *The Life and Works of A. K. Erlang.* Acta Polytechnica Scandinavica (Ap. 287, 1960). The Danish Academy of Technical Sciences, Copenhagen, Denmark.

Chelst, K., Tilles, A. Z., and Pipis, J. S. (1981). "A Coal-Unloader: A Finite Queueing System with Breakdowns." *Interfaces* 11(5), 12–25.

Edie, L. E. (1954). "Traffic Delays at Toll Booths." *J. Oper. Res. Soc. Amer.* 2, 107–138.

Gilliam, R. R. (1979). "An Application of Queueing Theory to Airport Passenger Security Screening." *Interfaces* 9(4), 117–123.

Gnedenko, B. V., and Kovalenko, I. N. (1968). *Introduction to Queueing Theory.* Jerusalem: Israel Program for Scientific Translations.

Khintchine, A. Y. (1960). *Mathematical Methods in the Theory of Queueing.* London: Charles Griffin.

Kolesar, P. (1970). "A Markovian Model for Hospital Administration Scheduling." *Management Sci.* 16, B384–B396.

Kolesar, P. (1979). "A Quick and Dirty Response to the Quick and Dirty Crowd; Particularly to Jack Byrd's 'The Value of Queueing Theory.'" *Interfaces* 9(1), 77–82. Includes a good set of references on applications.

Larson, R. C., and McKnew, M. A. (1982). "Police Patrol-Initiated Activities within a Systems Queueing Model." *Management Sci.* 28, 759–774. Includes a good set of references.

Lee, A. M. (1966). *Applied Queueing Theory.* London: Macmillan; New York: St. Martin's Press.

LeGall, P. (1962). *Les systèmes avec au sans attente et les processus stochastiques.* Vol. 1. Paris: Dunod.

McKeown, P. G. (1979). "An Application of Queueing Analysis to the New York State Child Abuse and Maltreatment Register Telephone Reporting System." *Interfaces* 9(3), 20–25.

Newell, G. F. (1971). *Applications of Queueing Theory.* London: Chapman and Hall.

Posner, M. J. M., and Watkin, M. (1981). "An Exponential Queueing Model for Determining Teller Requirements in Bank Branches." *INFOR* 19, 230–245.

Prabhu, N. U. (1965). *Queues and Inventories.* New York: Wiley.

Riordan, J. (1962). *Stochastic Service Systems.* New York: Wiley.

Rosenshine, M. (1975). "Queueing Theory: The State of the Art." *AIIE Trans.* 7, 257–267.

Saaty, T. L. (1961). *Elements of Queueing Theory.* New York: McGraw-Hill.

Saaty, T. L. (1966). "Seven More Years of Queues." *Nav. Res. Log. Quart.* 13, 447–476.

Syski, R. (1960). *Introduction to Congestion Theory in Telephone Systems.* London: Oliver and Boyd.

Takács, L. (1962). *Introduction to the Theory of Queues.* Oxford: Oxford University Press.

Takács, L. (1967). *Combinatorial Methods in the Theory of Stochastic Processes.* New York: Wiley.

Vogel, M. A. (1979). "Queueing Theory Applied to Machine Manning." *Interfaces* **9**(4), 1–8.

Yaffe, H. J. (1974). "A Model for Optimal Operation and Design of Solid Waste Transfer Stations." *Transportation Sci.* **8**, 265–306

Chapter 12
QUEUEING NETWORKS

12.1 INTRODUCTION

In Chapter 11 we identified the number of queues (servers) as one of the four basic characteristics of a queueing system. However, the systems we considered consisted of only one waiting line. In this chapter we discuss some of the simpler problems that arise when there are a number of queues to consider which can be arranged in series, parallel, or a combination of both configurations.

A network is a collection of nodes connected by a set of paths. In a queueing network servers or a group of servers are identified as nodes. Examples of queueing networks abound in this technology-dominated world: a telecommunications network, a job-shop scheduling system, and a data-processing system with multiprocessing and multiprogramming features are some of these.

A basic feature of a queueing network is the customer's need for service from more than one server. Many times a customer may have to return to the same server from whom service had been obtained earlier. (This phenomenon is known as feedback.) Thus the problem is one of analyzing system performance, taking the individual needs into account. Clearly we shall not be able to discuss the problem in its generality. We shall restrict ourselves to Markovian networks and some simple techniques of analyzing the underlying processes.

12.2 THE TANDEM QUEUE

Under Markovian assumptions (Poisson arrivals and exponential service times) the general model for a queueing network is given by the Markov population processes discussed in Section 7.9. A simple special case of such processes is the tandem queue in which single-server facilities are arranged strictly in series. Assembly lines in a factory, medical checkups in a clinic, a sequence of traffic intersections, are some common examples.

Consider a simple model of a tandem queue with only two servers in series. Customers arrive in a Poisson process and get served from the first and second servers in that order, on a first-come, first-served basis. Queues are allowed to build up in front of both servers. Let the service times of customers at the two servers have exponential distributions with rates μ_1 and μ_2, respectively.

Let $Q_1(t)$ and $Q_2(t)$ be the number of customers in the first and second queues, respectively, at time t. As a consequence of the Poisson assumptions just made, $\{Q_1(t), Q_2(t)\}$ is a vector Markov process with state space $\{(n_1, n_2): n_1, n_2 = 0, 1, 2, 3, \dots\}$. Write Q_1 and Q_2 for the long-run queue length values, and define

$$p_{n_1 n_2} = P(Q_1 = n_1, Q_2 = n_2) \tag{12.2.1}$$

which exists when $\rho_1 = \lambda/\mu_1 < 1$ and $\rho_2 = \lambda/\mu_2 < 1$. The steady-state equations for $p_{n_1 n_2}$ are

$$\lambda p_{00} = \mu_2 p_{01}$$

$$(\lambda + \mu_2) p_{0, n_2} = \mu_1 p_{1, n_2 - 1} + \mu_2 p_{0, n_2 + 1} \qquad n_2 > 0$$

$$(\lambda + \mu_1) p_{n_1 0} = \mu_2 p_{n_1 1} + \lambda p_{n_1 - 1, 0} \qquad n_1 > 0$$

$$(\lambda + \mu_1 + \mu_2) p_{n_1 n_2} = \mu_1 p_{n_1 + 1, n_2 - 1} + \mu_2 p_{n_1 n_2 + 1}$$

$$+ \lambda p_{n_1 - 1, n_2} \qquad n_1, n_2 > 0 \tag{12.2.2}$$

Suppose the solution is of the form (see R. R. P. Jackson, 1954, 1956)

$$p_{n_1 n_2} = \rho_1^{n_1} \rho_2^{n_2} p_{00} \tag{12.2.3}$$

which satisfies equations (12.2.2). Using Theorem 7.8.1 and the normalizing condition

$$\sum_{n_1 = 0}^{\infty} \sum_{n_2 = 0}^{\infty} p_{n_1 n_2} = 1 \tag{12.2.3}$$

we get

$$p_{00} = (1 - \rho_1)(1 - \rho_2)$$

$$p_{n_1 n_2} = (1 - \rho_1)(1 - \rho_2) \rho_1^{n_1} \rho_2^{n_2} \tag{12.2.4}$$

The tandem queue can be used to illustrate the use of local balance equations in deriving the limiting distribution as discussed in Section 7.9.

Comparing parameters and notations used for the tandem queue and those in Section 7.9, we write $p_{n_1 n_2} = p_1(n_1)$ for the same n_2, $p_{n_1 n_2} = p_2(n_2)$ for the same n_1 and note that, in using equation (7.9.21), we have

$$a_1 = \lambda \qquad b_1 = 0 \qquad g_{12} = \mu_1$$

$$a_2 = 0 \qquad b_2 = \mu_2 \qquad g_{21} = 0$$

Now applying equation (7.9.21), we get

$$\mu_1 \rho_1 = \lambda \tag{12.2.5}$$

giving $\rho_1 = \lambda/\mu_1$ and

$$\mu_2 \rho_2 = \mu_1 \rho_1 = \lambda \tag{12.2.6}$$

giving

$$\rho_2 = \frac{\lambda}{\mu_2}$$

The result (12.2.4) now follows from (7.9.18) and (7.9.17), after noting that $\phi_r(n) \equiv 1$ in this case.

It is interesting to observe the implications of equations (12.2.5) and (12.2.6) on the steady-state equations (12.2.2) which represent global balance. Noting that from (7.9.19) we have

$$\rho_1 = \frac{p_1(n_1 + 1)}{p_1(n_1)} \quad \text{and} \quad \rho_2 = \frac{p_2(n_2 + 1)}{p_2(n_2)}$$

a sufficient set of conditions for equations (12.2.5) and (12.2.6) to hold is

$$\lambda p_{n_1 n_2} = \mu_1 p_{n_1+1, n_2} \qquad n_1 = 0, 1, 2, \ldots$$

$$\lambda p_{n_1 n_2} = \mu_2 p_{n_1, n_2+1} \qquad n_2 = 0, 1, 2, \ldots \tag{12.2.7}$$

Using equations (12.2.7), the global balance equation (12.2.2) can be broken up into

$$\lambda p_{n_1 n_2} = \mu_2 p_{n_1, n_2+1} \qquad n_1 = 0, 1, 2, \ldots, n_2 = 0, 1, 2, \ldots$$

$$\lambda p_{n_1-1, n_2} = \mu_1 p_{n_1 n_2} \qquad n_2 = 0, 1, 2, \ldots, n_1 = 1, 2, \ldots$$

$$\mu_2 p_{n_1 n_2} = \mu_1 p_{n_1+1, n_2-1} \qquad n_1 = 0, 1, 2, \ldots, n_2 = 1, 2, 3, \ldots$$

$$\tag{12.2.8}$$

The last equation in (12.2.8) follows from combining the two equations in (12.2.7). The three equations in (12.2.8) balance three pairs of transitions into and out of state $(n_1 n_2)$ as shown in the following figure:

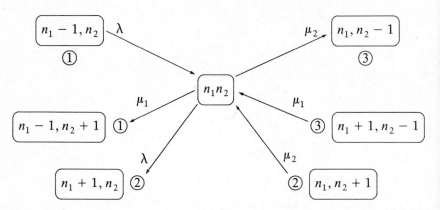

Hence these equations can be considered as equations of local balance. Together they represent equations of global balance, and therefore any solution satisfying this set of local balance equations should lead us to the final solution. Since local balance equations relate two state probabilities at a time, they lend themselves convenient for a recursive solution. Thus from the first two equations of (12.2.8) we have

$$p_{n_1 n_2} = \rho_1 p_{n_1 - 1, n_2}$$

giving

$$p_{n_1 n_2} = \rho_1^{n_1} p_{0 n_2} \qquad n_2 = 0, 1, 2, \ldots \qquad (12.2.9)$$

and

$$p_{n_1, n_2 + 1} = \rho_2 p_{n_1 n_2}$$

giving

$$p_{n_1 n_2} = \rho_2^{n_2} p_{n_1 0} \qquad n_1 = 0, 1, 2, \ldots \qquad (12.2.10)$$

Combining (12.2.9) and (12.2.10), we get

$$p_{n_1 n_2} = \rho_1^{n_1} \rho_2^{n_2} p_{00} \qquad (12.2.11)$$

which is equation (12.2.3). The unknown p_{00} is determined using the normalizing condition

$$\sum_{n_1} \sum_{n_2} p_{n_1 n_2} = 1$$

We get

$$p_{n_1 n_2} = (1 - \rho_1)(1 - \rho_2)\rho_1^{n_1}\rho_2^{n_2} \qquad (12.2.12)$$

Let $E(Q)$ be the mean number of customers in the system. From (12.2.12) we get

$$E(Q) = \frac{\rho_1}{1 - \rho_1} + \frac{\rho_2}{1 - \rho_2} \qquad (12.2.13)$$

Let Q_i $(i = 1, 2)$ be the limiting queue length in an $M/M/1$ queueing system with arrival and service parameters λ and μ_i, respectively. Then

$$P(Q_i = n) = (1 - \rho_i)\rho_i^n \qquad (12.2.14)$$

Clearly in (12.2.12) we have the product of such probabilities for $i = 1$ and 2. Hence we can conclude that in the tandem queue the limiting queue lengths Q_1 and Q_2 are independent of each other. This property also has been independently and more generally established by Burke (1956), who shows that the output process of an $M/M/s$ queue with an arrival parameter λ is again a Poisson process with the same parameter λ.

The results derived here can be easily extended for tandem queues with $s(\geq 2)$ servers. Let $\{Q_1, Q_2, \ldots, Q_s\}$ be the queue lengths corresponding to s servers in the limiting case. Also define $\rho_i = \lambda/\mu_i$, where μ_i is the service rate at the ith server. Then we get

$$p_{n_1 n_2 \cdots n_s} = P(Q_1 = n_1, Q_2 = n_2, \ldots, Q_s = n_s)$$

$$= \prod_{i=1}^{s} (1 - \rho_i)\rho_i^{n_i} \qquad n_1, n_2, \ldots, n_s \geq 0 \qquad (12.2.15)$$

12.3 OPEN QUEUEING NETWORKS

The product form of equation (12.2.15) establishes that the queue lengths $\{Q_i, i = 1, 2, \ldots, s\}$ are independent random variables. In the case of the tandem queue the arrival process at each node is also Poisson, and the limiting queue lengths are independent of each other. But as pointed out in Section 7.9, the independence of limiting queue lengths is not necessary for the product form solution, and under some special system structures relatively simple analysis techniques provide results that exhibit these properties.

Consider an open network of waiting lines studied by J. R. Jackson (1957, 1963; also see Whittle, 1968). The network consists of k nodes with node i serviced by s_i servers. Let the arrival of customers from outside the system be in Poisson processes, λ_i being the rate at node i. Let the service rate at node i be μ_i and the service time distribution be exponential. After service at node i, the customer moves to node j with probability α_{ij} and leaves the system with probability $1 - \sum_{j=1}^{k} \alpha_{ij}$. Let $p(n_1, n_2, \ldots, n_k)$ be the limiting distribution of the number of customers in the system at the various nodes (n_i at the ith node, $i = 1, 2, \ldots, k$). Under these assumptions J. R. Jackson has shown that the limiting distribution is given by

$$p(n_1, n_2, \ldots, n_k) = p_1(n_1) p_2(n_2) \cdots p_k(n_k) \qquad (12.3.1)$$

where

$$p_i(n_i) = \begin{cases} p_i(0) \dfrac{(\gamma_i/\mu_i)^r}{r!} & r = 0, 1, 2, \ldots, s_i \\[3mm] p_i(0) \dfrac{(\gamma_i/\mu_i)^r}{s_i!(s_i)^{r-s_i}} & r = s_i, s_i + 1, \ldots \end{cases} \qquad (12.3.2)$$

and

$$\gamma_i = \lambda_i + \sum_j \alpha_{ji} \gamma_j \qquad i = 1, 2, \ldots, k \qquad (12.3.3)$$

Clearly γ_i ($i = 1, 2, \ldots, k$) which can be determined from the set of equations (12.3.3) gives the effective arrival rate at the ith node, and $p_i(n_i)$ is the limiting distribution at the ith node, as given by equation (11.4.4). Consequently the results imply that the marginal distributions of the number of customers in the system at the various nodes are independent of each other and the queue length in each node behaves like the queue length of an $M/M/s$ queue in equilibrium. Unfortunately this interpretation cannot be carried through to saying that as $t \to \infty$, the arrival process at the ith node is a Poisson process with rate γ_i ($i = 1, 2, \ldots, k$) as implied by the $M/M/s$ queue structure. In fact for an $M/M/1$ queue with feedback, Burke (1976) has proved that the arrival process is characterized by a marginal interarrival time distribution which is a mixture of two exponentials.

The results (12.3.2) and (12.3.3) can be also obtained by appealing to the solution technique outlined in Section 7.9. We shall illustrate this procedure by considering a two-node network, with only one server at each node. For convenience we shall write $\alpha_{10} = 1 - \alpha_{12}$ and $\alpha_{20} = 1 - \alpha_{21}$.

Comparing parameters with those used in (7.9.21), we have

$$a_1 = \lambda_1 \qquad b_1 = \mu_1\alpha_{10} \qquad g_{12} = \mu_1\alpha_{12}$$

$$a_2 = \lambda_2 \qquad b_2 = \mu_2\alpha_{20} \qquad g_{21} = \mu_2\alpha_{21}$$

Corresponding to equations (7.9.21), we get

$$(\mu_1\alpha_{10} + \mu_1\alpha_{12})\rho_1 = \lambda_1 + \mu_2\alpha_{21}\rho_2$$

$$(\mu_2\alpha_{20} + \mu_2\alpha_{21})\rho_2 = \lambda_2 + \mu_1\alpha_{12}\rho_1 \qquad (12.3.4)$$

Solving for ρ_1 and ρ_2, we find

$$\rho_1 = \frac{\lambda_1 + \lambda_2\alpha_{21}}{\mu_1(1 - \alpha_{12}\alpha_{21})}$$

$$\rho_2 = \frac{\lambda_2 + \lambda_1\alpha_{12}}{\mu_2(1 - \alpha_{12}\alpha_{21})} \qquad (12.3.5)$$

It may be noted here that using the solution given by equation (12.3.3), ρ_1 and ρ_2 may be written as $\rho_1 = \gamma_1/\mu_1$ and $\rho_2 = \gamma_2/\mu_2$, which are the effective traffic intensities at the two nodes. The limiting distribution of customers in the system is now obtained as

$$p(n_1, n_2) = C\rho_1^{n_1}\rho_2^{n_2}$$

where C is determined using the normalizing condition $\sum_{n_1, n_2} p(n_1, n_2) = 1$. Thus we get the complete solution as

$$p(n_1, n_2) = (1 - \rho_1)(1 - \rho_2)\rho_1^{n_1}\rho_2^{n_2} \qquad (12.3.6)$$

We shall not venture into writing down the balance equations and solving them in this case since even when there are only two nodes, they become too cumbersome to be meaningful.

12.4   CLOSED QUEUEING NETWORKS

If in the network considered in the last section we eliminate the arrival
process from outside the network (exogenous arrivals) and consider a fixed
number of customers moving around and getting served from the k nodes,
we have a closed network studied by Gordon and Newell (1967) and
Whittle (1967), who identifies the system as a nonlinear migration process.
As discussed in Section 7.9, even though the various nodes of the network
are not independent of each other, in equilibrium the limiting distribution
exhibits the product form common to independent systems.

Let $N(N < \infty)$ be the total number of customers in the k nodes of the
system, and therefore if n_i is the number of customers at the ith node, we
have $\sum_{i=1}^{k} n_i = N$. Also let the ith node be served by s_i servers with service
rate μ_i and an exponential distribution. After service at node i, we assume
that the customer moves to node j with probability α_{ij}, so that $\sum_{j=1}^{k} \alpha_{ij} = 1$.
Using the results derived in Section 7.9 directly, we get

$$p(n_1, n_2, \ldots, n_k) = C \prod_{i=1}^{k} \frac{\rho_i^{n_i}}{\phi_i(1)\phi_i(2) \cdots \phi_i(n_i)} \qquad (12.4.1)$$

where

$$\phi_j(n) = \min(n, s_j)$$

and ρ_r satisfies the relation

$$\rho_r \sum_s \mu_r \alpha_{rs} = \sum_s \mu_s \alpha_{sr} \rho_s \qquad r = 1, 2, \ldots, k \qquad (12.4.2)$$

It should be noted that from the system of equations (12.4.2), ρ_r ($r =$
$1, 2, \ldots, k$) can be determined to within a multiplicative constant. But when
combined with C of (12.4.1), the arbitrariness can be eliminated by using the
normalizing condition

$$\sum_{n_1, n_2, \ldots, n_k} p(n_1, n_2, \ldots, n_k) = 1$$

As an illustration consider a cyclic queue with three nodes and one server
at each node. In a cyclic queue (see Koenigsberg, 1958) customers are served
in a cyclic manner, customers moving to node $i + 1$ from node i ($i =$
$1, 2, \ldots, k - 1$) and to node 1 from node k. Equations corresponding to

(12.4.2) have the following form:

$$\mu_1 \rho_1 = \mu_k \rho_k$$

$$\mu_2 \rho_2 = \mu_1 \rho_1$$

$$\mu_3 \rho_3 = \mu_2 \rho_2$$

$$\vdots$$

$$\mu_k \rho_k = \mu_{k-1} \rho_{k-1} \tag{12.4.3}$$

From these we get

$$\rho_2 = \frac{\mu_1}{\mu_2} \rho_1$$

$$\rho_3 = \frac{\mu_2}{\mu_3} \rho_2 = \frac{\mu_1}{\mu_3} \rho_1$$

$$\rho_4 = \frac{\mu_3}{\mu_3} \rho_3 = \frac{\mu_1}{\mu_4} \rho_1$$

$$\vdots$$

$$\rho_k = \frac{\mu_1}{\mu_k} \rho_1 \tag{12.4.4}$$

Consequently the limiting distribution takes the form

$$p(n_1, n_2, \ldots, n_k) = \left(C \rho_1^N \right) \prod_{i=2}^{k} \left(\frac{\mu_1}{\mu_i} \right)^{n_i} \tag{12.4.5}$$

The constant $(C\rho_i^N)$ is determined through normalization. For instance, the state space for $(N = 3, \ k = 3)$ can be identified as

$$(n_1, n_2, n_3) = \{(3,0,0), (2,1,0), (2,0,1), (1,2,0),$$

$$(1,0,2), (1,1,1), (0,3,0), (0,2,1),$$

$$(0,1,2), (0,0,3)\} \tag{12.4.6}$$

Writing $\mu_1/\mu_i = \nu_i$, the constant $(C\rho_1^N)$ is now obtained as

$$\left(C\rho_1^N \right) = \left[1 + \nu_2 + \nu_3 + \nu_2^2 + \nu_3^2 + \nu_2\nu_3 + \nu_2^3 + \nu_2^2\nu_3 + \nu_2\nu_3^2 + \nu_3^3 \right]^{-1}$$

$$\tag{12.4.7}$$

Given next is an illustration of the balance equation technique to solve this problem. Using the state space identified in (12.4.6) and using the notation $p_{n_1 n_2 n_3}$ in place of $p(n_1, n_2, n_3)$, we have the following steady-state equations of balance:

$$\mu_1 p_{300} = \mu_3 p_{201}$$

$$(\mu_1 + \mu_2) p_{210} = \mu_1 p_{300} + \mu_3 p_{111}$$

$$(\mu_1 + \mu_3) p_{201} = \mu_2 p_{210} + \mu_3 p_{102}$$

$$(\mu_1 + \mu_2) p_{120} = \mu_1 p_{210} + \mu_3 p_{021}$$

$$\vdots$$

$$\mu_3 p_{003} = \mu_2 p_{012} \tag{12.4.8}$$

With 10 unknown variables, obtaining a direct solution is cumbersome. Instead, we can break up the global balance equations into equations of local balance (i.e., between neighboring states). These are

$$\mu_1 p_{300} = \mu_3 p_{20}$$

$$\mu_2 p_{210} = \mu_1 p_{300}$$

$$\mu_1 p_{210} = \mu_3 p_{111}$$

$$\vdots$$

$$\mu_3 p_{003} = \mu_2 p_{012} \tag{12.4.9}$$

These equations can be written by simple observation regarding the transitions into and out of the system, or by using equations (12.4.4) which in this case turn out to be

$$\rho_2 = \frac{\mu_1}{\mu_2} \rho_1$$

$$\rho_3 = \frac{\mu_2}{\mu_3} \rho_2 = \frac{\mu_1}{\mu_3} \rho_1 \tag{12.4.10}$$

Recalling that ρ_r $(r = 1, 2, 3)$ are defined as (using the notation from Section 7.9)

$$\rho_r = \frac{p_r(i + 1)}{p_r(i)}$$

local balance equations can be written down as (writing ν_i instead of ρ_i)

$$\nu_2 = \frac{\mu_1}{\mu_2} = \frac{p_{210}}{p_{300}} = \frac{p_{021}}{p_{111}} = \frac{p_{030}}{p_{120}} = \frac{p_{012}}{p_{102}} = \frac{p_{111}}{p_{201}} = \frac{p_{120}}{p_{210}}$$

$$\nu_3 = \frac{\mu_1}{\mu_3} = \frac{p_{201}}{p_{300}} = \frac{p_{111}}{p_{210}} = \frac{p_{102}}{p_{201}} = \frac{p_{201}}{p_{120}} = \frac{p_{003}}{p_{102}} = \frac{p_{012}}{p_{111}} \quad (12.4.11)$$

These equations can be written down using the following rule: For μ_i/μ_j $(i, j = 1, 2, 3)$ the jth subscript in the numerator is one more than the denominator, the ith subscript is varied to cover all cases, and the subscript in position other than i and j remains constant. From equations (12.4.11) the following relations emerge:

$$p_{210} = \nu_2 p_{300} \qquad p_{120} = \nu_2^2 p_{300} \qquad p_{030} = \nu_2^3 p_{300} \qquad p_{201} = \nu_3 p_{300}$$

$$p_{111} = \nu_2 \nu_3 p_{300} \qquad p_{102} = \nu_3^2 p_{300} \qquad p_{003} = \nu_3^3 p_{300} \qquad p_{012} = \nu_2 \nu_3^2 p_{300}$$

$$p_{021} = \nu_2^2 \nu_3 p_{300} \qquad\qquad\qquad\qquad\qquad\qquad (12.4.12)$$

Now writing the general solution to be of the form

$$p_{n_1 n_2 n_3} = C \nu_1^{n_1} \nu_2^{n_2} \nu_3^{n_3} \quad (12.4.13)$$

with $n_1, n_2, n_3 = 0, 1, 2, 3$ and $n_1 + n_2 + n_3 = 3$, we find that the constant C has the form given by (12.4.7).

12.5 REMARKS

Analysis of stochastic processes occurring in more complex queueing networks is beyond the scope of our discussion. Interested readers may refer to review articles such as Baskett (1973) and Disney (1975) or books Kleinrock (1964, 1975, 1976), Beneš (1965), and Kelly (1979). Due to the complexity of the resulting processes, methods have been proposed to determine performance measures through approximations. One approach that has found wide usage in the literature on computer systems uses Norton's theorem for electrical networks in simplifying a queueing network. If one is interested in the behavior of a subsystem, in this approach the remainder of the network is collapsed into a single node with an equivalent composite set of traffic characteristics. Because of the product form solution characteristics of systems that exhibit local balance, the application of

Norton's theorem provides exact results in such cases. In systems that do not exhibit such properties, the degree of accuracy of approximation is in some way directly proportional to the system's departure from local balance. For a general discussion of this approach, the readers are referred to Chandy, Herzog, and Woo (1975a, 1975b).

Articles listed under the title "For Further Reading" give interested readers enough leads to prepare a comprehensive bibliography on the subject. It should also be noted that the bulk of the research on queueing networks has appeared in articles published in journals such as *Bell System Technical Journal, Computer Networks, Computer Surveys, IBM System Technical Journal,* and *IEEE Transactions on Communications.*

REFERENCES

Baskett, F. (1973). "Networks of Queues." *Proc. Seventh Annual Princeton Conference on Information Sciences and Systems.* Princeton University, pp. 428–434.

Beneš, V. E. (1965). *Mathematical Theory of Connecting Networks and Telephone Traffic.* New York: Academic Press.

Burke, P. J. (1956). "The Output of a Queueing System." *Oper. Res.* **4**, 699–714.

Burke, P. J. (1976). "Proof of a Conjecture on the Inter-arrival Time Distribution in an $M/M/1$ Queue with Feedback." *IEEE Trans. Comm.* **COM-24**, 575–576.

Chandy, K. M., Herzog, U., and Woo, L. (1975a). "Parametric Analysis of Queueing Networks." *IBM J. Res. Dev.* **19**, 36–42.

Chandy, K. M., Herzog, U., and Woo, L. (1975b) "Approximate Analysis of General Queueing Networks." *IBM J. Res. Dev.* **19**, 43–49.

Disney, R. L. (1975). "Random Flow in Queueing Networks: A Review and Critique." *AIIE Trans.* **7**, 268–288.

Gordon, W. J., and Newell, G. F. (1967). "Closed Queueing Systems with Exponential Servers." *Oper. Res.* **15**, 254–265.

Jackson, J. R. (1957). "Networks of Waiting Lines." *Oper Res.* **5**, 518–521.

Jackson, J. R. (1963). "Jobshop Like Queueing Systems." *Management Sci.* **10**, 131–142.

Jackson, R. R. P. (1954). "Queueing Processes with Phase Type Service." *Oper. Res. Quart.* **5**, 109–120.

Jackson, R. R. P. (1956). "Queueing Processes with Phase Type Service." *J. Roy. Stat. Soc.* **B18**, 129–132.

Kelly, F. P. (1979). *Reversibility and Stochastic Networks.* New York: Wiley.

Kleinrock, L. (1964). *Communication Nets: Stochastic Message Flow and Delay.* New York: McGraw-Hill.

Kleinrock, L. (1975). *Queueing Systems.* Vol. 1: *Theory.* New York: Wiley.

Kleinrock, L. (1976). *Queueing Systems.* Vol. 2: *Computer Applications.* New York: Wiley.

Koenigsberg, E. (1958). "Cyclic Queues." *Oper. Res. Quart.* **9**, 22–35.

Whittle, P. (1967). "Nonlinear Migration Processes." *Bull. Int. Stat. Inst.* **42**, 642–646.
Whittle, P. (1968). "Equilibrium Distributions for an Open Migration Process." *J. Appl. Prob.* **5**, 567–571.

FOR FURTHER READING

Baskett, F., Chandy, K. M., Muntz, R. R. and Palacios, F. G. (1975). "Open, Closed and Mixed Network of Queues with Different Classes of Customers." *J. ACM* **22**, 248–260.
Boyse, J. W., and Warn, D. R. (1975). "A Straightforward Model for Computer Performance Prediction." *Comp. Surveys* **7**, 73–93.
Buzen, J. P. (1973). "Computational Algorithms for Closed Queueing Networks with Exponential Servers." *Comm. ACM* **16**, 527–531.
Chandy, K. M., and Sauer, C. H. (1978). "Approximate Methods for Analyzing Queueing Network Models of Computing Systems." *Comp. Surveys* **10**, 281–317. Also see references cited in this paper.
Denning, P. J., and Buzen, J. P. (1978). "The Operational Analysis of Queueing Network Models." *Comp. Surveys* **10**, 225–261. Also see references cited in this paper.
Dukhovny, I. M., and Koenigsberg, E. (1981). "Invariance Properties of Queueing Networks and Their Application to Computer Communication Systems." *INFOR* **19**, 185–204.
Forys, L. J., and Messerli, E. (1975). "Analysis of Trunk Groups Containing Short Holding Time Trunks." *Bell Syst. Tech. J.* **54**, 1127–1153.
Georganas, N. D. (1980). "Modeling and Analysis of Message Switched Computer-Communication Networks with Multilevel Flow Control." *Comp. Networks* **4**, 285–294. Also see references cited in this paper.
Heffes, H. (1980). "A Class of Data Traffic Processes—Covariance Function Characterization and Related Queueing Results." *Bell Syst. Tech. J.* **59**, 897–929.
Inose, H., and Saito, T. (1978). "Theoretical Aspects in the Analysis and Synthesis of Packet Communication Networks." *Proc. IEEE* **66**, 1409–1422.
Kamoun, F., and Kleinrock, L. (1979). "Stochastic Performance Evaluation of Hierarchical Routing for Large Networks." *Comp. Networks* **3**, 337–353.
Kimbleton, S. R., and Schneider, G. M. (1975). "Computer Communication Networks: Approaches, Objectives and Performance Considerations." *Comp. Surveys* **7**, 129–173.
Kleinrock, L., and Kamoun, F. (1977). "Hierarchical Routing for Large Networks: Performance Evaluation and Optimization." *Comp. Networks* **1**, 155–174.
Krell, B. E., and Arminio, M. (1982). "Queueing Theory Applied to Data Processing Networks." *Interfaces* **12**(4), 21–33.
Lemoine, A. J. (1977). "Networks of Queues–A Survey of Equilibrium Analysis." *Management Sci* **24**, 464–481.
Lemoine, A. J. (1978). "Networks of Queues—A Survey of Weak Convergence Results." *Management Sci* **24**, 1175–1193.

Morris, R. J. T. (1981). "Priority Queueing Networks." *Bell Syst. Tech. J.* **60**, 1745–1769.

Reiser, M. (1976). "Interactive Modeling of Computer Systems." *IBM Syst. J.* **4**, 309–327.

Spaniol, O. (1979). "Modeling of Local Computer Networks." *Comp. Networks* **3**, 315–326.

Thomasian, A., and Kanakia, H. (1979). "Performance Study of Loop Networks Using Buffer Insertion." *Comp. Networks* **3**, 419–425.

Tobagi, F. A., Gerla, M., Peebles, R. W., and Manning, E. G. (1978). "Modeling and Measurement Techniques in Packet Communication Networks." *Proc. IEEE* **66**, 1423–1447.

Trivedi, K. S. (1982). *Probability and Statistics with Reliability, Queueing and Computer Science Applications.* Englewood Cliffs, N.J.: Prentice-Hall.

Wallace, V. L. (1973). "Algebraic Techniques for Numerical Solution of Queueing Networks." *Proc. Math. Methods in Queueing Theory* **98**. Lecture Notes in Economics and Math. Systems. New York: Springer-Verlag, pp. 295–305.

Chapter 13

SOME MARKOV MODELS IN COMMUNICATION AND INFORMATION SYSTEMS

13.1 INTRODUCTION

Time-shared computer systems applied the concept of resource sharing in information processing. As a natural extension networks of such systems are being used in a wide variety of areas. Data exchange facilities of government agencies, remote data-processing systems used by finance institutions, information banks of medical and health institutions, and computer communication systems of the defense establishments are only a few such examples.

Two methods are used in sending messages through a network. In the line-switched method the message route is identified before starting the transmission of the message. A good example is the telephone network. But in a message-switched method (or packet-switched method) a message (e.g., data) is routed through the network from node to node, queueing up whenever necessary at some nodes. The tasks such as switching messages from one route to another, coding them whenever necessary, and buffering them (placing them in a buffer to wait for their turn) for an orderly transmission are performed by microprocessors in the network. An efficient design of the network should aim at the reduction of queueing delay of messages at various stages of the network. Since the underlying processes are mostly stochastic, stochastic process techniques have been extensively used to help design better systems.

In the next few sections we shall discuss some simple probability models that can be used in the analysis and design of systems. Problems specific to networking have been discussed in Chapter 12. Here we shall concentrate on the use of queueing and other related probability models for buffering and transmission at a single terminal. We shall also analyze a specific message transfer network, the results of which could be useful in the design of such systems.

The general area of applied stochastic processes in communication systems is vast and cannot be summarized in a few pages. What we intend to provide is a sampler of simple problems that can be discussed with the techniques studied in earlier chapters. For more material interested readers are referred to books such as Kleinrock (1976) and Schwarz (1977) and articles published in journals such as the *Journal of the Association for Computing Machinery* and *IEEE Transactions on Communications*, to name only two of the major ones. Specifically, the two articles by Chu and Konheim (1972) and Kobayashi and Konheim (1977) provide an extensive bibliography on major topics.

13.2 STORAGE REQUIREMENTS FOR UNPACKED MESSAGES*

Consider a storage device with infinite capacity to which messages arrive in a Poisson process with parameter λ. It is used to store messages that will be transmitted through s channels. The message length is assumed to be exponential with mean $1/\mu_1$. Assuming a rate of transmission r, the transmission time (service time) can be assumed to be exponential with mean $1/\mu = 1/r\mu_1$. The messages are segmented for storage in fixed-size bins or buffers. An individual message may require several buffers, but no buffer contains data for more than one message. Each buffer consists of a data field and a certain amount of overhead for control and identification purposes. Let b be the size of the data field and a the overhead. Since reducing the data field size results in the increase of overhead in processing time, the selection of the optimum data field size is a significant problem in effecting a high data field utilization.

Let N be the number of buffers required by a message, and let $\{ q_n, n = 1, 2, \dots \}$ be its distribution. Noting that the message length is exponential with mean $1/\mu_1$, we get

$$q_n = \int_{(n-1)b}^{nb} \mu_1 e^{-\mu_1 x}\, dx \qquad n = 1, 2, \dots$$

$$= e^{-nb\mu_1}\left(e^{b\mu_1} - 1\right)$$

$$= e^{-n\beta}\left(e^{\beta} - 1\right) \qquad\qquad (13.2.1)$$

*Our treatment follows generally that of Pedersen and Shah (1972).

where we have written $b\mu_1 = \beta$. By direct simplification we also get

$$E(N) = \frac{1}{1 - e^{-\beta}}$$
(13.2.2)

Incorporating the overhead factor, expected total buffer space per message, S, can be obtained as

$$S = \frac{a + b}{1 - e^{-\beta}}$$
(13.2.3)

To determine a and b that minimize S, we differentiate (13.2.3) with respect to b and equate it to zero. We get

$$\frac{\partial S}{\partial b} = \frac{(1 - e^{-\beta}) - (a + b)(-e^{-\beta})(-\mu_1)}{(1 - e^{\beta})^2}$$

$$= 0$$

that is,

$$1 - e^{-\beta} - (a\mu_1 + b\mu_1)e^{-\beta} = 0$$
(13.2.4)

Writing $a\mu_1 = \alpha$, from (13.2.4) we get the relation

$$\alpha = e^{\beta} - 1 - \beta$$
(13.2.5)

a result due to Wolman (1965).

Let L be the length of the message. Then the following three utilization factors may be relevant:

$$U_1 = E \text{ (Utilization of allocated data field storage)} = E\left(\frac{L}{bN}\right)$$

$$U_2 = E \text{ (Utilization of the total allocated storage by the data)}$$

$$= E\left[\frac{L}{(a + b)N}\right]$$

$$U_3 = E \text{ (Utilization of the total allocated storage)} = E\left[\frac{L + aN}{(a + b)N}\right]$$

(13.2.6)

In all three factors the basic term is $E(L/N)$ which can be obtained in a straight-forward manner as

$$E\left(\frac{L}{N}\right) = \int_0^b x\mu_1 e^{-\mu_1 x}\,dx + \int_b^{2b} \frac{x}{2}\mu_1 e^{-\mu_1 x}\,dx$$

$$+ \int_{2b}^{3b} \frac{x}{3}\mu_1 e^{-\mu_1 x}\,dx + \cdots$$

$$= \sum_{k=1}^{\infty} \int_{(k-1)b}^{kb} \frac{x}{k}\mu_1 e^{-\mu_1 x}\,dx$$

$$= \frac{1}{\mu_1}\sum_{k=1}^{\infty}\left[\beta e^{-k\beta}(e^\beta - 1) + \frac{e^{-k\beta}}{k}(e^\beta - \beta e^\beta - 1)\right]$$

$$= \frac{1}{\mu_1}\left[\beta + (1 - \beta e^\beta - e^\beta)\ln(1 - e^{-\beta})\right] \qquad (13.2.7)$$

In simplifications we have used the results

$$\int_0^t \mu^2 x e^{-\mu x}\,dx = 1 - e^{-\mu t} - e^{-\mu t}\mu t$$

and

$$-\ln(1 - x) = x + \frac{x^2}{2} + \frac{x^3}{3} + \cdots \qquad |x| < 1$$

Thus we get

$$U_1 = 1 + \frac{1}{\beta}(1 - \beta e^\beta - e^\beta)\ln(1 - e^{-\beta})$$

$$U_2 = \frac{1}{\alpha + \beta}\left[\beta + (1 - \beta e^\beta - e^\beta)\ln(1 - e^{-\beta})\right] = \frac{\beta}{\alpha + \beta}U_1$$

$$U_3 = U_2 + \frac{a}{a + b} \qquad (13.2.8)$$

To determine storage requirements, it is necessary to determine the distribution of the number of message in storage. Note that the process is the standard $M/M/s$ queue, for which results are available in Section 11.4. Let $\{p_n\}$ be the steady-state probability distribution under the condition $\lambda/s\mu < 1$. (Note that $\mu = r\mu_1$, when r is the rate of transition.) Writing $\gamma = \lambda/\mu$, from (11.4.3) we have

$$p_n = \frac{\gamma^n}{n!}p_0 \qquad 0 \le n \le s$$

$$= \frac{\gamma^n}{s!s^{n-s}}p_0 \qquad n \ge s$$

where p_0 is obtained as

$$p_0 = \left[\sum_{j=0}^{s-1} \frac{\gamma^j}{j!} + \left(\frac{s}{s-\gamma} \right) \frac{\gamma^s}{s!} \right]^{-1} \tag{13.2.9}$$

The number of buffers required by a message is a random variable with distribution $\{q_n\}$ given by (13.2.1). Since we are interested in a random sum of discrete random variables, the derivation of its distribution is simpler if we use probability generating functions. Let

$$\Pi(z) = \sum_{n=0}^{\infty} p_n z^n$$

$$\nu(z) = \sum_{n=1}^{\infty} q_n z^n \tag{13.2.10}$$

Let $B(z)$ be the probability generating function for the number of buffers required. Following the discussion in Section 8.5, we get

$$B(z) = \Pi[\nu(z)] \tag{13.2.11}$$

From (13.2.1) we get

$$\nu(z) = \frac{(1 - e^{-\beta})z}{1 - ze^{-\beta}} = \frac{cz}{1 - dz} \tag{13.2.12}$$

where we have written $1 - e^{-\beta} = c$ and $e^{-\beta} = d$. Also we write

$$\Pi(z) = \sum_{n=0}^{\infty} p_n z^n$$

$$= p_0 \left\{ 1 + \sum_{n=1}^{\infty} \xi_n z^n \right\} \tag{13.2.13}$$

where $\{\xi_n\}$ are defined as

$$\xi_n = \frac{\gamma^n}{n!} \qquad n = 1, 2, \ldots, s$$

$$= \frac{\gamma^n}{s! s^{n-s}} \qquad n = s + 1, s + 2, \ldots$$

Substituting from (13.2.12) and (13.2.13) in (13.2.11), we get

$$B(z) = p_0\left\{1 + \sum_{n=1}^{\infty} \xi_n \left(\frac{cz}{1 - dz}\right)^n\right\} \qquad (13.2.14)$$

which is in a convenient form for expansion as a power series in z. For explicit expressions for the distribution of the number of occupied buffers and its mean and variance, the readers are referred to Pedersen and Shah (1972).

13.3 BUFFER BEHAVIOR FOR BATCH ARRIVALS*

In the previous section we considered buffer occupancy by messages of exponential length. In many data communication systems, however, the input traffic is in bursts (strings of characters) rather than in single characters. Also the discreteness of the input traffic makes the analysis of the previous section unsuitable. If, however, we consider the time to transmit a character as a unit service interval, it is possible to use an embedded Markov chain analysis at epochs of unit service completion.

In the design of a buffer its size should be determined based on the overflow probability which is given by the average fraction of the total number of arriving data rejected by the buffer. To determine the overflow probability, a model with a finite buffer size is necessary.

Investigations on operating computer systems have shown (see Fuchs and Jackson, 1969) that the burst arrivals (of batches of characters) can be approximated to follow a Poisson process and the burst length (batch size) can be approximated as being geometrically distributed. Based on these observations, we shall make the following assumptions:

1. Let G be a random variable representing batch size with the distribution

$$P(G = j) = g_j = \theta(1 - \theta)^{j-1} \qquad j = 1, 2, 3, \ldots \qquad (13.3.1)$$

2. Let $t_1, t_2, t_3, \ldots, t_n, \ldots$ be the unit service completion epochs.
3. Let X_{n+1} be the total number of characters arriving during (t_n, t_{n+1}).

Assuming that input bursts occur in a Poisson process with rate λ per unit

*Our treatment follows generally that of Chu (1970).

time, we get

$$P(X_{n+1} = j) = k_j = \sum_{r=0}^{j} e^{-\lambda} \frac{\lambda^r}{r!} g_j^{(r)*} \qquad (13.3.2)$$

where $g_j^{(r)*}$ is the r-fold convolution of $\{g_j\}$ with itself. Since g_j is a geometric distribution, we get

$$k_j = e^{-\lambda} \qquad \text{if } j = 0$$

$$= \sum_{r=1}^{j} \binom{j-1}{r-1} \theta^r (1 - \theta)^{j-r} e^{-\lambda} \frac{\lambda^r}{r!} \qquad j = 1, 2, 3, \ldots \quad (13.3.3)$$

where we have used the expression (7.2.39) for the negative binomial probability.

The queueing model is now given by the finite waiting room queue $M/G/1$ discussed in Chapter 11. The embedded Markov chain of this system has a transition probability matrix given by equation (11.6.35), and its limiting distribution can be determined using the recursive method outlined in (11.6.36)–(11.6.40). Let π_0 be the probability of an idle buffer as obtained from the embedded Markov chain analysis. The offered load β per unit service interval is obtained as λ/θ [θ^{-1} is the mean batch size in (13.3.1)]. The carried load α is obtained from π_0 as

$$\alpha = 1 - \pi_0 \qquad (13.3.4)$$

The overflow probability now follows as

$$P\text{ (Overflow)} = \frac{\text{Offered load } - \text{ Carried load}}{\text{Offered load}} = 1 - \frac{\alpha}{\beta} \quad (13.3.5)$$

A finite buffer model discussed by Mansfield and Tran-Gia (1982) assumes Poisson batch arrivals, exponential transmission times, and multiple channel use. The analysis follows in a direct manner using balance equations in steady state.

13.4 LOOP TRANSMISSION SYSTEMS*

One of the configurations of computer communications networks studied in the literature is the loop transmission system. In such networks the devices sending messages to one another are connected in a loop. Data are

*Our treatment follows generally that of Spragins (1972).

transmitted around the loop passing through each device on the way. A loop transmission system is an example of multiplexing networks in which some variant of time division multiplexing is used to share communication channels. In time-division multiplexing the time axis is divided into contiguous intervals of fixed length, called slots or frames. Normally a message is transmitted between a host computer and the terminals. For a good discussion of the various queueing problems arising out of such networks, the readers are referred to Kobayashi and Konheim (1977) or Schwarz (1977).

As an example, we shall consider a simple loop system with the following specifications: (1) The slots are of fixed duration, and they originate at the central control station and pass through the loop positions. (2) Each station on the loop has infinite buffering capacity. (3) Time to transmit signals around the loop is zero. (4) Each terminal generates requests for service and provides a buffer for requests that have not yet received service.

Using the queue analogy, the loop can be considered the server with the duration of slot as the service time. At the end of the service time another slot is available for use. Konheim and Meister (1972) consider a more general model in which a number C of such slots are available at every service. Then it can be thought of as a bus of capacity C that makes regular circuits around the loop, picking up passengers at the terminals until it is full and then discharging them at the central station.

Let $Q_i^{(n)}$ be the number of requests waiting for service at the ith terminal soon after the nth slot has left the terminal. Let $X_i^{(n)}$ be the number of requests arriving at the ith terminal during the nth service period. We assume that $\{ X_i^{(n)} \}$ are independent and identically distributed random variables. Further let

$$R_i(n) = \sum_{k=1}^{i} Q_k(n)$$

$$Y_i(n) = \sum_{k=1}^{i} X_i(n) \tag{13.4.1}$$

Since a request from terminal i can be serviced only if the slot has not been used by terminals $1, 2, \ldots, i - 1$, which come earlier in the loop, for the total number of requests in the first i terminals, we have the relation

$$R_i(n + 1) = \begin{cases} R_i(n) - 1 + Y_i(n + 1) & \text{if } R_i(n) > 0 \\ Y_i(n + 1) & \text{if } R_i(n) = 0 \end{cases}$$

$$\tag{13.4.2}$$

In (13.4.2) it should be noted that $\{Y_i(n)\}$ are independent and identically distributed random variables by virtue of our earlier assumption on $\{X_i(n)\}$.

The relation (13.4.2) is exactly in the form of the relation (11.6.10) for the queue $M/G/1$, and therefore the total expected number of requests in the first i terminals, as $n \to \infty$, is given by the queue length formula (11.6.25). Thus we have (dropping n to indicate that the random variables are considered as $n \to \infty$)

$$E[R_i] = \frac{1}{2}\left\{ E(Y_i) + \frac{V(Y_i)}{1 - E(Y_i)} \right\} \qquad (13.4.3)$$

Assuming that the arrival processes in different terminals are independent of each other, we get $E(Y_i) = \sum_{k=1}^{i} E(X_k)$ and $V(Y_i) = \sum_{k=1}^{i} V(X_k)$. Therefore

$$E[Q_i] = \frac{1}{2}\left\{ E(X_i) + \frac{\sum_{k=1}^{i} V(X_k)}{1 - \sum_{k=1}^{i} E(X_k)} - \frac{\sum_{k=1}^{i-1} V(X_k)}{1 - \sum_{k=1}^{i-1} E(X_k)} \right\}$$

$$(13.4.4)$$

The assumption that $\{X_i(n)\}$ are independent and identically distributed random variables imposes the condition that the arrival process for the requests be Poisson. Since the slot length is a constant, say T, we have

$$E(X_k) = \lambda_k T$$

$$V(X_k) = \lambda_k T$$

Let $\sum_{k=1}^{i} \lambda_k = \Lambda_i$. Then (3.14.4) simplifies to

$$E(Q_i) = \frac{1}{2}\lambda_i T\left\{ 1 + \frac{1}{(1 - \Lambda_i T)(1 - \Lambda_{i-1} T)} \right\} \qquad (13.4.5)$$

Using these results, it is possible to derive expressions for the mean delay experienced by a request. Since it involves properties of queueing systems beyond those covered in this book, interested readers are referred to Spragins (1972) and Konheim and Meister (1972). Incidentally, if one is interested in the limiting distribution of the number of requests waiting for service, it is readily obtained from the limiting queue length distribution in the $M/G/1$ queue derived in Chapter 11.

13.5 A PROBABILISTIC MODEL FOR HIERARCHICAL MESSAGE TRANSFER*

Consider a network with information centers $C_1, C_2, \ldots, C_N$ and a message arriving at an arbitrary center C_{i_0}. If the message cannot be satisfied at C_{i_0}, it can be referred to one or more of the remaining centers for action. Let c_{ij} be the cost of referring a message from center C_i to center C_j, and let h_i be the probability that a message will be satisfied at center C_i ($i = 1, 2, \ldots, N$). Now it can be shown (see Baker, 1969) that the best strategy with the least expected cost for selecting a referral path is one in which center $C_{i_{k+1}}$ is chosen after C_{i_k} so as to minimize the cost/probability ratio $c_{i_k j}/h_j$ with regard to the information centers not covered by the message so far.

Consistent with this strategy, consider a strictly hierarchical message transfer network consisting of information centers $C_1, C_2, \ldots, C_N$ and the referral path $C_N \to C_{N-1} \to C_{N-2} \to \cdots \to C_1$. Associated with the network, in order to complete the model, also consider two additional centers C_0 and C_R to represent the two outcomes: satisfaction and rejection of the message at each center. Rejection occurs when a center decides that the message cannot be processed anywhere in the network. Let p_{ij} ($i, j = 1, 2, \ldots, N$) be the referral probability from center C_i to center C_j, and let p_{i0} and p_{iR} ($i = 1, 2, \ldots, N$) be the probabilities of satisfaction and rejection at center C_i. Consider the $N + 2$ centers $\{C_R, C_0, C_1, \ldots, C_N\}$ as $N + 2$ states of a Markov chain, with p_{ij} ($i, j = R, 0, 1, 2, \ldots, N$) as the corresponding one-step transition probabilities. With the strictly hierarchical network structure the transition probability matrix can be represented as

$$
P = \begin{array}{c} \\ R \\ 0 \\ 1 \\ 2 \\ 3 \\ \vdots \\ N \end{array}
\begin{array}{c} \begin{array}{ccccccccc} R & \quad 0 & \quad 1 & \quad 2 & \quad 3 & \cdots & N-1 & \quad N \end{array} \\
\left[\begin{array}{cc|ccccccc}
1 & 0 & 0 & 0 & 0 & \cdots & 0 & 0 \\
0 & 1 & 0 & 0 & 0 & \cdots & 0 & 0 \\ \hline
p_{1R} & p_{10} & 0 & 0 & 0 & \cdots & 0 & 0 \\
p_{2R} & p_{20} & p_{21} & 0 & 0 & \cdots & 0 & 0 \\
p_{3R} & p_{30} & 0 & p_{32} & 0 & \cdots & 0 & 0 \\
\vdots & \vdots & \vdots & \vdots & \vdots & & \vdots & \vdots \\
p_{NR} & p_{N0} & 0 & 0 & 0 & \cdots & p_{N,N-1} & 0
\end{array} \right]
\end{array}
$$

$$(13.5.1)$$

*Our treatment follows generally that of Bhat, Nance, and Korfhage (1975). Significant portions of the article are reprinted from *Information Sciences* **9** (1975), 169–184, with the permission of the Elsevier Science Publishing Company.

For convenience we shall partition p as indicated:

$$P = \left[\begin{array}{c|c} I & 0 \\ \hline R & Q \end{array}\right] \qquad (13.5.2)$$

Suppose we are interested in developing a model for network cost. At this stage we assume that the costs c_{ij} are given. These costs may be the actual cost of message transfer or delay encountered in a transaction or a combination of the two. If the costs are assumed random, let γ_{ij} and η_{ij} be the mean and variance of c_{ij}.

Let K be the total network cost in a given length of time. This is comprised of costs of messages originating at different centers. Let m_i be the cost associated with a message originating at center C_i before it is eventually either rejected or filled by one of the centers of the network. Specific costs at center i $(i = 1, 2, \ldots, N)$ are (1) c_{iR} for rejecting, (2) c_{i0} for satisfying, and (3) $c_{i,i-1}$ for referring the message to center C_{i-1}. Also the cost m_i can be expressed as

$$m_i = \sum_{j=1}^{i} m_{ij} \qquad (13.5.3)$$

where for $j > i$

$$m_{ij} = 0$$

for $j = i$

$$m_{ii} = \begin{cases} c_{iR} & \text{with probability } p_{iR} \\ c_{i0} & \text{with probability } p_{i0} \\ c_{i,i-1} & \text{with probability } p_{i,i-1} \end{cases}$$

and for $j < i$

$$m_{ij} = \begin{cases} 0 & \text{with probability } p_{iR} + p_{i0} \\ m_{i-1,j} & \text{with probability } p_{i,i-1} \end{cases} \qquad (13.5.4)$$

Let

$$\mu_{ij} = E(m_{ij})$$

$$\sigma_{ij}^2 = V(m_{ij}) \qquad (13.5.6)$$

for $i, j = 1, 2, \ldots, N$, and denote matrices with μ_{ij} and σ_{ij}^2 as elements as M and S, respectively.

Taking expectations of (13.5.4), we get

$$\mu_{ii} = p_{iR}\gamma_{iR} + p_{i0}\gamma_{i0} + p_{i,i-1}\gamma_{i,i-1}$$

$$\mu_{ij} = p_{i,i-1}\mu_{i-1,j} \qquad j < i \tag{13.5.7}$$

Let

$$M_D = \begin{bmatrix} \mu_{11} & & & \bigcirc \\ & \mu_{22} & & \\ & & \ddots & \\ \bigcirc & & & \mu_{NN} \end{bmatrix} \tag{13.5.8}$$

Expressing (13.5.7) in matrix notations, we get

$$M = M_D + QM$$

which gives

$$M = (I - Q)^{-1}M_D \tag{13.5.9}$$

provided $(I - Q)^{-1}$ exists. It is easy to show that

$$(I - Q)^{-1} = \begin{bmatrix} 1 & 0 & \cdot & \cdot & 0 \\ p_{21} & 1 & \cdot & \cdot & 0 \\ p_{32}p_{21} & p_{32} & \cdot & \cdot & 0 \\ \vdots & \vdots & & & \vdots \\ p_{N,N-1}p_{N-1,N-2} \cdots p_{21} & & \cdot & \cdot & 1 \end{bmatrix} \tag{13.5.10}$$

which completely determines (13.5.9).

Let n_i be the number of messages originating at center C_i $(i = 1, 2, \ldots, N)$ during a given period. We then have

$$E(K) = \sum_{i=1}^{N} E(n_i) \sum_{j=1}^{N} \mu_{ij} \tag{13.5.11}$$

To derive the corresponding variances of costs, we proceed in the same manner as in Section 4.1. Squaring both sides of (13.5.4) and taking expectations, we get

$$\|E(m_{ij}^2)\| = (I - Q)^{-1}H \tag{13.5.12}$$

where

$$\eta_i = p_{iR}\eta_{iR} + p_{i0}\eta_{i0} + p_{i,i-1}\eta_{i,i-1} \tag{13.5.13}$$

and

$$H = \begin{bmatrix} \eta_1 & & & \bigcirc \\ & \eta_2 & & \\ & & \ddots & \\ \bigcirc & & & \eta_N \end{bmatrix} \tag{13.5.14}$$

Again writing

$$M_2 = \begin{bmatrix} \mu_{11}^2 & \mu_{12}^2 & \cdots & \mu_{1N}^2 \\ \vdots & \vdots & & \vdots \\ \mu_{N1}^2 & \mu_{N2}^2 & \cdots & \mu_{NN}^2 \end{bmatrix} \tag{13.5.15}$$

for the variance matrix S we get

$$S = (I - Q)^{-1}H - M_2 \tag{13.5.16}$$

To derive the variance of the total cost K, we also need $E(m_i^2)$. We have

$$m_i = \begin{cases} c_{iR} & \text{with probability } p_{iR} \\ c_{i0} & \text{with probability } p_{i0} \\ c_{i,i-1} + m_{i-1} & \text{with probability } p_{i,i-1} \end{cases} \tag{13.5.17}$$

Squaring (13.5.17), taking expectations, noting that due to the hierarchical network structure $c_{i,i-1}$ and m_{i-1} can be assumed to be independent, and writing

$$\sum_{j=1}^{N} \mu_{ij} = \mu_i = E(m_i) \tag{13.5.18}$$

and

$$\Gamma = \begin{bmatrix} 0 & 0 & 0 & \cdot & & \cdot & 0 \\ p_{21}\gamma_{21} & 0 & 0 & \cdot & & \cdot & 0 \\ 0 & p_{32}\gamma_{32} & 0 & \cdot & & \cdot & 0 \\ \cdot & \cdot & \cdot & \cdot & & \cdot & \cdot \\ \cdot & \cdot & \cdot & \cdot & & \cdot & \cdot \\ 0 & 0 & 0 & \cdot & p_{N,N-1}\gamma_{N,N-1} & & 0 \end{bmatrix} \tag{13.5.19}$$

we get

$$
\begin{pmatrix} E(m_1^2) \\ E(m_2^2) \\ \vdots \\ E(m_N^2) \end{pmatrix} = (I - Q)^{-1} \left[\begin{pmatrix} \eta_1 \\ \eta_2 \\ \vdots \\ \eta_N \end{pmatrix} + 2\Gamma \begin{pmatrix} \mu_1 \\ \mu_2 \\ \vdots \\ \mu_N \end{pmatrix} \right] \qquad (13.5.20)
$$

Let n_i be the number of messages originating at center C_i, as assumed. We assume that $E(n_i)$ and $E(n_i^2)$ exist and are known. Let $m_i^{(1)}, m_i^{(2)}, \ldots, m_i^{(n_i)}$ be the costs associated with these n_i messages. Then K_i, the total cost of messages originating at center C_i, is given by

$$
K_i = m_i^{(1)} + m_i^{(2)} + \cdots + m_i^{(n_i)} \qquad (13.5.21)
$$

Assuming that costs $m_i^{(r)}$ $(r = 1, 2, \ldots, n_i)$ are independent of the number of messages n_i, we get

$$
V(K_i) = V(n_i) [E(m_i)]^2 + E(n_i) V(m_i) \qquad (13.5.22)
$$

The term $V(m_i)$ in (13.5.22) is obtained from (13.5.20) by noting that $V(m_i) = E(m_i^2) - \mu_i^2$.

For the total cost K the variance expression becomes quite cumbersome:

$$
V(K) = V\left(\sum_{r=1}^{n_1} m_1^{(r)} + \sum_{r=1}^{n_2} m_2^{(r)} + \cdots + \sum_{r=1}^{n_N} m_N^{(r)} \right)
$$

$$
= \sum_{i=1}^{N} V(K_i) + 2 \sum\sum_{i<j} \mathrm{Cov}\left(\sum_{r=1}^{n_i} m_i^{(r)}, \sum_{k=1}^{n_j} m_j^{(k)} \right) \qquad (13.5.23)
$$

When n_i $(i = 1, 2, \ldots, N)$ are constants, (13.5.23) simplifies to

$$
V(K) = \sum_{i=1}^{N} n_i V(m_i) + 2 \sum\sum_{i<j} n_i n_j \sum_{l=1}^{N} q_{il} q_{jl} (\eta_l - \mu_{ll}^2) \qquad (13.5.24)
$$

where

$$
q_{il} = \begin{cases} p_{i,i-1} p_{i-1,i-2} \cdots p_{l+1,l} & l < i \\ 0 & \text{otherwise} \end{cases}
$$

13.6   REMARKS

In the last few sections we have discussed two different types of problems related to networks. In Sections 13.2–13.4 we considered problems arising within a network node in the course of information processesing and in Section 13.5 the message transfer problem over the entire network. Together with the preceding chapter on queueing networks, these topics provide a glimpse of the variety of probability models that arise in the analysis of such systems. Our intention is not to give a comprehensive treatment of network problems but some examples of probability models occurring in them. For more examples the readers are referred to books and articles cited earlier as well as those listed under the heading "For Further Reading."

REFERENCES

Baker, N. R. (1969). "Optional User Search Sequences and Implications for Information Systems Operation." *Amer. Documentation* **20**(3), 203–212.

Bhat, U. N., Nance, R. E., and Korfhage, R. R. (1975). "Information Networks: A Probabilistic Model for Hierarchical Message Transfer." *Information Sci.* **9**, 169–184.

Chu, W. W. (1970). "Buffer Behavior for Batch Poisson Arrivals and Single Constant Output." *IEEE Trans. Comm.* **COM-18**(5), 613–618.

Chu, W. W. and Konheim, A. G. (1972). "On the Analysis of Modeling of a Class of Computer Communication Systems." *IEEE Trans. Comm.* **COM-20**(3), 645–660.

Fuchs, E. and Jackson, P. E. (1969). "Estimates of Distributions of Random Variables for Certain Computer Communication Traffic Models." *Proc. ACM Symposium on Problems in the Optimization of Data Communications Systems*, Pine Mountain, Ga., pp. 201–227.

Kleinrock, L. (1976). *Queueing Systems.* Vol. 2: *Computer Applications.* New York: Wiley.

Kobayashi, H., and Konheim, A. G. (1977). "Queueing Models for Computer Communications System Analysis." *IEEE Trans. Comm.* **COM-25**(1), 2–29.

Konheim, A. G., and Meister, B. (1972). "Service in a Loop System." *J. ACM* **19**, 92–108.

Manfield, D. R., and Tran-Gia, P. (1982). "Analysis of a Finite Storage System with Batch Input Arising out of Message Packetization." *IEEE Trans. Comm.* **COM-30**(3), 456–463.

Pedersen, R. D., and Shah, J. C. (1972). "Multiserver Queue Storage Requirements with Unpacked Messages." *IEEE Trans. Comm.* **COM-20**(3), 462–465.

Schwartz, M. (1977). *Computer Communications Network Design and Analysis.* Englewood Cliffs, N.J.: Prentice-Hall.

Spragins, J. D. (1972). "Loop Transmission Systems–Mean Value Analysis." *IEEE Trans. Comm* **COM-20**(3), 592–602.

Wolman, E. (1965). "A Fixed Optimum Cell-size for Records of Various Lengths." *J. ACM*, **12**, 53–70.

FOR FURTHER READING

Abramson, N. (1973). "Packet Switching with Satellites." *AFIPS Proc. Nat. Comp. Conf.* **42**, 695–702.

Abramson, N., and Kuo, F. F. (eds.) (1973). *Computer Communication Networks.* Englewood Cliffs, N.J.: Prentice-Hall.

Bhat, U. N., and Nance, R. E. (1979). "An Evaluation of CPU Efficiency under Dynamic Quantum Allocation." *J. ACM* **26**, 761–778.

Cavers, J. K. (1975) "Cutset Manipulations for Communication Network Reliability Estimation." *IEEE Trans. Comm.* **COM-23**(6), 569–575.

Chang, W. (1970). "Single Server Queueing Processes in Computer Systems." *IBM Syst. J.* **9**, 36–71.

Coffman, E. G. (1967). "Studying Multiprogramming Systems with the Queueing Theory." *Datamation* **13**, 47–54.

Coffman, E. G. (1969). "Analysis of a Drum Input/Output Queue under Scheduled Operation in a Paged Computer System." *J. ACM* **16**, 73–90.

Coffman, E. G. (1969). "Markov Chain Analysis of Multiprogramming Computer Systems." *Nav. Res. Log. Quart.* **16**, 175–197.

Fischer, M. J., and Harris, T. C. (1976). "A Model for Evaluating the Performance of an Integrated Circuit—and Packet-Switched Multiplex Structure." *IEEE Trans. Comm* **COM-24**(2), 195–202.

Gaver, D. P. (1967) "Probability Models for Multiprogramming Computer Systems." *J. ACM* **14**, 423–438.

Gaver, D. P., and Lewis, P. A. W. (1971). "Probability Models for Buffer Storage Allocation Problems." *J. ACM* **18**, 186–198.

Haddad, A. H., Tsai, S., Goldberg, B., and Ranieri, G. C. (1975). "Markov Gap Models for Real Communication Channels." *IEEE Trans. Comm.* **COM-23** (11), 1189–1197.

Hayes, F. J., and Serman D. N. (1971). "Traffic Analysis of a Ring Switched Data Transmission System." *Bell Syst. Tech. J.* **50**, 2947–2978.

Kleinrock, L. (1964). *Communication Nets: Stochastic Message Flow and Delay.* New York: McGraw-Hill.

Kleinrock, L. (1969). "Models for Computer Networks." Proc. IEEE International Conference on Communications, Boulder, Colo., 21-9 to 21-16, June 9–11.

Kleinrock, L. and Lam, S. (1973). "Packet-Switching in a Slotted Satellite Channel." *AFIPS Proc. Nat. Comp. Conf.* **42**, 703–710.

Kleinrock, L. and Lam, S. (1975). "Packet-Switching in a Multi-Access Broadcast Channel: Performance Evaluation." *IEEE Trans. Comm.* **COM-23**(4), 410–423.

Kleinrock, L., and Tobagi, F. A. (1975). "Packet Switching in Radio Channels. Part I—Carrier Sense Multiple Access Modes and Their Throughput Delay Characteristics." *IEEE Trans. Comm.* **COM-23**(12), 1400–1416.

Konheim, A. G., and Meister, B. (1972). "Waiting Lines in Multiple Loop Systems." *J. Math. Anal. App.* **39**, 527–540.

Konheim, A. G., and Meister, B. (1973). "Distributions of Queue Lengths and Waiting Times in a Loop with Two-Way Traffic." *J. Comput. Syst. Sci.* **7**, 506–521.

McKinney, J. M. (1969). "A Survey of Analytical Time Sharing Models." *Comp. Reviews* **1**, 105–116.

Nance, R. E., and Bhat, U. N. (1978). "A Processor Utilization Model for a Multiprocessor Computer System." *Oper. Res.* **26**, 881–895.

Pennotti, M. C., and Schwartz, M. (1975). "Congestion Control in Store and Forward Tandem Links." *IEEE Trans. Comm* **COM-23**(12), 1434–1443.

Rubin, I. (1974). "Communication Networks: Message Path Delays." *IEEE Trans. Inform. Theory* **20**, 738–745.

Rubin, I. (1975). "Message Path Delays in Packet-Switching Communication Networks." *IEEE Trans. Comm.* **COM-23**(4), 186–192.

Tobagi, F. A., and Kleinrock, L. (1975). "Packet-Switching in Radio Channels. Part II—The Hidden Terminal Problem in Carrier Sense Multiple Access and the Busy-Tone Solution." *IEEE Trans. Comm.* **COM-23**(12), 1417–1433.

Chapter 14
SOME STOCHASTIC INVENTORY AND STORAGE PROCESSES

14.1 INTRODUCTION

An inventory system is a facility at which items of merchandise and materials are stocked in order to meet demands. There are two constituent processes underlying such a system: a demand process which is usually stochastic and a process of replenishment of stock which could be deterministic or stochastic based on the structure of the system. There are many similarities between inventory and queueing processes. However, because of the emphasis on the timely fulfillment of demand for products, inventory theorists have concentrated on the decision problems using the structural properties of the underlying processes and realistic costs associated with them.

An inventory problem is essentially a decision problem. Therefore it is not enough merely to provide the behavior of the process, it is also necessary to derive an operational policy incorporating the behavioral properties known for the system. Since the emphasis of this book is the study of stochastic processes and their applications, we shall not get into the existence and other fundamental problems related to optimal operational policies. We shall assume the existence of specific optimal policies and illustrate the use of system properties in their derivation.

In an inventory problem, broadly speaking, there are two modes of decision making, one with periodic review and the second with continuous review. A small retailer such as a newspaper vendor has to anticipate the demand level for a specified period, say a day, and start the period with enough inventory to satisfy the demands. If the inventory is more than the demand, costs will be incurred in holding them for the following period or in having excess obsolete items as in the case of the newspaper. On the other hand if the retailer runs out of inventory before satisfying the demands for the period, penalty costs are likely to be incurred. Customers not finding the needed items at one place many times start patronizing vendors who are more likely to carry items of their need.

In order to identify the structure of the problem, consider the costs involved during a period. Suppose the retailer starts the period with S items. Let the demand for the period be a random variable X with the probability distribution

$$P(X = n) = p_n \quad n = 0, 1, 2, \ldots$$

and mean δ. Let $\$h$ be the holding cost per item incurred by not being able to sell the item. Also let $\$u$ be the underage (penalty) cost per item for being short on inventory. Clearly the total cost per period to the retailer can be given as

$$C(S) = h \sum_{n=0}^{S} (S - n) p_n + u \sum_{n=S+1}^{\infty} (n - S) p_n$$

$$= (h + u) \sum_{n=0}^{S} (S - n) p_n + u(\delta - S) \qquad (14.1.1)$$

In order to investigate the properties of the cost function $C(S)$ as a function of S, consider the difference $\Delta C(S) = C(S + 1) - C(S)$. We have

$$\Delta C(S) = (h + u) \left[\sum_{n=0}^{S+1} (S + 1 - n) p_n - \sum_{n=0}^{S} (S - n) p_n \right] - u$$

$$= (h + u) \sum_{n=0}^{S} p_n - u$$

Also

$$\Delta^2 C(S) = \Delta C(S + 1) - \Delta C(S)$$

$$= (h + u) \left[\sum_{n=0}^{S+1} p_n - \sum_{n=0}^{S} p_n \right]$$

$$= (h + u) p_{S+1} > 0 \qquad (14.1.2)$$

The result (14.1.2) indicates that $C(S)$ is a convex function in S and therefore has a unique minimum. The minimum is attained for the value of S^* of S, such that

$$C(S^* - 1) - C(S^*) > 0$$

and

$$C(S^* + 1) - C(S^*) > 0$$

In a continuous review policy the cost considerations of the type illustrated here is made with respect to a reordering cycle which is dependent on the demand process. A well-known operational policy for a continuous review case is the so-called (s, S) policy. Under this policy the number of items stored varies between 0 and S, and the replenishment of stock up to the level S is ordered whenever the stock falls to the level s.

Under the usual cost structure involving some fixed costs, penalty costs for being unable to meet the demand and holding costs for storing items when not sold, it can be shown that a stationary (s, S) policy exists when the operation is under stable conditions. The examples considered in this chapter belong essentially to this class with varied degrees of randomness associated with the underlying processes. For the sake of completeness we start with the simple deterministic case and derive the well-known economic order quantity (EOQ) formula (also known as Wilson's formula).

Under appropriate assumptions the inventory level can be modeled as a Markov process. An embedded Markov chain can be identified either at demand arrival epochs or reorder points. Thus in the process of setting up an objective function for optimization (minimization if only costs are used and maximization if net profit is used in the objective function), the properties of the underlying stochastic process can be used in two types of models: (1) steady-state models, in which mean value characteristics are used under the assumption that the process is in steady state, and (2) transient models, in which the transition characteristics are built into the objective function in an evolutionary manner. Naturally the steady-state models are much simpler to analyze, and the essential steps consist of identifying the properties of the objective function relative to the optimization goal (convexity for minimization and concavity for maximization) and using an appropriate optimization technique. Transient models under a Markovian setting result in Markov decision models, an introduction to which was provided in Chapter 10.

In Sections 12.2 and 12.3 we shall consider only steady-state models based on the mean value characteristics of the inventory process. Some of these models can be considered classical by this time and can be found in books such as Whitin (1953), Morse (1958), Arrow, Karlin, and Scarf (1958), Hadley and Whitin (1963), and Sivazlian and Stanfel (1975).

As discussed in Chapter 10, the emphasis in Markov decision processes is the decision process itself rather than the behavior of the stochastic process. We consider that topic beyond the scope of this book, and readers interested in the study of dynamic models may look up references on Markov decision processes. Also it is not our intention to cover all aspects of the study of inventory systems in this chapter. Our major objective is the illustration of the application of stochastic processes in modeling problems.

Storage systems considered in the literature are similar to inventory systems but with less emphasis on associated costs. Storage processes have been studied mainly in connection with water storage in dams with an emphasis on the identification of behavioral properties of the basic processes. We shall discuss a simple storage process in the last section of this chapter following the work of Moran (1954). The significance of storage processes has been in their similarity to waiting time processes in queueing theory and risk reserve processes in insurance risk problems.

14.2 INVENTORY SYSTEMS WITH INSTANTANEOUS REPLENISHMENT OF STOCK

In this section we consider systems in which no delay is encountered between the time an order is placed and its fulfillment. Two different patterns of demand are used.

Deterministic Demands

Consider an inventory system with the following description:

1. Demand for items of merchandise arrive in a deterministic manner with a constant rate λ.
2. It is proposed to use an (s, S) inventory policy in which orders for Q $(= S - s)$ items are placed and instantaneously received when the inventory level falls to s.
3. The procurement cost of the replenishment of stock has a fixed cost of K dollars per order and a variable cost of c dollars per item in the order. Thus every order costs $K + cQ$ dollars.
4. The inventory cost is h dollars per item per unit time for holding it in storage.

The objective is to determine the best values for s and S so that the total cost is minimized.

Consider the time period from one order point to the next as a cycle. With a steady demand rate of λ the length of the cycle can be given as Q/λ. To determine the inventory-holding cost per unit time, we need the average inventory level during the cycle.

At the start of a cycle the inventory level starts at $S = s + Q$ and steadily decreases by Q to s at the end of the cycle. Thus the average inventory level is $s + Q/2$, giving the holding cost per unit time as $h(s + Q/2)$.

Let $C(s, Q)$ be the total cost per unit time. We have

$$C(s, Q) = (K + cQ)\frac{1}{Q/\lambda} + h\left(s + \frac{Q}{2}\right)$$

$$= c\lambda + \frac{K\lambda}{Q} + hs + \frac{hQ}{2}$$

$$= (c\lambda + hs) + \left(\frac{K\lambda}{Q} + \frac{hQ}{2}\right) \tag{14.2.1}$$

In order to minimize $C(s, Q)$, we note that it is a separable function in s and Q with $s \geq 0$ and $Q > 0$. Let $C_1(s) = c\lambda + hs$ and $C_2(Q) = K\lambda/Q + hQ/2$ so that

$$C_1(s) + C_2(Q) = C(s, Q)$$

Clearly $C_1(s)$ is minimized when $s = 0$, indicating that with the given cost structure, a threshold stock of $s > 0$ is unnecessary. This conclusion would have been obvious considering the deterministic nature of the demand. The function $C_2(Q)$ is convex, and the optimal value of $Q = Q^*$ is obtained by differentiating $C_2(Q)$ with respect to Q, equating the derivative to zero, and solving for Q. We get

$$\frac{dC_2(Q)}{dQ} = \frac{-K\lambda}{Q^2} + \frac{h}{2} = 0$$

which gives

$$Q^* = \sqrt{\frac{2K\lambda}{h}} \tag{14.2.2}$$

The result (14.2.2) is commonly known as the economic order quantity (EOQ) formula (or Wilson's formula) in inventory theory.

For smaller values of Q, because it is an integer, minimizing $C_2(Q)$ through differentiation is not appropriate. Then one should argue that Q^* should be such that

$$C_2(Q^* - 1) - C_2(Q^*) \geq 0$$

$$C_2(Q^*) - C_2(Q^* + 1) \leq 0 \qquad (14.2.3)$$

Using the first equation in (14.2.3), we get

$$\frac{K\lambda}{Q^* - 1} + \frac{h(Q^* - 1)}{2} - \frac{K\lambda}{Q^*} - \frac{hQ^*}{2} \geq 0$$

that is,

$$K\lambda\left(\frac{1}{Q^* - 1} - \frac{1}{Q^*}\right) \geq \frac{h}{2}$$

and

$$\frac{2K\lambda}{h} \geq Q^*(Q^* - 1) \qquad (14.2.4)$$

Using similar arguments on the second equation in (14.2.3), we get

$$\frac{2K\lambda}{h} \leq Q^*(Q^* + 1) \qquad (14.2.5)$$

Combining (14.2.4) and (14.2.5), we find

$$Q^*(Q^* - 1) \leq \frac{2K\lambda}{h} \leq Q^*(Q^* + 1) \qquad (14.2.6)$$

When the difference between $Q^* + 1$ and $Q^* - 1$ is small relative to Q^*, clearly we have the EOQ formula (14.2.2).

Stochastic Demands

Consider an inventory system with the same description as the deterministic system discussed previously except for the following modification to the demand arrival process: Demand for items of merchandise arrive in a renewal process at time points $t_0, t_1, t_2, \ldots$. Let $t_n - t_{n-1}$ $(n = 1, 2, \ldots)$ be independent and identically distributed random variables with mean $1/\lambda$ (the demand rate will be λ).

Let $I(t)$ be the inventory level at time t. The stochastic process $I(t)$ has a state space $\{s + 1, s + 2, \ldots, s + Q\}$. Let $p_n = \lim_{t \to \infty} P[I(t) = n]$. The stochastic process $I(t)$ visits states $s + Q, s + Q - 1, \ldots, s + 1$ in a cyclic fashion, and therefore we may use the results for the cyclically alternating renewal process given in Section 8.4. In particular, using the probability given in (8.4.9), we have

$$p_n = \frac{1/\lambda}{Q(1/\lambda)} = \frac{1}{Q} \qquad n = s + 1, \ldots, s + Q \qquad (14.2.7)$$

Thus we get

$$E[I] = \sum_{n=s+1}^{s+Q} n p_n$$

$$= s + \frac{1}{Q} \sum_{n=1}^{Q} n$$

$$= s + \frac{Q + 1}{2} \qquad (14.2.8)$$

During a reorder cycle Q demands arrive. Therefore

$$E(\text{length of a reorder cycle}) = \frac{Q}{\lambda}$$

Using these results in a cost function, we get

$$C(s, Q) = (K + cQ)\frac{1}{(Q/\lambda)} + hE[I]$$

$$= c\lambda + \frac{K\lambda}{Q} + h\left[s + \frac{Q + 1}{2}\right]$$

$$= \left(c\lambda + \frac{h}{2} + hs\right) + \left(\frac{k\lambda}{Q} + \frac{Q}{2}\right) \qquad (14.2.9)$$

The cost function (14.2.9) is in the same form as (14.2.1), and the optimal values for s and Q are determined in a similar manner [$s = 0$ and the optimal value Q^* is given by (14.2.2)].

14.3 INVENTORY SYSTEMS WITH LEAD TIMES

In the previous section we assumed that there is no delay between the time an order is placed and its fulfillment. In many situations it is an unnatural assumption. The primary emphasis in this section is to relax this assumption and to consider cases where a lead time is provided for order fulfillment. In the first two examples of this section we shall also assume that the lead times have exponential distributions.

When lead times are incorporated in the model, there is a positive probability for the arrival of demand to occur before the replenishment of stock. To incorporate such events, we consider two cases, one with complete backlogging and one without backlogging. When backlogging is allowed, no demand goes unfilled however long it has to wait.

An Inventory System with Backlogging*

Consider an inventory system with the following description:

1. Demand arrivals are in a Poisson process with parameter λ, and all demands are met either from the available stock or by ordering new items whenever necessary.

2. It is proposed to use an (s, S) inventory policy in which orders for $Q(= S - s)$ items are placed when the inventory falls to the level s.

3. The replenishment of stock is therefore in batches of Q items, and it is assumed that the amount of time needed to supply each batch has an exponential distribution with mean $1/\mu$.

4. The following four different costs are assumed:

   K—a fixed cost for placing an order.

   h— cost per unit time for holding an item in inventory.

   b— back order cost per unit per unit time.

   u— underage cost per unit for being short on inventory.

Let $C(S, Q)$ be the expected cost per unit time for operating the inventory policy with decision variables S and Q. We can consider two related stochastic processes in this system: $I(t)$ the on-hand inventory at time t and $X(t)$ the inventory on order plus the number of units demanded since the last order was placed. Clearly $I(t) + X(t) = S$, where negative values for $I(t)$ indicate the number of items on back order.

*Our treatment follows generally that of Gross and Harris (1973).

Let p_n be the steady-state probability that $X(t) = n$. Using p_n ($n = 0, 1, 2, \ldots$), the cost function $C(S, Q)$ can be given the following expression:

$$C(S, Q) = \frac{K\lambda}{Q} + h \sum_{n=0}^{S} (S - n)p_n + b \sum_{n=S}^{\infty} (n - S)p_n + u\lambda \sum_{n=S}^{\infty} p_n \quad (14.3.1)$$

Any use of cost function (14.3.1) in a decision process depends on the determination of the steady-state probabilities $\{ p_n \}$ of the process $X(t)$.

If we look at the $X(t)$ process by itself, it is clear that it is a queueing process in which arrivals occur in a Poisson process and the service is provided for batches of Q customers at a time, the service times having an exponential distribution. The steady-state equations for this process can be written down in the following manner:

$$\lambda p_0 = \mu p_Q$$

$$\lambda p_n = \lambda p_{n-1} + \mu p_{Q+n} \quad 1 \le n \le Q - 1$$

$$(\lambda + \mu) p_n = \lambda p_{n-1} + \mu p_{Q+n} \quad n \ge Q \quad (14.3.2)$$

In its present general form the set of equations (14.3.2) cannot be solved explicitly without going through generating functions. Multiplying both sides of the equations with appropriate powers z and defining $P(z) = \sum_{i=0}^{\infty} p_i z^i$, $|z| \le 1$, we get

$$\mu \left[P(z) - \sum_{i=0}^{Q-1} p_i z^i \right] + \lambda P(z) = \lambda z P(z) + \frac{\mu}{z^Q} \left[P(z) - \sum_{i=0}^{Q-1} p_i z^i \right]$$

which on simplification gives

$$P(z) = \frac{\mu(z^Q - 1) \sum_{i=0}^{Q-1} p_i z^i}{-\lambda z^{Q+1} + (\mu + \lambda) z^Q - \mu} \quad (14.3.3)$$

The probability generating function (14.3.3) can be completely specified by determining the unknowns $\{ p_0, p_1, \ldots, p_{Q-1} \}$. This is done by considering the roots of the denominator (it can be shown that there are Q distinct roots in $|z| \le 1$) and noting that the numerator should vanish at these points. For a discussion of this procedure, the readers are referred to Gross and Harris (1973) where a slightly more general system is analyzed.

When $P(z)$ is completely specified, we can express it as a power series in z and determine the individual probabilities $\{p_n\}$, which can then be substituted in the cost function (14.3.1). However, we should note that the probabilities $\{p_n\}$ determined in this manner are for specific values of Q and the only way we can carry out a decision process is through a search procedure, by trying out different values of Q. Therefore from a practical viewpoint the general model is likely to be untractable.

In the special case of $Q = 1$ (see Gross and Harris, 1971) the system of equations (14.3.2) shows that the process $X(t)$ is in fact the same as the queue length process in the queue $M/M/1$ discussed in Chapter 11, for which explicit expressions are available. We then have

$$p_n = \left(1 - \frac{\lambda}{\mu}\right)\left(\frac{\lambda}{\mu}\right)^n \qquad n = 0, 1, 2, \ldots \tag{14.3.4}$$

When $Q = 1$, the inventory system is such that every demand triggers an order, and therefore the decision variable is only S, the number of items to be held in stock. In the cost function of (14.3.1) we drop the first term $K\lambda$, since it does not involve the decision variable, and write

$$C(S) = h \sum_{n=0}^{S} (S - n)p_n + b \sum_{n=S}^{\infty} (n - S)p_n + u\lambda \sum_{n=S}^{\infty} p_n \tag{14.3.5}$$

The optimal value S^* is such that $C(S^*) = \min_{S \geq 0} C(S)$ with S being an integer. Clearly S^0 is locally optimal when

$$C(S^0) \leq C(S^0 + 1) \qquad \text{if } S^0 = 0$$

$$\left.\begin{array}{l} C(S^0) \leq C(S^0 + 1) \\ C(S^0) \leq C(S^0 - 1) \end{array}\right\} \qquad \text{if } S^0 > 0 \tag{14.3.6}$$

Let $\Delta C(S) = C(S + 1) - C(S)$. Then (14.3.6) can be written as

$$\Delta C(S^0) \geq 0 \qquad \text{if } S^0 = 0$$

$$\Delta C(S^0 - 1) \leq 0 \leq \Delta C(S^0) \qquad \text{if } S^0 > 0 \tag{14.3.7}$$

From (14.3.5) we get

$$\Delta C(S) = h\left[\sum_{n=0}^{S+1}(S+1-n)p_n - \sum_{n=0}^{S}(S-n)p_n\right]$$

$$+b\left[\sum_{n=S+1}^{\infty}(n-S-1)p_n - \sum_{n=S}^{\infty}(n-S)p_n\right]$$

$$+u\lambda\left[\sum_{n=S+1}^{\infty}p_n - \sum_{n=S}^{\infty}p_n\right]$$

$$= (h+b)\sum_{n=0}^{S}p_n - u\lambda p_S - b \qquad (14.3.8)$$

Using the value of p_n from (14.3.4), we have

$$\Delta C(S) = (h+b)(1-\rho^{S+1}) - u\lambda(1-\rho)\rho^S - b \qquad (14.3.9)$$

where we have written $\lambda/\mu = \rho$. If the function $C(S)$ is convex, then the optimum we get from (14.3.6) or (14.3.7) is also the global optimum. To show that $C(S)$ is convex, we consider the second-order difference

$$\Delta^2 C(S) = \Delta C(S+1) - \Delta C(S)$$

$$= (h+b)\rho^{S+1}(1-\rho) + u\lambda(1-\rho)^2\rho^S \qquad (14.3.10)$$

$$> 0$$

for any $h, b, u > 0$ since $\rho < 1$. Thus the global optimality of S^0 is clearly established. The optimal value $S^* = S^0$ is obtained by evaluating $\Delta C(S)$ for different values of S and identifying the value of S at which $\Delta C(S)$ becomes nonnegative.

For a numerical illustration of this procedure, the reader is referred to Gross and Harris (1971) where the authors have considered a system in which the lead times have exponential distributions with state dependent parameters.

An Inventory System without Backlogging*

As just seen, backlogging in an inventory system results in an underlying stochastic process which is similar to a batch service queue. A simpler model has been considered by Morse (1958) in which no backlogging is allowed.

*Our treatment follows generally that of Morse (1958).

Consider an inventory system with the following description:

1. Demand arrivals are in a Poisson process with parameter λ. Demands arriving when the on-hand inventory level is zero leave the system without having their demand satisfied.

2. It is proposed to use an (s, S) inventory policy in which orders for $Q(= S - s)$ items are placed when the inventory falls to the level s and additional orders for s items are placed when the level goes to zero. The second possibility arises because of the lead time between placing an order and getting it filled. When there are two outstanding orders—one for Q items and the other for s items $(s < Q)$—we assume that the order for s items gets filled before the order for Q items is satisfied.

3. The lead times for both orders are assumed to be random variables with the same exponential distribution with mean $1/\mu$.

4. The following three different costs are assumed:

   K— a fixed cost for placing an order.

   h— cost per unit time for holding an item in inventory.

   r— reorder cost per unit time when the inventory falls below the threshold level. It should be noted that since the system does not allow backlogging no back order cost is necessary in this case.

Let $C(s, Q)$ be the expected cost per unit time for operating the inventory policy with decision variables s and Q. As before, we can consider two related stochastic processes: $I(t)$ the on-hand inventory at time t and $X(t)$ the inventory on order plus the number of units demanded since the last order was placed. Clearly $I(t) + X(t) = S$. In order to be consistent with the earlier discussion, we shall work with the stochastic process $X(t)$ and define the steady-state probability that $X(t) = n$ as $p_n (n = 0, 1, 2, \ldots, S)$. The probability distribution of $I(t)$ easily follows from the relation

$$P[I(t) = n] = P[X(t) = S - n] = p_{S-n} \qquad (14.3.11)$$

For purposes of clarity the correspondence between the two processes can be exhibited as follows:

$I(t)$	0	1	2	$\ldots$	s	$\ldots$	Q	$\ldots$	S	
$X(t)$	S	$S-1$	$S-2$	$\ldots$	Q	$\ldots$	s	$\ldots$	0	(14.3.12)

As per (14.3.11) the probability that a demand will not be met by the system is given by p_S.

It is easy to visualize the system as going through periods when $I(t) > 0$ and $I(t) = 0$ alternately. Let $E(> 0)$ and $E(0)$ be the expected intervals of time for these periods, respectively. The lead time for filling an order is exponential with mean $1/\mu$, and therefore because of the memoryless property of the distribution, we have

$$E(0) = \frac{1}{\mu} \tag{14.3.13}$$

Referring back to results on alternating renewal processes, we also have

$$\lim_{t \to \infty} P[I(t) = 0] = \frac{E(0)}{E(> 0) + E(0)} \tag{14.3.14}$$

that is,

$$p_S = \frac{1}{\mu[E(> 0) + (1/\mu)]}$$

$$= \frac{1}{\mu E(> 0) + 1}$$

$$E(> 0) = \frac{1 - p_S}{\mu p_S} \tag{14.3.15}$$

In order to find an expression for the fixed cost of placing an order per unit time, we proceed as follows: Consider a cycle containing two adjacent periods during which $I(t) > 0$ and $I(t) = 0$. The average number of orders placed during such a cycle can be given as

$$\frac{\lambda E(> 0) - s}{Q} + 1 \tag{14.3.16}$$

When $\$K$ is the cost per unit time for placing an order, we have that component given by

$$\frac{K}{[E(0) + E(> 0)]}\left[\frac{\lambda E(> 0) - s}{Q} + 1\right] = K\mu p_S\left[\frac{\lambda E(> 0) - s}{Q} + 1\right]$$

$$= \frac{K\lambda}{Q}(1 - p_S) + K\mu p_S\left(1 - \frac{s}{Q}\right)$$

Other components of the cost $C(s, Q)$ are directly obtained. Thus we have

$$C(s, Q) = \frac{K\lambda}{Q}(1 - p_s) + K\mu p_S\left(1 - \frac{s}{Q}\right)$$

$$+ h \sum_{n=0}^{S}(S - n)p_n + r\left(2p_S + \sum_{n=Q}^{S-1} p_n\right) \quad (14.3.17)$$

The probabilities $\{p_n\}$ are determined using the steady-state equations for the process $X(t)$. Considering possible transitions, we have

$$\lambda p_0 = \mu p_Q$$
$$\lambda p_n = \lambda p_{n-1} + \mu p_{n+Q} \quad 0 < n \leq s$$
$$\lambda p_n = \lambda p_{n-1} \quad s < n < Q$$
$$(\lambda + \mu)p_Q = \lambda p_{Q-1} + \mu p_S$$
$$(\lambda + \mu)p_n = \lambda p_{n-1} \quad Q < n < S$$
$$2\mu p_S = \lambda p_{S-1} \quad (14.3.18)$$

Equations (14.3.18) can be solved recursively as follows (note that it is simpler to start with the last equation in this case): We get

$$p_n = \left(\frac{\lambda + \mu}{\lambda}\right)^{S-1-n}\left(\frac{2\mu}{\lambda}\right)p_S \quad Q \leq n < S$$

$$p_n = \left[\left(\frac{\lambda + \mu}{\lambda}\right)^{s}\left(\frac{2\mu}{\lambda}\right) - \frac{\mu}{\lambda}\right]p_S \quad s \leq n \leq Q - 1$$

$$p_n = \left(\frac{2\mu}{\lambda}\right)\left[\left(\frac{\lambda + \mu}{\lambda}\right)^{s} - \left(\frac{\lambda + \mu}{\lambda}\right)^{s-1-n}\right]p_S \quad 0 \leq n \leq s - 1$$

$$(14.3.19)$$

Using the normalizing condition $\sum_{n=0}^{S} p_n = 1$ in (14.3.19), we determine p_S. We get

$$p_S^{-1} = 1 + \sum_{n=Q}^{S-1}\left(\frac{\lambda + \mu}{\lambda}\right)^{S-1-n}\left(\frac{2\mu}{\lambda}\right) + \sum_{n=s}^{Q-1}\left[\left(\frac{\lambda + \mu}{\lambda}\right)^{s}\left(\frac{2\mu}{\lambda}\right) - \frac{\mu}{\lambda}\right]$$

$$+ \frac{2\mu}{\lambda}\sum_{n=0}^{s-1}\left[\left(\frac{\lambda + \mu}{\lambda}\right)^{s} - \left(\frac{\lambda + \mu}{\lambda}\right)^{s-1-n}\right]$$

which after simplification gives

$$p_S = \frac{\rho^{s+1}}{2Q(1 + \rho)^{s} + [\rho + s - Q]\rho^{s}} \quad (14.3.20)$$

where we have written $\lambda/\mu = \rho$.

For the cost function (14.3.17) we need the expected value $\sum_{n=0}^{S}(S-n)p_n$. Only major steps in the determination of the expected value are given here:

$$\sum_{n=0}^{S}(S-n)p_n = S - p_S\sum_{n=1}^{S}\frac{np_n}{p_S}$$

$$= S - p_S\left\{S + \sum_{n=Q}^{S-1}n\left(\frac{\lambda+\mu}{\lambda}\right)^{S-1-n}\left(\frac{2\mu}{\lambda}\right)\right.$$

$$+ \sum_{n=s}^{Q-1}n\left[\left(\frac{\lambda+\mu}{\lambda}\right)^{s}\left(\frac{2\mu}{\lambda}\right)-\frac{\mu}{\lambda}\right]$$

$$\left.+ \frac{2\mu}{\lambda}\sum_{n=0}^{s-1}n\left[\left(\frac{\lambda+\mu}{\lambda}\right)^{s}-\left(\frac{\lambda+\mu}{\lambda}\right)^{s-1-n}\right]\right\}$$

$$= p_S\left\{\sum_{n=1}^{s}n\left(\frac{\lambda+\mu}{\lambda}\right)^{n-1}\left(\frac{2\mu}{\lambda}\right)\right.$$

$$+ \sum_{n=s+1}^{Q}n\left[\left(\frac{\lambda+\mu}{\lambda}\right)^{s}\left(\frac{2\mu}{\lambda}\right)-\frac{\mu}{\lambda}\right]$$

$$\left.+ \frac{2\mu}{\lambda}\sum_{n=Q+1}^{S}n\left[\left(\frac{\lambda+\mu}{\lambda}\right)^{s}-\left(\frac{\lambda+\mu}{\lambda}\right)^{n-Q-1}\right]\right\}$$

$$= p_S\left[(S^2-s^2+Q)\left(\frac{\mu}{\lambda}\right)\left(\frac{\lambda+\mu}{\lambda}\right)^{s}\right.$$

$$\left.-(Q^2+Q-s^2-s)\left(\frac{\mu}{2\lambda}\right)-\frac{2Q\mu}{\lambda}\sum_{n=1}^{s}\left(\frac{\lambda+\mu}{\lambda}\right)^{n-1}\right]$$

$$= \frac{p_S}{\rho^{s+1}}\left[(S^2-s^2+Q)(1+\rho)^{s}-(Q^2+Q-s^2-s)\frac{\rho^s}{2}\right.$$

$$\left.-2Q\rho(1+\rho)^{s}+2Q\rho^{s+1}\right]$$

$$= \frac{Q(Q+2s+1-2\rho)(1+\rho)^{s}+\frac{1}{2}(4Q\rho+s^2+s-Q^2-Q)\rho^{s}}{2Q(1+\rho)^{s}+(\rho+s-Q)\rho^{s}}$$

$$(14.3.21)$$

In a similar fashion the probability terms in the fourth term of the cost function (14.3.17) simplify to

$$2p_S + \sum_{n=Q}^{S-1} p_n = \frac{2\rho(1 + \rho)^s}{2Q(1 + \rho)^s + (\rho + s - Q)\rho^s} \qquad (14.3.22)$$

The cost function (14.3.17) is thus completely determined. However, because of the complexity of the cost function we can only suggest a search procedure to determine the best values for s and Q of the inventory policy. Nevertheless, we quote here some simplified results derived by Morse (1958, pp. 151–152) by ignoring the fixed cost K and including a profit term in the objective function.

Let \$$g$ per unit be the gross profit realized by the sale. Expected number of items sold per unit time can be given as $\rho(1 - p_S)$. Now we can write a profit function $\phi(s, Q)$ as

$$\phi(s, Q) = g\rho(1 - p_S) - h \sum_{n=0}^{S} (s - n)p_n - r\left(2p_S + \sum_{n=Q}^{S-1} p_n\right)$$

$$(14.3.23)$$

Suppose the threshold value s^* for s is determined using the condition that the probability of being out of stock should be no larger than some amount $p_{\min}$. We get (using the notation $\simeq$ to indicate approximate equivalence)

$$s^* \simeq \frac{\ln(\rho/2Qp_{\min})}{\ln(1 + 1/\rho)} \qquad (14.3.24)$$

Using this value of s^* in (14.3.23) and maximizing, one gets

$$Q^* \simeq q + \frac{1}{2}qR^S\left[\frac{g\rho}{2r} + \frac{1}{2}\left(\frac{s + \rho}{\rho}\right)\right] \qquad (14.3.25)$$

where $q = (2\rho r/h)^{1/2}$ and $R = \rho/(1 + \rho)$.

An Inventory System with a Markov Chain as the Underlying Process

In the two examples discussed earlier in this section we have used a continuous parameter Markov process as the model for the underlying process. Consequently the demand arrivals have been assumed to be

Poisson, and the lead times have been assumed to be exponential. Alternately, we could have used a discrete parameter Markov process (Markov chain) as the model for the underlying process with the demand arrival points as epochs of observation. With such a model we can consider a renewal process for the demand arrivals and a geometric distribution for the lead times. The cost function and the analysis in this case are quite similar to the continuous time case. We illustrate the procedure using a modification of the system considered by Morse (1958).

Consider an inventory system with the following description:

1. Demand arrivals occur in a renewal process at time points $t_0, t_1, t_2, \ldots$. Demands arriving when the on-hand inventory level is zero leave the system without having their demand satisfied.

2. It is proposed to use an (s, S) inventory policy in which orders for Q ($= S - s$) items are placed when the inventory falls to the level s. It is also assumed that when the orders are received, the supplier will have enough additional items to bring the inventory level to S.

3. The probability that the replenishment of stock (from a level $\leq s$ to a level S) will take place within one demand period (time interval between two successive demands) is assumed to be a constant q ($0 < q < 1$). Consequently the lead time has a geometric distribution in terms of the number of demand periods.

4. The following two different costs are assumed (we ignore the fixed cost of placing an order in order to simplify the problem):

    h—cost per unit time for holding an item in inventory.

    u— underage cost per unit when a demand goes unsatisfied.

    It should be noted that a demand period is the unit of time when the underlying process is modeled as a Markov chain.

Let $C(s, Q)$ be the expected cost per unit time for operating the inventory policy, with decision variables s and Q. Let I_n be the inventory level just before the arrival of the nth demand. We write $\lim_{n \to \infty} I_n = I$. Let $\pi_n = P(I = n)$, the limiting distribution of the inventory level. Using π_n ($n = 0, 1, \ldots, S$), the cost function $C(s, Q)$ can be expressed as

$$C(s, Q) = h \sum_{n=1}^{S} n\pi_n + u\pi_0 \qquad (14.3.26)$$

An explicit expression for $C(s, Q)$ is obtained using the properties of the inventory process.

Clearly $\{I_n, n = 0, 1, 2, \ldots\}$ is a Markov chain. Its transition probability matrix P is given by (only nonzero elements are indicated):

$$
P = \begin{array}{c} \\ 0 \\ 1 \\ 2 \\ \vdots \\ s-1 \\ s \\ s+1 \\ s+2 \\ \vdots \\ S \end{array}
\begin{array}{cccccccc}
0 & 1 & \cdots & s-1 & s & s+1 & \cdots & S-1 \quad S \\
\left[\begin{array}{cccccccc}
1-q & & & & & & & q \\
1-q & & & & & & & q \\
& 1-q & & & & & & q \\
& & \ddots & & & & & \vdots \\
& & & 1-q & & & & q \\
& & & & 1-q & & & q \\
& & & & & 1 & & \\
& & & & & & \ddots & \\
& & & & & & & 1
\end{array}\right]
\end{array}
$$

$$(14.3.27)$$

The limiting distribution of the inventory level is obtained by solving the set of equations $\pi_j = \sum_i \pi_i P_{ij}$ ($j = 0, 1, 2, \ldots, S$) and $\sum_{n=0}^{S} \pi_n = 1$. We get

$$\pi_0 = (1-q)(\pi_0 + \pi_1)$$

$$\pi_n = (1-q)\pi_{n+1} \qquad 1 \le n \le s$$

$$\pi_n = \pi_{n+1} \qquad s+1 \le n \le S \qquad (14.3.28)$$

Solving (14.3.28), we obtain

$$\pi_n = \left(\frac{q}{1-q}\right)\left(\frac{1}{1-q}\right)^{n-1} \pi_0 \qquad 1 \le n \le s+1$$

and therefore

$$\pi_0^{-1} = 1 + \left(\frac{q}{1-q}\right)\sum_{n=1}^{s}\left(\frac{1}{1-q}\right)^{n-1} + Q\left(\frac{q}{1-q}\right)\left(\frac{1}{1-q}\right)^{s}$$

$$= \frac{1 + q(Q-1)}{(1-q)^{s+1}}$$

$$\pi_0 = \frac{(1-q)^{s+1}}{1 + q(Q-1)} \qquad (14.3.29)$$

Using π_0 in the solutions for $\pi_n = (n = 1, 2, \ldots, S)$ and after simplifications, we find

$$\sum_{n=1}^{S} n\pi_n = \frac{1}{1 + q(Q-1)} \left\{ q(1-q)^s \left[\sum_{n=1}^{s-1} n\left(\frac{1}{1-q}\right)^n \right] \right.$$

$$\left. + (1-q)\left[1 - (1-q)^s\right] + \frac{qQ}{2}(Q + 2s + 1) \right\} \quad (14.3.30)$$

Results (14.3.29) and (14.3.30) provide an explicit expression for the cost function (14.3.26). Optimal values for s and Q can now be determined by a convenient method such as a search procedure or an optimization technique.

14.4 STORAGE PROCESSES

In storage theory problems related to situations such as storing water in a reservoir are considered. These problems are similar to inventory problems since the basic feature is storing a commodity until it is released to or claimed by another party. However, a major distinction is that in a storage process, the control is generally on the release rule (output) where as in an inventory process the control is over stock replenishment (input). Also the resulting stochastic processes in storage theory are quite complex and are not amenable for easy manipulations in a decision procedure such as optimization. Therefore the control is based more on the behavioral characteristics of the system rather than on an optimizing procedure using an objective function. The practical approach is similar to that of queueing systems. We shall illustrate the basic structure of storage problems with the help of a dam model developed by Moran (1954) which is formulated in discrete time.

Let $t_0, t_1, t_2, \ldots$ be discrete points in time at which a dam is observed. The dam is assumed to have a finite capacity K. Let X_n be the amount of water (stated to the closest integer) flowing into the dam during the period (t_n, t_{n+1}) $(n = 0, 1, 2, \ldots)$. We assume that $X_0, X_1, X_2, \ldots$ are independent and identically distributed random variables with

$$P(X_n = j) = a_j \qquad j = 0, 1, 2, \ldots \qquad (14.4.1)$$

During each period an amount m of water is released from the dam. If the dam does not contain that amount, the entire amount is released. For purposes of mathematical formulation we shall assume that the release takes place at epochs t_n $(n = 1, 2, \ldots)$.

Let S_n be the amount of water in the dam at t_n soon after release. Because of the finite capacity of the dam excess water is allowed to overflow. Clearly the amount of water before release at t_{n+1} is $\min(K, S_n + X_n)$, and the amount released at t_{n+1} is $\min(m, S_n + X_n)$. Thus we have

$$S_{n+1} = \min(K, S_n + X_n) - \min(m, S_n + X_n) \qquad (14.4.2)$$

The process $\{S_n, n = 0, 1, 2, \ldots\}$ is a Markov chain with state space $\{0, 1, 2, \ldots, K\}$. Let

$$P_{ij} = P(S_{n+1} = j | S_n = i) \qquad i, j = 0, 1, \ldots, K$$

and define $A_j = a_0 + a_1 + \cdots + a_j$ and $\overline{A}_j = 1 - A_j$. The transition probability matrix $\boldsymbol{P} = \|P_{ij}\|$ is given as

$\boldsymbol{P} =$

↗	0	1	2	$\cdots$	$K - 2m$	$\cdots$	$K - m - 1$	$K - m$
0	A_m	a_{m+1}	a_{m+2}	$\cdots$	a_{K-m}	$\cdots$	a_{K-1}	$\overline{A}_{K-1}$
1	A_{m-1}	a_m	a_{m+1}	$\cdots$	a_{K-m-1}	$\cdots$	a_{K-2}	$\overline{A}_{K-2}$
.	.	.	.		.		.	.
$m-1$	A_1	a_2	a_3	$\cdots$	a_{K-2m+1}	$\cdots$	a_{K-m}	$\overline{A}_{K-m}$
m	a_0	a_1	a_2	$\cdots$	a_{K-2m}	$\cdots$	a_{K-m-1}	$\overline{A}_{K-m-1}$
$m+1$	0	a_0	a_1	$\cdots$	a_{K-2m-1}	$\cdots$	a_{K-m-2}	$\overline{A}_{K-m-2}$
.	.	.	.		.		.	.
$K - m$	0	0	0	$\cdots$	a_0	$\cdots$	a_{m-1}	$\overline{A}_{m-1}$

$$(14.4.3)$$

The matrix (14.4.3) is irreducible, aperiodic, and finite. The limiting distribution and other properties of the Markov chain can be determined in the usual manner.

When $m = 1$, a simple recursive approach can be suggested in deriving the limiting distribution of the Markov chain. In this case we have

$$\boldsymbol{P} = \begin{array}{c} \\ 0 \\ 1 \\ 2 \\ \vdots \\ K-1 \end{array} \begin{array}{cccccc} 0 & 1 & 2 & \cdots & K-2 & K-1 \\ \left[\begin{array}{cccccc} A_1 & a_2 & a_3 & \cdots & a_{K-1} & \overline{A}_{K-1} \\ a_0 & a_1 & a_2 & \cdots & a_{K-2} & \overline{A}_{K-2} \\ & a_0 & a_1 & \cdots & a_{K-3} & \overline{A}_{K-3} \\ & & & & \vdots & \vdots \\ & & & & a_0 & \overline{A}_0 \end{array}\right] \end{array} \qquad (14.4.4)$$

Note the similarity of the structure of this matrix and the transition probability matrix of the queueing system $M/G/1$ with a finite waiting room discussed in Section 11.6. The only difference lies in the elements of the first row. Following the algorithmic recursive method described therein, we can determine the limiting distribution of the storage level $(\pi_0, \pi_1, \pi_2, \ldots, \pi_{K-1})$ by first determining

$$\nu_i = \frac{\pi_i}{\pi_0} \qquad i = 1, 2, \ldots, K - 1$$

$$\nu_0 = 1 \tag{14.4.5}$$

and using equations

$$\nu_1 = \frac{1 - a_0 - a_1}{a_0}$$

$$\nu_2 = \frac{1 - a_1}{a_0} \nu_1 - \frac{a_2}{a_0}$$

$$\vdots$$

$$\nu_n = \frac{1 - a_1}{a_0} \nu_{n-1} - \frac{a_2}{a_0} \nu_{n-2} - \cdots - \frac{a_{n-1}}{a_0} \nu_1 - \frac{a_n}{a_0}$$

$$\nu_{K-1} = \frac{1 - a_1}{a_0} \nu_{k-2} - \frac{a_2}{a_0} \nu_{k-3} - \cdots - \frac{a_{K-2}}{a_0} \nu_1 - \frac{a_{K-1}}{a_0} \tag{14.4.6}$$

Now consider

$$\sum_{i=0}^{K-1} \nu_i = 1 + \frac{1}{\pi_0} \sum_{i=1}^{K-1} \pi_i$$

$$= \frac{1}{\pi_0} \sum_{i=0}^{K-1} \pi_i = \frac{1}{\pi_0}$$

Hence

$$\pi_0 = \frac{1}{\displaystyle\sum_{i=0}^{K-1} \nu_i}$$

$$\pi_i = \nu_i \pi_0 = \frac{\nu_i}{\displaystyle\sum_{i=0}^{K-1} \nu_i} \tag{14.4.7}$$

which completely determine the limiting distribution.

An analytical solution to this problem has been given by Moran (1956) as follows: Let

$$A(z) = \sum_{j=0}^{\infty} a_j z^j \qquad |z| < 1$$

$$V(z) = \sum_{i=0}^{\infty} v_i z^i \qquad |z| < 1$$

Then Moran has shown (also see Prabhu, 1965, p. 195) that

$$V(z) = \frac{a_0(1 - z)}{A(z) - z} \qquad (14.4.8)$$

which can be expanded as an infinite series in z to obtain v_i ($i = 1, 2, \ldots,$ $K - 1$). Nevertheless, the computational algorithm given by (14.4.7) is much more convenient and straightforward in practical applications.

In problems of water storage the occurrence of dry periods (when the dam is empty) is of major concern. Dry periods alternate with wet periods. Assuming a wet period starts with a certain amount of water in the dam, the wet period is the same as the first passage of the Markov chain to state zero. A similar concept in a queueing system is the busy period. As discussed in Section 4.2 in order to determine the characteristics of the wet period, we define $\tilde{P}$ by converting state 0 into an absorbing state.

$$\tilde{P} = \begin{bmatrix} 1 & 0 & 0 & \cdots & 0 & 0 \\ \hline a_0 & a_1 & a_2 & \cdots & a_{K-2} & \overline{A}_{K-2} \\ & a_0 & a_1 & \cdots & a_{K-3} & \overline{A}_{K-3} \\ & & & \vdots & \vdots & \vdots \\ & & & & a_0 & \overline{A}_0 \end{bmatrix} \qquad (14.4.9)$$

Partition $\tilde{P}$ as indicated:

$$\tilde{P} = \begin{bmatrix} I & 0 \\ \hline R & Q \end{bmatrix} \qquad (14.4.10)$$

where Q is a $(K - 1) \times (K - 1)$ matrix. As discussed in Chapter 4, the mean first passage times of the Markov chain to state 0 are given by the row sums of the matrix $(I - Q)^{-1}$. Using the distribution of initial water level at the start of a wet period as weights, a weighted average of the row sums gives the mean wet period.

If we are interested in the distribution of the wet period, we follow the algorithmic procedure developed in Section 4.2. Let T_i be the wet period of the dam if it starts with an initial water level of i. Define

$$f_{i0}^{(n)} = P(T_i = n) \qquad (14.4.11)$$

and let $F^{(n)}$ be the column vector with $f_{i0}^{(n)}(i = 1, 2, \ldots, K - 1)$ as elements. Then we get

$$F^{(n)} = Q^{n-1}F^{(1)} \qquad (14.4.12)$$

where $f_{i0}^{(1)}(i = 1, 2, \ldots, K - 1)$, the elements of $F^{(1)}$, are obtained as

$$f_{i0}^{(1)} = a_0 \qquad i = 1$$

$$= 0 \qquad i > 1 \qquad (14.4.13)$$

For a comprehensive analysis of problems in storage theory, the readers are referred to Moran (1959) and Prabhu (1965).

14.5 REMARKS

As mentioned earlier, our emphasis has been the study of some simple stochastic processes that occur in inventory systems. Interested readers can develop a comprehensive bibliography on the subject using the articles and books cited earlier and those listed under the heading "For Further Reading." It should also be noted that articles on inventory theory appear mostly in journals on operations research, management science, and industrial engineering. Articles on storage processes may be found in journals such as *Applied Probability, Advances in Applied Probability* and *Stochastic Processes and Applications*.

REFERENCES

Arrow, K. J., Karlin, S., and Scarf, H. (1958). *Studies in the Mathematical Theory of Inventory and Production*. Stanford: Stanford University, Press.

Gross, D., and Harris, C. M. (1971). "On One-for-One Ordering Inventory Policies with State Dependent Lead Times." *Oper. Res.* **19**, 735–760.

Gross, D., and Harris, C. M. (1973). "Continuous Review (s, S) Inventory Models with State Dependent Lead Times." *Management Sci.* **19**, 567–574.

Hadley, G., and Whitin, T. M. (1963). *Analysis of Inventory Systems*. Englewood Cliffs, N.J.: Prentice-Hall.

Moran, P. A. P. (1954). "A Probability Theory of Dams and Storage Systems." *Australian J. Appl. Sci.* **5**, 116–124.
Moran, P. A. P. (1959). *The Theory of Storage.* London: Methuen.
Morse, P. M. (1958). *Queues, Inventories and Maintenance.* New York: Wiley.
Prabhu, N. U. (1965). *Queues and Inventories.* New York: Wiley.
Sivazlian, B. D., and Stanfel, L. E. (1975). *Analysis of Systems in Operations Research.* Englewood Cliffs, N.J.: Prentice-Hall.
Whitin, T. M. (1953). *The Theory of Inventory Management.* Princeton: Princeton University Press.

FOR FURTHER READING

Allen, S. G., and D'Esopo, D. A. (1968). "An Ordering Policy for Repairable Stock Items." *Oper. Res.* **16**, 669–674.
Archibald, B. C. (1981). "Continuous Review (s, S) Policies with Lost Sales." *Management Sci.* **27**, 1171–1177.
Caine, G. J., and Plaut, R. H. (1976). "Optimal Inventory Policy When Stockouts Alter Demand." *Nav. Res. Log. Quart.* **23**, 1–13.
Das, C. (1977). "The $(S-1, S)$ Inventory Model under Time Limit on Backorders." *Oper. Res.* **25**, 835–850.
Graves, S. C. (1982). "The Application of Queueing theory to Continuous Perishable Inventory Systems." *Management Sci.* **28**, 400–406.
Gross, D. (1982). "On the Ample Service Assumption of Palm's Theorem in Inventory Modeling." *Management Sci.* **28**, 1065–1079.
Muckstadt, J. A., and Isaac, M. H. (1981). "An Analysis of Single Item Inventory Systems with Returns." *Nav. Res. Log. Quart.* **28**, 237–254.
Nahmias, S. (1982). "Perishable Inventory Theory: A Review." *Oper. Res.* **30**, 680–708. Includes a comprehensive set of references.
Nahmias, S., and Demmy, W. S. (1981). "Operating Characteristics of an Inventory System with Rationing." *Management Sci.* **27**, 1236–1245.
Pierskalla, W. P. (1969). "An Inventory Problem with Obsolescence." *Nav. Res. Log. Quart.* **16**, 217–228.
Posner, M. J. M., and Yansouni, B. (1972). "A Class of Inventory Models with Customer Impatience." *Nav. Res. Log. Quart.* **19**, 483–493.
Ravindran, A. (1972). "Management of Seasonal Style–Goods Inventories." *Oper. Res.* **20**, 265–275.
Sahin, I. (1979). "On the Stationary Analysis of Continuous Review (s, S) Inventory Systems with Constant Lead Times." *Oper. Res.* **27**, 717–729.

Chapter 15
STOCHASTIC PROCESSES
IN RELIABILITY THEORY

15.1 INTRODUCTION

In the pursuit of maximum efficiency in the operation of large-scale engineering systems, the complexity of equipment and configurations has been an ever-increasing phenomenon. Reliability theory is composed of the systematic procedures based on analytical techniques employed to ensure the operational efficiency of such complex systems. As a quantitative measure, *reliability* is defined as the probability that a system will adequately perform its intended purpose for a given period of time under stated environmental conditions.

Each component of a system has performance and reliability thresholds, and normally an increase in performance results in a decrease in reliability. When a large number of components are put together to form a single system, a careful analysis of the performance and reliability of the composite system is essential. In many cases it is possible to achieve a reliable system from less reliable components.

Because of the uncertainties of system behavior the underlying processes are basically stochastic (random) in nature. Therefore it is only to be expected that some of the theoretical results discussed in Chapters 2–10 should prove useful in reliability theory.

In Sections 15.2–15.6 we shall discuss some reliability problems that can be treated as problems in simple stochastic processes. Our treatment is not intended to be extensive or exhaustive; we shall only try to indicate the types of problems that arise and that can be modeled as simple Markov or renewal processes.

In the study of the reliability of systems the following basic characteristics are of major interest:

1. *Reliability:* Probability that the system or the device performs satisfactorily over a given time interval from the start of its operation.

2. *Pointwise availability:*   Probability that the system will be able to operate within certain tolerances at a given instant of time

3. *Interval reliability:*   Probability that the system continues to operate during a given interval of time

4. *Interval availability:*   Expected fraction of a given interval of time that the system will be able to operate within the tolerances. This quantity is also called *efficiency.*

5. *Limiting* values of quantities 2–4 for systems that have been in operation for a long period of time ($t \rightarrow \infty$).

6. *Distribution of time to failure and its mean:*   Failure distribution of the entire system.

7. *System down time:*   Fraction of time the system is inoperative.

These basic system characteristics depend largely on the life distributions of components and the structural configurations of systems. In many cases such as in the engines of an aircraft the components operate independently of each other; then for the efficient operation of the system a minimum number of active components would be necessary. But, if the components are in a series configuration, failure of even a single component renders the system inoperative. In a mixed configuration there can be several subsets of components, some of which are essential for the system to be operative.

Systems can also be classified as maintained or nonmaintained. In maintained systems there are provisions for repair of failed components, whereas in nonmaintained systems only replacements can be made with the available components.

15.2   GENERAL FAILURE DISTRIBUTION FOR COMPONENTS

The probability distribution of the length of life of a component or a device is called its *failure distribution.* Some common failure distributions used in practice are exponential, gamma, Weibull, and log-normal. Let $F(t)$ ($t \geq 0$) be the failure distribution, and let $f(t)$ be its probability density function. In this discussion we shall assume that $f(t)$ exists. Consider the ratio

$$h(t) = \frac{f(t)}{1 - F(t)} \qquad (15.2.1)$$

Probabilistically $f(t)\, dt$ is approximately the probability that the component fails during $(t, t + dt]$, and $1 - F(t)$ is the probability that it is at least of age t. Thus $h(t)\, dt$ represents approximately the probability that the component fails during $(t, t + dt]$, given that it is of age t. The function $h(t)$ is the instantaneous failure rate and is usually called simply the *failure rate*. It is also known as the *hazard rate* in the literature. Now let

$$R(t) = 1 - F(t) \tag{15.2.2}$$

Clearly $R(t)$ gives the probability that the component survives at least until t and hence is considered a measure of reliability. A convenient representation of $R(t)$ based on the failure rate $h(t)$ is obtained as follows: We have

$$h(t) = \frac{f(t)}{1 - F(t)} = -\frac{d}{dt}\ln[1 - F(t)] \tag{15.2.3}$$

giving $(d/dt)\ln[1 - F(t)] = -h(t)$, which on integration gives

$$\ln R(t) = -\int_0^t h(x)\, dx \tag{15.2.4}$$

Note that $R(0) = 1$ and $R(\infty) = 0$. Denoting $\int_0^t h(x)\,dx = H(t)$, from (15.2.4), we get

$$R(t) = e^{-H(t)} \tag{15.2.5}$$

Going back to (15.2.1), we can also write

$$f(t) = h(t)e^{-\int_0^t h(x)dx} \tag{15.2.6}$$

which is a generalized exponential form for any continuous probability density function.

A useful failure rate is that of the Weibull distribution. Let $h(t) = \lambda t^{\alpha - 1}$ $(\lambda, \alpha > 0; t \geq 0)$. Then

$$H(t) = \frac{\lambda t^\alpha}{\alpha}$$

$$R(t) = e^{-(\lambda/\alpha)t^\alpha}$$

$$f(t) = \lambda t^{\alpha - 1} e^{-(\lambda/\alpha)t^\alpha} \tag{15.2.7}$$

The mean and variance of the Weibull distribution (15.2.8) with parameters λ and α can be given as

$$\text{Mean} = \left(\frac{\alpha}{\lambda}\right)^{1/\alpha} \left| \overline{\left(\frac{\alpha + 1}{\alpha}\right)} \right.$$

$$\text{Variance} = \left(\frac{\alpha}{\lambda}\right)^{2/\alpha} \left[\left| \overline{\left(\frac{\alpha + 2}{\alpha}\right)} - \left(\left| \overline{\left(\frac{\alpha + 1}{\alpha}\right)} \right. \right)^2 \right. \right]$$

Setting $\alpha = 1$ in (15.2.8), we get

$$f(t) = \lambda e^{-\lambda t}$$

the exponential distribution whose failure rate is λ.

Finally it should be noted that since component life is a nonnegative random variable, expected life of the component can be conveniently expressed as

$$E[L] = \int_0^\infty R(t)\, dt \qquad (15.2.8)$$

A Replacement Problem

Let the failure distribution of a component be $F(t)$ with $h(t)$ as the failure rate. As soon as the component fails, a replacement is made with a similar component (with the same failure distribution). Let $P_n(t)$ be the probability that n components have been replaced during $(0, t]$. Since $h(t)\Delta t$ gives the probability of component failure during $(t, t + \Delta t]$ after it has attained the age t, we may write the forward Kolmogorov equations

$$P_0(t + \Delta t) = P_0(t)[1 - h(t)\Delta t] + o(\Delta t)$$

$$P_n(t + \Delta t) = P_n(t)[1 - h(t)\Delta t] + P_{n-1}(t)h(t)\Delta t + o(\Delta t) \qquad n > 0$$

$$(15.2.9)$$

and hence

$$P_0'(t) = -h(t)P_0(t)$$

$$P_n'(t) = -h(t)P_n(t) + h(t)P_{n-1}(t) \qquad n > 0 \qquad (15.2.10)$$

Solving these equations as in Section 7.2, we get

$$P_n(t) = e^{-H(t)}\frac{[H(t)]^n}{n!} \qquad (15.2.11)$$

Thus, when the failure distribution is Weibull as defined by (15.2.7), we get

$$P_n(t) = e^{-(\lambda/\alpha)t^\alpha}\left(\frac{\lambda t^\alpha}{\alpha}\right)^n \frac{1}{n!} \qquad (15.2.12)$$

Some General Remarks

As can be noted in problems using the failure rate $h(t)$, the properties of this function are crucial in the study of system reliability. Based on the increasing or decreasing nature of $h(t)$, we can classify failure distributions as having increasing or decreasing failure rates, and several important properties can be derived for these classes of distributions. This classification also helps in arriving at optimal policies of replacement or maintenance of items. Interested readers may refer to Barlow and Proschan (1965, 1975) for an excellent treatment of these topics.

In replacement problems of the type discussed before, a direct application of renewal theory also gives useful results. Replacement of failed components is a renewal process, and therefore we have

$$E[\text{Number of replacements in time } t] \quad = \sum_{n=1}^{\infty} F_n(t)$$

$$\text{Probability density function of a replacement} = \sum_{n=1}^{\infty} f_n(t)$$

$$\begin{array}{l}\text{Probability density of age of item}\\ \text{being used after system has been in}\\ \text{operation for a long time}\end{array} = \frac{1 - F(t)}{\mu}$$

Here we have defined $f_n(t)$ and $F_n(t)$ as the n-fold convolutions of $f(t)$ and $F(t)$, respectively, and μ as the expected lifetime of a component. These results are found in Chapter 8.

15.3 SERIES AND PARALLEL SYSTEMS

Systems are composed of one or more components. Depending on the system structure, system reliability can be determined when component reliabilities are known. Two basic structures in systems are series and parallel structures. For instance, for the success of a system, if all the components should operate satisfactorily, it is as though all of them are connected in series. If, however, a system has some duplicate devices used for the purpose of redundancy, then a parallel structure model is appropriate.

Consider M independent components connected in series. Let p_i be the reliability of the ith component during the operation of the system. If one wishes to express p_i $(i = 1, 2, \ldots, M)$ in terms of the life distribution of the components, using the results of the last section for the constant time

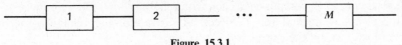

Figure 15.3.1

horizon T of operation, we have

$$p_i = R_i(T) = e^{-\int_0^T h_i(x)\,dx} \qquad (15.3.1)$$

where $h_i(x)$ is the failure rate of the ith component.

Figure 15.3.1 represents a series system with M components. Since for the successful operation of the system all components are needed, the system reliability R is obtained as

$$R = \prod_{i=1}^{M} p_i$$

Using (15.3.1), we may write this as

$$R = R(T) = \prod_{i=1}^{M} \left[e^{-\int_0^T h_i(x)\,dx} \right] \qquad (15.3.2)$$

In a parallel system two or more components are connected in parallel so that for the successful operation of the system only some of them are needed. For instance, consider a two-component parallel system in which at least one component is necessary for successful operation. Figure 15.3.2 represents this structure. Let p be the reliability of each component. The reliability R of the system can be obtained as

$$R = p^2 + 2p(1 - p)$$

which can also be obtained by noting that reliability R is the complement of the unreliability of the system.

$$R = 1 - (1 - p)^2 \qquad (15.3.3)$$

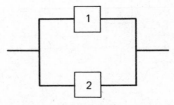

Figure 15.3.2

When there are M components, each with reliability p, and a minimum of k components is needed for a successful operation, the reliability of the system is given by the expression

$$R = \sum_{r=k}^{M} \binom{M}{r} p^r (1 - p)^{M-r}$$

$$= 1 - \sum_{r=0}^{k-1} \binom{M}{r} p^r (1 - p)^{M-r} \tag{15.3.4}$$

Consider an M-component parallel system with reliability p_i [given by (15.3.1)] for the ith component. Assume that the system can successfully operate even with one component. Then the reliability of the system can be given as

$$R = 1 - \prod_{i=1}^{M} (1 - p_i)$$

$$= 1 - \prod_{i=1}^{M} \left(1 - e^{-\int_0^T h_i(x)\,dx} \right). \tag{15.3.5}$$

Systems do not come as purely series or parallel structures. Most often they are mixtures of both. In determining the reliability of such systems, a combination of the principles employed in (15.3.2) and (15.3.5) may be necessary. For instance, consider a system with the structure given in Figure 15.3.3. Subsystem I is a parallel structure with reliability

$$R_{\mathrm{I}} = 1 - (1 - p_1)(1 - p_2) \tag{15.3.6}$$

Subsystem II is a parallel structure of two series structures. Hence its reliability is given as

$$R_{\mathrm{II}} = 1 - (1 - p_3 p_4)(1 - p_5 p_6) \tag{15.3.7}$$

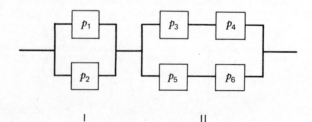

I   II

Figure 15.3.3

The entire system is a series structure composed of subsystems I and II. Hence the system reliability can be given as

$$R = R_I R_{II}$$

$$= \left[1 - (1 - p_1)(1 - p_2)\right]\left[1 - (1 - p_3 p_4)(1 - p_5 p_6)\right] \quad (15.3.8)$$

System structures can get much more complex due to such factors as dependent components and multiple failure modes. Even though the basic concept for the determination of system reliability remains the same, more time-saving methods are needed for the efficient computation of reliability. Event space methods identify events favorable to the success of operation and help compute the probabilities of these events. Path-tracing methods simplify the event space methods by tracing successful paths through subsystems. When the system contains only independent components cut-set and tie-set methods treat the system as a graph and use graph theoretic concepts and algorithms in the computation of reliability. Since our interest is primarily the underlying stochastic processes, we shall not get into their discussion here. Interested readers are referred to books such as Rau (1970) and Shooman (1968) where these and other methods are discussed in detail.

15.4 EXPONENTIAL FAILURE DISTRIBUTIONS: GENERAL MODEL

In Section 15.2 we discussed replacement problems for components with general failure distributions $F(t)$. On-line service was not considered on any of the failed components because that would have introduced further complexity. But if we specialize $F(t)$ to an exponential distribution, we can compensate the loss in generality by being able to consider several additional factors such as on-line service and redundancy. Furthermore, if we restrict all our distributions to exponentials, the resulting models are Markov processes with fairly straightforward analyses. We shall not comment here on the suitability of exponential distributions in these problems. We shall assume that conditions of the problem are such that these assumptions are valid.

A general Markov model for an M-component system can be given as follows: Let a system be composed of M components, interacting in such a way that, given the number of active components, the failure distribution for any component is exponential with a parameter depending only on that number. Let the on-line service have the same property. In particular, when the number of failed components is n, let λ_n be the failure rate for any component, and let μ_n be the rate of service. We naturally assume that the probability of occurrence of more than one event such as failure or service completion has a negligible probability. We shall also assume that at least k ($\leq M$) components should be in working order for the entire system to be operative. When $M - k + 1$ components fail, the system stops its operation until one of the components is repaired.

Let the number of failed components signify the states of the system. Clearly the state $M - k + 1$ represents the inoperative state of the system. Let $P_n(t)$ be the probability that the state of the system at time t is n ($0 \leq n \leq M - k + 1$). Because of the Markov property of the underlying distributions and processes, the generalized birth and death processes of Section 7.6 can be used as a mathematical model in this case. The forward Kolmogorov equations governing such a model have the form

$$P_0'(t) = -\lambda_0 P_0(t) + \mu_1 P_1(t) \qquad (15.4.1)$$

$$P_n'(t) = -(\lambda_n + \mu_n)P_n(t) + \lambda_{n-1}P_{n-1}(t)$$

$$+ \mu_{n+1}P_{n+1}(t) \qquad 0 < n \leq M - k \qquad (15.4.2)$$

$$P_{M-k+1}'(t) = -\mu_{M-k+1}P_{M-k+1}(t) + \lambda_{M-k}P_{M-k}(t) \qquad (15.4.3)$$

with the initial condition $P_0(0) = 1$. For specific forms of λ_n and μ_n these difference differential equations can be solved to give $P_n(t)$ for $n = 0, 1, 2, \ldots, M - k + 1$ and $0 < t < \infty$. If it is convenient, transform methods can be used in the solution of these equations. The expected length of time during which the system is inoperative in an interval $(0, T]$ is given by

$$\int_0^T P_{M-k+1}(t)\, dt \qquad (15.4.4)$$

As $t \to \infty$, the limiting distribution of the state of the system can be obtained from the steady-state equations: Let $p_n = \lim_{t \to \infty} P_n(t)$. Then

$$\mu_1 p_1 = \lambda_0 p_0$$

$$(\lambda_n + \mu_n)p_n = \lambda_{n-1}p_{n-1} + \mu_{n+1}p_{n+1} \qquad 0 < n \leq M - k$$

$$(15.4.5)$$

$$\mu_{M-k+1}p_{M-k+1} = \lambda_{M-k}p_{M-k}$$

$$\sum_{n=0}^{M-k+1} p_n = 1 \qquad (15.4.6)$$

Solving these recursively, or using results from (7.8.14) and (7.8.15), we get

$$p_n = \frac{\lambda_0 \lambda_1 \cdots \lambda_{n-1}}{\mu_1 \mu_2 \cdots \mu_n} p_0 \qquad 0 < n \le M - k + 1 \qquad (15.4.7)$$

$$p_0 = \left[1 + \frac{\lambda_0}{\mu_1} + \frac{\lambda_0 \lambda_1}{\mu_1 \mu_2} + \cdots + \frac{\lambda_0 \lambda_1 \cdots \lambda_{M-k}}{\mu_1 \mu_2 \cdots \mu_{M-k+1}} \right]^{-1} \qquad (15.4.8)$$

The fraction of time the system would be inoperative (system down time) is given by p_{M-k+1}, which is

$$p_{M-k+1} = \left[1 + \frac{\mu_{M-k+1}}{\lambda_{M-k}} + \frac{\mu_{M-k+1} \mu_{M-k}}{\lambda_{M-k} \lambda_{M-k-1}} + \cdots + \frac{\mu_{M-k+1} \mu_{M-k} \cdots \mu_1}{\lambda_{M-k} \lambda_{M-k-1} \cdots \lambda_0} \right]^{-1}$$

$$(15.4.9)$$

For the reliability and expected lifetime of the system we need the first passage probabilities of the underlying Markov process. This is obtained by considering state $M - k + 1$ as an absorbing state and deriving the proba-. bilities $_*P_n(t)$ which satisfy the following forward Kolmogorov equations:

$$_*P_0'(t) = -\lambda_0 {}_*P_0(t) + \mu_1 {}_*P_1(t)$$

$$_*P_n'(t) = -(\lambda_n + \mu_n) {}_*P_n(t) + \lambda_{n-1} {}_*P_{n-1}(t)$$

$$+ \mu_{n+1} {}_*P_{n+1}(t) \qquad 0 < n < M - k$$

$$_*P_{M-k}'(t) = -(\lambda_{M-k} + \mu_{M-k}) {}_*P_{M-k}(t)$$

$$+ \lambda_{M-k-1} {}_*P_{M-k-1}(t)$$

$$_*P_{M-k+1}'(t) = \lambda_{M-k} {}_*P_{M-k}(t) \qquad (15.4.10)$$

Note that the last two equations of (15.4.10) indicate that $M - k + 1$ is an absorbing state. The probability $_*P_n(t)$ ($n = 0, 1, 2, \ldots, M - k$), derived as a solution to (15.4.10), therefore gives the probability that the number of failed components is n at time t, with t being such that the system has been in continuous operation during $(0, t]$. Clearly the system reliability $R(t)$ can be given as

$$R(t) = \sum_{n=0}^{M-k} {}_*P_n(t) \qquad (15.4.11)$$

Using the relation

$$F(t) = 1 - R(t)$$

for the system failure distribution, the mean time to system failure $E(L)$ can be written

$$E(L) = \int_0^\infty [1 - F(t)] \, dt$$

$$= \int_0^\infty R(t) \, dt \tag{15.4.12}$$

If we define the Laplace transform

$${}_*\phi_n(\theta) = \int_0^\infty e^{-\theta t} {}_*P_n(t) \, dt \qquad \mathrm{Re}(\theta) > 0$$

we may write (15.4.12) as

$$E(L) = \sum_{n=0}^{M-k} \lim_{\theta \to 0} {}_*\phi_n(\theta) \tag{15.4.13}$$

When transform methods are used in the solution of (15.4.10), the last expression for $E(L)$ may prove to be convenient.

Suppose the M components belong to c distinct classes, and components belonging to the first class are essential for the system to be operative. The Markov model described before can be modified to include this general case by considering n as a vector:

$$n = (n_1, n_2, \ldots, n_c)$$

where n_i $(i = 1, 2, \ldots, c)$ is the number of failed components in the ith class. The states $\{M - k + 1, \ M - k + 2, \ldots, M\}$ will now include all components belonging to the first class that are essential for system operation. The modification given here is also relevant to situations where standby redundant components are maintained.

We shall close this section by discussing a Markov chain model for the M-component system when time can be considered a discrete parameter. As before, let the number of failed components signify the state of the system. Suppose the failure distributions of the components are such that the transitions of states between any two consecutive time points have the stationary transition probability matrix

$$
P = \begin{bmatrix}
P_{M-k+1,M-k+1} & P_{M-k+1,M-k} & \cdots & P_{M-k+1,0} \\
P_{M-k,M-k+1} & P_{M-k,M-k} & \cdots & P_{M-k,0} \\
\vdots & \vdots & & \vdots \\
P_{0,M-k+1} & P_{0,M-k} & \cdots & P_{00}
\end{bmatrix}
$$

$$(15.4.14)$$

where we have assumed that k components are essential for the system to be operative. The fraction of time the system will be inoperative in the long run is given by

$$
\pi_{M-k+1} \tag{15.4.15}
$$

where $\{\pi_n, \; n = 0, 1, 2, \ldots, M - k + 1\}$ is the limiting distribution of the Markov chain.

Following the methods described in Chapter 4, the failure distribution for the system can be derived by considering state $M - k + 1$ as an absorbing state and the states $\{0, 1, \ldots, M - k\}$ as belonging to a transient class. Let

$$
Q = \begin{bmatrix}
P_{M-k,M-k} & P_{M-k,M-k-1} & \cdots & P_{M-k,0} \\
\vdots & \vdots & & \vdots \\
P_{0,M-k} & P_{0,M-k-1} & \cdots & P_{00}
\end{bmatrix} \tag{15.4.16}
$$

which is a substochastic matrix whose elements give the probabilities of transitions of the Markov chain within the transient class. Let $Q_{ij}^{(n)}$ ($i, j = 0, 1, 2, \ldots, M - k$) be the elements of the nth power of the matrix Q. Let the operation of the system start with all the components working, and let L be the random variable representing the number of transitions that elapse before the first system failure ($M - k$ components fail). It is easy to see that

$$
P(L > n) = Q_{00}^{(n)} + Q_{01}^{(n)} + \cdots + Q_{0,M-k}^{(n)} \tag{15.4.17}
$$

But

$$
P(L = N) = P(L > n - 1) - P(L > n)
$$

$$
= Q_{00}^{(n-1)} + Q_{01}^{(n-1)} + \cdots + Q_{0,M-k}^{(n-1)}
$$

$$
- \left[Q_{00}^{(n)} + Q_{01}^{(n)} + \cdots + Q_{0,M-k}^{(n)} \right] \tag{15.4.18}
$$

The expected value of L, $E(L)$, can be directly obtained from the fundamental matrix

$$(I - Q)^{-1} \qquad (15.4.19)$$

Let μ_{ij} $(i, j = 0, 1, \ldots, M - k)$ be the elements of the fundamental matrix $(I - Q)^{-1}$ (see Chapter 4). Then

$$E(L) = \mu_{00} + \mu_{01} + \cdots + \mu_{0, M-k} \qquad (15.4.20)$$

Information thus derived is comparable to that available in the continuous parameter case.

In Sections 15.5 and 15.6 we shall discuss some Markov process models with specific values for the parameters λ_n and μ_n. We shall distinguish between two cases: (1) nonmaintained systems ($\mu_n = 0$, $n \geq 0$) and (2) maintained systems ($\mu_n > 0$ for at least some n).

15.5 EXPONENTIAL FAILURE DISTRIBUTIONS: NONMAINTAINED SYSTEMS

Service parameter μ_n ($n \geq 0$) of the general model discussed in Section 15.4 will be assumed to have the value zero in this section. Three specific structures for λ_n will be considered.

Systems with Redundant Components

To ensure efficient operation of a system, it is common to have more than one of each component available so that failed components can be replaced. Suppose a component has an exponential failure distribution with mean $1/\lambda$. If there are N identical standby components, clearly the time until the system failure (i.e., until $N + 1$ components fail in succession, assuming that replacements are instantaneous) has a gamma distribution with mean $(N + 1)/\lambda$. But in many cases even standby equipment may be put to some use and thus have a failure rate when it is not used on line. To illustrate this problem, consider a single equipment system with one redundant component as a standby. Let λ_1 and λ_2 ($\lambda_1 > \lambda_2$) be the failure rates of these components. Let $P_n(t)$ ($n = 0, 1, 2$) be the probability that the number of failed components in time $(0, t]$ is n. As there is no repair of the failed items, the reliability function $R(t)$ is given by

$$R(t) = P_0(t) + P_1(t) \qquad (15.5.1)$$

The number of failed components is a three-state Markov process whose forward Kolmogorov equations can be obtained as follows:

$$P_0(t + \Delta t) = P_0(t)\left[1 - (\lambda_1 + \lambda_2)\Delta t + o(\Delta t)\right]$$

$$P_1(t + \Delta t) = P_1(t)\left[1 - \lambda_1\Delta t + o(\Delta t)\right]$$

$$+ P_0(t)\left[\lambda_1\Delta t + \lambda_2\Delta t + o(\Delta t)\right]$$

$$P_2(t + \Delta t) = P_2(t) + P_1(t)\lambda_1\Delta t + o(\Delta t) \qquad (15.5.2)$$

These give

$$P_0'(t) = -(\lambda_1 + \lambda_2)P_0(t)$$

$$P_1'(t) = -\lambda_1 P_1(t) + (\lambda_1 + \lambda_2)P_0(t)$$

$$P_2'(t) = \lambda_1 P_1(t) \qquad (15.5.3)$$

with $P_0(0) = 1$. Proceeding as in Section 7.2, these equations can be directly solved to give

$$P_0(t) = e^{-(\lambda_1 + \lambda_2)t}$$

$$P_1(t) = \frac{\lambda_1 + \lambda_2}{\lambda_2}\left[e^{-\lambda_1 t} - e^{-(\lambda_1 + \lambda_2)t}\right]$$

$$P_2(t) = 1 + \frac{\lambda_1}{\lambda_2}e^{-(\lambda_1 + \lambda_2)t} - \frac{\lambda_1 + \lambda_2}{\lambda_2}e^{-\lambda_1 t} \qquad (15.5.4)$$

Another way of solving these and similar equations is through Laplace transforms (see Appendix B). This method will be illustrated later in Section 15.6. In the present case, however, we may use direct probabilistic arguments to derive the results (15.5.4).

When the failure rate is constant, the life distribution of the component has an exponential distribution. Therefore the probability that a component with hazard rate λ_i ($i = 1, 2$) has not failed during $(0, t]$ is given by $e^{-\lambda_i t}$. Consequently we have

$$P_0(t) = \left(e^{-\lambda_1 t}\right)\left(e^{-\lambda_2 t}\right) = e^{-(\lambda_1 + \lambda_2)t} \qquad (15.5.5)$$

When one component has failed, it could be the standby component or the one in operation. If the latter, the standby component takes the place of the failed component immediately. Thus we get

$$P_1(t) = (e^{-\lambda_1 t})(1 - e^{-\lambda_2 t})$$

$$+ \int_{\tau=0}^{t} \lambda_1 e^{-\lambda_1 \tau} e^{-\lambda_2 \tau} e^{-\lambda_1(t-\tau)} \, d\tau$$

$$= \frac{\lambda_1 + \lambda_2}{\lambda_2} [e^{-\lambda_1 t} - e^{-(\lambda_1 + \lambda_2)t}] \tag{15.5.6}$$

Finally, $P_2(t) = 1 - P_0(t) - P_1(t)$ which on simplification gives the last equation in (15.5.4).

For the reliability $R(t)$ of the system we have

$$R(t) = P_0(t) + P_1(t)$$

$$= \frac{\lambda_1 + \lambda_2}{\lambda_2} e^{-\lambda_1 t} - \frac{\lambda_1}{\lambda_2} e^{-(\lambda_1 + \lambda_2)t} \tag{15.5.7}$$

The expected lifetime of the system, $E(L)$, can now be obtained [using (15.4.12)] as

$$E(L) = \int_0^\infty R(t) \, dt = \frac{\lambda_1 + \lambda_2}{\lambda_2} \int_0^\infty e^{-\lambda_1 t} \, dt - \frac{\lambda_1}{\lambda_2} \int_0^\infty e^{-(\lambda_1 + \lambda_2)t} \, dt$$

$$= \frac{\lambda_1 + \lambda_2}{\lambda_1 \lambda_2} - \frac{\lambda_1}{\lambda_2(\lambda_1 + \lambda_2)}$$

$$= \frac{1}{\lambda_1} + \frac{1}{\lambda_1 + \lambda_2} \tag{15.5.8}$$

When $\lambda_1 = \lambda_2 = \lambda$, we get

$$E(L) = \tfrac{3}{2}\lambda \tag{15.5.9}$$

A Four-Engine Aircraft

An aircraft has four engines, each of which has a failure rate λ. For a successful flight at least two engines should be operating (Dick, 1963). Let $P_n(t)$ be the probability that at time t there are n failed engines. The forward Kolmogorov equations for the process can be given as

$$P_0'(t) = -4\lambda P_0(t)$$

$$P_1'(t) = -3\lambda P_1(t) + 4\lambda P_0(t)$$

$$P_2'(t) = -2\lambda P_2(t) + 3\lambda P_1(t) \tag{15.5.10}$$

These equations can be solved using Laplace transforms, as illustrated in the next section. In the present case it is simpler to use the method outlined in Section 15.3.

When the failure rate for an engine is a constant λ, its reliability during $(0, t]$ is given by $e^{-\lambda t}$. Using the formula (15.3.5) with $p = e^{-\lambda t}$, we have

$$R(t) = \sum_{r=2}^{4} \binom{4}{r} (e^{-\lambda t})^r (1 - e^{-\lambda t})^{4-r}$$

$$= e^{-4\lambda t} + 4e^{-3\lambda t}(1 - e^{-\lambda t})$$

$$+ 6e^{-2\lambda t}(1 - e^{-\lambda t})^2 \tag{15.5.11}$$

The three terms on the right-hand side of (15.5.11) are, respectively, $P_0(t)$, $P_1(t)$, and $P_2(t)$ of equations (15.5.10). Simplifying (15.5.11), one gets

$$R(t) = 3e^{-4\lambda t} - 8e^{-3\lambda t} + 6e^{-2\lambda t} \tag{15.5.12}$$

The expected lifetime $E(L)$ of the aircraft can be easily obtained as

$$E(L) = \int_0^\infty R(t)\, dt = \frac{3}{4\lambda} - \frac{8}{3\lambda} + \frac{6}{2\lambda} = \frac{13}{12\lambda} \tag{15.5.13}$$

Suppose the aircraft is such that it needs at least one engine on either side for a successful flight. To accommodate this restriction, we may consider the aircraft as a series system of two parallel component subsystems (e.g., as in Figure 15.3.3 but with only one component in each branch of subsystem II and $p_i = p = e^{-\lambda t}$ for all i). The reliability $R(t)$ of such a system can be given as

$$R(t) = \left[1 - (1 - e^{-\lambda t})^2\right]^2$$

$$= [2e^{-\lambda t} - e^{-2\lambda t}]^2$$

$$= e^{-4\lambda t} - 4e^{-3\lambda t} + 4e^{-2\lambda t} \tag{15.5.14}$$

and

$$E(L) = \frac{11}{12\lambda}.$$

An M-Component System

Consider M components of a system, each having a failure rate λ and failing independently of each other. Suppose, for the successful operation of the system, at least k of the M components should be operative. Let $R_i(t)$ be the reliability of the ith component; because of the independence of the failure distributions of components, the reliability $R_i(t)$ is given by

$$R_i(t) = e^{-\lambda t} \qquad i = 1, 2, \ldots, M \tag{15.5.15}$$

Proceeding as in (15.3.5), we get the system reliability

$$
\begin{aligned}
R(t) &= \sum_{r=k}^{M} \binom{M}{r} [e^{-\lambda t}]^r [1 - e^{-\lambda t}]^{M-r} \\
&= \sum_{r=k}^{M} \binom{M}{r} e^{-r\lambda t} [1 - e^{-\lambda t}]^{M-r} \\
&= \sum_{r=k}^{M} \sum_{l=0}^{M-r} \frac{M!}{r! l! (M - r - l)!} (-1)^l e^{-(r+l)\lambda t} \tag{15.5.16}
\end{aligned}
$$

and

$$E(L) = \sum_{r=k}^{M} \sum_{l=0}^{M-r} \frac{M!}{r! l! (M - r - l)!} \frac{(-1)^l}{(r + l)\lambda} \tag{15.5.17}$$

A simpler expression for $E(L)$ can be obtained by considering the system as a pure death process with an initial population size of M. As the components fail, total death rate decreases from $M\lambda$ to $(M - 1)\lambda$ to $(M - 2)\lambda, \ldots,$ to $k\lambda$. The period during which the total death rate is $r\lambda$ $(r = M, M - 1, \ldots, k)$ has an exponential distribution with mean $1/r\lambda$. Adding such periods, we get

$$E(L) = \sum_{r=k}^{M} \frac{1}{r\lambda} \tag{15.5.18}$$

15.6 EXPONENTIAL FAILURE DISTRIBUTIONS: MAINTAINED SYSTEMS

In this section we shall discuss some systems in which maintenance and repair service are available for components.

Single Equipment with No Redundancy

Consider equipment that has an exponential failure distribution with parameter λ. When it fails, the service time needed to bring it back to working order has an exponential distribution with parameter μ. Let 1 be the state when the equipment is being repaired, and let 0 be the state when it is working. Clearly the necessary mathematical model is a two-state Markov process with states 0 and 1, and all the results derived in Section 7.5 are applicable. For instance, depending on the initial state of the equipment, its availability at time t is given by

$$P_{00}(t) = \frac{\mu}{\lambda + \mu} + \frac{\lambda}{\lambda + \mu} e^{-(\lambda + \mu)t}$$

$$P_{01}(t) = \frac{\lambda}{\lambda + \mu} - \frac{\lambda}{\lambda + \mu} e^{-(\lambda + \mu)t}$$

$$P_{10}(t) = \frac{\mu}{\lambda + \mu} - \frac{\mu}{\lambda + \mu} e^{-(\lambda + \mu)t}$$

$$P_{11}(t) = \frac{\lambda}{\lambda + \mu} + \frac{\mu}{\lambda + \mu} e^{-(\lambda + \mu)t} \tag{15.6.1}$$

As $t \to \infty$, the equipment can be found to be in states 0 and 1 with probabilities

$$p_0 = \frac{\mu}{\lambda + \mu} \quad \text{and} \quad p_1 = \frac{\lambda}{\lambda + \mu} \tag{15.6.2}$$

respectively. These also give the expected fraction of time the system (single equipment) is inoperative and operative in the long run. But for finite t the expected length of time the system will be inoperative depends on the initial state of the system and is given by

$$\mu_{11}(t) = \frac{\lambda t}{\lambda + \mu} + \frac{\mu}{(\lambda + \mu)^2} [1 - e^{-(\lambda + \mu)t}]$$

$$\mu_{01}(t) = \frac{\lambda t}{\lambda + \mu} - \frac{\lambda}{(\lambda + \mu)^2} [1 - e^{-(\lambda + \mu)t}] \tag{15.6.3}$$

The distribution of the length of time the system will be inoperative has been given by Barlow and Hunter (1961) by specializing a result due to Takács (1957). Let $D(t)$ be the total length of time during $(0, t]$ that the system has been inoperative. Then

$$P[D(t) \leq x] = e^{-\lambda(t-x)}\left[1 + \sqrt{\lambda\mu(t-x)}\right.$$

$$\left. \times \int_0^x e^{-\mu y} y^{-1/2} I_1\left[2\sqrt{\lambda\mu(t-x)y}\right] dy\right] \quad (15.6.4)$$

where $I_1(x)$ is the Bessel function of order 1 for the imaginary argument, defined by

$$I_1(x) = \sum_{j=0}^{\infty} \frac{[(1/2)x]^{2j+1}}{j!(j+1)!} \quad (15.6.5)$$

Standard tables of Bessel functions are available. See, for example, *Handbook of Mathematical Functions with Formulas, Graphs and Tables* (Eds: M. Abramowitz and I. A. Stegun), Applied Mathematics Series 55, National Bureau of Standards, U.S. Department of Commerce, 1964.

For larger values of t the asymptotic distribution of $D(t)$ will be more useful. Specializing Takács' (1957) results, we have the random variable

$$\frac{D(t) - \lambda t/(\lambda + \mu)}{\sqrt{2\lambda\mu t/(\lambda + \mu)^3}} \quad (15.6.6)$$

distributed asymptotically as a standard normal variate with mean 0 and variance 1.

Single Equipment with Redundancy

Consider equipment that has an exponential failure distribution with parameter λ and standby redundant equipment with the same failure distribution. When the on-line equipment fails, the service time needed to bring it back to working order has an exponential distribution with parameter μ. We shall assume instantaneous replacement for the failed unit if a redundant unit is available. The system is said to fail when both items are out of order. Let $P_n(t)$ be the probability that the number of failed items at time t is n. Clearly the reliability function $R(t)$ is given by

$$R(t) = P_0(t) + P_1(t) \quad (15.6.7)$$

Under these conditions the number of failed items at time t is a Markov process determined by the forward Kolmogorov equations:

$$P_0'(t) = -\lambda P_0(t) + \mu P_1(t)$$

$$P_1'(t) = -(\lambda + \mu)P_1(t) + \lambda P_0(t)$$

$$P_2'(t) = \lambda P_1(t) \tag{15.6.8}$$

The probabilities $P_n(t)$ can be determined by solving these equations using transforms. Let

$$\phi_n(\theta) = \int_0^\infty e^{-\theta t} P_n(t)\, dt \qquad \text{Re}(\theta) > 0$$

We have

$$-1 + \theta\phi_0(\theta) = \int_0^\infty e^{-\theta t} P_0'(t)\, dt \tag{15.6.9}$$

$$\theta\phi_n(\theta) = \int_0^\infty e^{-\theta t} P_n'(t)\, dt \qquad n = 1, 2 \tag{15.6.10}$$

Using (15.6.9) and (15.6.10) in (15.6.8), we get

$$\theta\phi_0(\theta) = 1 - \lambda\phi_0(\theta) + \mu\phi_1(\theta)$$

$$\theta\phi_1(\theta) = -(\lambda + \mu)\phi_1(\theta) + \lambda\phi_0(\theta)$$

$$\theta\phi_2(\theta) = \lambda\phi_1(\theta) \tag{15.6.11}$$

Solving (15.6.11), we get

$$\phi_0(\theta) = \frac{\theta + \lambda + \mu}{\theta^2 + \theta(2\lambda + \mu) + \lambda^2} \tag{15.6.12}$$

$$\phi_1(\theta) = \frac{\lambda}{\theta^2 + \theta(2\lambda + \mu) + \lambda^2} \tag{15.6.13}$$

$$\phi_2(\theta) = \frac{\lambda}{\theta} \cdot \frac{\lambda}{\theta^2 + \theta(2\lambda + \mu) + \lambda^2} \tag{15.6.14}$$

Let ξ_1 and ξ_2 be the roots of the equation

$$\theta^2 + \theta(2\lambda + \mu) + \lambda^2 = 0 \tag{15.6.15}$$

such that

$$\xi_1 = \frac{-(2\lambda + \mu) + \sqrt{4\lambda\mu + \mu^2}}{2}$$

$$\xi_2 = \frac{-(2\lambda + \mu) - \sqrt{4\lambda\mu + \mu^2}}{2} \tag{15.6.16}$$

and

$$\xi_1 + \xi_2 = -(2\lambda + \mu)$$

$$\xi_1\xi_2 = \lambda^2 \tag{15.6.17}$$

Thus, expressing the denominator of (15.6.12) as $(\theta - \xi_1)(\theta - \xi_2)$, we get

$$\phi_0(\theta) = \frac{\theta + \lambda + \mu}{(\theta - \xi_1)(\theta - \xi_2)} \tag{15.6.18}$$

Expressing the right-hand side of (15.6.18) in partial fractions (see Appendix B),

$$\phi_0(\theta) = \frac{\lambda + \mu + \xi_1}{\xi_1 - \xi_2} \cdot \frac{1}{\theta - \xi_1} - \frac{\lambda + \mu + \xi_2}{\xi_1 - \xi_2} \cdot \frac{1}{\theta - \xi_2} \tag{15.6.19}$$

Inverting $1/(\theta - \xi_1)$ and $1/(\theta - \xi_2)$, we get

$$P_0(t) = \frac{(\lambda + \mu + \xi_1)e^{\xi_1 t} - (\lambda + \mu + \xi_2)e^{\xi_2 t}}{\xi_1 - \xi_2} \tag{15.6.20}$$

Proceeding in a similar manner, from (15.5.13) we get

$$\phi_1(\theta) = \frac{\lambda}{\xi_1 - \xi_2}\left[\frac{1}{\theta - \xi_1} - \frac{1}{\theta - \xi_2}\right]$$

giving

$$P_1(t) = \frac{\lambda}{\xi_1 - \xi_2}[e^{\xi_1 t} - e^{\xi_2 t}] \tag{15.6.21}$$

We therefore have the reliability

$$R(t) = P_0(t) + P_1(t)$$

$$= \frac{\xi_1 e^{\xi_2 t} - \xi_2 e^{\xi_1 t}}{\xi_1 - \xi_2} \tag{15.6.22}$$

The expected life of the system $E(L)$ can be easily obtained by noting that

$$E(L) = \int_0^\infty R(t)\, dt = \lim_{\theta \to 0}\left[\phi_0(\theta) + \phi_1(\theta)\right]$$

$$= \frac{\lambda + \mu}{\lambda^2} + \frac{\lambda}{\lambda^2} = \frac{2\lambda + \mu}{\lambda^2} \tag{15.6.23}$$

The probability density of the lifetime is given by

$$F'(t) = \frac{d}{dt}\left[1 - R(t)\right] = P_2'(t)$$

$$= \frac{\lambda^2}{\xi_1 - \xi_2}\left[e^{\xi_1 t} - e^{\xi_2 t}\right] \tag{15.6.24}$$

As in the case of a second car, in some situations even the redundant unit is liable to fail, though with a lesser rate. To account for this failure, let λ_1 and λ_2 be the failure rates of the on-line and off-line equipment, respectively, with $\lambda_1 \geq \lambda_2$. Then the forward Kolmogorov equations corresponding to (15.6.8) are

$$P_0'(t) = -(\lambda_1 + \lambda_2)P_0(t) + \mu P_1(t)$$

$$P_1'(t) = -(\lambda_1 + \mu)P_1(t) + (\lambda_1 + \lambda_2)P_0(t)$$

$$P_2'(t) = \lambda_1 P_1(t) \tag{15.6.25}$$

These equations can be solved as before to yield

$$P_0(t) = \frac{(\lambda_1 + \mu + \xi_1)e^{\xi_1 t} - (\lambda_1 + \mu + \xi_2)e^{\xi_2 t}}{\xi_1 - \xi_2} \tag{15.6.26}$$

$$P_1(t) = \frac{\lambda_1 + \lambda_2}{\xi_1 - \xi_2}\left[e^{\xi_1 t} - e^{\xi_2 t}\right] \tag{15.6.27}$$

and

$$R(t) = \frac{\xi_1 e^{\xi_2 t} - \xi_2 e^{\xi_1 t}}{\xi_1 - \xi_2} \qquad (15.6.28)$$

where

$$\xi_1 = \frac{-(2\lambda_1 + \lambda_2 + \mu) + \sqrt{\lambda_2^2 + \mu^2 - 2\lambda_2\mu - 4\lambda_1\mu}}{2} \qquad (15.6.29)$$

$$\xi_2 = \frac{-(2\lambda_1 + \lambda_2 + \mu) - \sqrt{\lambda_2^2 + \mu^2 - 2\lambda_2\mu - 4\lambda_1\mu}}{2} \qquad (15.6.30)$$

Consequently the expected lifetime of the system will be obtained as

$$E(L) = \frac{2\lambda_1 + \lambda_2 + \mu}{\lambda_1(\lambda_1 + \lambda_2)} \qquad (15.6.31)$$

If units are repaired as soon as they fail with rate μ, the long-term availability of the system is given by the limiting distribution of the states of the system. Let $p_n = \lim_{t \to \infty} P_n(t)$, $n = 0, 1, 2$. The steady-state equations are

$$(\lambda_1 + \lambda_2)p_0 = \mu p_1$$

$$(\lambda_1 + \mu)p_1 = (\lambda_1 + \lambda_2)p_0 + 2\mu p_2$$

$$2\mu p_2 = \lambda_1 p_1 \qquad (15.6.32)$$

Solving these with the help of the additional condition

$$\sum_0^2 p_n = 1$$

we get

$$p_0 = \frac{2\mu^2}{2\mu^2 + (2\mu + \lambda_1)(\lambda_1 + \lambda_2)} \qquad (15.6.33)$$

$$p_1 = \frac{2\mu(\lambda_1 + \lambda_2)}{2\mu^2 + (2\mu + \lambda_1)(\lambda_1 + \lambda_2)} \qquad (15.6.34)$$

$$p_2 = \frac{\lambda_1(\lambda_1 + \lambda_2)}{2\mu^2 + (2\mu + \lambda_1)(\lambda_1 + \lambda_2)} \qquad (15.6.35)$$

Incidentally it may be noted that p_2 of (15.6.35) gives the long-run expected fraction of time the system will be inoperative (downtime).

A Phased-Array Radar System*

Modern phased-array radars are usually complex systems containing many hundreds of radiating and receiving elements. The availability of such a system is related to a minimum number of operable elements. Thus with every system a discrete threshold can be set. Let M be the total number of elements and k be the least number of elements needed for the system to be operative. (At most $M - k$ can be out.) Suppose the elements have an exponential life distribution with a failure rate λ. Let s (≥ 1) simultaneous repairs be possible at any time, and let the repair time on each element have an exponential distribution with mean $1/\mu$. Let $N(t)$ be the number of elements that are inoperative at time t, and let $P_n(t)$ be the probability that $N(t)$ is equal to n at time t.

Clearly this is a special case of the general model described in Section 15.4, and all the results follow by specializing that model:

$$\lambda_n = (M - n)\lambda \qquad 0 \leq n \leq M$$

$$\mu_n = n\mu \qquad 1 \leq n \leq s$$

$$\mu_n = s\mu \qquad n \geq s \qquad (15.6.36)$$

It should also be pointed out that this problem is identical to the machine interference problem discussed in Section 11.7. Proceeding as in that section for

$$p_n = \lim_{t \to \infty} P_n(t)$$

we get

$$p_n = \binom{M}{n}\rho^n p_0 \qquad 0 \leq n \leq s$$

$$= \binom{M}{n}\frac{n!}{s!s^{n-s}}\rho^n p_0 \qquad s \leq n \leq M - k + 1 \qquad (15.6.37)$$

where we have assumed that when there are only $k - 1$ components working, the operation of the system is stopped until one of the $M - k + 1$

*See Hevesh and Harrahy (1966).

components is repaired. Also we have written $\rho = \lambda/\mu$ as the utilization factor. The unknown p_0 in (15.6.37) is obtained by the usual assumption

$$\sum_{n=0}^{M-k+1} p_n = 1 \tag{15.6.38}$$

In particular, when $s = 1$, we get

$$p_0 = \left[1 + \frac{M!}{(M-1)!}\rho + \frac{M!}{(M-2)!}\rho^2 + \cdots + \frac{M}{(k-1)!}\rho^{M-k+1} \right]^{-1} \tag{15.6.39}$$

$$p_n = \frac{M!}{(M-n)!}\rho^n p_0 \qquad 1 \le n \le M - k + 1 \tag{15.6.40}$$

In the long run the fraction of system downtime will be given by p_{M-k+1}.

To obtain the system reliability and the expected lifetime, we have to analyze the transient behavior based on the appropriate forward Kolmogorov equations. Clearly this is beyond the scope of our discussion.

More General Models

More general forms of these and other Markov models can be obtained in reliability theory by simply relaxing some basic assumptions. Every system is characterized by the following four major characteristics:

1. Number of components.
2. System configuration.
3. Failure distributions.
4. Repair distributions.

Numerous interesting and useful models can be structured with these basic features under varying conditions. Only in some simple cases, as illustrated in this chapter, do such models remain Markovian. When models become non-Markovian, either more general solution techniques must be employed or the models must be made Markovian by the use of supplementary variables. Even then the solution techniques required are complex.

Numerous papers investigating the properties of complex systems have appeared in *IEEE Transactions on Reliability*, *Proceedings of National Symposia of Reliability and Quality Control*, *Naval Research Logistics*

Quarterly, *Operations Research*, and *Opsearch*. A sampling of such papers is included under the heading "For Further Reading." A wide coverage of stochastic processes occurring in reliability problems can also be found in books such as Barlow and Proschan (1965, 1975), Gnedenko (1968), Lawless (1982), Mann et al. (1974), Martz and Waller (1982), Nelson (1982), Rau (1970), Sandler (1963), Shooman (1968), Trivedi (1982), Tsokos and Shimi (1977), and Zelen (1963).

REFERENCES

Barlow, R. E., and Hunter, L. C. (1961). "Reliability Analysis of a One Unit System." *Oper. Res.* **9**(2), 200–208.

Barlow, R. E., and Proschan, F. (1965). *Mathematical Theory of Reliability*. New York: Wiley.

Barlow, R. E., and Proschan, F. (1975). *Statistical Theory of Reliability and Life Testing*. New York: Holt, Rinehart and Winston.

Dick, R. S. (1963). "The Reliability of Repairable Complex Systems." Part B: The Dissimilar Machine Case *IEEE Trans. Reliability* **R-12**(1), 1–8.

Gnedenko, B. V. (1968). *Mathematical Methods in Reliability Theory*. New York: Academic Press (English trans.; Russian ed., 1965).

Hevesh, A. H., and Harrahy, D. J. (1966). "Effects of Failure on Phased-Array Radar Systems." *IEEE Trans. Reliability* **R-15**(1), 22–31.

Lawless, J. F. (1982). *Statistical Models and Methods for Lifetime Data*. New York: Wiley.

Mann, N. R., Schafer, R. E., and Singpurwalla, N. D. (1974). *Methods for Statistical Analysis of Reliability and Life Data*. New York: Wiley.

Martz, H. F., and Waller, R. A. (1982). *Bayesian Reliability Analyses*. New York: Wiley.

Nelson, W. (1982). *Applied Life Data Analysis*. New York: Wiley.

Rau, J. G. (1970). *Optimization and Probability in Systems Engineering*. New York: Van Nostrand.

Sandler, G. H. (1963). *System Reliability Engineering*. Englewood Cliffs, N.J.: Prentice-Hall.

Shooman, M. L. (1968). *Probabilistic Reliability, an Engineering Approach*. New York: McGraw-Hill.

Takács, L. (1957). "On Certain Sojourn Time Problems in the Theory of Stochastic Processes." *Acta Math.* (Academiae Scientiarum Hungaricae), **8**, 169–191.

Trivedi, K. S. (1982). *Probability and Statistics with Reliability, Queueing and Computer Science Applications*. Englewood Cliffs, N.J.: Prentice-Hall.

Tsokos, C. P., and Shimi, I. N., eds. (1977). *The Theory and Applications of Reliability*. Vol. 1. New York: Academic Press.

Zelen, M., ed. (1963). *Statistical Theory of Reliability*, Proceedings of an Advanced Seminar, May 8–10, 1962. Madison: University of Wisconsin Press.

FOR FURTHER READING

The first four papers list more than 200 articles covering the topic. Also articles appearing in IEEE Transactions on Reliability consider mostly stochastic models.

Kumar, A., and Agarwal, M. (1980). "A Review of Standby Redundant Systems." *IEEE Trans. Reliability* **R-29**, 290–294.

Lie, C. H., Hwang, C. L., and Tillman, F. A. (1977). "Availability of Maintained Systems: A State-of-the-Art Survey." *AIIE Trans.* **9**, 247–259. Includes a good bibliography.

Osaki, S., and Nakagawa, T. (1976). "Bibliography for Reliability and Availability of Stochastic Systems." *IEEE Trans. Reliability* **R-25**, 284–286.

Tillman, F. A., Hwang, C. L. and Kuo, W. (1980). "System Effectiveness Models: An Annotated Bibliography." *IEEE Trans. Reliability* **R-29**, 295–304.

Arnold, T. F. (1973). "The Concept of Coverage and its Effect on the Reliability Model of a Repairable System." *IEEE Trans. Computers* **C-22**, 251–254.

Ashar, K. G. (1960). "Probabilistic Model of System Operation with a Varying Degree of Spares and Services Facilities." *Oper. Res.* **8**, 707–718.

Barlow, R. E., and Hunter, L. C. (1960). "Mathematical Models for System Reliability." *The Sylvania Technologist* **13**, (1, 2).

Bollinger, R. C., and Salvia, A. A. (1982). "Consecutive *k*-out-of-*n*: F Networks." *IEEE Trans. Reliability* **R-31**, 53–56. Includes a good set of references.

Derman, C., Lieberman, G. J., and Ross, S. M. (1982). "On the Consecutive *k*-out-of-*n*: F System." *IEEE Trans. Reliability*, **R-31**, 57–63.

Epstein, B., and Hosford, J. (1960). "Reliability of Some Two Unit Redundant Systems." *Proc. Sixth National Symposium on Reliability and Quality Control*, 469–476.

Jain, H. C. (1967). "Reliability of a Complex System with Types of Subsystems." *Logistic Rev.* **3**, 21–37.

Jain, H. C. (1969). "System Effectiveness with Pre-emptive Resume Repair Policy." *Advancing Frontiers in Operations Research*. New Delhi: Hindustan, pp. 295–305.

Kaufmann, A., Groucho, D., and Cruon, R. (1977). *Mathematical Models for the Study of the Reliability System*. New York: Academic Press.

Kodama, M. (1976). "Probabilistic Analysis of a Multicomponent Series-Parallel System under Preemptive Repeat Repair Discipline." *Oper. Res.* **24**, 500–515.

Malhotra, N. K. (1969). "Complex System Reliability." *Advancing Frontiers in Operations Research*. New Delhi: Hindustan, pp. 283–294.

Mathur, F. P., and Avizienis, A. (1970). "Reliability Analysis and Architechure of a Hybrid-Redundant Digital System: Generalized Triple Modular Redundancy with Repair." *Proc. AFIPS, SJCC*. **36**, 375–383.

Mohan, C., Garg, R. C., and Singal, P. R. (1962). "Dependability of a Complex System." *Oper. Res.* **10**, 310–313.

Natarajan, R. (1967). "Assignment of Priority in Improving System Reliability." *IEEE Trans. Reliability* **R-16**(3), 104–110.

Natarajan, R. (1969). "Reliability of a Single Unit System with Spares and Single Repair Facility." *Advancing Frontiers in Operations Research*. New Delhi: Hindustan, pp. 227–238.

Shogan, A. W. (1976). "Sequential Bounding of the Reliability of a Stochastic Network." *Oper. Res.* **24**, 1027–1044.

Smith, C. O. (1976). *Introduction to Reliability in Design*, New York: McGraw-Hill.

Stiffler, J. J., Bryant, L. A., and Guccione, I. (1979). "Care-III Final Report, Phase I." NASA Contractor Report 159122, Langley Research Center; Hampton, Virginia.

Weinstock, G. D. (1962). "Mathematical Analysis of Redundant Systems–Solving the Problem with Matrix Algebra." *Proc. Eighth National Symposium on Reliability and Quality Control*.

Wilkin, D. R., and Langford, E. S. (1966). "Failure Probability Formulas for Systems with Spares." *Oper. Res.* **14**, 731–732.

Chapter 16

MARKOVIAN
COMBAT MODELS

16.1 INTRODUCTION

Many attempts have been made to understand combat operations using mathematical models. Initial attempts were mostly deterministic, the classical one being the analysis made by F. W. Lanchester in 1916. Assuming different loss rates, he developed differential equations dealing with the losses on opposing sides with large numbers of combatants on both sides. When combats are between two contestants we call them duels. Duels that involve probabilistic assumptions on firing time and/or the success of hits are called stochastic duels. For a discussion on the development of this topic readers are referred to Ancker (1967).

In this chapter we shall discuss three specific Markovian combat models: a combat between two opposing forces of varied strength, a duel between two combatants with exponential firing times, and fencing, which is essentially the sports version of a Markov duel. The similarity of the underlying stochastic processes results in similar solution techniques for the three models.

16.2 LANCHESTER'S COMBAT MODEL

Consider a combat between two opposing forces, say Red and Blue, starting with initial numbers R and B combatants, respectively. Using differential equation models, Lanchester (1916) has hypothesized the following well-known linear and square laws for the attrition of combatants. Let $R(t)$ and $B(t)$ be the surviving combatants at time t. Then the linear law is of the form

$$\frac{dR(t)}{dt} = -\alpha \quad \text{and} \quad \frac{dB(t)}{dt} = -\beta$$

and the square law is of the form

$$\frac{dR(t)}{dt} = -\alpha B(t) \quad \text{and} \quad \frac{dB(t)}{dt} = -\beta R(t)$$

A probabilistic analysis of these laws has been given by Morse and Kimball (1951) who also discuss the relative merits of the deterministic and the probabilistic forms as well as the shortcomings of the two laws to represent real combat situations. Apart from using a probabilistic model, one could generalize the deterministic law to include more general attrition rates. This is the approach used by Brackney (1959) who suggests the incorporation of several factors in the attrition rate: the sizes of the opposing forces, probabilities of kill in individual encounters, area (space) occupied by contending forces, and the time required by the combatants to fire their weapons. Giving a recurrence relation technique in a discrete time probabilistic analysis, Smith (1965) provides a general setting to include some of the factors suggested by Brackney. However, the emphasis by Smith is on an analytical closed form expression, and therefore he is able to include only a simple mixed law to give an explicit solution for the probability of survival. Furthermore the discrete time analysis does not give the state of combat for a given time. A variation of Smith's approach has been used by Jain and Nagabhushanam (1974) to investigate a correlated model for combat (also see Kisi and Hirose, 1966).

As indicated by the detailed discussion of an example given by Morse and Kimball (1951, p. 70), improvements are needed in the combat models to bring them closer to reality. It is very desirable that such a model be flexible so that different attrition components (e.g., those suggested by Brackney, 1959), including those without any particular analytical form, can be tested for purposes of validation. Efforts to extend earlier models have been restricted due to the need for closed form analytical expressions such as those given by Smith (1965) for use in applications. However, now because of the advent of the computers with ever increasing computational power, algorithmic solutions have become satisfactory. The Markov process model that follows discusses the problem with a large measure of generality and provides explicit solutions for relatively small problems. For larger problems procedures are outlined for an algorithmic solution that is convenient for computational purposes.

The Markov Process Model and the Transform Solution

Let $R(t)$ and $B(t)$ be the surviving combatants after time t in the Red and Blue forces, respectively. We assume $R(0) = R$ and $B(0) = B$. Define

$$P_{ij}(t) = P[R(t) = i, B(t) = j] \tag{16.2.1}$$

The state of the process is given by the pair $[R(t), B(t)]$. Assume that during a small interval of time $(t, t + \Delta t]$, the state changes occur with the following probabilities:

$$(i, j) \rightarrow (i - 1, j) \text{ with probability } \alpha_{ij}\Delta t + o(\Delta t)$$

$$(i, j) \rightarrow (i, j - 1) \text{ with probability } \beta_{ij}\Delta t + o(\Delta t)$$

where $o(\Delta t)$ is such that $o(\Delta t)/\Delta t \rightarrow 0$ as $\Delta t \rightarrow 0$.

All other changes are assumed to have a probability of the order $o(\Delta t)$. Under these assumptions $[R(t), B(t)]$ is a vector Markov process which satisfies the following forward Kolmogorov equations:

$$P'_{RB}(t) = -(\alpha_{RB} + \beta_{RB})P_{RB}(t)$$

$$P'_{Rj}(t) = -(\alpha_{Rj} + \beta_{Rj})P_{Rj}(t) + \beta_{R,j+1}P_{R,j+1}(t) \qquad 0 < j < B$$

$$P'_{iB}(t) = -(\alpha_{iB} + \beta_{iB})P_{iB}(t) + \alpha_{i+1,B}P_{i+1,B}(t) \qquad 0 < i < R$$

$$P'_{ij}(t) = -(\alpha_{ij} + \beta_{ij})P_{ij}(t) + \alpha_{i+1,j}P_{i+1,j}(t)$$
$$+ \beta_{i,j+1}P_{i,j+1}(t) \qquad 0 < i < R, 0 < j < B$$

$$P'_{0j}(t) = \alpha_{1j}P_{1j}(t) \qquad 0 < j \leq B$$

$$P'_{i0}(t) = \beta_{i1}P_{i1}(t) \qquad 0 < i \leq R \qquad (16.2.2)$$

These equations can be solved using Laplace transforms. Let

$$\phi_{ij}(\theta) = \int_0^\infty e^{-\theta t}P_{ij}(t)\,dt \qquad \text{Re}(\theta) > 0 \qquad (16.2.3)$$

Taking transforms of equations in (16.2.2), and using the boundary condition $R(0) = R$ and $B(0) = B$, we get

$$\phi_{RB}(\theta) = \frac{1}{\theta + \alpha_{RB} + \beta_{RB}}$$

$$\phi_{Rj}(\theta) = \left(\frac{1}{\theta + \alpha_{Rj} + \beta_{Rj}}\right)\beta_{R,j+1}\phi_{R,j+1}(\theta)$$

$$\phi_{iB}(\theta) = \left(\frac{1}{\theta + \alpha_{iB} + \beta_{iB}}\right)\alpha_{i+1,B}\phi_{i+1,B}(\theta)$$

$$\phi_{ij}(\theta) = \left(\frac{1}{\theta + \alpha_{ij} + \beta_{ij}}\right)\left[\alpha_{i+1,j}\phi_{i+1,j}(\theta) + \beta_{i,j+1}\phi_{i,j+1}(\theta)\right]$$

$$(16.2.4)$$

Clearly equations (16.2.4) are in a suitable form for a recursive solution. In giving a general solution procedure, we use the following simplifying notations:

$$a_{ij} \equiv a_{ij}(\theta) = \frac{\alpha_{ij}}{\theta + \alpha_{ij} + \beta_{ij}}$$

$$b_{ij} \equiv b_{ij}(\theta) = \frac{\beta_{ij}}{\theta + \alpha_{ij} + \beta_{ij}}$$

$$c_{ij} \equiv c_{ij}(\theta) = \frac{1}{\theta + \alpha_{ij} + \beta_{ij}} \qquad (16.2.5)$$

Now solving (16.2.4) recursively, we get

$$\phi_{RB}(\theta) = c_{RB}$$

$$\phi_{R,B-1}(\theta) = b_{RB} c_{R,B-1}$$

$$\phi_{Rj}(\theta) = \left(\prod_{l=j+1}^{B} b_{Rl} \right) c_{Rj}$$

and

$$\phi_{iB}(\theta) = \left(\prod_{l=i+1}^{R} a_{lB} \right) c_{iB} \qquad (16.2.6)$$

Using the last equation in (16.2.4), first we get $\phi_{R-1,j}(\theta)$, for $j = B - 1$, $B - 2, B - 3, \ldots$, then $\phi_{R-2,j}(\theta)$, for $j = B - 1, B - 2, B - 3, \ldots$, and so on, in a recursive manner. For instance,

$$\phi_{R-1,B-1}(\theta) = c_{R-1,B-1}[\alpha_{R,B-1}\phi_{R,B-1}(\theta) + \beta_{R-1,B}\phi_{R-1,B}(\theta)]$$

in which expressions are available from previous results. Thus we get

$$\phi_{R-1,B-1}(\theta) = c_{R-1,B-1}[a_{R,B-1}b_{RB} + b_{R-1,B}a_{RB}]$$

$$\phi_{R-1,B-2}(\theta) = c_{R-1,B-2}[\alpha_{R,B-2}\phi_{R,B-2}(\theta) + \beta_{R-1,B-1}\phi_{R-1,B-1}(\theta)]$$

$$(16.2.7)$$

which can be simplified using expressions from (16.2.6) and (16.2.7). We get

$$\phi_{R-1,B-2}(\theta) = c_{R-1,B-2}(a_{R,B-2}b_{R,B-1}b_{RB} + b_{R-1,B-1}a_{R,B-1}b_{RB}$$

$$+ b_{R-1,B-1}b_{R-1,B}a_{RB}) \qquad (16.2.8)$$

An examination of the terms in (16.2.7) and (16.2.8) and the structure of the recurrence relation of equation (16.2.4) reveals that $\phi_{ij}(\theta)$ can be expressed as

$$\phi_{ij}(\theta) = c_{ij} \sum_{r=1}^{n_{ij}} A_r$$

where $A_1, A_2, \ldots, A_{n_{ij}}$ correspond to the n_{ij} distinct paths leading to state (i, j) from state (R, B). As can be seen from Figure 16.2.1, the number of distinct paths leading to state (i, j) from state (R, B) is given by the binomial coefficient $\binom{R-i+B-j}{R-i}$. Each path involves $R - i$ left movements (steps) and $B - j$ right movements (steps) for a total of $(R - i + B - j)$ steps. With each left step we associate an a function, and with each right step we associate a b function; these are Laplace transforms of

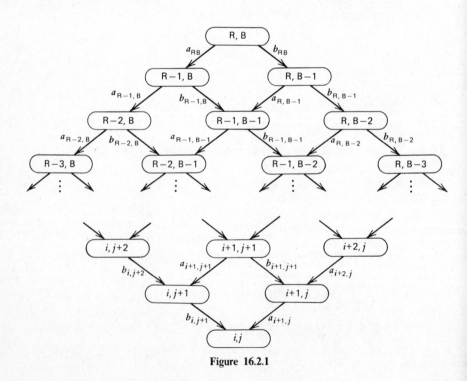

Figure 16.2.1

the corresponding transition densities. For instance, consider the state $(R - 1, B - 2)$. The three distinct paths leading to this state are

Path 1: $(R, B) \rightarrow (R, B - 1) \rightarrow (R, B - 2) \rightarrow (R - 1, B - 2)$

Path 2: $(R, B) \rightarrow (R, B - 1) \rightarrow (R - 1, B - 1) \rightarrow (R - 1, B - 2)$

Path 3: $(R, B) \rightarrow (R - 1, B) \rightarrow (R - 1, B - 1) \rightarrow (R - 1, B - 2)$

$$(16.2.9)$$

Now consider the transition $(R, B - 1) \rightarrow (R, B - 2)$. For the corresponding b function we have

$$
b_{R, B-1} = \frac{\beta_{R, B-1}}{\theta + \alpha_{R, B-1} + \beta_{R, B-1}}
$$

$$
= \left(\frac{\beta_{R, B-1}}{\alpha_{R, B-1} + \beta_{R, B-1}} \right) \left(\frac{\alpha_{R, B-1} + \beta_{R, B-1}}{\theta + \alpha_{R, B-1} + \beta_{R, B-1}} \right) \quad (16.2.10)
$$

In the expression on the right-hand side of (16.2.10), the first term is the probability that state $(R, B - 2)$ follows state $(R, B - 1)$, and the second term is the Laplace transform of the exponential distribution which is the probability density of the time needed for this transition. These follow from known properties of Markov processes [see (7.2.31) and (7.6.34)]. Thus the three terms in parentheses in equation (16.2.8) correspond to Laplace transforms for the transition densities on the three paths identified in (16.2.9). It should also be noted that $c_{R-1, B-2}$ is the Laplace transform of $\exp[-(\alpha_{R-1, B-2} + \beta_{R-1, B-2})t]$ which gives the probability that the process is still in state $(R - 1, B - 2)$ at time t.

In general, for any state $(i, j), \phi_{ij}(\theta)$ is determined by identifying all paths leading to (i, j) from (R, B) and using the appropriate transforms in the product corresponding to each path. The transforms to be used are

$$
a_{kl} \equiv a_{kl}(\theta) \qquad \text{if transition is from } (k, l) \rightarrow (k - 1, l)
$$

$$
b_{kl} \equiv b_{kl}(\theta) \qquad \text{if transition is from } (k, l) \rightarrow (k, l - 1)
$$

In a computer algorithm this can be done by tracing paths in a systematic manner. Starting from a node (i, j), one would trace back the nodes $(i + 1, j)$ with an associated transform $a_{i+1, j}$ and $(i, j + 1)$ with an associated transform $b_{i, j+1}$. These two nodes lead to two paths, one of which can be followed initially, while stacking the other one for consideration later. At every stage this procedure is repeated until the initial node (R, B) is reached. This procedure is illustrated in the following example.

Example: State $(R - 1, B - 2)$

State	Transform			Stacked paths Path no.	State	Transform
Path 1:						
$(R, B - 2)$	$a_{R, B-2}$			2	$(R - 1, B - 1)$	$b_{R-1, B-1}$
$(R, B - 1)$	$a_{R, B-2}$	$b_{R, B-1}$				
(R, B)	$a_{R, B-2}$	$b_{R, B-1}$	b_{RB}			
Path 2:						
$(R - 1, B - 1)$	$b_{R-1, B-1}$					
$(R, B - 1)$	$b_{R-1, B-1}$	$a_{R, B-1}$		3	$(R - 1, B)$	$b_{R-1, B-1}$ $b_{R-1,}$
(R, B)	$b_{R-1, B-1}$	$a_{R, B-1}$	b_{RB}			
Path 3:						
$(R - 1, B)$	$b_{R-1, B-1}$	$b_{R-1, B}$				
(R, B)	$b_{R-1, B-1}$	$b_{R-1, B}$	a_{RB}			

Determination of Probabilities $P_{ij}(t)$

The probabilities $P_{ij}(t)$ are determined by inverting the Laplace transform $\phi_{ij}(\theta)$. Let n_{ij} be the number of paths leading to (i, j) from (R, B), and let $A_1(\theta), A_2(\theta), \ldots, A_{n_{ij}}(\theta)$ be the corresponding product transform functions as determined by tracing the paths. It should be noted that $A_r(\theta)$ is of the form

$$A_r(\theta) = \left(\frac{\alpha_{i_1 j_1}}{\theta + \alpha_{i_1 j_1} \beta_{i_1 j_1}} \right) \left(\frac{\alpha_{i_2 j_2}}{\theta + \alpha_{i_2 j_2} + \beta_{i_2 j_2}} \right) \quad \cdots \quad (R - i \text{ terms})$$

$$\times \left(\frac{\beta_{i_k j_k}}{\theta + \alpha_{i_k j_k} + \beta_{i_k j_k}} \right) \left(\frac{\beta_{i_{k+1} j_{k+1}}}{\theta + \alpha_{i_{k+1} j_{k+1}} + \beta_{i_{k+1} j_{k+1}}} \right) \quad \cdots \quad (B - j \text{ terms})$$

$$(16.2.11)$$

The number of terms on the right-hand side of (16.2.11) is $(R - i) +$

$(B - j)$. The transform $\phi_{ij}(\theta)$ is given as

$$\phi_{ij}(\theta) = c_{ij}(\theta) \sum_{r=1}^{n_{ij}} A_r(\theta) \qquad (16.2.12)$$

It should be noted that each term on the right-hand side of $A_r(\theta)$ is the transform of an exponential term; so also is $c_{ij}(\theta)$. Hence the $P_{ij}(t)$ can be determined from $\phi_{ij}(\theta)$ by inverting $\phi_{ij}(\theta)$ with the help of partial fraction expansion. This procedure is illustrated using $\phi_{R-1, B-2}(\theta)$ for $R = 2$, $B = 3$. Specializing (16.2.8), we have

$$\phi_{11}(\theta) = c_{11}(\theta)[b_{23}(\theta)b_{22}(\theta)a_{21}(\theta) + b_{23}(\theta)a_{22}(\theta)b_{12}(\theta)$$

$$+ a_{23}(\theta)b_{13}(\theta)b_{12}(\theta)]$$

$$= c_{11}(\theta)[A_1(\theta) + A_2(\theta) + A_3(\theta)] \qquad (16.2.13)$$

where

$$c_{11}(\theta) = \frac{1}{\theta + \alpha_{11} + \beta_{11}}$$

$$a_{ij}(\theta) = \frac{\alpha_{ij}}{\theta + \alpha_{ij} + \beta_{ij}}$$

$$b_{ij}(\theta) = \frac{\beta_{ij}}{\theta + \alpha_{ij} + \beta_{ij}}$$

Using partial fraction expansions, we get

$c_{11}(\theta)A_1(\theta)$

$$= \left(\frac{1}{\theta + \alpha_{11} + \beta_{11}} \right) \lim_{\theta \to -(\alpha_{11}+\beta_{11})} [A_1(\theta)]$$

$$+ \left(\frac{1}{\theta + \alpha_{23} + \beta_{23}} \right) \lim_{\theta \to -(\alpha_{23}+\beta_{23})} [(\theta + \alpha_{23} + \beta_{23})c_{11}(\theta)A_1(\theta)]$$

$$+ \left(\frac{1}{\theta + \alpha_{22} + \beta_{22}} \right) \lim_{\theta \to -(\alpha_{22}+\beta_{22})} [(\theta + \alpha_{22} + \beta_{22})c_{11}(\theta)A_1(\theta)]$$

$$+ \left(\frac{1}{\theta + \alpha_{21} + \beta_{21}} \right) \lim_{\theta \to -(\alpha_{21}+\beta_{21})} [(\theta + \alpha_{21} + \beta_{21})c_{11}(\theta)A_1(\theta)]$$

$$= k_1 \left(\frac{1}{\theta + \alpha_{11} + \beta_{11}} \right) + k_2 \left(\frac{1}{\theta + \alpha_{23} + \beta_{23}} \right)$$

$$+ k_3 \left(\frac{1}{\theta + \alpha_{22} + \beta_{22}} \right) + k_4 \left(\frac{1}{\theta + \alpha_{21} + \beta_{21}} \right)$$

where we have used k_1, k_2, k_3, and k_4 to represent constants determined in the previous step. Using L^{-1} to denote the inverse, we get

$$L^{-1}\left[c_{11}(\theta)A_1(\theta)\right] = k_1 e^{-(\alpha_{11}+\beta_{11})t} + k_2 e^{-(\alpha_{23}+\beta_{23})t}$$

$$+ k_3 e^{-(\alpha_{22}+\beta_{22})t} + k_4 e^{-(\alpha_{21}+\beta_{21})t} \quad (16.2.14)$$

The inverse transforms of the other two terms in $\phi_{11}(\theta)$ can be determined in a similar manner. The probability $P_{ij}(t)$ is now obtained as

$$P_{ij}(t) = L^{-1}\left[c_{11}(\theta)A_1(\theta)\right] + L^{-1}\left[c_{11}(\theta)A_2(\theta)\right] + L^{-1}\left[c_{11}(\theta)A_3(\theta)\right]$$

$$(16.2.15)$$

The Probability of Survival and the Expected Duration of Combat

The probability of survival of either of the two forces (Red and Blue) is easily obtained from the transforms. We have

$P(\text{Red wins before time } t \text{ with } i \text{ surviving combatants}) = P_{i0}(t)$

$$= \int_0^t P'_{i0}(\tau)\,d\tau$$

$$= \beta_{i1}\int_0^t P_{i1}(\tau)\,d\tau \quad (16.2.16)$$

$P(\text{Ultimately Red wins with } i \text{ surviving combatants})$

$$= \beta_{i1}\int_0^\infty P_{i1}(\tau)\,d\tau$$

$$= \beta_{i1}\phi_{i1}(0) \quad (16.2.17)$$

Similarly for the Blue force we get

$P(\text{Ultimately Blue wins with } j \text{ surviving combatants}) = \alpha_{1j}\phi_{1j}(0)$

$$(16.2.18)$$

Let C_{RB} be the duration of the combat. Clearly

$$P(C_{RB} > t) = \sum\sum_{ij>0} P_{ij}(t) \quad (16.2.19)$$

Also

$$E[C_{RB}] = \int_0^\infty P(C_{RB} > t)\, dt$$

$$= \sum_{i=1}^{R} \sum_{j=1}^{B} \phi_{ij}(0) \qquad (16.2.20)$$

A Numerical Example

Let $R = 2$, $B = 3$, and the attrition rates be given as follows:

$$\alpha_{23} = 1.8 \qquad \beta_{23} = 2.5$$
$$\alpha_{22} = 1.5 \qquad \beta_{22} = 3.0$$
$$\alpha_{21} = 0.5 \qquad \beta_{21} = 4.0$$
$$\alpha_{13} = 3.0 \qquad \beta_{13} = 0.5$$
$$\alpha_{12} = 2.0 \qquad \beta_{12} = 1.0$$
$$\alpha_{11} = 1.0 \qquad \beta_{11} = 2.0 \qquad (16.2.21)$$

In this example we shall determine the probabilities of survival and the expected duration of the combat. For notational convenience we shall write $\phi_{ij}(0) = \phi_{ij}$ and

$$a_{ij} = a_{ij}(0) = \frac{\alpha_{ij}}{\alpha_{ij} + \beta_{ij}}$$

$$b_{ij} = b_{ij}(0) = \frac{\beta_{ij}}{\alpha_{ij} + \beta_{ij}} = 1 - a_{ij}$$

$$c_{ij} = c_{ij}(0) = \frac{1}{\alpha_{ij} + \beta_{ij}}$$

Substituting values from (16.2.21), we get

$$c_{23} = 0.2326 \qquad a_{23} = 0.4187 \qquad b_{23} = 0.5813$$
$$c_{22} = 0.2222 \qquad a_{22} = 0.3333 \qquad b_{22} = 0.6667$$
$$c_{21} = 0.2222 \qquad a_{21} = 0.1111 \qquad b_{21} = 0.8889$$
$$c_{13} = 0.2857 \qquad a_{13} = 0.8571 \qquad b_{13} = 0.1429$$
$$c_{12} = 0.3333 \qquad a_{12} = 0.6667 \qquad b_{12} = 0.3333$$
$$c_{11} = 0.3333 \qquad\qquad (16.2.22)$$

Specializing the expressions derived earlier for $R = 2$ and $B = 3$ and substituting appropriate values from (16.2.22), we get

$$\phi_{23} = c_{23} = 0.2326$$

$$\phi_{22} = b_{23}c_{22} = 0.1292$$

$$\phi_{21} = b_{22}b_{23}c_{21} = 0.0861$$

$$\phi_{13} = a_{23}c_{13} = 0.1196$$

$$\phi_{12} = c_{12}[a_{22}b_{23} + b_{13}a_{23}] = 0.0845$$

$$\phi_{11} = c_{11}[a_{21}b_{22}b_{23} + b_{12}a_{22}b_{23} + b_{12}b_{13}a_{23}] = 0.0425$$

Let R_i be the event that the Red force wins with i surviving combatants ($i = 1, 2$); similarly define the event B_j. Using expressions given in equations (16.2.17) and (16.2.18), we get

$$P(R_1) = \beta_{11}\phi_{11} = 0.0850$$

$$P(R_2) = \beta_{21}\phi_{21} = 0.3444$$

$$P(B_1) = \alpha_{11}\phi_{11} = 0.4250$$

$$P(B_2) = \alpha_{12}\phi_{12} = 0.1690$$

$$P(B_3) = \alpha_{13}\phi_{13} = 0.3588$$

Therefore

$$P(\text{Red force wins ultimately}) = 0.4294$$

$$P(\text{Blue force wins ultimately}) = 0.5703$$

The expected duration of the combat follows from equation (16.2.20):

$$E[C_{23}] = \sum_{i=1}^{2} \sum_{j=1}^{3} \phi_{ij} = 0.6945 \text{ time units}$$

Conclusion

The combat model discussed here is quite general as far as the form of the attrition rate is concerned, provided the attrition is not time dependent. In the numerical example we have purposely chosen rates that do not come from any analytical expressions. Even though it is possible to generalize the process to time-dependent attrition rates, we do not believe such an effort is worthwhile from a practical viewpoint. The expressions will be too cumbersome to be amenable to numerical computations.

A limitation of the model is the Markovian assumption for the process. If a deterministic model does not represent reality (as discussed by Morse and Kimball, 1951), a Markov process model should do considerably better. Even though ideally it should be possible to set up a better probabilistic model than the Markovian one given here, the computational complexities of such a model are likely to make such an effort purely academic. The Markov process model is an approximation that is likely to provide the best results one can get using analytical models. Inclusion of other complicating factors can be done best using a mixture of mathematical and simulation models.

16.3   A MARKOV DUEL

The combat situation considered in Lanchester's model involved two opposing forces with varying numbers of combatants on each side. A stochastic duel refers to a combat in which two combatants face each other, and the associated events are random outcomes. Various problems arising out of a duel are discussed in Ancker (1967). As developed by Ancker and his associates (see references cited in his papers), the theory of stochastic duels is based on the determination of the probability of a series of events rather than the study of the properties of the underlying stochastic processes. A Markov process formulation of the duel (hence we may call it a Markov duel), has been given by Bhashyam and Singh (1967). In this section we shall take their formulation and derive the properties of the underlying Markov process using the procedure developed in the last section for the Lanchester model. This procedure is much simpler and can be used even under more general conditions than those assumed by Bhashyam and Singh.

Suppose two deulists A and B are engaged in a duel. They fire, each with the objective of killing the opponent. Initially we assume that they have an unlimited amount of ammunition. When A and B are ready to fire at each other at time t, the probability that A will fire at B during $(t, t + \Delta t]$ is assumed to be $\lambda_A \Delta t + o(\Delta t)$. The probability of a similar event in the case of B is assumed to be $\lambda_B \Delta t + o(\Delta t)$. Let $p_A(m)$ be the probability that A scores the kill in A's mth round and $p_B(n)$ be the probability that B scores the kill in B's nth round. We write $q_A(m) = 1 - p_A(m)$ and $q_B(n) = 1 - p_B(n)$. Clearly $q_A(m)$ is the probability that B escapes A's fire during A's mth round, and similarly $q_B(n)$ is the probability that A escapes the fire. Let $A(t)$ and $B(t)$ be the number of rounds fired without success by A and B, respectively. Under the preceding assumptions $[A(t), B(t)]$ is a vector Markov process. Define the following transition probabilities with the initial conditions $A(0) = 0$ and $B(0) = 0$:

$$P_{mn}(t) = P[A(t) = m, B(t) = n]$$

$$A_{mn}(t) = P[\text{A scores the kill at the } m\text{th round before } t,\ B(t) = n]$$

$$B_{mn}(t) = P[\text{B scores the kill at the } n\text{th round before } t,\ A(t) = m]$$

$$(16.3.1)$$

It should be noted that under these assumptions the firing time for each duelist has an exponential distribution.

The forward Kolmogorov equations for the process $[A(t), B(t)]$ are as follows:

$$P'_{00}(t) = -(\lambda_A + \lambda_B)P_{00}(t)$$

$$P'_{0n}(t) = -(\lambda_A + \lambda_B)P_{0n}(t) + \lambda_B q_B(n)P_{0,n-1}(t) \qquad n > 0$$

$$P'_{m0}(t) = -(\lambda_A + \lambda_B)P_{m0}(t) + \lambda_A q_A(m)P_{m-1,0}(t) \qquad m > 0$$

$$P'_{mn}(t) = -(\lambda_A + \lambda_B)P_{mn}(t) + \lambda_A q_A(m)P_{m-1,n}(t)$$

$$+\lambda_B q_B(n)P_{m,n-1}(t) \qquad m > 0, n > 0$$

$$A'_{mn}(t) = \lambda_A p_A(m)P_{m-1,n}(t)$$

$$B'_{mn}(t) = \lambda_B p_B(n)P_{m,n-1}(t) \qquad (16.3.2)$$

Define the Laplace transform

$$\phi_{mn}(\theta) = \int_0^\infty e^{-\theta t}P_{mn}(t)\,dt \qquad (\text{Re}(\theta) > 0)$$

Taking transforms, the first four equations of (16.3.2) give

$$\phi_{00}(\theta) = \frac{1}{\theta + \lambda_A + \lambda_B}$$

$$\phi_{0n}(\theta) = \frac{\lambda_B q_B(n)}{\theta + \lambda_A + \lambda_B} \phi_{0, n-1}(\theta)$$

$$\phi_{m0}(\theta) = \frac{\lambda_A q_A(m)}{\theta + \lambda_A + \lambda_B} \phi_{m-1, 0}(\theta)$$

$$\phi_{mn}(\theta) = \frac{1}{\theta + \lambda_A + \lambda_B} \left[\lambda_A q_A(m) \phi_{m-1, n}(\theta) + \lambda_B q_B(n) \phi_{m, n-1}(\theta) \right]$$

$$(16.3.3)$$

Solving the first three equations of (16.3.3) recursively, we get

$$\phi_{0n}(\theta) = \frac{\lambda_B^n Q_B(n)}{(\theta + \lambda_A + \lambda_B)^{n+1}} \qquad (16.3.4)$$

$$\phi_{m0}(\theta) = \frac{\lambda_A^m Q_A(m)}{(\theta + \lambda_A + \lambda_B)^{m+1}} \qquad (16.3.5)$$

where we have written $q_B(1) q_B(2) \cdots q_B(n) = Q_B(n)$ and $q_A(1) q_A(2) \cdots q_A(m) = Q_A(m)$. For a recursive solution of the last equation of (16.3.3) we note the transition structure of the $[A(t), B(t)]$ process as in Figure 16.3.1. As described in the last section, the term within brackets in the last equation of (16.3.3) corresponds to the transform corresponding to transitions through all paths for the process from state $(0, 0)$ to state (m, n). For each path the transform can be given as

$$\frac{\prod_{i=0}^{m} [\lambda_A q_A(i)] \prod_{j=1}^{n} [\lambda_B q_B(j)]}{(\theta + \lambda_A + \lambda_B)^{m+n}} \qquad (16.3.6)$$

Note that the transform functions for each path is the same for this model as compared to the Lanchester model of the last section. This is due to our assumption that q_A's and q_B's are dependent only on their respective

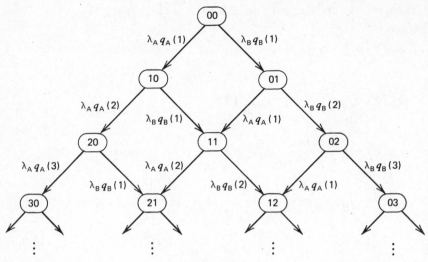

Figure 16.3.1

processes. If we assume that they depend on both $A(t)$ and $B(t)$, we will have expressions similar to those from the last section. However, such generalizations do not seem realistic in this case.

The number of paths leading to (m, n) from $(0, 0)$ is clearly given by the binomial coefficient $\binom{m+n}{m}$. Thus the solution for the last equation in (16.3.3) is completely determined as

$$\phi_{mn}(\theta) = \binom{m+n}{m} \frac{\lambda_A^m \lambda_B^n Q_A(m) Q_B(n)}{(\theta + \lambda_A + \lambda_B)^{m+n+1}} \tag{16.3.7}$$

The transition probability $P_{mn}(t)$ can be easily obtained from (16.3.7) by inversion if we note that the Laplace transform of the gamma distribution

$$f(y) = e^{-\lambda y} \frac{\lambda^n y^{n-1}}{(n-1)!} \tag{16.3.8}$$

is given by

$$\int_0^\infty e^{-\theta t} f(t)\, dt = \left(\frac{\lambda}{\theta + \lambda} \right)^n \tag{16.3.9}$$

Note that $f(y)$ in (16.3.8) is the probability density of the sum of n

exponential random variables (Section 7.2). Now writing (16.3.7) as

$$\phi_{mn}(\theta) = \binom{m+n}{m} \frac{\lambda_A^m \lambda_B^n Q_A(m) Q_B(n)}{(\lambda_A + \lambda_B)^{m+n+1}} \left(\frac{\lambda_A + \lambda_B}{\theta + \lambda_A + \lambda_B} \right)^{m+n+1}$$

we get after simplifications

$$P_{mn}(t) = \frac{\lambda_A^m \lambda_B^n Q_A(m) Q_B(n)}{m! n!} t^{m+n} e^{-(\lambda_A + \lambda_B)t} \qquad (16.3.10)$$

To determine $A_{mn}(t)$ and $B_{mn}(t)$, the probability of win before t for A and B, respectively, we proceed as follows: We have

$$A_{mn}(t) = \int_0^t A'_{mn}(\tau) \, d\tau$$

$$= \lambda_A p_A(m) \int_0^t P_{m-1,n}(\tau) \, d\tau \qquad (16.3.11)$$

From the identity (7.2.34) we have

$$\int_0^t e^{-(\lambda_A + \lambda_B)\tau} \frac{(\lambda_A + \lambda_B)^{m+n}}{(m+n-1)!} \tau^{m+n-1} d\tau$$

$$= \sum_{r=m+n}^{\infty} e^{-(\lambda_A + \lambda_B)t} \frac{(\lambda_A + \lambda_B)^r t^r}{r!} \qquad (16.3.12)$$

Thus we get

$$A_{mn}(t) = \binom{m+n-1}{n} \frac{\lambda_A^m \lambda_B^n p_A(m) Q_A(m-1) Q_B(n)}{(\lambda_A + \lambda_B)^{m+n}}$$

$$\times \sum_{r=m+n}^{\infty} e^{-(\lambda_A + \lambda_B)t} \frac{(\lambda_A + \lambda_B)^r t^r}{r!} \qquad (16.3.13)$$

Let $P(A)$ be the probability of eventual win by A. This is obtained by letting $t \to \infty$ in (16.3.13) and summing over all values of m and n. Noting that the integral on the left-hand side of (16.3.12) goes to 1 as $t \to \infty$, we get

$$P(A) = \sum_{m=1}^{\infty} \sum_{n=0}^{\infty} \binom{m+n-1}{n} \frac{\lambda_A^m \lambda_B^n p_A(m) Q_A(m-1) Q_B(n)}{(\lambda_A + \lambda_B)^{m+n}}$$

$$(16.3.14)$$

Further simplifications are possible for specific forms of kill probabilities $p_A(\cdot)$ and $p_B(\cdot)$.

Alternately, $P(A)$ can be directly obtained from the transforms $\phi_{mn}(\theta)$ if we note that

$$P(A) = \sum_{m=1}^{\infty} \sum_{n=0}^{\infty} \lambda_A p_A(m) \int_0^{\infty} P_{m-1,n}(t)\, dt$$

$$= \sum_{m=1}^{\infty} \sum_{n=0}^{\infty} \lambda_A p_A(m) \phi_{m-1,n}(0) \qquad (16.3.15)$$

Similarly we get

$$B_{mn}(t) = \binom{m+n-1}{m} \frac{\lambda_A^m \lambda_B^n p_B(n) Q_A(m) Q_B(n-1)}{(\lambda_A + \lambda_B)^{m+n}}$$

$$\times \sum_{r=m+n}^{\infty} e^{-(\lambda_A + \lambda_B)t} \frac{(\lambda_A + \lambda_B)^r t^r}{r!} \qquad (16.3.16)$$

and

$$P(B) = \sum_{m=0}^{\infty} \sum_{n=1}^{\infty} \binom{m+n-1}{m} \frac{\lambda_A^m \lambda_B^n p_B(n) Q_A(m) Q_B(n-1)}{(\lambda_A + \lambda_B)^{m+n}}$$

$$(16.3.17)$$

Let D be the duration of the duel. We have

$$E[D] = \int_0^{\infty} \sum_{m=0}^{\infty} \sum_{n=0}^{\infty} P_{mn}(t)\, dt$$

$$= \sum_{m=0}^{\infty} \sum_{n=0}^{\infty} \phi_{mn}(0)$$

$$= \sum_{m=0}^{\infty} \sum_{n=0}^{\infty} \binom{m+n}{m} \frac{\lambda_A^m \lambda_B^n Q_A(m) Q_B(n)}{(\lambda_A + \lambda_B)^{m+n+1}} \qquad (16.3.18)$$

Suppose the amount of ammunition at the disposal of the duelists is limited. Assume that A and B have enough ammunition for M and N rounds, respectively. Clearly there is the possibility of eventual draw in this case.

The forward Kolmogorov equation (16.3.2) should now be modified to take into account the following restriction on the state space of the process: The first four equations hold only for $m < M$ and $n < N$. We also have

$$P'_{Mn}(t) = -\lambda_B P_{Mn}(t) + \lambda_A q_A(M) P_{M-1,n}(t) + \lambda_B q_B(n) P_{M,n-1}(t)$$

$$0 < n < N$$

$$P'_{mN}(t) = -\lambda_A P_{nN}(t) + \lambda_A q_A(m) P_{m-1,N}(t) + \lambda_B q_B(N) P_{m,N-1}(t)$$

$$0 < m < M$$

$$P'_{MN}(t) = \lambda_A q_A(M) P_{M-1,N}(t) + \lambda_B q_B(N) P_{M,N-1}(t) \qquad (16.3.19)$$

Taking transforms as before, we get

$$\phi_{Mn}(\theta) = \frac{1}{\theta + \lambda_B} \left[\lambda_A q_A(M) \phi_{M-1,n}(\theta) + \lambda_B q_B(n) \phi_{M,n-1}(\theta) \right]$$

$$(16.3.20)$$

$$\phi_{mN}(\theta) = \frac{1}{\theta + \lambda_A} \left[\lambda_A q_A(m) \phi_{m-1,N}(\theta) + \lambda_B q_B(N) \phi_{m,N-1}(\theta) \right]$$

$$(16.3.21)$$

Solving (16.3.20) recursively for $n = 0, 1, 2, \ldots$, we get the general expression

$$\phi_{Mn}(\theta) = \left(\frac{1}{\theta + \lambda_B} \right) \frac{\lambda_A^M \lambda_B^n Q_A(M) Q_B(n)}{(\theta + \lambda_A + \lambda_B)^M}$$

$$\times \sum_{j=0}^{n} \binom{N+j-1}{j} \frac{1}{(\theta + \lambda_A + \lambda_B)^j} \frac{1}{(\theta + \lambda_B)^{n-j}}$$

$$(16.3.22)$$

Similarly we obtain

$$\phi_{mN}(\theta) = \left(\frac{1}{\theta + \lambda_A} \right) \frac{\lambda_A^m \lambda_B^N Q_A(m) Q_B(N)}{(\theta + \lambda_A + \lambda_B)^N}$$

$$\times \sum_{j=0}^{m} \binom{M+j-1}{j} \frac{1}{(\theta + \lambda_A + \lambda_B)^j} \frac{1}{(\theta + \lambda_A)^{m-j}}$$

$$(16.3.23)$$

If one is interested in the finite time probabilities $P_{Mn}(t)$ and $P_{mN}(t)$, the transforms in (16.3.22) and (16.3.23) can be inverted, after using partial fraction expansions as described in the last section. The probability that A or B wins eventually is now given as

$$P(A) = \sum_{m=1}^{M} \sum_{n=0}^{N} \lambda_A p_A(m) \phi_{m-1,n}(0)$$

and

$$P(B) = \sum_{m=0}^{M} \sum_{n=1}^{N} \lambda_B p_B(n) \phi_{m,n-1}(0) \tag{16.3.24}$$

In (16.3.24) substitutions of appropriate expressions should be made from (16.3.7), (16.3.22), and (16.3.23). Finally, the probability of eventual draw is obtained as

$$P(\text{Draw}) = \lambda_A q_A(M) \phi_{M-1,N}(0) + \lambda_B q_B(N) \phi_{M,N-1}(0) \tag{16.3.25}$$

16.4 FENCING

Fencing is a worldwide sport derived from the ancient art of offense and defense with a sword. It is practiced with three different weapons, the foil, the épée (or "dueling sword") and the saber.[*] Touches are scored when a player touches valid targets on the opponent's body with the weapon; these targets differ for the three kinds of weapons used. In competitions the player who scores a predetermined number of touches first is the winner. The president of the jury starts the play by ordering the players to fence; but the players may take some time to advance, trying to judge the tactics of the other. The one who advances first will be on the offensive, and the other on the defensive. Depending on the offensive or defensive nature of the play, the player's chances of scoring a touch may change. As soon as one scores a touch or gets into an inconclusive situation, the players return to their positions and start again.

For the sake of modeling we shall consider only the play time; the time spent in getting ready to play will be ignored. We shall also assume that the play will be decided as soon as one of the players makes a move. Under these restrictions let $\lambda \Delta t + o(\Delta t)$ be the probability that player A will go

[*]*Encyclopaedia Britannica* (1967), Vol. 9, pp. 162–169.

on the offensive during $(t, t + \Delta t]$, and let $\mu \Delta t + o(\Delta t)$ be the corresponding probability for player B, where $o(\Delta t)$ is such that $o(\Delta t)/\Delta t \to 0$ as $\Delta t \to 0$. We assume that these events are independent of each other and also independent of the events that have occurred earlier. When A is on the offensive, let α_1, β_1, and γ_1 be the probabilities that A will score a touch, B will score a touch, and none will score; and when B is on the offensive, let the corresponding probabilities be α_2, β_2, and γ_2, respectively. Thus during $(t, t + \Delta t]$ the probability that A will score a touch is given by $(\alpha_1 \lambda + \alpha_2 \mu)\Delta t + o(\Delta t)$, and the probability that B will score a touch is given by $(\beta_1 \lambda + \beta_2 \mu)\Delta t + o(\Delta t)$. The probability that none will score during $(t, t + \Delta t]$ is given by $1 - (\lambda + \mu)\Delta t + (\lambda \gamma_1 + \mu \gamma_2)\Delta t + o(\Delta t)$. We shall further assume that the player who scores k points first will be declared the winner.

Let $P_{ij}(t)$ be the probability that at time t, A's score is i and B's score is j. We shall assume $P_{ij}(0) = 1$ only if $i = j = 0$ and $P_{ij}(0) = 0$ otherwise. For $0 < i, j < k$ we can therefore write

$$P_{ij}(t + \Delta t) = P_{ij}(t)\left[1 - (\lambda + \mu)\Delta t + (\lambda \gamma_1 + \mu \gamma_2)\Delta t\right] + P_{i-1,j}(t)$$

$$\times [\alpha_1 \lambda + \alpha_2 \mu]\Delta t + P_{i, j-1}(t)[\beta_1 \lambda + \beta_2 \mu]\Delta t + o(\Delta t)$$

$$(16.4.1)$$

Now write

$$\alpha_1 \lambda + \alpha_2 \mu = a \qquad \beta_1 \lambda + \beta_2 \mu = b \qquad \gamma_1 \lambda + \gamma_2 \mu = c$$

so that

$$a + b + c = \lambda + \mu$$

and

$$\lambda + \mu - \lambda \gamma_1 - \mu \gamma_2 = a + b$$

Rearranging (16.4.1), dividing by Δt, and letting $\Delta t \to 0$, we get

$$P_{ij}'(t) = -(a + b)P_{ij}(t) + aP_{i-1,j}(t) + bP_{i, j-1}(t) \qquad 0 < i, j < k$$

$$(16.4.2)$$

By similar arguments we also get

$$P_{00}'(t) = -(a + b)P_{00}(t)$$

$$P_{i0}'(t) = -(a + b)P_{i0}(t) + aP_{i-1,0}(t) \qquad 0 < i < k$$

$$P_{0j}'(t) = -(a + b)P_{0j}(t) + bP_{0,j-1}(t) \qquad 0 < j < k$$

$$P_{kj}'(t) = aP_{k-1,j}(t) \qquad\qquad\qquad\qquad j < k$$

$$P_{ik}'(t) = bP_{i,k-1}(t) \qquad\qquad\qquad\qquad i < k \qquad (16.4.3)$$

These equations may be solved by the use of Laplace transforms. Let

$$\phi_{ij}(\theta) = \int_0^\infty e^{-\theta t} P_{ij}(t)\, dt \qquad \mathrm{Re}(\theta) > 0 \qquad (16.4.4)$$

Taking transforms of equations (16.4.3), we get

$$\phi_{00}(\theta) = \frac{1}{\theta + a + b}$$

$$\phi_{i0}(\theta) = \frac{a}{\theta + a + b}\phi_{i-1,0}(\theta) \qquad\qquad 0 < i < k$$

$$\phi_{0j}(\theta) = \frac{b}{\theta + a + b}\phi_{0,j-1}(\theta) \qquad\qquad 0 < j < k$$

$$\phi_{ij}(\theta) = \frac{1}{\theta + a + b}\left[a\phi_{i-1,j}(\theta) + b\phi_{i,j-1}(\theta)\right] \qquad 0 < i, j < k$$

$$\phi_{kj}(\theta) = \frac{a}{\theta}\phi_{k-1,j}(\theta) \qquad\qquad\qquad\qquad j < k$$

$$\phi_{ik}(\theta) = \frac{b}{\theta}\phi_{i,k-1}(\theta) \qquad\qquad\qquad\qquad i < k \qquad (16.4.5)$$

Solving these equations recursively, we get

$$\phi_{00}(\theta) = \frac{1}{\theta + a + b}$$

$$\phi_{i0}(\theta) = \frac{a^i}{(\theta + a + b)^{i+1}} \qquad i < k$$

$$\phi_{0j}(\theta) = \frac{b^j}{(\theta + a + b)^{j+1}} \qquad j < k$$

$$\phi_{ij}(\theta) = \binom{i+j}{i} \frac{a^i b^i}{(\theta + a + b)^{i+j+1}} \qquad i, j < k \qquad (16.4.6)$$

It may be noted that (16.4.6) has the preceding three expressions as special cases. The Laplace transform appearing in (16.4.6) is in the standard form of the Laplace transform of a gamma function, as given in (16.3.9). Comparing (16.4.6) with (16.3.9), we can write

$$P_{ij}(t) = \frac{a^i b^j}{i! j!} e^{-(a+b)t} t^{i+j} \qquad i, j < k \qquad (16.4.7)$$

The probability that A has scored a win before t, when B's score is only $j < k$ is given by $P_{kj}(t)$. We have from (16.4.3)

$$P_{kj}(t) = \int_0^t P'_{kj}(\tau) \, d\tau$$

$$= a \int_0^t P_{k-1, j}(\tau) \, d\tau \qquad (16.4.8)$$

giving

$$P_{kj}(t) = \int_0^t \frac{a^k b^j}{(k-1)! j!} e^{-(a+b)\tau} \tau^{k-1+j} d\tau \qquad j < k \qquad (16.4.9)$$

and

$$P_{ik}(t) = \int_0^t \frac{a^i b^k}{i!(k-1)!} e^{-(a+b)\tau} \tau^{k-1+i} d\tau \qquad i < k \qquad (16.4.10)$$

If we let $t \to \infty$ in (16.4.9) and (16.4.10), we obtain the probabilities of ultimate win for A and B, respectively. Let $W_j^{(A)}$ be the probability of A's ultimate win when B's score is j and let $W_i^{(B)}$ be the probability of

B's ultimate win when A's score is i. We get

$$W_j^{(A)} = \frac{a^k b^j}{(k-1)! j!} \int_0^\infty e^{-(a+b)\tau} \tau^{k-1+j} d\tau$$

$$= \binom{k+j-1}{j} \left(\frac{a}{a+b}\right)^k \left(\frac{b}{a+b}\right)^j \qquad (16.4.11)$$

and

$$W_i^{(B)} = \frac{a^i b^k}{(k-1)! i!} \int_0^\infty e^{-(a+b)\tau} \tau^{k-1+i} d\tau$$

$$= \binom{k+i-1}{i} \left(\frac{a}{a+b}\right)^i \left(\frac{b}{a+b}\right)^k \qquad (16.4.12)$$

which may be recognized as negative binomial probabilities. $W_j^{(A)}$ and $W_i^{(B)}$ can be obtained directly from the transforms (16.4.6) if we note that

$$W_j^{(A)} = a \int_0^\infty P_{k-1,j}(t) \, dt = a\phi_{k-1,j}(0)$$

and

$$W_i^{(B)} = b \int_0^\infty P_{i,k-1}(t) \, dt = b\phi_{i,k-1}(0)$$

The ultimate probabilities in each case can be obtained from $\sum_{j=0}^{k-1} W_j^{(A)}$ and $\sum_{i=0}^{k-1} W_i^{(B)}$, respectively. Finally, consider the duration of the game. Let G be the random variable representing the length of the game. Then we have

$$P(G > t) = \sum_{i=0}^{k-1} \sum_{j=0}^{k-1} P_{ij}(t) \qquad (16.4.13)$$

Referring to (15.4.12), we can write

$$E[G] = \int_0^\infty P(G > t) \, dt$$

$$= \sum_{i=0}^{k-1} \sum_{j=0}^{k-1} \phi_{ij}(0)$$

$$= \sum_{i=0}^{k-1} \sum_{j=0}^{k-1} \binom{i+j}{i} \frac{a^i b^j}{(a+b)^{i+j+1}} \qquad (16.4.14)$$

REFERENCES

Ancker, C. J., Jr. (1967). "The Status of Development in the Theory of Stochastic Duels–II." *Oper. Res.* **15**, 388–406. This paper includes an annotated bibliography.

Bhashyam, N., and Singh, N. (1967). "Stochastic Duels with Varying Single Shot Kill Probabilities." *Oper. Res.* **15**, 233–244.

Brackney, H. (1959). "Dynamics of Military Combat." *Oper. Res.* **7**, 30–44.

Jain, G. C., and Nagabhushanam, A. (1974). "A Two-State Markovian Correlated Combat." *Oper. Res.* **22**, 440–444.

Kisi, T., and Hirose, T. (1966). "Winning Probability in an Ambush Engagement." *Oper. Res.* **14**, 1137–1138.

Lanchester, F. W. (1916). *Aircraft in Warfare: The Dawn of the Fourth Arm.* London: Constable.

Morse, P. M., and Kimball, G. E. (1951). *Methods of Operations Research.* New York: Technology Press of MIT and Wiley.

Smith, D. G. (1965). "The Probability Distribution of the Number of Survivors in a Two-Sided Combat Situation." *Oper. Res. Quart.* **16**, 429–437.

FOR FURTHER READING

Ancker, C. J., Jr. (1964). "Stochastic Duels with Limited Ammunition Supply." *Oper. Res.* **12**, 38–50.

Ancker, C. J., Jr. (1982). *One-on-One Stochastic Duels.* Military Applications Section, ORSA.

Ancker, C. J., Jr., and Gafarian A. V. (1965). "The Distribution of the Time Duration of Stochastic Duels." *Nav. Res. Log. Quart.* **12**, 275–294.

Ancker, C. J., Jr., and Williams, T. (1965). "Some Discrete Processes in the Theory of Stochastic Duels." *Oper. Res.* **13**, 202–216.

Barfoot, C. B. (1969). "The Attrition-Rate Coefficient: Some Comments on Seth Bonder's Paper and a Suggested Alternative Model." *Oper. Res.* **17**, 888–894.

Barefoot, C. B. (1974). "Markov Duels." *Oper. Res.* **22**, 318–330.

Bhashyam, N. (1970). "Stochastic Duels with Non-Repairable Weapons." *Nav. Res. Log. Quart.* **17**, 121–129.

Bonder, S. (1967). "The Lanchester Attrition-Rate Coefficient." *Oper. Res.* **15**, 221–232.

Dolansky, L. (1964). "Present State of the Lanchester Theory of Combat." *Oper. Res.* **12**, 344–358.

Grubbs, F. E., and Shuford, J. H. (1973). "A New Formulation of Lanchester Combat Theory." *Oper. Res.* **21**, 926–941.

Kimbleton, S. R. (1971). "Attrition Rates for Weapons with Markov Dependent Fire." *Oper. Res.* **19**, 698–706.

Nagabhushanam, A., and Jain, G. C. (1972). "Stochastic Duels with Damage." *Oper. Res.* **20**, 350–356.

Taylor, J. G. (1971). "A Note on the Solution to Lanchester-Type Equations with Variable Coefficients." *Oper. Res.* **19**, 709–712.

Taylor, J. G. (1980). "Theoretical Analysis of Lanchester Type Combat between Two Homogeneous Forces with Supporting Fires." *Nav. Res. Log. Quart.* **27**, 109–121.

Williams, T., and Ancker, C. J., Jr. (1963). "Stochastic Duels." *Oper. Res.* **11**, 803–817.

Chapter 17

SOCIAL AND
BEHAVIORAL PROCESSES

17.1 INTRODUCTION

The uncertainties associated with human behavior compel analytically oriented social and behavioral scientists to appeal routinely to mathematical models based on probability theory and stochastic processes. Several excellent papers and books have appeared on this area of applications of stochastic processes, and it is not our intention here to summarize all these developments. In this chapter we shall only try to point out the highlights of a few specific models, so as to indicate the scope of possible applications. For more extensive discussions of specific and general investigative areas, reference can be made to Anderson (1954), Bartholomew (1973), Blumen et al. (1955), Coleman (1964a, 1964b), Lazarsfeld (1951), Schuessler (1978, 1980), Singer and Spilerman (1976), Wasserman (1978, 1980), Young and Vassiliou (1974), and other articles cited in these works and the References given at the end of the chapter. Bartholomew (1973) gives a good account of the various models that will be of interest to us. The models we shall discuss are social mobility of individuals, industrial mobility of workers, educational systems, labor force size, the diffusion of information, and social networks. In all these cases we shall concentrate on one or two relevant research papers and discuss the problems as posed by the authors.

Despite the stochastic nature of social processes, their inherent unstructured nature renders realistic modeling not an easy task. In most cases the state space is finite, so a continuous parameter would be more realistic than a discrete one. However, for the sake of simplicity, in the first four examples we shall use Markov chain models even when a Markov process model is more realistic. Relaxation of such assumptions can be made by the reader in future investigations.

17.2   SOCIAL MOBILITY

In the study of social structure the influence of social status of a parent on that of the children as they grow up is a major consideration. Sons tend to adopt their fathers' professions more frequently in a conservative society than in a progressive and developing one. For this kind of influence a Markov model with generations as the time periods seem to be an ideal one. In fact a survey conducted in England and Wales during 1949 (as reported by Glass, 1954) justifies the use of such a model. Some conclusions of this survey will be mentioned later.

Consider a society consisting of m ($< \infty$) distinct social classes, such as professional, managerial, skilled, semiskilled, unskilled. We assume that in this society the male head of household alone decides the social class of the family, and his social status is dependent only on his father's. Under these assumptions, each person can be considered to be occupying one of the m states, and the social status of a family (considered as a continuing unit through sons and their sons, etc.) in a generation can be modeled as a finite Markov chain.

Most of the results derived for finite Markov chains in Chapter 4 are relevant to this model. To be specific, the limiting distribution of the Markov chain gives the structure of the society after it has existed for several generations. The convergence rates of the n-step transition probabilities measure the rates at which the society achieves stability in structure. Occupation times represent the pattern of movement of individuals in different generations through different social classes, and recurrence times of states indicate the influence of certain classes to the entire social structure. In addition the underlying process of social mobility needs further investigation. One major aspect is the problem of measuring social mobility itself. Prais (1955), in modeling the social statuses of fathers and sons in England and Wales for 1949 as a Markov chain on the basis of the survey data, suggests a measure for social mobility based on the time families would remain in different classes through generations.

Let α_j be the number of generations that a family occupies the jth social class at one stretch before moving out into another class. Let the transition probability matrix of the Markov chain be

$$P = \begin{bmatrix} P_{11} & P_{12} & \cdots & P_{1m} \\ \vdots & \vdots & & \vdots \\ P_{m1} & P_{m2} & \cdots & P_{mm} \end{bmatrix} \qquad (17.2.1)$$

Then P_{jj} is the probability that a family will remain in state j from one generation to the next. Thus we have

$$P(\alpha_j = n) = P_{jj}^{n-1}(1 - P_{jj})$$ (17.2.2)

where P_{jj}^n denotes the nth power of P_{jj}. This is a geometric distribution and the mean and variance of α_j are given by

$$E[\alpha_j] = \frac{1}{1 - P_{jj}}$$ (17.2.3)

$$V[\alpha_j] = \frac{P_{jj}}{(1 - P_{jj})^2}$$ (17.2.4)

Clearly $E[\alpha_j]$ is the average number of generations spent by a continuing family in a social class j, and if it can be compared with a similar measure for an ideal society, we have a measure for the social mobility of the present society. For the ideal society Prais (1955) uses one whose transition probability matrix of the social classes is given by the limiting distribution of the Markov chain under investigation. He calls it the "perfectly mobile" society. Let

$$\boldsymbol{\pi} = [\pi_1, \pi_2, \ldots, \pi_m]$$ (17.2.5)

be the limiting distribution of the Markov chain whose transition probability matrix is given by (17.2.1). Then Prais' perfectly mobile society has the transition probability matrix

$$\Pi = \begin{bmatrix} \pi_1 & \pi_2 & \cdots & \pi_m \\ \pi_1 & \pi_2 & \cdots & \pi_m \\ \cdot & \cdot & & \cdot \\ \cdot & \cdot & & \cdot \\ \pi_1 & \pi_2 & \cdots & \pi_m \end{bmatrix}$$ (17.2.6)

The Markov chain whose transition probability matrix is given by (17.2.6) has the property that the probability that the Markov chain will occupy any state after transition is independent of the present state. However, it should be noted that the probabilities of occupying different states are not the same. From this viewpoint, to call it a "stable but independent" society rather than a perfectly mobile society would be more appropriate. This would imply that the social structure is stable (the proportions of families in different classes remain the same), but social mobility is independent of the present state. For a perfectly mobile society in the strict sense, a doubly stochastic transition probability matrix with all elements equal to $1/m$ would be appropriate.

Now, for the Markov chain with a transition probability matrix given by (17.2.6), the occupation time α_j^c has the mean given by

$$E\left[\alpha_j^c\right] = \frac{1}{1 - \pi_j} \tag{17.2.7}$$

The ratio of mean occupation times (17.2.3) and (17.2.7) can be taken to be a measure of social mobility. This can be given as

$$\frac{E\left[\alpha_j\right]}{E\left[\alpha_j^c\right]} = \frac{1 - \pi_j}{1 - P_{jj}} \tag{17.2.8}$$

In a similar way the ratio

$$\frac{P_{jj}}{\pi_j} \tag{17.2.9}$$

can be considered to be a measure of immobility.

A different kind of measure of social mobility has been given by Bartholomew (1973) based on the expected number of class boundaries crossed in one transition. Assuming that the society is in a state of equilibrium and the limiting distribution of the classes is given by $\{\pi_j, j = 1, 2, \ldots, m\}$, the expected number of class boundaries crossed in one transition is

$$D = \sum_{i=1}^{m} \sum_{j=1}^{m} \pi_i P_{ij} |i - j| \tag{17.2.10}$$

Even though this measure describes the phenomenon of social mobility better than the measure based on occupation times and the comparison Markov chain (whose choice is rather arbitrary), its usefulness is limited. It cannot be used for comparing societies with different class structures. For $m = 2$, and the transition probability matrix P given by

$$P = \begin{bmatrix} 1 - a & a \\ b & 1 - b \end{bmatrix} \tag{17.2.11}$$

we get

$$D = \frac{2ab}{a + b} \tag{17.2.12}$$

which gives $D = 1$ if $a = 1$, $b = 1$, which refers to a perfectly unstable society. When $a = 0$ and $b = 0$, we have a perfectly stable society for which, by going back to (17.2.10), we can find $D = 0$.

As an illustration of the application of the Markov model to social mobility data, we quote some results appearing in Prais' article for the sample survey data which were published in Glass (1954). The random sample for the survey consisted of 3500 males resident in England and Wales and aged 18 years and over. Social statuses of fathers and sons were noted in these 3500 cases, and Prais' classification of statuses are as follows:

1. Professional and high administrative.
2. Managerial and executive.
3. Higher-grade supervisory and nonmanual.
4. Lower-grade supervisory and nonmanual.
5. Skilled manual and routine nonmanual.
6. Semiskilled manual.
7. Unskilled manual.

For the father–son status relationship, the following transition probability matrix was obtained:

Father \ Son	1	2	3	4	5	6	7
1	0.388	0.146	0.202	0.062	0.140	0.047	0.015
2	0.107	0.267	0.227	0.120	0.206	0.053	0.020
3	0.035	0.101	0.188	0.191	0.357	0.067	0.061
4	0.021	0.039	0.112	0.212	0.430	0.124	0.062
5	0.009	0.024	0.075	0.123	0.473	0.171	0.125
6	0.000	0.013	0.041	0.088	0.391	0.312	0.155
7	0.000	0.008	0.036	0.083	0.364	0.235	0.274

Considering social status as a finite Markov chain with the seven states as given here, Prais derives the limiting distribution from the transition probabilities for comparison with the actual distributions of social classes as can be estimated from the survey. These results are presented in Table 17.2.1. For the same data the measures of social mobility given by (17.2.8) are presented in Table 17.2.2. Looking at these tables, we can see that the Markov chain model fits remarkably well with the realistic situation, and the

Table 17.2.1

| | | | Actual distribution | |
Class		Limiting distribution	Father	Sons
1	Professional	0.023	0.037	0.029
2	Managerial	0.042	0.043	0.046
3	Higher-grade nonmanual	0.088	0.098	0.094
4	Lower-grade nonmanual	0.127	0.148	0.131
5	Skilled manual	0.409	0.432	0.409
6	Semiskilled manual	0.182	0.131	0.170
7	Unskilled manual	0.129	0.111	0.121

lower five social classes have somewhat similar mobility factors as compared to a stable but independent society. For other conclusions the readers are referred to Glass (1954) and Prais (1955).

In Chapter 4, when discussing the characteristics of a two-state Markov chain, we derived the means and variances of the number of processes occupying any one of the states after n ($n = 1, 2, 3, \ldots$) transitions, starting with an initial distribution. An extension of those results to the m-state Markov chain can be used to derive more information about the social structure at any given generation. One major inherent difficulty in interpreting such results would be in approximating the generation counts for different families by actual time. Assuming that such difficulties can be ironed out with good enough approximations, we shall derive a recurrence relation for the variances and covariances of the number of families in a given state during two consecutive generations. These results are due to Pollard (1966).

Table 17.2.2

Class		$E[\alpha_j]$	$E[\alpha_j^c]$	$E[\alpha_j]/E[\alpha_j^c]$
1	Professional	1.63	1.02	1.59
2	Managerial	1.36	1.04	1.30
3	Higher-grade nonmanual	1.23	1.10	1.12
4	Lower-grade nonmanual	1.27	1.15	1.11
5	Skilled manual	1.90	1.69	1.12
6	Semiskilled manual	1.45	1.22	1.19
7	Unskilled manual	1.38	1.15	1.20

Let $C_j^{(n)}$ be the number of families in social class j ($j = 1, 2, \ldots, m$) at the nth generation. We shall assume that $\sum_{j=1}^{m} C_j^{(n)} = C$, a constant for $n = 0, 1, 2, \ldots$. For the sake of convenience we shall consider the variance and covariance terms separately. We have, for $i \neq j$,

$$\text{Cov}\left\{ C_i^{(n+1)}, C_j^{(n+1)} \right\} = E\left[C_i^{(n+1)} C_j^{(n+1)} \right] - E\left[C_i^{(n+1)} \right] E\left[C_j^{(n+1)} \right]$$

(17.2.13)

Let $C_{ri}^{(n)}$ be the number of families that have moved from state r to state i during the nth generation. We have

$$C_i^{(n+1)} = \sum_{r=1}^{m} C_{ri}^{(n)} \qquad n \geq 0 \tag{17.2.14}$$

and

$$E\left[C_i^{(n+1)} \right] = \sum_{r=1}^{m} E\left[C_{ri}^{(n)} \right]$$

$$= \sum_{r=1}^{m} E\left[C_r^{(n)} \right] P_{ri} \tag{17.2.15}$$

Using (17.2.14) and (17.2.15) in (17.2.13), we get

$$\text{Cov}\left\{ C_i^{(n+1)}, C_j^{n+1} \right\} = E\left[\left(\sum_{r=1}^{m} C_{ri}^{(n)} \right) \left(\sum_{s=1}^{m} C_{sj}^{(n)} \right) \right]$$

$$- E\left[\sum_{r=1}^{m} C_{ri}^{(n)} \right] E\left[\sum_{s=1}^{m} C_{sj}^{(n)} \right]$$

$$= \sum_{r,s=1}^{m} E\left[C_{ri}^{(n)} C_{sj}^{(n)} \right] - \sum_{r,s=1}^{m} E\left[C_{ri}^{(n)} \right] E\left[C_{sj}^{(n)} \right]$$

(17.2.16)

But we also have

$$E\left[C_{ri}^{(n)} C_{sj}^{(n)} | C_r^{(n)}, C_s^{(n)} \right] = C_r^{(n)} C_s^{(n)} P_{ri} P_{sj} \qquad r \neq s \tag{17.2.17}$$

$$E\left[C_{ri}^{(n)} C_{rj}^{(n)} | C_r^{(n)} \right] = C_r^{(n)} \left(C_r^{(n)} - 1 \right) P_{ri} P_{rj} \tag{17.2.18}$$

Equation (17.2.18) follows from the fact that if X_1 and X_2 have a multinomial distribution with parameters n, p_1, and p_2, then

$$E[X_1 X_2] = n(n-1) p_1' p_2$$

Thus we write

$$\text{Cov}\left\{ C_i^{(n+1)}, C_j^{(n+1)} \right\}$$

$$= \sum_{r \neq s=1}^{m} P_{ri} P_{sj} E\left[C_r^{(n)} C_s^{(n)} \right] + \sum_{r=1}^{m} P_{ri} P_{rj} E\left[C_r^{(n)}\left(C_r^{(n)} - 1 \right) \right]$$

$$- \sum_{r,s=1}^{m} P_{ri} P_{sj} E\left[C_r^{(n)} \right] E\left[C_s^{(n)} \right]$$

$$= \sum_{r,s=1}^{m} P_{ri} P_{sj} \left[E\left(C_r^{(n)} C_s^{(n)} \right) - E\left(C_r^{(n)} \right) E\left(C_s^{(n)} \right) \right]$$

$$- \sum_{r=1}^{m} P_{ri} P_{rj} E\left[C_r^{(n)} \right]$$

$$= \sum_{r,s=1}^{m} P_{ri} P_{sj} \text{Cov}\left\{ C_r^{(n)}, C_s^{(n)} \right\} - \sum_{r=1}^{m} P_{ri} P_{rj} E\left[C_r^{(n)} \right] \qquad (17.2.19)$$

Proceeding in a similar manner, we get

$$V\left\{ C_i^{(n+1)} \right\} = \sum_{r=1}^{m} \sum_{s=1}^{m} P_{ri} P_{si} \text{Cov}\left\{ C_r^{(n)}, C_s^{(n)} \right\} + \sum_{r=1}^{m} P_{ri}(1 - P_{ri}) E\left[C_r^{(n)} \right]$$

$$(17.2.20)$$

The recurrence relations (17.2.19) and (17.2.20) can now be used to derive the variances and covariances of the number of families in different classes in successive generations. The recurrence relation (17.2.15) gives the expected values of such numbers in a similar way.

In particular, if $P_{ij} = \pi_j$, $i = 1, 2, \ldots, m$ (i.e., in Prais' perfectly mobile society),

$$E\left[C_i^{(n+1)} \right] = \sum_{r=1}^{m} E\left[C_r^{(n)} \right] \pi_i \qquad n \geq 0 \qquad (17.2.21)$$

giving

$$E\left[C_i^{(n+1)} \right] = \pi_i C \qquad (17.2.22)$$

We also have

$$\sum_{r,s} \text{Cov}\left[C_r^{(n)}, C_s^{(n)} \right] = 0$$

and hence,

$$\text{Cov}\left[C_i^{(n+1)}, C_j^{(n+1)}\right] = -\pi_i \pi_j C \qquad i \neq j \qquad (17.2.23)$$

$$V\left[C_i^{(n+1)}\right] = \pi_i (1 - \pi_i) C \qquad (17.2.24)$$

Evidently $C_j^{(n)}$ has a multinomial distribution with parameters C and $\{\pi_j\}_{j=1}^{m}$.

17.3 INDUSTRIAL MOBILITY OF LABOR

One of the social phenomena dependent on time is the mobility of the industrial labor force. If we select a suitable time scale, we can always find that a portion of the labor force in an industry has changed from one time period to another. On a small scale even part of the faculty of educational institutions exhibit the same kind of mobility as the industrial work force. The epochs of time individuals change jobs vary. This presents several problems in modeling, and therefore any attempt at considering the same time scale for all the individuals would at best be an approximation.

Let us assume a closed system of employers (called industries) and employees so that at a given time an employee would be at the service of any one of the given m industries. We do not rule out the possibility of unemployment being designated as one of the m industries. The number of employees in the entire system remains the same during the period of investigation, with the losses in work force being made good by the gains.

Suppose this industrial labor system is observed at regular intervals of time, and the number of employees changing their jobs during any such interval and their distributions are noted. Assuming that the present experience of the employees in one industry affects their choice of another industry, a Markov chain model seems to be a realistic approximation of the employment behavior of the work force. If no assumptions can be made about the time spent by an employee with an employer, more general models than a Markov process would be needed to represent employment behavior. A semi-Markov process would perhaps serve as a good approximation. But if we can realistically assume that an employee changes jobs in a Poisson fashion—the length of time this individual would be at the service of any employer has an exponential distribution—a Markov chain model is satisfactory.

Thus let $\{X_n, \, n = 0, 1, 2, \dots\}$ be the state of employment of an individual at the time of the nth observation, and let the state space of X_n be $\{1, 2, \dots, m\}$ representing the m industries. The transition probability matrix of $\{X_n, \, n = 0, 1, 2, \dots\}$ can be estimated from observed data in the usual manner, and useful information on the characteristics of industrial mobility can be obtained as in the case of social mobility. An illustration of this type of Markov chain model is provided by Blumen et al. (1955) in their study of the industrial mobility of labor, based on data derived from the Continuous Work History Sample (CWHS) of the Bureau of Old-Age and Survivors Insurance (Perlman and Mandel, 1944). The CWHS is a 1% sample of all workers who have been covered by employment insurance since the inception of the Social Security System in 1937. Blumen et al.'s study used a 10% sample of CWHS. Eleven types of industry classification were used in identifying the employers; these ranged from agricultural, construction, and printing and publishing to business professions and government. In addition unemployment was used as one of the eleven categories. Extensive tables based on quarterly data are given in Blumen et al. (1955), and a glance through the transition probability matrices and the characteristics derived in them will satisfy the reader of the validity of such models. As an example we shall quote from pp. 64–65 of the report in which the authors have combined the eleven classifications into five and obtained the limiting distribution of these classifications. Let the five classifications be designated as follows:

CDE   Household products, printing and publishing, metal equipment, miscellaneous manufacturing, and so forth.

   G   Wholesale and retail trade.

 FH   Transportation, communication, utilities, and business.

ABJK   Agriculture, construction, professional service, and government.

   U   Unemployment.

The estimated one-quarter transition probability matrix for males 20–24 years of age is

	CDE	G	FH	ABJK	U
CDE	0.832	0.033	0.013	0.028	0.095
G	0.046	0.788	0.016	0.038	0.112
FH	0.038	0.034	0.785	0.036	0.107
ABJK	0.054	0.045	0.017	0.728	0.156
U	0.082	0.065	0.023	0.071	0.759

The limiting distribution of this matrix is compared in Table 17.3.1 with the

Table 17.3.1

Classification	Limiting distribution	Actual distribution 1947–1949
CDE	0.27	0.282
G	0.18	0.170
FH	0.08	0.068
ABJK	0.15	0.137
U	0.32	0.343

actual distribution of workers in the age group 20–24 for these five industry groups for the period 1947–1949. The agreement of results in the last two columns of Table 17.3.1 is truly remarkable, thus justifying the validity of the Markov model for the phenomenon of industrial mobility.

An interesting modification suggested by Blumen et al. (1955) to account for the inclination of a certain portion of the labor force to stay with the same employer for a long time needs further investigation. They classify the labor force as consisting of stayers and movers. Thus, if there are m industries, consider a fraction s_i $(i = 1, 2, \ldots, m)$ of the labor staying with industry i continuously. Let

$$R = \begin{bmatrix} R_{11} & R_{12} & \cdots & R_{1m} \\ \vdots & \vdots & & \vdots \\ R_{m1} & R_{m2} & \cdots & R_{mm} \end{bmatrix} \tag{17.3.1}$$

be the transition probability matrix for the mover part of the labor force. The transition probability matrix P, for the entire mover-stayer model can then be represented by

$$P = \begin{bmatrix} s_1 + (1 - s_1)R_{11} & (1 - s_1)R_{12} & \cdots & (1 - s_1)R_{1m} \\ (1 - s_2)R_{21} & s_2 + (1 - s_2)R_{22} & \cdots & (1 - s_2)R_{2m} \\ \vdots & \vdots & & \vdots \\ (1 - s_m)R_{m1} & (1 - s_m)R_{m2} & \cdots & s_m + (1 - s_m)R_{mm} \end{bmatrix}$$

$$\tag{17.3.2}$$

$$= S + (I - S)R \tag{17.3.3}$$

where S is the diagonal matrix

$$S = \begin{bmatrix} s_1 & & & \\ & s_2 & & \bigcirc \\ & & \ddots & \\ & \bigcirc & & s_m \end{bmatrix} \tag{17.3.4}$$

For the n-step transition probabilities we assume that no transitions occur for the s_i $(i = 1, 2, \ldots, m)$ portion of the work force, and hence we get

$$\|\tilde{P}_{ij}^{(n)}\| = S + (I - S)R^n \tag{17.3.5}$$

Note that $\|P_{ij}^{(n)}\| \neq P^n$ as in the ordinary chain. As $n \to \infty$, let $\lim_{n \to \infty} R^n = \Pi'$, where Π' has identical rows, each row representing the limiting probability vector

$$[\pi_1', \pi_2', \cdots, \pi_m'] \tag{17.3.6}$$

for the mover portion of the labor. Let $\lim_{n \to \infty} \|P_{ij}^{(n)}\| = \Pi$. From (17.3.5) and (17.3.6), we then get

$$\Pi = S + (I - S)\Pi'$$

$$= \begin{bmatrix} s_1 + (1 - s_1)\pi_1' & (1 - s_1)\pi_2' & \cdots & (1 - s_1)\pi_m' \\ (1 - s_1)\pi_1' & s_2 + (1 - s_2)\pi_2' & \cdots & (1 - s_2)\pi_m' \\ \vdots & \vdots & & \vdots \\ (1 - s_m)\pi_1' & (1 - s_m)\pi_2' & \cdots & s_m + (1 - s_m)\pi_m' \end{bmatrix}$$

$$\tag{17.3.7}$$

It is clear from (17.3.7) that the limiting distribution of the work force in different industries is not independent of the initial state.

If the initial distribution of employees in different industries is known, the distribution of employees after n time periods can be obtained from (17.3.5) and (17.3.7) for $n < \infty$ as well as $n \to \infty$. Let $\{ p_j^{(n)} \}_{j=1}^m$ be the distribution of employees after n transitions. For the vector $p^{(n)} = (p_1^{(n)}, p_2^{(n)}, \ldots, p_m^{(n)})$, we then get

$$p^{(n)} = p^{(0)}S + p^{(0)}(I - S)R^n \tag{17.3.8}$$

and for $p^* = \lim_{n \to \infty} p^{(n)}$

$$p^* = p^{(0)}S + p^{(0)}(I - S)\Pi' \tag{17.3.9}$$

A major practical problem in this mover-stayer Markov model concerns the estimation of the elements of the transition probability matrix R and the fraction of movers and stayers in each industry. In a given period of time from industrial records it is not possible to identify the movers and stayers but only the overall number who stay with their present employers for another period and the number of those who move to other employers. With these data, by methods given in Chapter 5, it is possible to estimate the elements of the transition probability matrix

$$P = S + (I - S)R$$

If n_i is the number of employees in industry i at one period, out of whom $n_{ij}^{(1)}$ move into industry j ($j = 1, 2, \ldots, m$), then the (i, j)th element of matrix P has the following maximum likelihood estimate:

Maximum likelihood estimate $\left[s_i + (1 - s_i)R_{ii} \right] = \dfrac{n_{ii}^{(1)}}{n_i} = \hat{P}_{ii}$

$$i = 1, 2, \ldots, m$$

$$\tag{17.3.10}$$

Maximum likelihood estimate $\left[(1 - s_i)R_{ij} \right] = \dfrac{n_{ij}^{(1)}}{n_i} = \hat{P}_{ij} \qquad i \neq j$

$$\tag{17.3.11}$$

To derive separate estimates for s_i ($i = 1, 2, \ldots, m$) and R_{ij} ($i, j = 1, 2, \ldots, m$), additional information is needed. We shall give a method of estimation suggested by Goodman (1961). Let $f_i^{(k)}$ be the fraction of those workers in industry i ($i = 1, 2, \ldots, m$) initially who remained continuously employed in that industry for the next k periods. Clearly we have

$$\text{Estimate}\left[s_i + (1 - s_i)R_{ii}^k \right] = f_i^{(k)} \tag{17.3.12}$$

and hence

$$\text{Estimate}\left[(1 - s_i)(1 - R_{ii}^k) \right] = 1 - f_i^{(k)} \tag{17.3.13}$$

Note that in (17.3.12) and (17.3.13) R_{ii}^k is the kth power of the one-step transition probability R_{ii}. Combining (17.3.10) and (17.3.13), we get

$$\text{Estimate}\left[\frac{R_{ii} - R_{ii}^k}{1 - R_{ii}^k}\right] = \frac{\hat{P}_{ii} - f_i^{(k)}}{1 - f_i^{(k)}} = h_{ii} \qquad (17.3.14)$$

and combining (17.3.11) and (17.3.13), we also get

$$\text{Estimate}\left[\frac{R_{ij}}{1 - R_{ii}^k}\right] = \frac{\hat{P}_{ij}}{1 - f_i^{(k)}} = h_{ij} \qquad (17.3.15)$$

Rewriting (17.3.14),

$$\hat{R}_{ii} = h_{ii} + (1 - h_{ii})\hat{R}_{ii}^k \qquad (17.3.16)$$

which has a unique solution in the interval $0 \leq \hat{R}_{ii} < 1$ and $0 \leq h_{ii} < 1 - (1/k)$. The solution $\hat{R}_{ii}$ is such that

$$\hat{R}_{ii} - h_{ii} = (1 - h_{ii})R_{ii}^k \geq 0 \qquad (17.3.17)$$

When $h_{ii} \geq 1 - (1/n)$, we set $\hat{R}_{ii} = 1$. Equation (17.3.17) can be solved by graphical methods or successive approximation. Now going back to (17.3.15), R_{ij} can be estimated from

$$\hat{R}_{ij} = h_{ij}\frac{1 - \hat{R}_{ii}}{1 - h_{ii}} \qquad (17.3.18)$$

When the initial work force in industry i is large, the estimators obtained here are consistent. Once we have the estimators for R_{ii} and R_{ij}, estimators for s_i can be readily obtained from (17.3.10), for it can be rewritten as (with $\hat{s}_i$ as the estimate of s_i)

$$\hat{s}_i(1 - \hat{R}_{ii}) = \hat{P}_{ii} - \hat{R}_{ii}$$

and hence

$$\hat{s}_i = \frac{\hat{P}_{ii} - \hat{R}_{ii}}{1 - \hat{R}_{ii}} \qquad (17.3.19)$$

The estimators for the limiting probability vector $\boldsymbol{\pi}' = (\pi_1', \pi_2', \ldots, \pi_m')$ are found by simply solving

$$\boldsymbol{\pi}'\hat{\boldsymbol{R}} = \boldsymbol{\pi}' \qquad \sum_{i=1}^{m} \pi_i' = 1 \qquad (17.3.20)$$

An indirect method of estimation has been suggested by Blumen et al. (1955) in their initial report. This is based on observing the initial state and the state after a large number of transitions for every employee and assuming that, by that time, the process has settled down to equilibrium. Their estimates are obtained under the assumption that π_j' ($j = 1, 2, \ldots, m$) are small, such that $\pi_j'^2$ are negligible. When this assumption does not hold, estimates obtained are evidently biased. For a discussion of the estimation procedure as well as the properties see Goodman (1961), who also suggests some improvements for their method. In the Blumen-Kogan-McCarthy method, various estimates are obtained in the following sequence:

1. Maximum likelihood estimates of π_j' ($j = 1, 2, \ldots, m$) are derived under some restrictive assumptions.
2. Estimates of s_i are obtained from the relation $\pi_{ii} = s_i + (1 - s_i)\pi_i'$, where the maximum likelihood estimate of $P_{ii}^{(k)}$ for large k is considered to be the estimate of π_{ii}.
3. Estimates of R_{ij} ($i, j = 1, 2, \ldots, m$) are obtained from the one-step transition probabilities

$$P_{ii} = s_i + (1 - s_i)R_{ii} \qquad P_{ij} = (1 - s_i)R_{ij}$$

using the maximum likelihood estimates of P_{ij} ($i, j = 1, 2, \ldots, m$) in the process.

17.4   EDUCATIONAL ADVANCEMENT

Exercise 38 at the end of Chapter 4 required the reader to construct a Markov model for an educational system for predicting the number of graduates added to the work force. The basic assumption to be made for this model is that the advancement of a college student through the freshman, sophomore, junior, and senior years can be modeled as a Markov chain. A modified version of such a model was used by Gani (1963) in projecting enrollments and degrees awarded in Australian universities. Gani's model and results will be discussed in this section. This model can also be used to represent promotions in an organization; in view of this we shall first formulate it as a slightly more general problem. Let m be the number of stages a student has to pass through before graduating or an employee has to pass through in a promotion process. We assume that there are no demotions. Let P_{ij} ($j = i, i + 1, \ldots, m$) be the probability that an individual in state i at one period will be found in state j at the next period. We shall assume $\sum_{j=1}^{m} P_{ij} < 1$ so that with probability $1 - \sum_{j=1}^{m} P_{ij}$, the individual will be out of the system during the next period. In the educational process this may be due to dropping out, and in organizations this may be due to resignations, retirements, and so on.

Let $C_i^{(n)}$ be the number of individuals at state i after n periods, and let

$$E\left[C_i^{(n)}\right] = \gamma_i^{(n)} \tag{17.4.1}$$

Further let N_n be the number of new entrants to the system during the nth period, with $N_n \rho_i$ of these entering the ith state $(i = 1, 2, \ldots, m)$. Clearly the expected values $\gamma_i^{(n)}$ satisfy the recurrence relations

$$\gamma_i^{(n)} = \sum_{r=1}^{i} P_{ri} \gamma_r^{(n-1)} + N_n \rho_i \tag{17.4.2}$$

Let Q be the transpose of the transition probability matrix (which is substochastic) so that

$$Q = \begin{bmatrix} P_{11} & & & \\ P_{12} & P_{22} & & \bigcirc \\ P_{13} & P_{23} & & \\ \vdots & \vdots & \ddots & \\ P_{1m} & P_{2m} & \cdots & P_{mm} \end{bmatrix} \tag{17.4.3}$$

Also write

$$\gamma_n = \begin{bmatrix} \gamma_1^{(n)} \\ \gamma_2^{(n)} \\ \vdots \\ \gamma_m^{(n)} \end{bmatrix} \qquad \rho = \begin{bmatrix} \rho_1 \\ \rho_2 \\ \vdots \\ \rho_m \end{bmatrix} \tag{17.4.4}$$

In matrix notations (17.4.2) can be written

$$\gamma_n = Q\gamma_{n-1} + N_n \rho \tag{17.4.5}$$

which gives

$$\gamma_n = Q(Q\gamma_{n-2} + N_{n-1}\rho) + N_n \rho$$

$$= Q^2 \gamma_{n-2} + N_{n-1} Q\rho + N_n \rho \tag{17.4.6}$$

Proceeding in this manner, we can write

$$\gamma_n = Q^{k+1}\gamma_{n-k-1} + \sum_{j=0}^{k} N_{n-j}Q^j\rho \qquad (17.4.7)$$

Letting $k \to \infty$ and noting that $Q^k \to 0$ as $k \to \infty$ because of its substochastic nature, an alternate form can be obtained:

$$\gamma_n = \sum_{j=0}^{\infty} N_{n-j}Q^j\rho \qquad (17.4.8)$$

Knowing Q, ρ, and N_r $(r = 0, 1, 2, \ldots)$, (17.4.8) can be used to predict the expected number of individuals in every state after n periods.

For the educational system and most organizations it is reasonable to assume that $P_{ij} = 0$ for $j \neq i, i + 1$. In other words, the advancement of the individual through the system is at most one state (level) for any period. Then we get

$$Q = \begin{bmatrix} P_{11} & & & & \\ P_{12} & P_{22} & & \bigcirc & \\ & P_{23} & P_{33} & & \\ & & P_{34} & & \\ & \bigcirc & & \ddots & \\ & & & P_{m-1,m} & P_{mm} \end{bmatrix} \qquad (17.4.9)$$

Assuming that P_{ii}'s are all distinct for $i = 1, 2, \ldots, m$, $P_{11}, P_{22}, \ldots, P_{mm}$ are seen to be the eigenvalues of the Q matrix, and therefore

$$Q = A^{-1}\Lambda A \qquad (17.4.10)$$

where

$$\Lambda = \begin{bmatrix} P_{11} & & & \\ & P_{22} & \bigcirc & \\ & & \ddots & \\ \bigcirc & & & P_{mm} \end{bmatrix} \qquad (17.4.11)$$

and A and A^{-1} are lower triangular matrices with unit elements in the diagonal. The remaining elements are

$$a_{ij} = \prod_{k=j}^{i-1} P_{k,k+1} (P_{ii} - P_{kk})^{-1} \tag{17.4.12}$$

and

$$a_{ij}^{-1} = \prod_{k=j+1}^{i} P_{k-1,k} (P_{jj} - P_{kk})^{-1} \qquad i > j$$

respectively. With the use of the diagonalized matrix Q, an alternate form for γ_n can be given as

$$\gamma_n = \sum_{j=0}^{\infty} N_{n-j} A^{-1} \Lambda^j A \rho \tag{17.4.13}$$

The total enrollment at the start of the nth year can be obtained by summing over all elements of the column vector γ_n. This is accomplished by premultiplying γ_n by the unit row vector in which all elements are 1.

As a specific example Gani (1963) uses the estimates of transition probabilities of student advancement in Australian universities. The students pass through seven states: first year, second year, Bachelor's pass, Bachelor's honors, Master's, and two years of doctoral work. Based on the figures given in the *Murray Committee of Australian Universities* (1957, Canberra, Commonwealth Government Printer), Gani obtains the estimates of transition probabilities as

$$P_{11} = 0.15 \qquad P_{22} = 0.11 \qquad P_{33} = 0.10$$

$$P_{12} = 0.61 \qquad P_{23} = 0.71 \qquad P_{34} = 0.162$$

In addition it is found that 81% of the third-year students pass and become Bachelor's pass graduates (out of whom 20% proceed to the fourth year), 5% of the fourth-year students repeat, and 91% pass to become honors graduates. Estimates of the number of Bachelor's degrees (pass and honors) can be obtained:

$$B_n + H_n = (0.81)\gamma_3^{(n-1)} + (0.91)\gamma_4^{(n-1)}$$

where B_n and H_n represent the expected numbers of pass and honors graduates at the end of the nth year. For the years 1955–1960 the estimated

Table 17.4.1

Year	N_n	$B_n + H_n$ Estimated	$B_n + H_n$ Actual	Error (%)
1950	5208			
1951	5084			
1952	5033			
1953	4958			
1954	4951			
1955	5791	3114	2967	+ 5.0
1956	6881	3068	3089	− 0.7
1957	7401	3048	3025	+ 0.8
1958		3335	3382	− 1.4
1959		3876	3615	+ 7.2
1960		4313	4184	+ 3.1

figures obtained through these formulae and the actual figures as reported in the University Statistics of the Commonwealth Bureau of Census and Statistics, 1960, are given in Table 17.4.1. It should be noted that all new entrants are assumed to enter the system only through the first year. The closeness of the estimated results to the actual figures validates the use of Markov models for educational systems of this type. The model developed here is general enough to be applicable in most other similar situations. The reader is now advised to take a second look at Exercise 38 of Chapter 4.

17.5   LABOR FORCE PLANNING AND MANAGEMENT

The use of analytical models in labor force planning and management (also known as manpower planning and management in the literature) increased rapidly in the 1970s. Among these stochastic models have dominated due to the inherent stochastic nature of the social and behavioral processes. As well as more theoretical papers published in methodological journals, papers with case histories of applications have been appearing with increasing frequency in field journals. Also statistics is becoming more and more acceptable in court evidence on issues related to discrimination in hiring and promotion. In short, stochastic models have become an integral part of organizational decision making.

As illustrated in the last three sections, while using theoretical results, the need for extensions and modifications is quite common. In this section we shall illustrate one of the modifications that has been used to account for the various factors that influence the promotion process in an organization. The procedure outlined next is due to Young and Vassiliou (1974).

Consider a graded organization through which people move by recruitment, promotion, and wastage (retirement, retrenchment, termination, etc.). Let there be m grades: $1, 2, \ldots, m$, the movement through grades being $i \to j$, where $j < i$. Let $\eta_i^{(n)}$ be the expected number of people in grade i at the beginning of the nth period. Normally in an organization the number of people in a grade is smaller than the number of approved positions in that grade; only in exceptional circumstances, such as a period of contraction, the actual number is larger than the approved number. We shall identify the approved number of positions as the inherent profile, and we shall denote it by $\eta_i^{*(n)}$ for grade i at the beginning of the nth period. We shall also define the following additional symbols:

$l_i^{(n)} =$ the probability that an individual in grade i leaves the organization during period n

$p_{ij}^{(n)} =$ the probability that an individual moves from grade i to grade j during period n

$\rho_i^{(n)} =$ expected number of recruits to grade i during period n

In general $p_{ij}^{(n)} > 0$ for $j > i$ or $j < i$, where $j > i$ case indicates demotions. One of the significant aspects of this model is the determination of $p_{ij}^{(n)}$, using the interaction between the number of available suitable candidates and the number of vacancies for promotion. The existence of this interaction has been noted as the "ecological principle" by the authors. Let $\delta_j^{(n)}$ be the expected number of vacancies occurring in grade j during period n, and let $a_{ij}^{(n)}$ be the probability that an individual in grade i will be acceptable as a candidate for one of the positions in grade j during period n. Since there are $\delta_j^{(n)}$ available vacancies, we get

$$p_{ij}^{(n)} = a_{ij}^{(n)} \delta_j^{(n)} \qquad (17.5.1)$$

For purposes of determining $p_{ij}^{(n)}$, we may identify the expected number of people who remain in grade j at the end of period n, having been in the grade at the start of the period, as $\eta_j^{+(n)}$ and write

$$\eta_j^{+(n)} = \left[1 - l_j^{(n)} - \sum_{k=1}^{j-1} p_{j,j-k}^{(n)} \right] \eta_j^{(n)} \qquad (17.5.2)$$

Using this expression, the expected number of vacancies occurring in grade

j during period n can be given as

$$\delta_j^{(n)} = \eta_j^{*(n+1)} - \eta_j^{+(n)} \qquad (17.5.3)$$

The problem now is the estimation of the inherent profile. The authors use what can be called an "inertia principle" in its determination. It is generally accepted that organizational systems tend to continue in a uniform way until acted on by strong externally applied forces. Using this principle, one may argue that a grade's share of the inherent profile and the probabilities $a_{ij}^{(n)}$ remain fairly constant over the years and suggest a least squares method of estimation (for details, see Young and Vassiliou, (1974).

Using the probabilities $l_i^{(n)}$ and $p_{ij}^{(n)}$, the expected number of people in grade i during two consecutive periods n and $(n + 1)$ can be recursively related by the following equation:

$$\eta_i^{(n+1)} = \left[1 - l_i^{(n)} - \sum_{k=1}^{i-1} p_{i,i-k}^{(n)} \right] \eta_i^{(n)}$$

$$+ \sum_{j \neq i} p_{ji}^{(n)} \eta_j^{(n)} + \rho_i^{(n)} \qquad i = 1, 2, \ldots, m \qquad (17.5.4)$$

Using this model, the authors analyze data from a British firm for the period 1955–1966. They use the data for the first seven years for parameter estimation and the remaining years for validation of the model. When enough data were available, they conclude that the evidence suggests that "the model is resilient and responds well after sudden shocks occur in the system." For a detailed analysis, the readers are referred to the article.

The preceding study is but an example of how realities of assumptions have made researchers explore modifications and extensions to the basic model. For a comprehensive discussion of stochastic models in social and behavioral processes, the readers are referred to Bartholomew (1973). A selected bibliography is supplied at the end of the chapter under the heading "For Further Reading."

17.6 DIFFUSION OF INFORMATION

A social phenomenon that lends itself to stochastic modeling is the diffusion of information among members of a social group. After a news item or a rumor has originated, the individuals in the first group of "hearers" become "spreaders," and their "hearers" turn "spreaders," and so on, until the news diffuses through the population. In some cases a "spreader" may get "stifled" by trying to spread to one who has already heard about it. Several modifications can be made in this pattern, and the simplest of all is the one that can be modeled as a pure birth process.

This social phenomenon is very similar to the epidemic processes on which considerable analytic work has been carried out. For an extensive discussion of these, the reader is referred to Bailey (1957). For an epidemic process the disease is carried by "infectives" to "susceptibles," possibly after an "incubation period." Therefore, but for the incubation period, both processes have the same characteristics. The pure birth process is also a model for a simple epidemic, as demonstrated by Bailey (1957); Taga and Isii (1959) discuss it in the context of information diffusion. Here we give their formulation for the simplest case and indicate some of the information that can be derived from such a model. Bartholomew (1973, Chaps. 9, 10) gives an excellent exposition of this topic, modeled mainly as a modification of an epidemic process.

Suppose the social group consists of N individuals called members (or families or groups), and let the probabilities of events occurring in this group during a time interval $(t, t + \Delta t]$ be given as follows: Let S be the original source and let

$P\{$A member of the group will receive information from $S\}$
$$= \alpha \Delta t + o(\Delta t)$$
$P\{$Any of the n members in the group will receive information from $S\}$
$$= n\alpha \Delta t + o(\Delta t)$$
$P\{$No one among the n-member group will receive information from $S\}$
$$= 1 - n\alpha \Delta t + o(\Delta t)$$
$P\{$More than one member will receive information from $S\}$
$$= o(\Delta t)$$
$P\{$Any of the n members of the group will receive information from a spreader$\}$
$$= n\beta \Delta t + o(\Delta t)$$
$P\{$No one among the n members will receive information from the spreader$\}$
$$= 1 - n\beta \Delta t + o(\Delta t)$$
$P\{$More than one member will receive information from a spreader$\}$
$$= o(\Delta t)$$

Under these assumptions, when n out of N have already received the information, the probability that one more member of the group will receive information either from S or from one of the spreaders is given by $\lambda_n \Delta t + o(\Delta t)$, where

$$\lambda_n = (N - n)(\alpha + n\beta) \qquad n = 0, 1, 2, \dots, N - 1 \qquad (17.6.1)$$

The probability that during $(t, t + \Delta t]$ the number who have received information will remain at n is then given by $1 - \lambda_n \Delta t + o(\Delta t)$, and the probability that the number will increase by more than one is $o(\Delta t)$. In all these assumptions it should be recalled that $o(\Delta t)$ is such that $o(\Delta t)/\Delta t \to 0$ as $\Delta t \to 0$.

Let $X(t)$ be the number of individuals who are in possession of the information at time t, and let

$$P_n(t) = P[X(t) = n | X(0) = 0] \qquad (17.6.2)$$

Clearly, under the Markov assumptions made before, $X(t)$ is a pure birth process with parameter λ_n ($n = 0, 1, 2, \ldots, N$). Referring to Section 7.3, we get

$$P_n(t) = \sum_{\nu=0}^{n} A_n^{(\nu)} e^{-\lambda_\nu t} \qquad (17.6.3)$$

where

$$A_n^{(\nu)} = \frac{\lambda_0 \lambda_1 \cdots \lambda_{n-1}}{\Pi_{i=0, i \neq \nu}^{n} (\lambda_i - \lambda_\nu)} \qquad (17.6.4)$$

For small values of N these probabilities can be numerically evaluated. When N is large, their properties can only be derived through indirect means. For further discussion of such numerical and graphical methods, the reader is referred to Bailey (1957).

Taga and Isii also consider related stochastic processes such as the number of individuals who have received information but have not become spreaders and the number who have received the information directly from the original source. In addition to getting recurrence relations for the joint distributions of these stochastic processes, they derive useful relations to determine their expected values, variances, and estimates.

Another characteristic of importance in the preceding model is the time until the information gets through to a given number of individuals. Let W_n be the waiting time until the nth member receives the information. From the theory of general birth and death processes, we know that when the number of spreaders is k, the time until one or more individual joins the group of spreaders has an exponential distribution with mean $1/\lambda_k$. Thus W_n is the sum of n independent, exponential random variables, and hence

$$E[W_n] = \sum_{k=0}^{n} \frac{1}{\lambda_k}$$

$$= \sum_{k=0}^{n} \frac{1}{(N - k)(\alpha + k\beta)} \qquad (17.6.5)$$

which can be easily evaluated. We can also write

$$V[W_n] = \sum_{k=0}^{n} \frac{1}{\lambda_k^2}$$

$$= \sum_{k=0}^{n} \frac{1}{(N-k)^2(\alpha + k\beta)^2} \qquad (17.6.6)$$

A modification of this model suggested by Daley and Kendall (1965) is realistic for some communities where spreaders lose interest in spreading the news if they feel that the hearer already has the information. The population is divided into three groups: ignorants, spreaders, and stiflers. A spreader who contacts a stifler turns a stifler; if a spreader contacts another spreader, both turn stiflers. Let n_1 be the number of ignorants and n_2 be the number of spreaders at time t. Let m be the number of stiflers at that time. Then during $(t, t + \Delta t)$ the following transitions are possible:

$$(n_1, n_2) \to (n_1 - 1, n_2 + 1) \quad \text{with probability} \quad n_1 n_2 \beta \Delta t + o(\Delta t)$$

$$(n_1, n_2) \to (n_1, n_2 - 1) \quad \text{with probability} \quad n_2 m \beta \Delta t + o(\Delta t)$$

$$(n_1, n_2) \to (n_1, n_2 - 2) \quad \text{with probability} \quad (\tfrac{1}{2})n_2(n_2 - 1)\beta \Delta t + o(\Delta t)$$

Note that when there are n_2 spreaders, $(\tfrac{1}{2})n_2(n_2 - 1)$ pairs of them can make contact and turn stiflers. Based on these probabilities, equations can be written, and numerical results can be obtained. Because of the complexity of the method we shall not go into its details here.

17.7 STOCHASTIC MODELS FOR SOCIAL NETWORKS

A social network identifies the interpersonal relationships within a social group. Using graph theoretic concepts, Holland and Leinhardt (1977) have given a dynamic model for social networks. Their model has been extended by Wasserman (1980) to give a stochastic model that provides a structure to include interpersonal relationships such as reciprocity and popularity. The basic premise of the Wasserman approach is that a social group is a dynamic entity with a gradually evolving structure that eventually reaches a statistical equilibrium.

For graph theoretic concepts the readers are referred to Section 3.5. For a short introduction to the essential terminology reference can also be made to Wasserman (1978). In order to represent the various interrelationships between the nodes of a social network, a directed graph, called *digraph* for short, is used. The adjacency matrix of the digraph identifies possible transitions between the nodes. In a social network model the diagonal elements of the adjacency matrix are set to zero in order to avoid looping within a node. Let $X(t)$ be the adjacency matrix at time t. The elements are

$$X_{ij}(t) = \begin{cases} 1 & \text{if } i \to j \text{ transition is possible } (i \neq j) \\ 0 & \text{otherwise} \end{cases}$$

$$X_{ii}(t) = 0 \qquad \text{for all } i \text{ and } t$$

The following assumptions are made in the model:

1. $X(t)$ is a Markov process with transition probabilities

$$P_{xy}(s, t) = P[X(t) = y | X(s) = x] \qquad (17.7.1)$$

2. The probability that any two arcs act in collusion and change simultaneously during a small interval of time $(t, t + \Delta t]$ is of the order $o(\Delta t)$. Thus during a small interval of time $(t, t + \Delta t]$, a relationship between two individuals i and j can change only in the following two ways: (a) If there has not been a bond from $i \to j$ at time t, such a bond may develop by time $t + \Delta t$. The probability of this event is assumed to be

$$(\Delta t)\lambda_{0ij}(x, t) + o(\Delta t) \qquad (17.7.2)$$

    (b) If there has been a bond from $i \to j$ at time t, it may not exist by time $t + \Delta t$. The probability of this event is assumed to be

$$(\Delta t)\lambda_{1ij}(x, t) + o(\Delta t) \qquad (17.7.3)$$

Depending on the nature of relationship to be considered, different forms for the λ's in (17.7.2) and (17.7.3) can be identified. Two such special cases have been considered by Wasserman (1980).

The first model is for the reciprocity of friendship. Suppose the probability that a choice $i \to j$ will be made or withdrawn during an interval of length Δt (independent of t) depends only on the presence or absence of the reciprocated choice $j \to i$. In terms of λ's the infinitesimal transition rates can be written down as

$$\lambda_{0ij}(x, t) = \lambda_0 + \mu_0 x_{ji}$$

$$\lambda_{1ij}(x, t) = \lambda_1 + \mu_1 x_{ji} \qquad \lambda_0, \lambda_1 > 0; \ \mu_0 \geq 0, \ \mu_1 \geq -\lambda_1 \qquad (17.7.4)$$

If there are N individuals in a group, even though $N(N-1)$ pairs of transitions are possible, the same structure as in (17.7.4) is assumed for all pairs. Define

$$D_{ij}(t) = \left[X_{ij}(t), X_{ji}(t) \right]$$

as the dyad for the pair (i, j). The dyad is a Markov process with states $(0,0), (1,0), (0,1),$ and $(1,1)$. [State $(1,0)$ represents the existence of an $i \to j$ bond and the nonexistence of $j \to i$ bond at time t]. In terms of interpersonal relationships these states may be identified as: $(0,0)$—null, $(0,1)$ and $(1,0)$—asymmetric, and $(1,1)$—mutual.

For this four-state process the transition rate matrix has the form

$$
\begin{array}{c@{\quad}c@{\quad}c@{\quad}c@{\quad}c}
 & (0,0) & (1,0) & (0,1) & (1,1) \\
\begin{array}{c}(0,0)\\(1,0)\\(0,1)\\(1,1)\end{array} &
\left[\begin{array}{cccc}
-2\lambda_0 & \lambda_0 & \lambda_0 & 0 \\
\lambda_1 & -(\lambda_0 + \lambda_1 + \mu_0) & 0 & \lambda_0 + \mu_0 \\
\lambda_1 & 0 & -(\lambda_0 + \lambda_1 + \mu_0) & \lambda_0 + \mu_0 \\
0 & \lambda_1 + \mu_1 & \lambda_1 + \mu_1 & -2(\lambda_1 + \mu_1)
\end{array}\right]
\end{array}
$$

$$(17.7.5)$$

The steady-state distribution of the four states is determined by the usual methods. We get

$$p_{00} = \frac{\lambda_1(\lambda_1 + \mu_1)}{(\lambda_0 + \lambda_1)(\lambda_1 + \mu_1) + \lambda_0(\lambda_0 + \mu_0 + \lambda_1 + \mu_1)}$$

$$p_{10} = p_{01} = \frac{\lambda_0(\lambda_1 + \mu_1)}{(\lambda_0 + \lambda_1)(\lambda_1 + \mu_1) + \lambda_0(\lambda_0 + \mu_0 + \lambda_1 + \mu_1)}$$

$$p_{11} = \frac{\lambda_0(\lambda_0 + \mu_0)}{(\lambda_0 + \lambda_1)(\lambda_1 + \mu_1) + \lambda_0(\lambda_0 + \mu_0 + \lambda_1 + \mu_1)} \qquad (17.7.6)$$

The second model considered by Wasserman is the so-called popularity model. In this model individual i's choice or nonchoice of individual j depends only on the number of individuals who choose j. The last number, say x_{+j} is given by the in-degree of node j, that is, the column sum $X_{\cdot j}(t) = \Sigma_i X_{ij}(t)$ of the adjacency matrix at time t. In this case, of the $N(N-1)$ pairs of transitions in a group of size N, for a fixed j, $N-1$ pairs have identical transition rates. In terms of λ's of (17.7.2) and (17.7.3) the transition rates for this model can be written down as

$$\lambda_{0ij}(x,t) = \lambda_0 + \nu_0 x_{+j}$$

$$\lambda_{1ij}(x,t) = \lambda_1 + \nu_1 x_{+j} \qquad \lambda_0, \lambda_1 > 0;\ \nu_0 > 0,\ -\lambda_1/(N-1) < \nu_1 < 0$$

The last restriction is imposed in order to make the rate $\lambda_1 + (N-1)\nu_1 > 0$ and $\nu_1 < 0$.

Consider the jth column process of $X(t)$:

$$X_{\cdot j}(t) = \left[X_{1j}(t), X_{2j}(t), \ldots, X_{Nj}(t) \right]^T.$$

Each of the column processes has 2^{N-1} states consisting of all possible zero-one vectors of length N, with $X_{jj}(t)$ fixed at zero. Due to the transition rates defined in (17.7.7), $\{ X_{\cdot j}(t),\ j = 1, 2, \ldots, N \}$ are independent and identical Markov processes. For $N = 3$ the transition rate matrix can be given as

	$(0,0)$	$(1,0)$	$(0,1)$	$(1,1)$
$(0,0)$	$-2\lambda_0$	λ_0	λ_0	0
$(1,0)$	$\lambda_1 + \nu_1$	$-(\lambda_0 + \nu_0 + \lambda_1 + \nu_1)$	0	$\lambda_0 + \nu_0$
$(0,1)$	$\lambda_1 + \nu_1$	0	$-(\lambda_0 + \nu_0 + \lambda_1 + \nu_1)$	$\lambda_0 + \nu_0$
$(1,1)$	0	$\lambda_1 + 2\nu_1$	$\lambda_1 + 2\nu_1$	$-2(\lambda + 2\nu_1)$

In this case $X_{1j}(t)$ is represented by a vector of length 2, ignoring the single zero entry corresponding to $X_{jj}(t) = 0$. The steady-state probabilities are obtained in the usual manner. We get

$$p_{00} = \frac{A + 2B}{A + 2(B + C + D)}$$

$$p_{10} = p_{01} = \frac{2D}{A + 2(B + C + D)}$$

$$p_{11} = \frac{2C}{A + 2(B + C + D)}$$

where

$$A = 1 + \lambda_0 + \nu_0 + \lambda_1 + \nu_1$$

$$B = (1 + \lambda_1 + \nu_1)(\lambda_1 + 2\nu_1)$$

$$C = \lambda_0(\lambda_0 + \nu_0)$$

$$D = \lambda_0 + 2(\lambda_1 + 2\nu_1)$$

For further discussion on the model and methods for estimating model parameters, the readers are referred to Wasserman (1980).

REFERENCES

Anderson, T. W. (1954). "Probability Models for Analyzing Time Changes in Attitudes." In P. F. Lazarsfeld (ed.), *Mathematical Thinking in the Social Sciences*. Glencoe, Ill.: The Free Press, pp. 17–66.

Bailey, N. T. J. (1957). *The Mathematical Theory of Epidemics*. New York: Hafner.

Bartholomew, D. J. (1973). *Stochastic Models for Social Processes*. 2nd ed. New York: Wiley.

Blumen, I., Kogan, M., and McCarthy, P. J. (1955). *The Industrial Mobility of Labor as a Probability Process*. Ithaca, N.Y.: Cornell University Press.

Coleman, J. S. (1964a). *Models of Change and Response Uncertainty*. Englewood Cliffs, N.J.: Prentice-Hall.

Coleman, J. S. (1964b). *Introduction to Mathematical Sociology*. Glencoe, Ill.: The Free Press.

Daley, D. J., and Kendall, D. G. (1965). "Stochastic Rumors." *J. Inst. Math. Appl.* 42–55.

Gani, J. (1963). "Formulae for Projecting Enrollments and Degrees Awarded in Universities." *J. Roy. Stat. Soc.* **A126** 400–409.

Glass, D. V., ed. (1954). *Social Mobility in Britain*. London: Rutledge and Keegan Paul.

Goodman, L. A. (1961). "Statistical Methods for the 'Mover-Stayer' Model." *J. Amer. Stat. Assoc.* **56**, 841–868.

Holland, P. W., and Leinhardt, S. (1977). "A Dynamic Model for Social Networks." *J. Math. Sociol.* **5**, 5–20.

Lazarsfeld, P. F., ed. (1951). *Mathematical Thinking in the Social Sciences*. Glencoe, Ill.: The Free Press.

Perlman, J. and Mandel, B. (1944). "The Continuous Work History Sample under Old-Age and Survivors Insurance." *Social Security Bulletin*.

Pollard, J. H. (1966). "On the Use of the Direct Matrix Product in Analyzing Certain Stochastic Population Models." *Biometrika* **53**, 397–415.

Prais, S. J. (1955). "Measuring Social Mobility." *J. Roy. Stat. Soc.* **A118**, 56–66.

Schuessler, K. F., ed. (1978, 1980). *Sociological Methodology*. San Francisco: Jossey-Bass.

Singer, B., and Spilerman, S. (1976). "The Representation of Social Processes by Markov Models." *Amer. J. Sociol.*, **82**, 1–54.

Taga, Y., and Isii, K. (1959). "On a Stochastic Model Concerning the Pattern of Communication—Diffusion of News in a Social Group." *Ann. Inst. Stat. Math.* **11**, 25–43.

Wasserman, S. (1978). "Models for Binary Directed Graphs and Their Applications." *Adv. Appl. Prob.* **10**, 803–818.

Wasserman, S. (1980). "Analyzing Social Networks as Stochastic Processes." *J. Amer. Stat. Assoc.* **75**(370), 280–294.

Young, A., and Vassiliou, P. C. G. (1974). "A Non-Linear Model on the Promotion of Staff." *J. Roy. Stat. Soc.* A137, 584–595.

FOR FURTHER READING

Anderson, E. E. (1980). "A Stochastic Model of Industrial Fluctuations: An International Comparison." *Omega* 8, 219–226.

Bartholomew, D. J. (1969). "Renewal Theory Models for Manpower Systems." In N. A. B. Wilson (ed.) *Manpower Research* English Universities Press (now Hodder & Stoughton Educational) London: 120–128.

Bartholomew, D. J. (1971). "The Statistical Approach to Manpower Planning." *Statistician* 20, 3–26.

Bartholomew, D. J., and Smith, A. R., eds. (1971). *Manpower and Management Science*. Lexington, Ky.: Heath.

Davies, G. S. (1973). "Structural Control in a Graded Manpower System." *Management Sci.* 20, 76–84.

Eaton, B. C. (1970). "Studying Mass Layoff through Markov Chains." *Ind. Rel.* 9, 394–403.

Glen, J. J. (1977). "Length of Service Distributions in Markov Manpower Models." *Oper. Res. Quart.* 28, 975–985.

Hayne, W. J., and Marshall, K. T. (1977). "Two-Characteristic Markov-Type Manpower Flow Models." *Nav. Res. Log. Quart.* 24, 235–255.

Hodge, R. W. (1968). "Occupational Mobility as a Probability Process." *Demography* 3, 19–34.

Keeny, G. A., Morgan, R. W., and Ray, K. H. (1977). "An Analytical Model for Company Manpower Planning." *Oper. Res. Quart.* 28, 983–995.

Larson, G. W. (1979). "Wastage in a Hierarchical Manpower System." *J. Oper. Res. Soc.* 30, 341–348.

McClean, S. I. (1977). "The Steady State Behavior of a Manpower Planning Model in Which Class Corresponds to Length of Service." *Oper. Res. Quart.* 28, 305–311.

McClean, S. I., and Karageorgos, D. L. (1979). "An Age-Stratified Manpower Model Applied to the Educational System." *Statistician* 28, 9–18.

Morgan, R. W. (1979). "Some Models for a Hierarchical Manpower System." *J. Oper. Res. Soc.* 30, 727–736.

Nielsen, G. L., and Young, A. R. (1973). "Manpower Planning: A Markov Chain Application." *Public Personnel Management*, 133–143.

Price, W. L., Martel, A., and Lewis, K. A. (1980). "A Review of Mathematical Models in Human Resource Planning." *Omega* 8(6), 639–645.

Rowland, K. M., and Sovereign, M. G. (1969). "Markov Chain Analysis of Internal Manpower Supply," *Ind. Rel.* 9, 88–99.

Sales, P. (1971). "The Validity of the Markov Chain Model for a Class of the Civil Service." *Statistician* 20, 85–110.

Schinnar, A. P. (1978). "Mobility and Turnover Control of Economically Active Populations." *TIMS Studies Management Sci.* 8, 75–98.

Schinnar, A. P., and Stewman, S. (1978). "A General Class of Markov Models of Social Mobility with Duration Memory Patterns." *J. Math. Sociol.* **6**.

Spilerman, S. (1972). "Extensions of the Mover-Stayer Model." *Amer. J. Sociol.* **78**, 599–626.

Stewman, S. (1975). "Two Markov Models of Open System Occupational Mobility: Underlying Conceptualizations and Empirical Tests." *Amer. Sociol. Rev.* **40**, 298–321.

Stewman, S. (1976). "Markov Models of Occupational Mobility—Theoretical Development and Empirical Support, Parts I and II." *J. Math. Sociol.* **4**, 201–245, 247–278.

Stewman, S. (1978). "Markov and Renewal Models for Total Manpower Systems." *Omega* **6**, 341–351.

Uyar, K. M. (1972). "Markov Chain Forecasts of Employee Replacement Needs." *Ind. Rel.* **11**, 96–106.

Vassiliou, P. C. G. (1976). "A Markov Chain Model for Wastage in Manpower Systems." *Oper. Res. Quart.* **27**, 57–70.

Vassiliou, P. C. G. (1978). "A High Order Non-Linear Markovian Model for Promotion in Manpower Systems." *J. Roy. Stat. Soc.* **A141**, 86–94.

Vassiliou, P. C. G. (1981). "On the Asymptotic Behavior of Age Distributions in Manpower Systems." *J. Oper. Res. Soc.* **32**, 503–506.

Vroom, V. H., and MacCrimmon, K. R. (1968). "Toward a Stochastic Model of Managerial Careers." *Administrative Sci. Quart.* **13**, 26–46.

Young, A. (1965). "Models for Planning Recruitment and Promotion of Staff." *British J. Ind. Rel.* **3**, 301–310.

Young, A., and Almond, G. (1961). "Predicting Distributions of Staff." *Computer J.* **3**, 246–250.

Zanakis, S. H., and Marat, M. W. (1981). "A Markovian Goal Programming Approach to Aggregate Manpower Planning." *J. Oper. Res. Soc.* **32**, 55–63.

Zanakis, S. H., and Maret, M. W. (1980). "A Markov Chain Application to Manpower Supply Planning." *J. Oper. Res. Soc.* **31**, 1095–1102.

Chapter 18

MARKOV MODELS IN BIOLOGICAL SCIENCES

18.1 INTRODUCTION

The use of stochastic models in understanding biological phenomena has its beginning with the study of the probability of extinction of families by Watson and Galton (1874). Biological processes such as population growth have been the major motivating factors for advances in the study of branching, and birth and death processes. We have discussed these stochastic processes extensively in earlier chapters. Also appearing earlier has been an example from population genetics. In this chapter we shall discuss further the classical problem of genetics and concentrate on other diverse phenomena such as the recovery process, cell survival after irradiation, and the flow of particles through organs. The treatment is not comprehensive by any means. It is only indicative of the potential for the application of stochastic models in understanding biological phenomena. The literature is vast and applications are diverse. For additional reading interested readers are referred to books such as Bharucha-Reid (1960), Bailey (1964, 1967), Chiang (1968), Ewens (1969), and Iosifescu and Tăutu (1973) among others. The last reference is particularly useful for readers interested in specific models.

18.2 MARKOV MODELS IN POPULATION GENETICS

The genetic composition of populations has been the subject of investigations by several researchers, namely Fisher, Wright, Haldane, Kimura, and Moran. Here we shall present some stochastic models postulated by Fisher (1930), Wright (1931), and Moran (1958). First we shall consider two models introduced by Fisher and Wright where mutation is included. Long-term properties of this model are due to Feller (1951). The second model introduced by Moran uses a continuous time process, the properties of which have also been investigated by Karlin and McGregor (1964).

Consider a population of N haploid individuals which belong to either of two types: A or a. We assume that the next generation is formed by selecting with replacement N individuals from the parent generation under two hypotheses: random mating without mutation and with mutation.

Let i be the number of individuals of type A in the parent generation. Then $N - i$ individuals are of type a. When no factors such as mutation are assumed, the probability that an individual in the next generation is of type A is obtained as i/N, and the probability that it is of type a is $N - i/N$. Clearly the number of type A individuals is a Markov chain, with the one-step transition probability given by the binomial probability

$$P_{ij} = \binom{N}{j}\left(\frac{i}{N}\right)^j\left(1 - \frac{i}{N}\right)^{N-j} \tag{18.2.1}$$

The state space of the Markov chain is $\{0, 1, 2, \ldots, N\}$, with states 0 and N representing homozygotic states which are absorbing. Using eigenvalue techniques, Feller (1951) has shown that the rate of approach to absorption is given by $(1 - 1/N)^n$ (which is the rate at which Q^n approaches a null matrix, where Q is the matrix of transition probabilities among transient states).

Next let us assume that mutation is possible. Let α_1 be the probability that an individual of type A mutates to type a and α_2 be the probability that the mutation is from type a to A from one generation to another. Now p_A, the probability that the individual of the next generation is of type A, and p_a, the corresponding probability for type a, are given as

$$p_A = \frac{i(1 - \alpha_1) + (N - i)\alpha_2}{N}$$

$$p_a = \frac{i\alpha_1 + (N - i)(1 - \alpha_2)}{N} \tag{18.2.2}$$

The one-step transition probability P_{ij} is obtained by writing

$$P_{ij} = \binom{N}{j} p_A^j p_a^{N-j}$$

The Markov chain in this case is not absorbing, and therefore the limiting distribution exists. Again, using eigenvalue techniques, Feller (1951) has shown that the rate of convergence of $P_{ij}^{(n)}$ to its limiting value π_j is given by $(1 - \alpha_1 - \alpha_2)^n$.

Moran's model assumes that the next generation is formed when an individual duplicates and produces a progeny that replaces a second individual at random. Thus, when mutation is allowed, changes in the number of A-type individuals take place with the following probabilities:

$$p_{i,i+1} = \left(\frac{N-i}{N} \right) \left[\frac{i(1-\alpha_1) + (N-i)\alpha_2}{N} \right]$$

$$p_{i,i-1} = \left(\frac{i}{N} \right) \left[\frac{i\alpha_1 + (N-i)(1-\alpha_2)}{N} \right] \qquad (18.2.3)$$

A continuous time model is obtained by incorporating the distribution of time interval between successive events when changes of state occur. Let each individual initiate a change of state at intervals of time which have independent and identical exponential distributions with mean $1/\mu$. Combining this factor with the changes of state, we may set up a finite state birth and death process where the birth and death parameters λ_j and μ_j are given as

$$\lambda_j = \mu p_{j,j+1} \quad \text{and} \quad \mu_j = \mu p_{j,j-1} \qquad (18.2.4)$$

For a discussion of the properties of this process, the readers are referred to Karlin and McGregor (1962, 1964).

18.3   RECOVERY, RELAPSE, AND DEATH DUE TO DISEASE

Studies of hospital data have shown that the process of initial treatment of an individual for a disease like cancer, the patient's recovery from it, the relapse of the disease, and possible death due to the initial treatment (or operation) or the recurring relapses is controlled more by random causes than by the deterministic environment. A stochastic process therefore is a natural model for an analytically minded investigator. Here we shall discuss a simple Markov process model as set up by Fix and Neyman (1951) for recovery, relapse, and death due to cancer.

For the sake of simplicity, assume that an individual who has contracted cancer may go through the following stages:

0—Initial state of being under treatment for cancer.

1— State of being dead immediately following treatment for cancer (death from cancer or operative death).

2— State of recovery in which the patient is not under treatment but under observation.

3— State of being lost after recovery either through death not connected with cancer or difficulties of tracing the patient.

We also assume that when a transition occurs, only the following types of transitions are possible:

$$(0 \to 1) \qquad (0 \to 2) \qquad (2 \to 0) \qquad (2 \to 3)$$

This implies that death due to cancer is always preceded by treatment and the patient becomes untraceable only after recovery. We shall also assume that the time spent in states 0 and 2 are exponential random variables with the following transition rates:

λ_{01}— Transition rate for $0 \to 1$.
λ_{02}— Transition rate for $0 \to 2$.
λ_{20}— Transition rate for $2 \to 0$.
λ_{23}— Transition rate for $2 \to 3$.

Let the patient be initially in state 0. Define

$$P_n(t) = \text{Probability that the patient is found in state } n \text{ at time } t$$

Due to the assumption of exponential distributions for stay in a state, the state of the patient is a Markov process. Forward Kolmogorov equations for this process can be written as in Chapter 7 for the determination of probabilities $P_n(t)$. We get

$$P_0'(t) = -(\lambda_{01} + \lambda_{02})P_0(t) + \lambda_{20}P_2(t)$$

$$P_1'(t) = \lambda_{01}P_0(t)$$

$$P_2'(t) = -(\lambda_{20} + \lambda_{23})P_2(t) + \lambda_{02}P_0(t)$$

$$P_3'(t) = \lambda_{23}P_2(t) \tag{18.3.1}$$

Steady-state solutions of these equations do not provide additional information, since under the assumptions, a patient eventually has to enter one of the absorbing states 1 and 3. To solve these equations for finite time, we use the transform methods employed in Chapter 15 for similar problems in reliability theory. Let

$$\phi_n(\theta) = \int_0^\infty e^{-\theta t}P_n(t)\,dt \qquad \text{Re}(\theta) > 0$$

such that, under the assumption $P_0(0) = 1$ and $P_n(0) = 0$, we have

$$\int_0^\infty e^{-\theta t} P_0'(t)\, dt = \theta \phi_0(\theta) - 1$$

and

$$\int_0^\infty e^{-\theta t} P_n'(t)\, dt = \theta \phi_n(\theta) \qquad n = 1, 2, 3$$

Now from (18.3.1) we get

$$\theta \phi_0(\theta) - 1 = -(\lambda_{01} + \lambda_{02})\phi_0(\theta) + \lambda_{20}\phi_2(\theta)$$

$$\theta \phi_1(\theta) = \lambda_{01}\phi_0(\theta)$$

$$\theta \phi_2(\theta) = -(\lambda_{20} + \lambda_{23})\phi_2(\theta) + \lambda_{02}\phi_0(\theta)$$

$$\theta \phi_3(\theta) = \lambda_{23}\phi_2(\theta) \qquad\qquad (18.3.2)$$

Solving these equations, we get

$$\phi_2(\theta) = \frac{\lambda_{02}}{\theta + \lambda_{20} + \lambda_{23}} \phi_0(\theta) \qquad\qquad (18.3.3)$$

$$\phi_0(\theta) = \frac{\theta + \lambda_{20} + \lambda_{23}}{\theta^2 + (\lambda_{01} + \lambda_{02} + \lambda_{20} + \lambda_{23})\theta + \lambda_{01}\lambda_{20} + \lambda_{01}\lambda_{23} + \lambda_{02}\lambda_{23}}$$

$$(18.3.4)$$

The denominator of (18.3.4) is a quadratic in θ, and hence we can write

$$\phi_0(\theta) = \frac{\theta + \lambda_{20} + \lambda_{23}}{(\theta - \xi_1)(\theta - \xi_2)} \qquad\qquad (18.3.5)$$

where

$$\xi_1 = \frac{-B - \sqrt{B^2 - 4C}}{2} \qquad \xi_2 = \frac{-B + \sqrt{B^2 - 4C}}{2} \qquad (18.3.6)$$

with $\lambda_{01} + \lambda_{02} + \lambda_{20} + \lambda_{23} = B$ and $\lambda_{01}\lambda_{20} + \lambda_{01}\lambda_{23} + \lambda_{01}\lambda_{23} = C$. Expressing the right-hand side of (18.3.5) in partial fractions (see Appendix B), we write

$$\phi_0(\theta) = \frac{\xi_1 + \lambda_{20} + \lambda_{23}}{\xi_1 - \xi_2} \cdot \frac{1}{\theta - \xi_1} - \frac{\xi_2 + \lambda_{20} + \lambda_{23}}{\xi_1 - \xi_2} \cdot \frac{1}{\theta - \xi_2}$$

which when inverted gives

$$P_0(t) = \frac{1}{\xi_1 - \xi_2}\left[(\xi_1 + \lambda_{20} + \lambda_{23})e^{\xi_1 t} - (\xi_2 + \lambda_{20} + \lambda_{23})e^{\xi_2 t}\right]$$

$$(18.3.7)$$

Similarly $\phi_2(\theta)$ of (18.3.3) gives

$$\phi_2(\theta) = \frac{\lambda_{02}}{\xi_1 - \xi_2}\left[\frac{1}{\theta - \xi_1} - \frac{1}{\theta - \xi_2}\right]$$

and

$$P_2(t) = \frac{\lambda_{02}}{\xi_1 - \xi_2}\left[e^{\xi_1 t} - e^{\xi_2 t}\right] \tag{18.3.8}$$

Expressions for $P_1(t)$ and $P_3(t)$ can be obtained either directly from (18.3.1) by integrating both sides of these equations or from (18.3.2) by inverting their Laplace transforms. Ultimately we get

$$P_1(t) = \lambda_{01}\left[\frac{\lambda_{20} + \lambda_{23}}{\xi_1 \xi_2} + \frac{(\xi_1 + \lambda_{20} + \lambda_{23})e^{\xi_1 t}}{\xi_1(\xi_1 - \xi_2)} - \frac{(\xi_2 + \lambda_{20} + \lambda_{23})e^{\xi_2 t}}{\xi_2(\xi_1 - \xi_2)}\right]$$

$$(18.3.9)$$

$$P_3(t) = \lambda_{02}\lambda_{23}\left[\frac{1}{\xi_1 \xi_2} + \frac{\xi_2 e^{\xi_1 t} - \xi_1 e^{\xi_2 t}}{\xi_1 \xi_2(\xi_1 - \xi_2)}\right] \tag{18.3.10}$$

A quantitative measure for the net risk of death from cancer at time t is obtained by eliminating the risk of being lost after recovery or dying from other causes (i.e., state 3). Let the probabilities of the states of the patient after eliminating state 3 be $P_n^*(t)$, $n = 0, 1, 2$. These probabilities satisfy the differential equations:

$$P_0^{*'}(t) = -(\lambda_{01} + \lambda_{02})P_0^*(t) + \lambda_{20}P_2^*(t)$$

$$P_1^{*'}(t) = \lambda_{01}P_0^*(t)$$

$$P_2^{*'}(t) = -\lambda_{20}P_2^*(t) + \lambda_{02}P_0^*(t) \tag{18.3.11}$$

These equations have the same solutions as (18.3.1), except that $\lambda_{23} = 0$ in the present case. With this modification we find

$$\xi_1^* = \frac{-(\lambda_{01} + \lambda_{02} + \lambda_{20}) - \sqrt{(\lambda_{01} + \lambda_{02} + \lambda_{20})^2 - 4\lambda_{01}\lambda_{20}}}{2}$$

$$\xi_2^* = \frac{-(\lambda_{01} + \lambda_{02} + \lambda_{20}) + \sqrt{(\lambda_{01} + \lambda_{02} + \lambda_{20})^2 - 4\lambda_{01}\lambda_{20}}}{2}$$

$$(18.3.12)$$

$$P_0^*(t) = \frac{1}{\xi_1^* - \xi_2^*}\left[(\xi_1^* + \lambda_{20})e^{\xi_1^* t} - (\xi_2^* + \lambda_{20})e^{\xi_1^* t}\right]$$

$$P_1^*(t) = \lambda_{01}\left[\frac{\lambda_{20}}{\xi_1^*\xi_2^*} + \frac{(\xi_1^* + \lambda_{20})e^{\xi_1^* t}}{\xi_1^*(\xi_1^* - \xi_2^*)} - \frac{(\xi_2^* + \lambda_{20})e^{\xi_2^* t}}{\xi_2^*(\xi_1^* - \xi_2^*)}\right]$$

$$P_2^*(t) = \frac{\lambda_{02}}{\xi_1^* - \xi_2^*}\left[e^{\xi_1^* t} - e^{\xi_2^* t}\right] \qquad (18.3.13)$$

The net risk of death due to cancer at time t is given by $P_1^*(t)$.

The length of time a patient will survive after the treatment starts is also an important measure. Denoting this time by D, its expected value can be derived in a way similar to deriving the expected length of life of a system in reliability theory. We have

$$P(D > t) = P\{\text{Patient will survive at least until } t\}$$

$$= P_0^*(t) + P_2^*(t)$$

$$= \frac{(\xi_1^* + \lambda_{20} + \lambda_{02})e^{\xi_1^* t} - (\xi_2^* + \lambda_{20} + \lambda_{02})e^{\xi_2^* t}}{\xi_1^* - \xi_2^*}$$

$$(18.3.14)$$

$$E[D] = \int_0^\infty P(D > t)\,dt$$

$$= \frac{\lambda_{02} + \lambda_{20}}{\lambda_{01}\lambda_{20}} \qquad (18.3.15)$$

The model can be further extended to include a larger number of states depending on the nature of the disease and the recovery and relapse process.

Fix and Neyman (1951) have also given methods of estimating the transition rates based on available data; these estimates follow along the lines indicated in Chapter 7. The estimation of the transition rate matrix has been the subject of another study by Zahl (1955). His data consist of observing the patients at periodic intervals of time and classifying each into one of a finite number of states. Zahl also treats the large sample distribution of the estimates and tests of hypothesis connected with the model.

In the analysis of the recovery process a Markov process model has been used for two reasons. First, a more general model would have complicated the analysis, and consequently we would have had to settle for less information as the output. Second, when this process was studied in the 1950s, it was a pioneering effort, and more sophisticated models were not available. But while looking at it from the perspective of the 1980s, one is bound to conclude that a more general model such as a semi-Markov process would be realistic for the situation. In many recovery processes Markovian assumptions of transition may not be realistic. For instance, once the treatment starts, the patient might be required to remain under treatment for at least some amount of time. In such cases, along with the probabilities of transition, distributions of lengths of time between transitions should also be considered. A semi-Markov process, discussed in Chapter 8, provides the necessary mechanism to allow for the distributions of time intervals between transitions.

A semi-Markov process model for the recovery process can be analyzed much the same way as the analysis of a human fertility model carried out by Perrin and Sheps (1964) and Sheps and Perrin (1966). They set up the human reproduction process as a semi-Markov process with the following states: S_0—nonpregnant, fecund period; S_1—pregnancy; S_2—postpartum sterile period associated with abortion or fetal loss; S_3—postpartum sterile period associated with stillbirth; S_4—postpartum sterile period associated with live birth. Possible transitions are: $S_0 \to S_1$; $S_1 \to S_2$, S_3, or S_4; $S_2 \to S_0$; $S_4 \to S_0$. Using a semi-Markov process analysis, Sheps and Perrin (1966) study the following characteristics: (1) the distribution of intervals between successive pregnancy terminations of various types, (2) the distribution of the number of miscarriages, stillbirths, and livebirths per individual for a given period of time, and (3) the fertility rate.

18.4   CELL SURVIVAL AFTER IRRADIATION

Living cells are damaged when energetic radiation such as ionizing radiation passes through them. One theory in the field of radiobiology postulates that the damaged cell is likely to be repaired in the absence of further hits, given enough time. Assuming a Poisson process for radiation, Bansal and Gupta (1978) have proposed a model in which the repair time has a general continuous distribution of the type given in equation (15.2.7). This model is a refinement over an earlier model proposed by Gupta (1967) in which it is assumed that (1) the cell can be hit by radiation in n regions with different rates, (2) no additional hits are allowed when the cell is in a damaged state, and (3) the repair times are exponential.

Consider a living cell subject to radiation as being in any one of the three states: 0—normal undamaged state; 1—partially damaged and undergoing repair, and 2—irreparably damaged state. Let $P_i(t)$, $i = 0, 1, 2$, be the probability that the cell is in state i at time t. Under the assumption of a Poisson radiation process with parameter λ and exponential repair times with mean $1/\mu$, the state of the cell is a Markov process with 2 as the absorbing state. Forward Kolmogorov equations of the process take the form

$$P_0'(t) = -\lambda P_0(t) + \mu P_1(t)$$

$$P_1'(t) = -(\lambda + \mu) P_1(t) + \lambda P_0(t)$$

$$P_2'(t) = \lambda P_1(t) \tag{18.4.1}$$

which are exactly the same as equations (15.6.8) written down for the reliability problem of a single equipment with a standby redundant unit. Using the solution derived there, for the normal state of the cell, we get

$$P_0(t) = \frac{(\lambda + \mu + \xi_1) e^{\xi_1 t} - (\lambda + \mu + \xi_2) e^{\xi_2 t}}{\xi_1 - \xi_2} \tag{18.4.2}$$

where

$$\xi_1 = \frac{-(2\lambda + \mu) + \sqrt{4\lambda\mu + \mu^2}}{2}$$

$$\xi_2 = \frac{-(2\lambda + \mu) - \sqrt{4\lambda\mu + \mu^2}}{2} \tag{18.4.3}$$

Using experimental data on survival of human lymphocytes following ionizing radiations reported by Madhavanath (1971), Bansal and Gupta (1978) estimate λ and μ for three different doses of radiation and calculate $P_0(t)$ as given by equation (18.4.2) for various values of t. A comparison of these values with experimental values indicates that the Markov process model is in close agreement for all doses and small values of t, and for higher doses and all values of t.

The equations for the general case use the hazard function of the repair time distribution in place of μ and are beyond the scope of this book. Interested readers are referred to the 1978 paper of Bansal and Gupta.

An extension of this model to the n-region case considered by Gupta earlier but now with less restrictive assumptions can be made by identifying $n + 2$ states $\{0, 1, 2, \ldots, n, n + 1\}$ in which state 0 is the normal undamaged state and state $n + 1$ is the irreparably damaged state. We assume that, when the cell is in any one of the states $\{1, 2, \ldots, n\}$ which are now the damaged states, another hit at any region will take it to the irreparably damaged state $n + 1$. Let λ_i and μ_i be the hit and repair rates for region i, and write $\Sigma\lambda_i = \lambda$. Define $P_i(t)$, $(i = 0, 1, 2, \ldots, n + 1)$ as the probability that the cell is in state i at time t. The forward Kolmogorov equations for the process can be written down as

$$P_0'(t) = -\lambda P_0(t) + \sum_{i=1}^{n} \mu_i P_i(t)$$

$$P_i'(t) = -(\lambda + \mu_i)P_i(t) + \lambda_i P_0(t) \qquad i = 1, 2, \ldots, n$$

$$P_{n+1}'(t) = \lambda \sum_{i=1}^{n} P_i(t) \tag{18.4.4}$$

Defining the Laplace transform

$$\phi_i(\theta) = \int_0^{\infty} e^{-\theta t} P_i(t)\, dt \qquad \mathrm{Re}(\theta) > 0$$

and taking transforms of equations (18.4.4) as we did in the case of equations (15.6.8), we get [assuming $P_i(0) = 1$ if $i = 0$ and $= 0$ if $i \neq 0$]

$$-1 + \theta\phi_0(\theta) = -\lambda\phi_0(\theta) + \sum_{i=1}^{n} \mu_i \phi_i(\theta) \tag{18.4.5}$$

$$\theta\phi_i(\theta) = -(\lambda + \mu_i)\phi_i(\theta) + \lambda_i \phi_0(\theta) \tag{18.4.6}$$

The last equation gives

$$\phi_i(\theta) = \frac{\lambda_i}{\theta + \lambda + \mu_i} \phi_0(\theta) \tag{18.4.7}$$

which on substitution in (18.4.5) gives

$$\phi_0(\theta) = \left[\theta + \lambda - \sum_{i=1}^{n} \frac{\lambda_i \mu_i}{\theta + \lambda + \mu_i}\right]^{-1} \tag{18.4.8}$$

As in the case of reliability systems the expected life of a cell subject to irradiation can be obtained as

$$E[L] = \int_0^\infty \left[\sum_{i=0}^{n} P_i(t)\right] dt \tag{18.4.9}$$

$$= \lim_{\theta \to 0} \sum_{i=0}^{n} \phi_i(\theta)$$

$$= \frac{1 + \sum_{i=1}^{n} \lambda_i/(\lambda + \mu_i)}{\lambda - \sum_{i=1}^{n} \lambda_i \mu_i/(\lambda + \mu_i)} \tag{18.4.10}$$

When $\mu_1 = \mu_2 = \cdots = \mu_n = \mu$ (i.e., when the repair rate is the same for all regions), (18.4.10) simplifies to

$$E[L] = \frac{2\lambda + \mu}{\lambda^2} \tag{18.4.11}$$

which is the same results as the earlier single region case [see (15.6.23)].

18.5   COMPARTMENTAL ANALYSIS

Compartment models are used to represent the distribution of elements such as a drug, tracer, or a pollutant residing over time in distinct body (human or animal) units (blood, liver, tissue, etc.). The number of particles of a given molecule in a compartment is represented by a stochastic process. In the formulation of these models two different approaches have been adopted: differential equation models and stochastic models. In the former approach models are analyzed using linear and nonlinear systems theory, whereas in the latter case the use of Markov processes has been common. For an illustration of the relationship of these two approaches the readers are referred to Eisenfeld (1979). Also see Jacquez (1972) for a comprehensive treatment of differential equation models.

In this section we shall concentrate on some simple Markov models for a compartmental system. One of the initial attempts to represent the number of particles in a compartment by a Markov process is by Thakur et al. (1972). Consider a single compartment, and let $X(t)$ be the number of particles in the compartment at time t. It is assumed that each particle leaves the compartment with a constant rate μ [i.e., when $X(t) = n$, the departure rate is $n\mu$] and that input takes place in a nonhomogeneous Poisson process with rate $\lambda(t)$ at time t. Using a birth and death process model and writing

$$P_n(t) = P[X(t) = n] \qquad (18.5.1)$$

we get the forward Kolmogorov equation

$$P_n'(t) = -[n\mu + \lambda(t)]P_n(t) + \lambda(t)P_{n-1}(t) + (n+1)\mu P_{n+1}(t)$$

$$(18.5.2)$$

Distribution characteristics of $X(t)$ can now be determined using this equation.

A simpler approach to the determination of distribution characteristics of $X(t)$ in a slightly more general setting has been given by Purdue (1974a, 1974b). Consider a single compartment with the following assumptions: (1) Particles enter the compartment as before according to a nonhomogeneous Poisson process with rate $\lambda(t)$ at time t. (2) The probability that a particle that enters the compartment at time $x > 0$ is still present at time t is given by $1 - F(x, t)$. It may be noted that when $F(x, t) = 1 - e^{-\mu(t-x)}$, we get the model of Thakur et al. (1972).

Assuming that particles in the compartment behave independently of each other, the number $X(t)$ can be considered to be made up of two independent groups $Y(t)$ and $Z(t)$ defined as follows:

$Y(t)$ = number of particles present at time 0 that are still present at time t.

$Z(t)$ = number of particles that arrive during $(0, t]$ and are still present at time t.

Let $B_i(t)$ be an indicator variable, defined as

$$B_i(t) = \begin{cases} 1 & \text{if particle } i \text{ was present at time 0 and is still present at time } t \\ 0 & \text{otherwise} \end{cases}$$

Clearly

$$P[B_i(t) = 1] = 1 - F(0, t)$$

and

$$E[B_i(t)] = 1 - F(0, t)$$

$$V[B_i(t)] = F(0, t)[1 - F(0, t)] \tag{18.5.3}$$

Also

$$Y(t) = \sum_{i=1}^{X(0)} B_i(t)$$

showing that $Y(t)$ is a sum [random sum if $X(0)$ is random] of Bernoulli random variables. Therefore, when $X(0)$ is a constant, $Y(t)$ has a binomial distribution with probability of success $1 - F(0, t)$.

To determine the distribution of $Z(t)$, first we have to determine the distribution of the number of particles entering the compartment during $(0, t]$. Let $N(t)$ be this number. By extending the results of time homogeneous Poisson process to a nonhomogeneous Poisson process, we get

$$P[N(t) = n] = e^{-m(t)} \frac{[m(t)]^n}{n!} \qquad n = 0, 1, 2, \ldots \tag{18.5.4}$$

where $m(t) = \int_0^t \lambda(\tau) d\tau$. Of the $N(t)$ particles entering the system in $(0, t]$ only $Z(t)$ are present at time t. Now let $p(t)$ be the probability that a particle enters the system in $(0, t)$ and is still present at t. We have,

$$p(t) = \frac{1}{m(t)} \int_0^t \lambda(x)[1 - F(x, t)] \, dx$$

$$= \frac{h(t)}{m(t)} \tag{18.5.5}$$

where we define

$$h(t) = \int_0^t \lambda(x)[1 - F(x, t)] \, dx \tag{18.5.6}$$

Now we have

$$P[Z(t) = k] = \sum_{n=k}^{\infty} P[Z(t) = k|N(t) = n] P[N(t) = n]$$

$$= \sum_{n=k}^{\infty} \binom{n}{k} [p(t)]^k [1 - p(t)]^{n-k} e^{-m(t)} \frac{[m(t)]^n}{n!}$$

$$= e^{-m(t)} \sum_{n=k}^{\infty} \binom{n}{k} \left[\frac{h(t)}{m(t)} \right]^k \left[1 - \frac{h(t)}{m(t)} \right]^{n-k} \frac{[m(t)]^n}{n!}$$

$$= e^{-m(t)} \frac{[h(t)]^k}{k!} \sum_{n=k}^{\infty} \frac{[m(t)]^{n-k}}{(n-k)!} \left[1 - \frac{h(t)}{m(t)} \right]^{n-k}$$

$$= e^{-m(t)} \frac{[h(t)]^k}{k!} \sum_{n=0}^{\infty} \frac{[m(t) - h(t)]^n}{n!}$$

$$= e^{-h(t)} \frac{[h(t)]^k}{k!} \tag{18.5.7}$$

The mean and variance of $X(t)$ can be now easily determined using the means and variances of constituent processes. Let

$$E[X(0)] = \mu_0 \quad \text{and} \quad V[X(0)] = \sigma_0^2$$

Then we have

$$E[X(t)] = h(t) + \mu_0[1 - F(0, t)]$$

$$V[X(t)] = h(t) + \mu_0 F(0, t)[1 - F(0, t)] + \sigma_0^2 [1 - F(0, t)]^2$$

$$\tag{18.5.8}$$

When $X(0) = i$ (a known constant)

$$E[X(t)|X(0) = i] = h(t) + i[1 - F(0, t)]$$

$$V[X(t)|X(0) = i] = h(t) + iF(0, t)[1 - F(0, t)] \tag{18.5.9}$$

Under the constant output rate of Thakur et al. (1972) we get

$$E[X(t)|X(0) = i] = h(t) + ie^{-\mu t}$$

$$V[X(t)|X(0) = i] = h(t) + ie^{-\mu t}[1 - e^{-\mu t}] \qquad (18.5.10)$$

where $h(t)$ is given by

$$h(t) = \int_0^t \lambda(x)e^{-\mu(t-x)} dx$$

Purdue (1974b) extends this analysis to a two-compartment model.

While considering a two-compartment model, one of the factors that may be relevant is the possibility of the particle visiting a compartment more than once. Some results for this type of a model can be found in Purdue (1974b).

In the models considered here the emphasis has been on deriving information on the distribution characteristics for the model and its parameters. However, in practice, one is faced with the problem of obtaining estimates of parameters using the available data. In Saffer et al. (1976) a problem of this nature has been considered in the context of investigations into the medical evaluation of normal-abnormal liver function. Radioactive Rose Bengal is used as a tracer element in the study of biliary function in patients suffering from a hepatic disease. Rose Bengal injected into the blood is taken up by the liver, and ultimately it appears in the urine and feces. In modeling this system as a four-compartment model, because of the movement of Rose Bengal to and fro between the blood and the liver, these two form two communicating compartments, whereas urine and feces are absorbing ones. Schematically the compartments are shown in Figure 18.5.1.

Consider the four compartments as the four states of a Markov chain. The transition probability matrix of the Markov chain, arranged in the canonical form is given as follows:

$$\begin{array}{cccc} & B(2) & D(4) & A(1) & C(3) \end{array}$$

$$P = \begin{array}{c} B(2) \\ D(4) \\ A(1) \\ C(3) \end{array} \begin{bmatrix} 1 & 0 & 0 & 0 \\ 0 & 1 & 0 & 0 \\ P_{12} & 0 & P_{11} & P_{13} \\ 0 & P_{34} & P_{31} & P_{33} \end{bmatrix} \qquad (18.5.11)$$

Partitioning P in the usual manner as

$$P = \left[\begin{array}{c|c} I & 0 \\ \hline R & Q \end{array} \right]$$

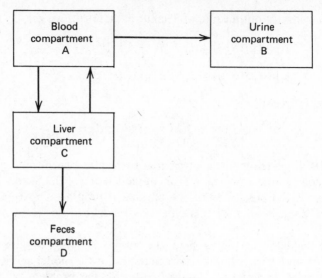

Figure 18.5.1

where

$$R = \begin{bmatrix} P_{12} & 0 \\ 0 & P_{34} \end{bmatrix} \qquad Q = \begin{bmatrix} P_{11} & P_{13} \\ P_{31} & P_{33} \end{bmatrix}$$

The elements of the first row of the matrix

$$F = (I - Q)^{-1} R \qquad\qquad (18.5.12)$$

give probabilities that Rose Bengal will ultimately appear in urine and feces. The elements of $(I - Q)^{-1}$ give the expected time needed for this eventual transition. This information can be used for diagnostic purposes in the case of different hepatic diseases.

The difficulty in estimating the elements of the transition probability matrix (18.5.11) arises due to our inability to measure the level of Rose Bengal in liver (state C). Let A_t, B_t, C_t, and D_t be the amount of Rose Bengal present in the four states at time t ($t = 0, 1, 2, \ldots$). Using the elements of the transition probability matrix P, for one unit of Rose Bengal injected into the blood at time 0, we get

$$A_0 = 1 \qquad B_0 = C_0 = D_0 = 0$$

$$A_{t+1} = P_{11}A_t + P_{31}C_t$$

$$B_{t+1} = B_t + P_{12}A_t$$

$$C_{t+1} = P_{13}A_t + P_{33}C_t$$

$$D_{t+1} = D_t + P_{34}C_t \qquad (18.5.13)$$

In the absence of data on C_t, Saffer et al. (1976) have given a nonlinear programming technique applied to the least squares procedure on equation (18.5.13), for the determination of the transition probabilities P_{ij} ($i, j = 1, 2, 3, 4$). (For another illustration of the use of the least squares procedure under similar circumstances, the readers are referred to Matis and Hartley, 1971.) An alternative method using IMSL (International Mathematical and Statistical Library) Version 5 Package for the minimization problem has been reported in Anderson et al. (1977). These authors have also given upper-bound estimates for the four transition probabilities P_{12}, P_{13}, P_{31}, and P_{34} by using the relationship between the absorption probabilities in the F matrix of (18.5.12) and the one-step transition probabilities.

REFERENCES

Anderson, D. H., Eisenfeld, J., Saffer, S. I., Reisch, J. S., and Mize, C. E. (1977). "The Mathematical Analysis of a Four Compartment Stochastic Model of Rose Bengal Transport through the Hepatic System." *Nonlinear Systems and Applications*, Proc. International Conference. New York: Academic Press, pp. 351–371.

Bailey, N. T. J. (1964). *The Elements of Stochastic Processes with Application to the Natural Sciences*. New York: Wiley.

Bailey, N. T. J. (1967). *The Mathematical Approach to Biology and Medicine*. New York: Wiley.

Bansal, B. N. L., and Gupta, P. C. (1978). "A Stochastic Model for Cell Survival after Irradiation." *Biometrics* **34**, 653–658.

Bharucha-Reid, A. T. (1960). *Elements of the Theory of Markov Processes and Their Applications*. New York: McGraw Hill.

Chiang, C. L. (1968). *Introduction to Stochastic Processes in Biostatistics*. New York: Wiley.

Eisenfeld, J. (1979). "Relationship between Stochastic and Differential Models of Compartmental Systems." *Math. Biosci.* **43**, 289–305.

Ewens, W. J. (1969). *Population Genetics*. London: Methuen.

Feller, W. (1951). "Diffusion Processes in Genetics." *Proc. Second Berkeley Symposium on Mathematical Statistics and Probability*. Berkeley: University of California Press, pp. 227–234.

Fisher, R. A. (1930). *The Genetical Theory of Natural Selection*. Oxford: Clarendon Press.

Fix, E., and Neyman, J. (1951). "A Simple Stochastic Model of Recovery, Relapse, Death and Loss of Patients." *Human Biol.* **23**, 205–241.

Gupta, P. C. (1967). "Probability of Survival of a Cell in a Complex System When Irradiated with Ionizing Radiations." *Bull. Math. Biophysics* **29**, 753–757.

Iosifescu, M., and Tăutu, P. (1973). *Stochastic Processes and Applications in Biology and Medicine*. Berlin: Springer-Verlag.

Jacquez, J. A. (1972). *Compartmental Analysis in Biology and Medicine*. Amsterdam: Elsevier.

Karlin, S., and McGregor, J. (1964). "On Some Stochastic Models in Genetics." In Gurland, J., ed., *Stochastic Models in Medicine and Biology*. Madison: University of Wisconsin Press, pp. 245–279.

Madhavanath, U. (1971). "Effects of Densely Ionizing Radiations on Human Lymphocytes Cultured *in Vitro*." Ph.D. dissertation. University of California, Berkeley.

Matis, J. H., and Hartley, H. O. (1971). "Stochastic Compartmental Analysis: Model and Least Squares Estimation from Time Series Data." *Biometrics* **27**, 77–102.

Moran, P. A. P. (1958). "Random Processes in Genetics." *Proc. Cambridge Phil. Soc.* **54**, 60–71.

Perrin, E. B., and Sheps, M. C. (1964). "Human Reproduction: A Stochastic Process." *Biometrics* **20**, 28–45.

Purdue, P. (1974a). "Stochastic Theory of Compartments." *Bull. Math. Biology* **36**, 305–309.

Purdue, P. (1974b). "Stochastic Theory of Compartments: One and Two Compartment Systems." *Bull. Math. Biology* **36**, 577–587.

Saffer, S. I., Mize, C. E., Bhat, U. N., and Szygenda, S. A. (1976). "Use of Nonlinear Programming and Stochastic Modeling in the Medical Evaluation of Normal-Abnormal Liver Function." *IEEE Trans. Biomed. Eng.* **BME-23**, 200–207.

Sheps, M. C., and Perrin, E. B. (1966). "Further Results from a Human Fertility Model with a Variety of Pregnancy Outcomes." *Human Biol.* **38**, 180–193.

Thakur, A. K., Rescigno, A., and Schafer, D. E. (1972). "On the Stochastic Theory of Compartments I. A Single Compartment System." *Bull. Math. Biophysics* **34**, 53–65.

Watson, H. W., and Galton, F. (1874). "On the Probability of the Extinction of Families." *J. Anthropol. Inst.* **4**, 138.

Wright, S. (1931). "Evolution in Mendelian Populations." *Genetics* **16**, 97–159.

Zahl, S. (1955). "A Markov Process Model for Follow-up Studies." *Human Biol.* **27**, 90–120.

FOR FURTHER READING

Alling, D. W. (1958). "The After History of Pulmonary Tuberculosis: A Stochastic Model." *Biometrics* **14**, 527–548.

Armitage, P., and Doll, R. (1961). "Stochastic Models for Carcinogenesis." *Proc. Fourth Berkeley Symposium on Mathematical Statistics and Probability*. Berkeley: University of California Press, pp. 19–38.

Bailey, N. T. J. (1961). *Introduction to the Mathematical Theory of Genetic Linkage*. Oxford: Clarendon Press.

Chiang, C. L. (1957). "An Application of Stochastic Processes to Experimental Studies on Flour Beetles." *Biometrics* 13, 79–97.

Chiang, C. L. (1971). "A Stochastic Model of Human Fertility." *Biometrics* 27, 345–356.

Dunn, J. K., and Hardy, R. J. (1980). "A Stochastic Model for the Occurrence of Transient Ischemic Attacks." *Biometrics* 36, 91–103.

Gani, J., and Jerwood, D. (1971). "Markov Chain Methods in Chain Binomial Epidemic Models." *Biometrics* 27, 591–603.

Greenhouse, S. W. (1961). "A Stochastic Process Arising in the Study of Muscular Contraction." *Proc. Fourth Berkeley Symposium on Mathematical Statistics and Probability*. Berkeley: University of California Press, pp. 257–265.

Kao, E. P. C. (1974). "Modeling the Movement of Coronary Patients within a Hospital by Semi-Markov Processes." *Oper. Res.* 22, 683–699.

Karlin, S., and McGregor, J. (1962). "On A Genetic Model of Moran." *Proc. Cambridge Phil. Soc.* 58, 299–311.

Lehoczky, J. P., and Gaver, D. P. (1977). "A Diffusion Approximation Analysis of a General n-Compartment System." *Math. Biosciences* 36, 127–148.

Lu, K. H. (1966). "A Path Probability Approach to Irreversible Markov Chains with an Application in Studying the Dental Caries Process." *Biometrics* 22, 791–809.

Matis, J. H., Wehrly, T. E., and Metzler, C. M. (1983). "On Some Stochastic Formulations and Related Statistical Moments of Pharmacokinetic Models." *J. Pharmacokinetics and Biopharmaceutics* 11, 77–92.

Rao, C. R., and Kshirasagar, A. M. (1978). "A Semi-Markovian Model for Predator-Prey Interactions." *Biometrics* 34, 611–619.

Shachtman, R. H., and Hogue, C. J. (1976). "Markov Chain Model for Events Following Induced Abortion." *Oper. Res.* 24, 916–932.

Shachtman, R. H., Schoenfelder, J. R., and Hogue, C. J. (1981). "Using a Stochastic Model to Investigate Time to Absorption Distributions." *Oper. Res.* 29, 589–603.

Sheps, M. C., and Menken, J. A. (1971). "A Model for Studying Birth Rates Given Time Dependent Changes in Reproductive Parameters." *Biometrics* 27, 325–343.

Trinkl, F. H. (1974). "A Stochastic Analysis of Programs for the Mentally Retarded." *Oper. Res.* 22, 1175–1191.

Chapter 19

SOME MARKOV MODELS IN BUSINESS MANAGEMENT

19.1 INTRODUCTION

Markov models in marketing and accounting have been introduced in earlier chapters (3, 4, 5, and 10) as examples. Here we shall discuss them in more detail and describe other models in topics such as the theory of interest, human resource management, and income determination.

19.2 CONSUMER BEHAVIOR

The basic phenomenon in marketing is the purchasing behavior of the individual consumer. Numerous efforts have been and are being made to arrive at analytical models that would adequately describe this behavior. The nature of the interacting forces underlying the marketing process is stochastic, and hence several stochastic models have been tried. One of the earlier models proposed for the brand-switching behavior of a consumer is a Bernoulli process in which the probability of buying a given brand does not depend on the brands purchased on previous occasions (see Brown, 1952; Cunningham, 1956). Even though the simplicity of the model is appealing, obviously it cannot meet all the realistic conditions of consumer behavior. A more natural model for this phenomenon is a Markov chain. Suppose there are m brands of a product in the market and there is reason to believe that consumer preference for a brand is influenced only by the brand of product an individual has been currently using. Let X_n be this individual's choice for the nth period, where X_n can take values $1, 2, \ldots, m$, corresponding to the m brands of the product. By observing the behavior of a sufficiently large number of customers, the transition probability matrix may be estimated by using the methods discussed in Chapter 5. Properties of the Markov chain can be derived in the usual manner. One of the earliest papers in this direction was by Maffei (1960), who used a two-state Markov chain to describe the brand preferences of consumers when there are only two brands of the product. Working with the expected number, stated erroneously as the actual number of consumers who buy these two products, he considers characteristics such as the expected cumulated gain or loss of consumers of a certain brand and convergence times for achieving stability of behavior. More sophisticated approaches have been made to the problem of applying Markov models to consumer behavior, and for an extensive bibliography on the topic, the reader is referred to Herniter and Howard (1964), Ehrenberg (1965), Carmen (1966), Sheth (1967), and Massey et al. (1970).

In reality it is hard to justify the use of a Markov chain with a stationary transition probability matrix for consumer behavior. However, defining the time periods suitably, it is possible to find intervals of time for which such transition probability matrices are applicable. So, even though the limiting results of such models may not be very relevant to the real world, transient results can be considered to be good approximations.

Standard techniques of analysis have been used for investigating the consumer behavior modeled as a time-homogeneous Markov chain in Lipstein (1959), Harary and Lipstein (1962), and Howard (1963). Harary and Lipstein used the mover-stayer model as developed by Blumen et al. (1955) for labor mobility, to separate the "hard-core" buyers of a brand from the "switchers." An appraisal of Markov brand-switching models has been made by Ehrenberg (1965), who has pointed out the several pitfalls in such a modeling process. One major criticism of Ehrenberg centers around the backward transition probabilities in Markov chains which we have identified while discussing reversed chains in Section 5.3. (Also see Ehrenberg, 1968, 1970; Massy and Morrison, 1968; and Charnes et al., 1971 for a lively discussion on the suitability of Markov models for consumer behavior.) For an application of the brand-switching model to the analysis of marketing data, see Draper and Nolin (1964), who use the *Chicago Tribune* consumer panel data (January 1959–December 1961) which cover a three-year listing of purchases of cake mixes of three brands and discuss the consumer attitudes based on their analysis.

An attempt to include the time-dependence of transition probabilities in the Markov model has been made by Lipstein (1965) in studying the effect of advertising on consumer attitudes. He defines a matrix A_n whose elements represent the effect of advertising during the nth time period and a reaction matrix R_n representing the consumer attitudes before advertising for that period. Clearly

$$R_{n+1} = R_n A_n \qquad n = 0, 1, 2, \ldots \qquad (19.2.1)$$

The elements of the matrix A_n are obtained by a regression analysis of the advertising expenditures and returns. Thus, if $(b_1^{(n)}, b_2^{(n)}, \ldots, b_m^{(n)})$ are the advertising costs for the m brands of the merchandise during the nth time period and $\beta_{ijr}^{(n)}$ $(r = 1, 2, \ldots, m)$ are the corresponding regression coefficients, then the element $a_{ij}^{(n)}$ of the matrix A_n can be obtained (after suitable normalization):

$$a_{ij}^{(n)} = \beta_{ij1}^{(n)}b_1^{(n)} + \beta_{ij2}^{(n)}b_2^{(n)} + \cdots + \beta_{ijm}^{(n)}b_m^{(n)} \qquad (19.2.2)$$

With these factors the n-step transition probability matrix can be expressed as

$$(R_0R_1R_2 \cdots R_n) = R_0(R_0A_0)(R_1A_1)(R_2A_2) \cdots (R_{n-1}A_{n-1})$$

$$= R_0(R_0A_0)(R_0A_0A_1)(R_0A_0A_1A_2) \cdots (R_0A_0A_1 \cdots A_{n-1})$$

$$(19.2.3)$$

It should be noted from (19.2.1) that R_n $(n = 0, 1, 2, \ldots)$ is a matrix whose elements can be estimated from standard techniques given in Chapter 5. Therefore an indirect way of determining the elements of A_n is from

$$A_n = R_n^{-1}R_{n+1} \qquad (19.2.4)$$

provided R_n^{-1} exists. As reported by Lipstein (1965), all R matrices of practical interest were found to have dominant diagonals. A matrix has a dominant diagonal if the diagonal element of a row is greater than the sum of the nondiagonal elements of that row, and this is a sufficient condition for the existence of the inverse.

In the study of consumer behavior one factor to be noted is the influence of availability and price of the product on purchasing behavior. To incorporate this factor, Lipstein modifies the transition probability for the nth period $(n = 1, 2, \ldots)$ as

$$R_{ij}^{(n)} = R_{ij}^{(n)}d_j^{(n)}\left[p_j^{(n)} + e_{ij}^{(n)}\right] \qquad (19.2.5)$$

where $R_{ij}^{(n)}$ is the (i, j)th element of R_n, $d_j^{(n)}$ is the availability factor for brand j, $p_j^{(n)}$ is the price factor for brand j, and $e_{ij}^{(n)}$ is some chance factor relative to the transition $i \to j$.

As is the case with many social phenomena consumer preferences do not change at discrete epochs of time. As a first step in generalization, a Markov process model can be used. This can be further generalized to include arbitrarily distributed interpurchase times with the help of semi-Markov processes, as suggested by Howard (1963). Hauser and Wisniewski (1982) provide a semi-Markov process model that incorporates consumer dynamics such as diffusion of innovation and response to new products and allows interdependence between purchase timing and brand selection. The transition probabilities in the model are dependent on managerial actions. The investigation includes estimation of parameters of the model as well as validation through simulation and empirical results.

Another approach to the brand selection problem is that of Kuehn (1958, 1962; also see Massy et al., 1970). Considering brand choice as a learning process, he has formulated a linear learning model that takes into account the effects of past purchases—going beyond the immediately last purchase. For a comparative treatment of this model with the Markov model, see Herniter and Howard (1964).

For other stochastic models of consumer behavior the readers are referred to Massy et al. (1970), Bass (1970), Chatfield and Goodhardt (1970), Bass et al. (1976), and Mahajan et al. (1982), and papers cited in them.

19.3 SELECTING A PORTFOLIO OF CREDIT RISKS

In Chapters 3 and 4 we discussed a problem of estimating the loss expectancy rates in accounting, using Markov chain models. The model discussed there was originally due to Cyert et al. (1962); their model made direct use of the theoretical results developed in Chapter 4 (originally due to Kemeny and Snell, 1960). An extension of these results to several classes of credit risks was made by Beranek (1962, pp. 319–320). Here we shall discuss a method suggested by Cyert and Thompson (1967) for selecting a portfolio of credit risks. It is a common practice with credit agencies to classify each credit applicant with several risk categories. This is usually done by a crude form of multiple regression model using "scoring functions" of relevant independent variables. Suppose there are c risk categories and $m + 2$ stages of the account, out of which two refer to the "paid up" and the "bad debt" states of an account. Assuming that the credit applicant passes through the states of the account based on a transition probability matrix, the management can determine the elements of c transition probability matrices, one for each risk category. This seems to be a more realistic assumption than the one based on a single transition probability matrix for all credit applicants.

The end objective of this procedure is to enable the firm to arrive at a portfolio of accounts receivable with different proportions of customers in the various risk categories. From the standpoint of total risk the firm then should be able to afford more people in higher-risk categories when it has relatively more in the low-risk categories. The criterion for this policy proposed by Cyert and Thompson is based on the coefficient of variation of the total expected receipts and acceptance of customers as long as this coefficient remains below a preset value. For numerical examples and the derivation of formulae in the discounted case (discounted due to inflation and other factors), the reader is referred to Cyert and Thompson (1967).

19.4 TERM STRUCTURE

The problem of term structure may be considered to be the special case of the problem of determining capital values under uncertainty. The payments stream in this problem may be taken as deterministic, but future interest rates are stochastic. To account for the stochastic nature of interest rates, it may be assumed that there are a finite number of interest rates ρ_i $(i = 1, 2, \ldots, m)$, and from period to period the interest rates vary as a Markov chain. We shall briefly describe the analysis of the term structure problem based on this Markov assumption as given by Pye (1966). Let P be the transition probability matrix for the interest rates ρ_i $(i = 1, 2, \ldots, m)$ where ρ_i's are such that $\rho_{i+1} > \rho_i$ for all i. Also let $\gamma_i^{(n)}$ be the expected price of an n-period bond when the current interest rate is ρ_i $(i = 1, 2, \ldots, m)$. Further let r be the coupon payment each period, and let γ_0 be the principal of the bond. We shall also assume that the conditional expected value of the bond plus the interest received in the next period is discounted by the factor $(1 + \rho_i)^{-1}$. Thus we have the recurrence relation

$$\gamma_i^{(n)} = \frac{1}{1 + \rho_i}\left(r + \sum_j P_{ij}\gamma_j^{(n-1)} \right) \qquad i = 1, 2, \ldots, m; n \geq 1 \qquad (19.4.1)$$

Define

$$\gamma^{(n)} = \begin{bmatrix} \gamma_1^{(n)} \\ \gamma_2^{(n)} \\ \vdots \\ \gamma_m^{(n)} \end{bmatrix} \qquad \xi = \begin{bmatrix} 1 \\ 1 \\ \vdots \\ 1 \end{bmatrix}$$

and

$$D = \begin{bmatrix} \dfrac{1}{1 + \rho_i} & & & \\ & \dfrac{1}{1 + \rho_2} & & \mathbf{O} \\ & & \ddots & \\ \mathbf{O} & & & \dfrac{1}{1 + \rho_m} \end{bmatrix} \qquad (19.4.2)$$

Then in matrix notations (19.4.1) can be written as

$$\gamma^{(n)} = D\left[\xi r + P\gamma^{(n-1)} \right] \qquad (19.4.3)$$

from which we can get

$$\gamma^{(1)} = D\xi(r + \gamma_0)$$

$$\gamma^{(2)} = Dr\xi + DPD\xi(r + \gamma_0)$$

$$\vdots$$

$$\gamma^{(n)} = \sum_{k=0}^{n-2} (DP)^k Dr\xi + (DP)^{n-1} D\xi(r + \gamma_0) \qquad (19.4.4)$$

The matrix DP is substochastic (there is at least one row whose row sum is less than 1), and hence $(DP)^k \to 0$ as $k \to \infty$. Thus, using result (4.1.2), we can write

$$\sum_{k=0}^{\infty} (DP)^k = (I - DP)^{-1} \qquad (19.4.5)$$

Therefore, letting $n \to \infty$ in (19.4.4), we get

$$\gamma^* = \lim_{n \to \infty} \gamma^{(n)} = (I - DP)^{-1} Dr\xi \qquad (19.4.6)$$

Corresponding to each $\gamma_i^{(n)}$ there is a yield $y_i^{(n)}$ which is the single rate of interest that discounts the payments stream of the bond to $\gamma_i^{(n)}$ (see Fisher, 1930, and Lutz, 1967, for a further discussion on the theory of interest). With this definition $\gamma_i^{(n)}$ and $y_i^{(n)}$ are related as

$$\gamma_i^{(n)} = z_i^{(n)} \gamma_0 \qquad (19.4.7)$$

where

$$z_i^{(n)} = \left[\frac{1}{1 + y_i^{(n)}} \right]^n \qquad (19.4.8)$$

Simple expressions for $y_i^{(n)}$ can be obtained when the interest rate in one year is independent of the interest rate in the preceding year and $r = 0$. Then the transition probability matrix P has identical rows $\alpha = (\alpha_1, \alpha_2, \ldots, \alpha_m)$. We then have

$$P = \xi\alpha \qquad (19.4.9)$$

Since $r = 0$, from (19.4.4) we get

$$\gamma^{(n)} = (D\xi\alpha)^{n-1} D\xi\gamma_0$$

$$= D\xi(\alpha D\xi)^{n-1} \gamma_0 \qquad (19.4.10)$$

But

$$\alpha D \xi = \sum_{j=1}^{m} \frac{\alpha_j}{1 + \rho_j}$$

and hence

$$\gamma^{(n)} = \left(\sum_{j=1}^{m} \frac{\alpha_j}{1 + \rho_j} \right)^{n-1} D \xi \gamma_0$$

$$= \left(\sum_{j=1}^{m} \frac{\alpha_j}{1 + \rho_j} \right)^{n-1} \begin{bmatrix} \dfrac{1}{1 + \rho_1} \\[2mm] \dfrac{1}{1 + \rho_2} \\[1mm] \vdots \\[1mm] \dfrac{1}{1 + \rho_m} \end{bmatrix} \gamma_0 \qquad (19.4.11)$$

giving

$$\gamma_i^{(n)} = \left(\sum_{j=1}^{m} \frac{\alpha_j}{1 + \rho_j} \right)^{n-1} \left(\frac{1}{1 + \rho_i} \right) \gamma_0 \qquad (19.4.12)$$

Comparing with (19.4.7), we can write

$$z_i^{(n)} = \left(\sum_{j=1}^{m} \frac{\alpha_j}{1 + \rho_j} \right)^{n-1} \left(\frac{1}{1 + \rho_i} \right) \qquad (19.4.13)$$

Therefore from (19.4.8) we get

$$y_i^{(n)} = \left(1 + \rho_i \right)^{1/n} \left(\sum_{j=1}^{m} \frac{\alpha_j}{1 + \rho_j} \right)^{(1/n)-1} - 1 \qquad (19.4.14)$$

In his paper Pye (1966) also discusses the properties of the yield curves and the behavior of the expected future one-period interest rates based on the properties of the transition probability matrix P.

19.5 HUMAN RESOURCE MANAGEMENT

One of the important resources of a firm is its employee group. The use of scientific techniques in the management of this human resource has been gaining increasing attention during the last few years. There are two aspects to the problem. For the benefit of the firm human resource management should take into account the economic value of the resource as well as human considerations. The first aspect considers what is good for the firm purely on economic terms, whereas the second one addresses problems of unfairness and discrimination among its employees.

We discuss here two investigations which have tackled these two problems using Markov chain models. Jaggi and Lau (1974) associated economic values for each class of employees and determined the total economic values of the employees to the firm over a period of time by following their progression through different classes using a Markov chain model. (For a discussion and refinement of the model, also see Morse, 1975, and Jaggi and Lau, 1975.) The second investigation is by Churchill and Shank (1975) who use Markov chain models to compare the progression of males and females through different ranks as well as identifying a long-run management mix for the firm.

Suppose there are k different ranks in a firm. The ranks of employees are observed at regular intervals, and using the data so generated, a transition probability matrix for the progression of employees through different ranks is obtained. Let the states of the Markov chain be the k ranks $\{1, 2, \ldots, k\}$, and add an additional state, say 0, to account for departures. Assuming that there would be no transitions to lower ranks, the transition probability matrix can be represented as follows:

$$
P = \begin{array}{c} \\ 0 \\ 1 \\ 2 \\ 3 \\ . \\ . \\ . \\ k \end{array}
\begin{array}{cccccc}
\nearrow\; 0 & 1 & 2 & 3 & \cdots & k \\
\left[\begin{array}{c|ccccc}
1 & 0 & 0 & 0 & \cdots & 0 \\
\hline
P_{10} & P_{11} & P_{12} & P_{13} & \cdots & P_{1k} \\
P_{20} & 0 & P_{22} & P_{23} & \cdots & P_{2k} \\
P_{30} & 0 & 0 & P_{33} & \cdots & P_{3k} \\
. & . & . & . & \cdots & . \\
. & . & . & . & \cdots\dagger & . \\
. & . & . & . & \cdots & . \\
P_{k0} & 0 & 0 & 0 & \cdots & P_{kk}
\end{array}\right]
\end{array}
$$

$$
= \left[\begin{array}{c|c} I & 0 \\ \hline R & Q \end{array}\right] \tag{19.5.1}
$$

From the theory of Markov chains we know that the elements of the fundamental matrix $(I - Q)^{-1}$ give the expected number of periods an employee spends in each of the ranks before leaving the firm.

Let V_i ($i = 1, 2, \ldots, k$) be the economic value of the employee to the firm while the employee occupies the ith rank. Also let N_i ($i = 1, 2, \ldots, k$) be the number of employees in the ith rank. Clearly the long-run expected economic value to the firm from its employees is given by

$$v = N^T(I - Q)^{-1}V \qquad (19.5.2)$$

where N and V are column vectors with N_i and V_i as elements.

If we are interested in determining the economic value in the short run, we replace $(I - Q)^{-1}$ by an appropriate sum of the powers of Q. For instance, if we are interested in determining the total value for R periods, we get

$$v_R = N^T\left(\sum_{n=0}^{R} Q^n\right)V \qquad (19.5.3)$$

Suppose we wish to consider future values at a discounted rate of r per period. Then formulae (19.5.3) and (19.5.2) take the form

$$v_R = N^T\left(\sum_{n=0}^{R} (rQ)^n\right)V \qquad (19.5.4)$$

$$v = N^T(I - rQ)^{-1}V \qquad (19.5.5)$$

The Markov model approach takes a more global view of the value of an employee to the firm than previous models given by Lev and Schwartz (1971) and Flamholtz (1971). Consequently it can be used to compare the economic value of employees between firms as well as to aggregate economic values of employees for all the firms in a sector of the economy. To do so, in addition to the expected values obtained in (19.5.2) and (19.5.5), we also need variances of the total value, which can be easily derived using results given in Chapter 4.

Churchill and Shank (1975) have used the Markov model to compare the employment policies of a corporation for male and female employees. Statistical tests applied to the transition probabilities for the progression of employees through different job categories indicate the difference between the two employee groups. They also suggest that the model can be used in a planning mode to determine the long-range management mix resulting from a given hiring policy and a transition probability matrix for promotions. If N_i is the number of new employees hired into rank i every year, and N is a column vector of N_i ($i = 1, 2, \ldots, k$), then the management mix M_n after n years is given by

$$M_n = B^TQ^n + N^T\sum_{r=0}^{n} Q^r \qquad (19.5.6)$$

where B is a column vector giving the number of employees in different job categories initially.

As $n \to \infty$, the long-range management mix M^* is obtained as

$$M^* = N^T(I - Q)^{-1} \tag{19.5.7}$$

Applying these formulae separately for males and females, one can arrive at desirable hiring policies.

19.6 INCOME DETERMINATION UNDER UNCERTAINTY

Shank (1971) has proposed a Markov model in the process of income determination in a business enterprise subject to uncertain events. One such firm is a small Christmas tree nursery with about 25,000 trees. The trees are grouped into 10 classes based on their height (0–1 ft, 1–2 ft,..., 9–10 ft). These are the 10 transient states of a Markov chain. Three recurrent states are included to represent two different prices the trees can bring when sold and the possibility that the tree dies or is destroyed because of blight, disease, or deformity. Let $\$C_1$ (for trees between 4–6 ft) and $\$C_2$ (for trees taller than 6 ft) be the two price levels. The nonzero elements of the transition probability matrix are identified as follows, the states being 1 ($\$C_1$), 2 ($\$C_2$), 3 (scrap), 4 (0–1 ft), 5 (1–2 ft) $\cdots$ 13 (9–10 ft):

$$
P = \begin{array}{c}
\begin{array}{cccccccc} \nearrow \quad 1 & 2 & 3 & 4 & 5 & 6 & \cdots & 12 \quad 13 \end{array} \\
\begin{array}{c} 1 \\ 2 \\ 3 \\ 4 \\ 5 \\ 6 \\ 7 \\ 8 \\ 9 \\ 10 \\ 11 \\ 12 \\ 13 \end{array}
\left[\begin{array}{ccc|cccccc}
1 & & & & & & & & \\
& 1 & & & & & & & \\
& & 1 & & & & & & \\
\hline
& & P_{43} & P_{44} & P_{45} & & & & \\
& & P_{53} & & P_{55} & P_{56} & & & \\
& & P_{63} & & & & & & \\
& & P_{73} & & & & \ddots & & \\
P_{81} & & P_{83} & & & & & & \\
P_{91} & & P_{93} & & & & & & \\
& P_{10,2} & P_{10,3} & & & & & & \\
& P_{11,2} & P_{11,3} & & & & & & \\
& P_{12,2} & P_{12,3} & & & & & P_{12,12} & P_{12,13} \\
& P_{13,2} & P_{13,3} & & & & & & P_{13,13}
\end{array} \right]
\end{array}
$$

$$
= \left[\begin{array}{c|c} I & 0 \\ \hline R & Q \end{array} \right] \tag{19.6.1}
$$

Let c_{ij} be the cost associated with a transition of a tree from state i to state j $(i, j = 3, 4, 5, \ldots, 13)$. Also let s_{ij} $((i = 8, \ldots, 13, \ j = 1, 2)$ be the net cash receipt realized from the sale of trees. Define v_i as the net realizable value of a tree that is in a transient state i. Then for $i = 4, 5, \ldots, 13$ we have

$$v_i = \sum_{j=1}^{2} P_{ij} s_{ij} - P_{i3} c_{i3} + \sum_{j=4}^{13} P_{ij}(v_j - c_{ij})$$

$$= \sum_{j=1}^{2} P_{ij} s_{ij} - P_{i3} c_{i3} - \sum_{j=4}^{13} P_{ij} c_{ij} + \sum_{j=4}^{13} P_{ij} v_j \qquad (19.6.2)$$

Writing

$$\sum_{j=1}^{2} P_{ij} s_{ij} - P_{i3} c_{i3} - \sum_{v=4}^{13} P_{ij} c_{ij} = h_i$$

and using V and H to represent column vectors with elements v_i and h_i, respectively, in matrix notation we may write (19.6.2) as

$$V = H + QV$$

giving

$$V = (I - Q)^{-1} H \qquad (19.6.3)$$

If one has to discount future values of the inventory, a discount factor of r per time period can be incorporated in (19.6.3) to obtain

$$V = (I - rQ)^{-1} H \qquad (19.6.4)$$

Let N_i $(i = 4, 5, \ldots, 13)$ be the number of trees in each state, and let N be the column vector of N_i's. Then the total inventory wealth is given by $N^T V$. For a discussion of the usefulness of the model, the readers are referred to Shank (1971).

REFERENCES

Bass, F. M. (1970). "The Theory of Stochastic Preference and Brand Switching." *J. Marketing Res.* **11**, 1–20.

Bass, F. M., Jeuland, A., and Wright, G. P. (1976). "Equilibrium Stochastic Choice and Market Penetration Theories: Derivations and Comparisons." *Management Sci.* **22**, 1051–1063.

Beranek, W. (1962). *Analysis for Financial Decisions.* Homewood, Ill.: Irwin.

Blumen, I., Kogan, M., and McCarthy, P. M. (1955). *The Industrial Mobility of Labor as a Probability Process.* Ithaca, N.Y.: Cornell University Press.

Brown, G. (1952). "Brand Loyalty—Fact or Fiction?" *Advertising Age* **23** (June), 9.

Carmen, J. C. (1966). "Brand Switching and Linear Learning Models." *J. Advertising Res.* **6**(2), 23.

Charnes, A., Cooper, W. W., DeVoe, J. K., and Learner, D. B. (1971). "Models of Fact and Laws of Behavior." *Management Sci.* **17**, 367–369.

Chatfield, C., and Goodhardt, G. J. (1970). "The Beta-Binomial Model for Consumer Purchasing Behavior." *Appl. Stat.* **19**, 240–250.

Churchill, N. C., and Shank, J. K. (1975). "Accounting for Affirmative Action Programs: A Stochastic Flow Approach." *Accounting Rev.* **50**(4), 643–656.

Cunningham, R. M. (1956). "Customer Loyalty to Store and Brand." *Harvard Bus. Rev.* **34** (January–February), 116–128.

Cyert, R. M., and Thompson, G. L. (1967). "Selecting a Portfolio of Credit Risks by Markov Chains." *Management Science Research Report* 82. Graduate School Industrial Administration, Carnegie Institute of Technology, Pittsburgh.

Cyert, R. M., Davidson, H. J., and Thompson, G. L. (1962). "Estimation of the Allowance for Doubtful Accounts by Markov Chains." *Management Sci.* **8**, 287–303.

Draper, J. C., and Nolin, L. H. (1964). "A Markov Chain Analysis of Brand Preferences." *J. Advertising Res.* **4**(3), 33–38.

Ehrenberg, A. S. C. (1964). "An Appraisal of Brand Switching Models." *J. Marketing Res.* **2**, 347–363.

Ehrenberg, A. S. C. (1968). "On Clarifying *M* and *M*." *J. Marketing Res.* **5**, 228–229.

Ehrenberg, A. S. C. (1970). "Models of Fact: Examples from Marketing." *Management Sci.* **16**, 435–445.

Fisher, I. (1930). *The Theory of Interest.* London: Macmillan.

Flamholtz, E. (1971). "A Model for Human Resource Valuation: A Stochastic Process with Service Rewards." *Accounting Rev.* **46**, 253–267.

Harary, F., and Lipstein, B. (1962). "The Dynamics of Brand Loyalty: A Markovian Approach." *Oper. Res.* **10**, 19–40.

Hauser, J. R., and Wisniewski, K. J. (1982). "Dynamic Analysis of Consumer Response to Marketing Strategies." *Management Sci.* **28**, 455–486.

Herniter, J. D., and Howard, R. A. (1964). "Stochastic Marketing Models," *Progress in Operations Research.* Vol. 2. (Eds. D. B. Hertz and R. T. Eddison), New York: Wiley, pp. 33–96.

Howard, R. A. (1963). "Stochastic Process Models of Consumer Behavior." *J. Advertising Res.* **3**, 35–42.

Jaggi, B., and Lau, H. S. (1974). "Toward a Model for Human Resource Valuation." *Accounting Rev.* **49**, 321–329.

Jaggi, B., and Lau, H. S. (1975). "Toward a Model for Human Resource Valuation: A Reply." *Accounting Rev.* **50**, 348–350.

Kemeny, J. G., and Snell, J. L. (1960). *Finite Markov Chains.* New York: Van Nostrand.

Kuehn, A. A. (1958). "An Analysis of the Dynamics of Consumer Behavior and Its Implications for Marketing Management." Ph.D. dissertation, Carnegie Technological Institute.

Kuehn, A. A. (1962). "Consumer Brand Choice as a Learning Process." *J. Advertising Res.* **2**, 10–17.

Lev, B., and Schwartz, A. (1971). "On the Use of the Economic Concept of Human Capital in Financial Statements." *Accounting Rev.* **46**, 103–111.

Lipstein, B. (1959). "The Dynamics of Brand Loyalty and Brand Switching." *Proc. Fifth Annual Conference of Advertising Research Foundation*, pp. 101–108.

Lipstein, B. (1965). "A Mathematical Model of Consumer Behavior." *J. Marketing Res.* **2**, 259–265.

Lutz, F. A., (1967). *The Theory of Interest.* Dordrecht-Holland: D. Reidel Pub. Co.

Maffei, R. B. (1960). "Brand Preferences and Simple Markov Processes." *Oper. Res.* **8**, 210–218.

Mahajan, V., Green, P. E., and Goldberg, S. M. (1982). "A Conjoint Model for Measuring Self- and Cross-Price/Demand Relationships." *J. Marketing Res.* **19**, 334–342.

Massy, W. F., and Morrison, D. G. (1968). "Comments on Ehrenberg's Appraisal of Brand-Switching Models." *J. Marketing Res.* **5**, 225–228.

Massy, W. F., Montgomery, D. B., and Morrison, D. G. (1970). *Stochastic Models of Buying Behavior.* Cambridge, Mass: The MIT Press.

Morse, W. J. (1975). "Toward a Model for Human Resource Valuation: A Comment." *Accounting Rev.* **50**, 345–347.

Pye, G. (1966). "A Markov Model of the Term Structure." *Quart. J. Economics* **80**, 60–72.

Shank, J. K. (1971). "Income Determination under Uncertainty: An Application of Markov Chains." *Accounting Rev.* **46**, 57–74.

Sheth, J. N. (1967). "A Review of Buyer Behavior." *Management Sci.* **13**(12), B718.

FOR FURTHER READING

Albright, S. C., and Winston, W. (1979). "A Birth–Death Model of Advertising and Pricing." *Adv. Appl. Prob.* **11**, 134–152.

Balachandran, K. R., Maschmeyer, R. A., and Livingstone, J. L. (1981). "Product Warranty Period: A Markovian Approach to Estimation and Analysis of Repair and Replacement Costs." *Accounting Rev.* **56**, 115–124.

Barta, T. A. (1976). "Placing a Value on Human Resources: A Markov Approach." *AIIE Trans.* **8**, 336–342. Includes a good set of references.

Bass, F. M. (1974). "The Theory of Stochastic Preference and Brand Switching." *J. Marketing Res.* **11**, 1–20.

Corcoran, A. W. (1978). "The Use of Exponentially Smoothed Transition Matrices to Improve Forecasting of Cash Flows from Accounts Receivable." *Management Sci.* **24**, 732–739.

Deshmukh, S., and Winston, W. (1977). "A Controlled Birth and Death Process Model of Optimal Product Pricing under Stochastically Changing Environment." *J. Appl. Probl.* **14**, 328–339.

Ehrenberg, A. S. C. (1972). *Repeat-Buying: Theory and Applications.* New York: American Elsevier.

Rao, A. G. (1970). *Quantitative Theories in Advertising.* New York: Wiley.

Tapiero, C. S. (1975). "Random Walk Models of Advertising and Their Diffusion Processes." *Ann. Econ. Soc. Measurement* **4**, 293–311.

Tapiero, C. S. (1975). "On Line and Adaptive Optimum Advertising Control by a Diffusion Approximation." *Oper. Res.* **23**, 890–908.

van Kuelen, J. A. M., Spronk, J., and Corcoran, A. W. (1981). "On the Cyert-Davidson-Thompson Doubtful Accounts Model." *Management Sci.* **27**, 108–112.

Vidale, M. L., and Wolfe, H. B. (1957). "An Operations Research Study of Sales Response to Advertising." *Oper. Res.* **5**, 370–381.

Chapter 20

STOCHASTIC MODELS IN TRAFFIC FLOW THEORY AND GEOLOGICAL SCIENCES

The application of stochastic models can be found in diverse areas under engineering, natural, physical, and the social sciences. Examples from several of these major areas have been described in Capters 11–19. In this chapter we shall identify two more areas of applications: the theory of traffic flow and geological sciences. In view of the lack of common theme between the topics, references have been provided at the end of each topic.

20.1 SOME TRAFFIC FLOW PROBLEMS

A variety of stochastic models arise in studying the behavior of traffic on a road system. A sampling of these is given in this section.

Pedestrian Traffic on a Sidewalk

Pedestrians get on and off a sidewalk at certain rates; this process can be modeled as a birth and death process. Suppose pedestrians get on the sidewalk in a Poisson process with parameter λ. Let the time spent by each pedestrian on the sidewalk be an exponential random variable with mean μ^{-1}. When there are n people on the sidewalk at time t, the probability that any one of them would leave in the infinitesimal interval $(t, t + \Delta t]$ will then be $n\mu \Delta t + o(\Delta t)$. It should be noted that we have assumed that the time t is represented by the length of the sidewalk; this assumes a constant speed for all the pedestrians.

Clearly we have a birth and death process with birth parameter λ and death parameter $\mu_n = n\mu$, and hence results from Section 7.6 can be directly applied. However, from realistic considerations the limiting distribution may not have much use in this case. Because of changes in the arrival and departure patterns within a short distance on the sidewalk, transient behavior of the volume of traffic would give more useful information. As we have not discussed much about the transient solution of birth and death processes, we shall not go into its details.

A realistic model for pedestrian traffic can be formulated as follows: Along a one-way street in the center of a city, suppose road crossings are to be marked at optimum locations. If we use a birth and death process model, when the traffic intensity is greater than 1 with a certain probability, we can get the minimum distance on the sidewalk so as to make the volume of traffic greater than a preassigned quantity. This quantity will have to be decided on the basis of some tangible as well as intangible costs connected with the traffic pattern.

Freeway Traffic

With certain simplifying assumptions, pure birth and pure death processes are suitable for commuter traffic before and after a working day. Non-Markovian stochastic processes would be more realistic in these cases.

Parking Lot Traffic

Suppose the parking lot of a supermarket has space for s cars. Assume the cars arrive in the parking lot in a Poisson process with parameter λ, and let the parking times of cars be exponential random variables with mean μ^{-1}. The number of cars in the lot behaves exactly as the queue length process in the loss queueing system $M/M/s$ of Section 11.5. System characteristics discussed for the queueing system are directly applicable in this situation.

Intersection Traffic

Consider the intersection of a major and a minor road. Traffic on the minor road has a stop sign and is supposed to give way to the traffic on the major road. The situation is the same when a pedestrian tries to cross a street. A two-state Markov chain model for the delay of a minor road vehicle can be formulated as follows: Gaps occur in the major road flow of traffic and for the minor road traffic to get into or go across the major road, a certain length of gap is necessary. Suppose the minor road traffic can move only if the gap is of length T or more. The consecutive gaps occurring in the major road traffic can be classified either as acceptable and or not acceptable, depending on their length. Let $(t_1, t_2, \ldots, t_n, \ldots)$ be the discrete points in time when these gaps occur. Suppose further that the distribution of the type of gaps can be approximated by a transition probability matrix as follows:

| | Gap length at t_{n+1} | |
Gap length at t_n	Acceptable	Not acceptable
Acceptable	p_1	q_1
Not acceptable	p_2	q_2

$$(20.1.1)$$

This transition probability matrix can be used to obtain information about the delay (in terms of time points, or an approximate length of time using mean distance between vehicles on the major road), the state of the system under equilibrium conditions, the fraction of time when acceptable gaps are found, and so on. These results can be used to decide whether to install traffic signals. A renewal theory formulation will be given later in this section.

The Markov chain model can also be used to measure congestion in signalized intersections. Roughly, such a model can be formulated as follows: Let the red period be of length R units of time and the green period of G units of time. Let the unit of time be the distance (in time) between two vehicles passing through the intersection. Therefore we can assume that the maximum number of vehicles passing through the intersection in a green period is G and the maximum number of vehicles arriving in a green-red cycle is $R + G = C$, for instance. In this model we shall assume that the amber period is merged with one of the red or green periods. Let $t_1, t_2, t_3, \ldots$ be the instants of time at which the consecutive red periods start. Let V_n be the number of vehicles waiting at the start of the nth red period (t_n), and let A_n be the number of vehicles arriving during the interval $(t_{n-1}, t_n]$. Also let

$$P(A_n = j) = a_j \qquad j = 0, 1, 2, \ldots, C \qquad (20.1.2)$$

For the sake of simplicity we shall assume that the random variables A_n are independent and identically distributed as (20.1.2). Clearly $\{V_n\}$ can now be represented as a Markov chain with state space $\{0, 1, 2, \ldots\}$. We can write

$$V_{n+1} = \begin{cases} V_n + A_{n+1} - G & \text{if } V_n + A_{n+1} - G > 0 \\ 0 & \text{if } V_n + A_{n+1} - G \leq 0 \end{cases} \qquad (20.1.3)$$

The transition probability matrix of this Markov chain has the form

$$
\begin{array}{c}
 \\
0 \\
1 \\
\vdots \\
G \\
G+1 \\
G+2 \\
\vdots
\end{array}
\begin{array}{ccc}
0 & 1 & 2 \\
\left[
\begin{array}{cccc}
\displaystyle\sum_{0}^{G} a_i & a_{G+1} & a_{G+2} & \cdots \\[2ex]
\displaystyle\sum_{0}^{G-1} a_i & a_G & a_{G+1} & \cdots \\[2ex]
\vdots & \vdots & \vdots & \vdots \\
a_0 & a_1 & a_2 & \cdots \\
0 & a_0 & a_1 & \cdots \\
0 & 0 & a_0 & \cdots \\
\vdots & \vdots & \vdots & \vdots
\end{array}
\right]
\end{array}
\qquad (20.1.4)
$$

It can be shown that this Markov chain is irreducible and aperiodic and the limiting distribution exists if $E(A_n)/G < 1$. To determine the limiting distribution and other characteristics, the generating function technique has to be employed.

By considering the distribution of the number of arrivals during green and red periods separately and extending this method further, optimum values can also be obtained for the length of red and green periods (see Uematu, 1957).

One of the major characteristics of interest to traffic engineers is the delay encountered by a vehicle in an intersection. Several analyses have appeared for intersection with fixed traffic signals. For a critical review of these analyses and an excellent bibliography the readers are referred to Allsop (1972) (see also Hutchinson (1972) for a numerical comparison of theoretical expressions). After recounting the various queue theoretic models, Allsop observes that the mathematical fascination of such analyses has caused them to be pursued to a level of intricacy far greater than is likely to find practical application. A more recent study of the delay problem can be found in Ohno (1978) who gives approximate expressions for the average delay. A general overview of delay problems for isolated intersections has been given by McNeil and Weiss (1974), who also provide an excellent bibliography.

Left-Turning Traffic

As mentioned earlier, there are several aspects of traffic behavior that make modeling difficult. One of these is the left-turning traffic. Consider a two-lane road—one northbound (N) and the other southbound (S)—at an intersection. Let us assume that passing does not occur at the intersection, and a left-turning vehicle is held up only when the opposing traffic is right-turning or going straight through. For the sake of simplicity we shall assume that the right-turning vehicles pose identical problems as those going straight; therefore we can categorize the traffic as those turning left and those not turning left. We shall obtain a Markov chain model for this traffic under the assumption that queues of vehicles do not vanish on either lane. This model can then be used to estimate the maximum capacity of the two lanes (see Newell, 1959).

Let p be the probability that an N-lane vehicle wishes to turn left, and let p' be the probability that an S-lane vehicle wishes to turn left. Let $(t_1, t_2, \ldots)$ be the discrete set of time points at which traffic on either lane moves into the intersection. At these points the states of the system are as follows:

0—Neither vehicle is held over.

N— Northbound vehicle is held over.

S— Southbound vehicle is held over.

Between two consecutive time points possible transitions are as given in Table 20.1.1.

Corresponding to these transitions, the transition probability matrix can be given as

$$
\boldsymbol{P} = \begin{bmatrix} pp' + (1-p)(1-p') & p(1-p') & p'(1-p) \\ p' & 1-p' & 0 \\ p & 0 & 1-p \end{bmatrix} \quad (20.1.5)
$$

Table 20.1.1

	0	N	S
0	Both vehicles turn left, or both do not turn left	N-lane vehicle wants to turn left, but S-lane vehicle does not turn left	S-lane vehicle wants to turn left, but N-lane vehicle does not turn left
N	S-lane vehicle also turns left	S-lane vehicle does not turn left	Not possible
S	N-lane vehicle also turns left	Not possible	N-lane vehicle does not turn left

Suppose we are interested in the number of vehicles cleared during time $(0, t_n)$ from the N and S lanes. We shall call these expected values the capacities of the lanes, and we shall denote them by C_N and C_S, respectively. Using arguments similar to those employed in deriving (5.2.63) and writing

$$\boldsymbol{P}^k = \begin{bmatrix} P_{00}^{(k)} & P_{0N}^{(k)} & P_{0S}^{(k)} \\ P_{N0}^{(k)} & P_{NN}^{(k)} & P_{NS}^{(k)} \\ P_{S0}^{(k)} & P_{SN}^{(k)} & P_{SS}^{(k)} \end{bmatrix} \qquad (20.1.6)$$

we have

$$C_N = \sum_{k=1}^{n} P_{00}^{(k)} + \sum_{k=1}^{n} P_{N0}^{(k)}$$

$$C_S = \sum_{k=1}^{n} P_{00}^{(k)} + \sum_{k=1}^{n} P_{S0}^{(k)} \qquad (20.1.7)$$

In particular, when $p = p'$ in (20.1.5), we get

$$\boldsymbol{P} = \begin{bmatrix} 1 - 2p + 2p^2 & p - p^2 & p - p^2 \\ p & 1 - p & 0 \\ p & 0 & 1 - p \end{bmatrix} \qquad (20.1.8)$$

Simplifying $\boldsymbol{P}^k$ by the eigenvalue method of Section 4.2, we finally obtain

$$C_N = C_S = \frac{n(2 - p)}{3 - 2p} + \frac{(1 - p)[1 - (1 - p)^n (1 - 2p)^n]}{p(3 - 2p^2)} \qquad (20.1.9)$$

These would give the maximum capacities of the two lanes.

Pedestrian Delay

While discussing the intersection traffic, we have seen a simple Markov chain model for the delay of a pedestrian intending to cross a street or a minor-road vehicle intending to get through the major-road traffic. Clearly the information provided by the Markov chain model is insufficient, and therefore we shall try to give a renewal theory formulation of the delay problem. The terminology used would be relative to the pedestrian delay problem.

The road traffic can be considered a renewal process with gaps between vehicles as renewal periods. (For the sake of convenience gaps would include the length of vehicles.) We shall further assume that the gaps have an exponential distribution with mean μ^{-1}. (This assumption ignores the length of a vehicle.) As before, we shall assume that the gaps are acceptable to the pedestrian only if their lengths exceed T.

Let W be the random variable representing the time a pedestrian has to wait (delay) to cross the road. Let $X_1, X_2, \ldots$ be the successive gaps. Clearly

$$P(W = 0) = P(X_1 > T)$$

$$= e^{-\mu T} \qquad (20.1.10)$$

Let $f(x)$ $(x > 0)$ be the probability density of W. If the pedestrian can accept only the $(n + 1)$th gap, then

$$W = X_1 + X_2 + \cdots + X_n \qquad (20.1.11)$$

and X_i $(i = 1, 2, \ldots, n)$ are such that $X_i \leq T$. The conditional probability density of these X_i's can be given as

$$h(x) = \frac{\mu e^{-\mu x}}{1 - e^{-\mu T}} = \frac{g(x)}{1 - e^{-\mu T}} \qquad 0 < x < T \qquad (20.1.12)$$

and therefore the conditional probability density of W, given that the pedestrian accepts the $(n + 1)$th gap, can be obtained as

$$f(x|n) = [h(x)]^{n*} = \frac{[g(x)]^{n*}}{(1 - e^{-\mu T})^n} \qquad (20.1.13)$$

where $n*$ denotes the n-fold convolution. Now the event that the road crossing takes place in the $(n + 1)$st period has a probability equal to

$$(1 - e^{-\mu T})^n e^{-\mu T} \qquad (20.1.14)$$

Combining (20.1.13) and (20.1.14), we get

$$f(x) = e^{-\mu T} \sum_{n=1}^{\infty} [g(x)]^{n*} \qquad (20.1.15)$$

Because of the truncated form of $g(x)$, it is convenient to use Laplace transforms in deriving an explicit expression for $f(x)$. Let

$$\gamma(\theta) = \int_0^T e^{-\theta x} g(x) \, dx \qquad \text{Re}(\theta) > 0$$

$$= \mu \int_0^T e^{-(\theta + \mu)x} \, dx$$

$$= \frac{\mu}{\theta + \mu} [1 - e^{-(\theta + \mu)T}] \qquad (20.1.16)$$

and

$$\psi(\theta) = \int_0^\infty e^{-\theta x} f(x) \, dx \qquad \text{Re}(\theta) > 0$$

$$= e^{-\mu T} + e^{-\mu T} \sum_{n=1}^\infty [\gamma(\theta)]^n$$

$$= \frac{(\theta + \mu) e^{-\mu T}}{\theta + \mu e^{-(\theta + \mu)T}} \qquad (20.1.17)$$

On inversion, this gives

$$f(x) = e^{-\mu T} \delta(x) + \mu e^{-\mu T} \sum_{r=0}^{n-1} [-e^{-\mu T}]^r$$

$$\times \left\{ \frac{[\mu(x - rT)]^{r-1}}{(r - 1)!} + \frac{[\mu(x - rT)]^r}{r!} \right\}$$

$$(n - 1)T \le x \le nT; \quad n = 1, 2, \ldots \qquad (20.1.18)$$

where $\delta(x) = 1$ if $x = 0$ and $\delta(x) = 0$ otherwise. Differentiating $\psi(\theta)$ and setting $\theta = 0$, we also get

$$E(W) = -\psi'(0) = \frac{1}{\mu} (e^{\mu T} - 1 - \mu T) \qquad (20.1.19)$$

$$V(W) = \psi''(0) - [\psi'(0)]^2 = \frac{1}{\mu^2} [e^{2\mu T} - 2\mu T e^{\mu T} - 1] \qquad (20.1.20)$$

Headway Distribution

Headway is the space that separates vehicles in traffic. The simplest model is to consider vehicles as points occurring in a Poisson process along a line. This ignores the reality that headways generally belong to two classes: follower headways and nonfollower headways. Follower headways are those created when a vehicle follows another, and such headways can be considered to have a different distribution than the nonfollower ones. A simple distribution model can be set up as follows (Branston, 1976).

Let $f(x)$ be the probability density function (p.d.f.) of all headways, with $g(x)$ being the p.d.f. of follower headways and $h(x)$ the p.d.f. of nonfollower headways. Also let α be the proportion of follower vehicles. Then we may write

$$f(x) = \alpha g(x) + (1 - \alpha)h(x) \qquad (20.1.21)$$

With an analogy to a queueing situation one could identify the follower headway as the service time in a single-server queue. Suppose additionally we assume that a nonfollower headway has the distribution that is a convolution of $g(x)$ and the exponential density $\lambda e^{-\lambda x}$, then the general headway distribution can be considered the distribution of time intervals between departures in an $M/G/1$ queue in steady state. With this interpretation α will be the traffic intensity, which also gives the probability that the system is busy. Thus we get

$$f(x) = \alpha g(x) + (1 - \alpha)\int_0^x g(y)\lambda e^{-\lambda(x-y)}\,dy \qquad (20.1.22)$$

The headway model described here is a modification of the original model given earlier by Buckley (1968), which he called a semi-Poisson model. His model has a conditional exponential distribution for $h(x)$, so as to make the nonfollower headway longer than the follower headway. Thus Buckley's form for $h(x)$ is given by

$$h(x) = \frac{\lambda e^{-\lambda x}\int_0^x g(y)\,dy}{\int_0^\infty \lambda e^{-\lambda x}\int_0^x g(y)\,dy\,dx} \qquad (20.1.23)$$

Using the notation $a^*(\theta) = \int_0^\infty e^{-\theta x}a(x)\,dx$ for the Laplace transform, the denominator in (20.1.23) may be written as

$$\int_0^\infty \lambda e^{-\lambda x}\int_0^x g(y)\,dy\,dx = g^*(\lambda)$$

Thus we have

$$f(x) = \alpha g(x) + (1 - \alpha)\big[g^*(\lambda)\big]^{-1}\lambda e^{-\lambda x}G(x) \qquad (20.1.24)$$

Buckley has applied this model with a gamma headway distribution to data from a freeway in suburban Sydney, Australia, and found the model to have a good fit.

Comparing his generalized queueing model (20.1.22) with Buckley's semi-Poisson model (20.1.24) on M4 motorway data from West London, Canada, Branston concludes that the generalized queueing model with a log-normal distribution of follower headways gives the better fit.

Buckley's semi-Poisson model has been generalized by Wasielewski (1974, 1979) to give an integral equation form, which is also convenient for the determination of the follower headway distribution from the data on the overall headway distribution.

For a discussion of various models of traffic flow, the readers are referred to Edie (1974), which also includes an excellent bibliography on the topic.

REFERENCES

Allsop, R. E. (1972). "Delay at a Fixed Time Traffic Signal. I: Theoretical Analysis." *Transportation Sci* **6**, 260–285. Includes a good set of references.

Branston, D. (1976). "Models of Single Lane Time Headway Distributions." *Transportation Sci*. **10**, 125–148.

Buckley, D. J. (1968). "A Semi-Poisson Model of Traffic Flow." *Transportation Sci*. **2**, 107–133.

Edie, L. C. (1974). "Flow Theories." In D. C. Gazis, ed., *Traffic Science*. New York: Wiley, pp. 1–108.

Hutchinson, T. P. (1972). "Delay at a Fixed Time Traffic Signal. II: Numerical Comparisons of Some Theoretical Expressions." *Transportation Sci*. **6**, 286–305.

McNeil, D. R., and Weiss, G. H. (1974). "Delay Problems for Isolated Intersections." In D. C. Gazis, ed., *Traffic Science*. New York: Wiley, pp. 109–174.

Newell, G. F. (1959). "The Effect of Left Turns on the Capacity of a Traffic Intersection." *Quart. Appl. Math*. **17**(1), 67–76.

Ohno, K. (1978). "Computational Algorithm for a Fixed Cycle Traffic Signal and New Approximate Expressions for Average Delay." *Transportation Sci*. **12**, 29–47.

Uematu, T. (1957). "On the Traffic Control at an Intersection Controlled by the Repeated Fixed Cycle Traffic Light." *Ann. Inst. Stat. Math., Tokyo* **9**, 87–107.

Wasielewski, P. (1974). "An Integral Equation for the Semi-Poisson Headway Distribution Model." *Transportation Sci*. **8**, 237–247.

Wasielewski, P. (1979). "Car Following Headways On Freeways Interpreted by the Semi-Poisson Headway Distribution Model." *Transportation Sci*. **13**, 36–55.

FOR FURTHER READING

Ashton, W. D. (1966). *The Theory of Road Traffic Flow*. New York: Wiley.

Blumenfeld, D. E., and Weiss, G. H. (1971). "Merging from an Acceleration Lane." *Transportation Sci*. **5**, 161–168.

Cowan, R. J. (1975). "Useful Headway Models." *Trans. Res.* **9**, 371–375.

Drew, D. R. (1968). *Traffic Flow Theory and Control.* New York: McGraw-Hill.

Dunne, M. C., Rothery, R. W., and Potts, R. B. (1968). "A Discrete Markov Model of Vehicular Traffic." *Transportation Sci.* **2**, 233–251.

Gazis, D. C., ed. (1974). *Traffic Science.* New York: Wiley.

Haight, F. A. (1963). *Mathematical Theories of Traffic Flow.* New York: Academic Press.

Hauer, E., and Templeton, J. G. C. (1972). "Queuing in Lanes." *Transportation Sci.* **6**, 247–259.

Holland, H. J. (1967). "A Stochastic Model for Multilane Traffic Flow." *Transportation Sci.* **1**, 184–205.

Miller, A. J. (1960). "An Empirical Model for Multilane Road Traffic." *Transportation Sci.* **4**, 164–186. Includes a good set of references.

Newell, G. F. (1969). "Properties of Vehicle Actuated Signals. I: One-Way Streets." *Transportation Sci.* **3**, 30–52. Includes a good set of references.

Newell, G. F., and Osuna, E. E. (1969). "Properties of Vehicle-Actuated Signals. II: Two-way Streets." *Transportation Sci* **3**, 99–125.

20.2 SEDIMENTATION AND STRATIGRAPHY

Stratigraphy is concerned with the development of the earth's rocky framework through successive geologic ages, and sedimentation has played a major role in that process. Sedimentary rocks are formed as layers in a stratigraphic sequence. These bedding types, or lithologies (rock formations), can be classified based on their nature and identified as states of a stochastic process. Sandstone, shale, and limestone are some examples of such states. A vertical section of a stratigraphic sequence can be considered a realization of a stochastic process, and an appropriate model can be developed.

The earliest published work on the use of stochastic models for stratigraphic data seems to be Vistelius (1949). More recent applications have come from Carr et al. (1966), Potter and Blakely (1967), Gingerich (1969), Schwarzacher (1969), Miall (1973), and others. For some of the theoretical problems related to geological applications, the readers are referred to Krumbein (1967, 1968), Krumbein and Dacey (1969), Dacey and Krumbein (1970), Harbaugh and Bonham-Carter (1970, chap. 4), and Schwarzacher (1972).

Markov models are used in stratigraphy to understand geological processes. Recognizing that the sedimentation process is governed by random phenomena, a Markov model is developed for a stratigraphic sequence of lithologies. The validity of the model is checked using simulation and tests of hypotheses. When first-order Markov models are found to be unsatisfactory, higher-order models are tried. We give two examples of studies indicating the nature of use made of Markov models.

Synthesized Stratigraphic Sections

A Markov model is used to synthesize stratigraphic sections and bedding types with the hope that if the depositional process is better understood, better predictions are possible. As an example of this process Carr et al. (1966) simulate a stratigraphic section, with the Mississippian Chester Series of the Illinois Basin as a model. Observations of the series resulted in the following characteristics of the model: state 1—sandstone; state 2—shale; state 3—limestone. The resulting transition probability matrix was

$$P = \begin{bmatrix} 0.1 & 0.8 & 0.1 \\ 0.4 & 0.2 & 0.4 \\ 0.1 & 0.8 & 0.1 \end{bmatrix} \qquad (20.2.1)$$

The thickness distributions of the three types of layers were also obtained taking appropriate measurements. Normally the distributions used for this purpose are, log-normal, exponential, and chi-square.

Simulated stratigraphic sections are obtained by the following procedure involving four steps:

1. Choose an initial state i at random.
2. Determine the thickness of the ith lithology from the thickness distribution.
3. Using the transition probabilities P_{ij}, choose the jth lithology as the following state.
4. Repeat steps (2) and (3) and build the stratigraphic sections.

The synthetic stratigraphic section generated by simulation can be compared with the natural section for its properties. If they match to a substantial extent, we have a model by which further information on the stratigraphic section can be derived.

Using the foregoing transition probability, Carr et al. (1966) develop a 3093-ft section with two hundred lithologic units to illustrate the procedure.

An Ancient Alluvial Plain Succession as a Markov Chain

In Miall (1973) data from the western side of the Boothia Uplift of the Peel Sound Formation of Prince of Wales Island, Arctic Canada, are used for a Markov model. Boothia Uplift became very active in the Early Devonian age and deposited various lithofacies such as conglomerate, sandstone, siftstone, shale, and dolomite. For the purpose of our discussion we shall consider a section of conglomerate–sandstone facies with the

Table 20.2.1

↗	1	2	3	4	
1	—	3	11	1	15
2	5	—	1	8	14
3	4	0	—	15	19
4	5	13	6	—	24
	14	16	18	24	72

following lithofacies as distinct states: 1—conglomerate; 2—pebbly sandstone; 3—course to medium sandstone; 4—fine sandstone. Using a vertical sequence stratigraphic section, the transition count matrix (Table 20.2.1) is obtained for the four states. The transition epochs are identified as the points at which the lithofacies change from one to another, regardless of the thickness of the lower bed. Therefore the transitions are always from one state to some other state; as a result the diagonal elements of the matrix are zeros. The corresponding models have been identified as embedded Markov chains in the geosciences literature, indicating that they are in fact the underlying Markov chain of a semi-Markov process. These data provide the following estimates for the transition probabilities:

$$\hat{P} = \begin{bmatrix} — & 0.20 & 0.73 & 0.07 \\ 0.36 & — & 0.07 & 0.57 \\ 0.21 & 0.00 & — & 0.79 \\ 0.21 & 0.54 & 0.25 & — \end{bmatrix} \qquad (20.2.2)$$

The objective of this analysis is to provide a pattern of sedimentation using a mathematical model. If a Markov model is suitable, looking at the transition probabilities, it is possible to infer properties of the depositional mechanism of the sedimentation process. Thus as a first step a χ^2 test is used to establish the validity of the model.

For purposes of testing, the null hypothesis used is that the observations are attributable to a process of independent trials. Under the assumption of independence, as illustrated in Section 5.4, the transition probabilities are obtained by considering the proportion of occurrences of a state without regard to the previous state. Thus typically in the case of Markov chains in which all transitions are possible, an independent trials matrix has identical probability row vectors. However, in the present case, where transitions occur from a state only to some other state, approximate probabilities are obtained by adjusting the divisor for each row. This is illustrated as follows using notations employed earlier in Chapter 5. For an exact method under similar situations the readers are referred to Goodman (1968). [We are using the column sums instead of row sums of the transition count matrix to be consistent with estimates derived in (5.4.31). Miall uses row sums in his analysis. See Naylor and Woodcock, 1977, Hiscott, 1981, and Basava and Bhat, 1985, for further elaboration on this change.]

Case 1: Unrestricted Transitions

The transition count matrix is

	1	2	$\cdots$	m	
1	n_{11}	n_{12}	$\cdots$	n_{1m}	n_1
2	n_{21}	n_{22}	$\cdots$	n_{2m}	n_2
$\vdots$	$\vdots$	$\vdots$		$\vdots$	$\vdots$
m	n_{m1}	n_{m2}	$\cdots$	n_{mm}	n_m
	$n_{.1}$	$n_{.2}$		$n_{.m}$	n

Then the estimated transition probability matrix is

$$\hat{P}^0 = \begin{bmatrix} \dfrac{n_{.1}}{n} & \dfrac{n_{.2}}{n} & \cdots & \dfrac{n_{.m}}{n} \\ \dfrac{n_{.1}}{n} & \dfrac{n_{.2}}{n} & \cdots & \dfrac{n_{.m}}{n} \\ \vdots & \vdots & & \vdots \\ \dfrac{n_{.1}}{n} & \dfrac{n_{.2}}{n} & \cdots & \dfrac{n_{.m}}{n} \end{bmatrix}$$

Case 2: Restricted Transitions

The transition count matrix is

	1	2	3	$\cdots$	m	
1	—	n_{12}	n_{13}	$\cdots$	n_{1m}	n_1
2	n_{21}	—	n_{23}	$\cdots$	n_{2m}	n_2
$\vdots$	$\vdots$	$\vdots$	$\vdots$		$\vdots$	$\vdots$
m	n_{m1}	n_{m2}	n_{m3}	$\cdots$	—	n_m
	$n_{.1}$	$n_{.2}$	$n_{.3}$		$n_{.m}$	n

Then the estimated transition probability matrix is

$$
\hat{P}^0 = \begin{bmatrix}
- & \dfrac{n_{.2}}{n - n_{.1}} & \dfrac{n_{.3}}{n - n_{.1}} & \cdots & \dfrac{n_{.m}}{n - n_{.1}} \\[2ex]
\dfrac{n_{.1}}{n - n_{.2}} & - & \dfrac{n_{.3}}{n - n_{.2}} & \cdots & \dfrac{n_{.m}}{n - n_{.2}} \\[2ex]
\vdots & \vdots & \vdots & & \vdots \\[2ex]
\dfrac{n_{.1}}{n - n_{.m}} & \dfrac{n_{.2}}{n - n_{.m}} & \dfrac{n_{.3}}{n - n_{.m}} & \cdots & -
\end{bmatrix}
$$

For the data from Table 20.2.1 the transition probability matrix under null hypothesis is obtained as follows (Case 2):

$$
\hat{P}^0 = \begin{bmatrix}
- & 0.28 & 0.31 & 0.41 \\
0.25 & - & 0.32 & 0.43 \\
0.26 & 0.30 & - & 0.44 \\
0.29 & 0.33 & 0.38 & -
\end{bmatrix}
\tag{20.2.3}
$$

The χ^2 statistic of (5.4.32) is obtained as 33.35 with $16 - 4 - 4 - 3 = 5$ d.f. From χ^2 tables, for the corresponding distribution we have

$$
P(\chi^2 > 16.7) = 0.005
$$

Alternately, we could use the likelihood ratio statistic of (5.4.33) for this test. In this case we get $-2\ln\Lambda = 38.8061$. Under the null hypothesis this statistic should have a χ^2 distribution with $9 - 4 = 5$ d.f. Thus both tests lead us to the conclusion that the assumption of independence can be rejected even at the 0.5% level of significance. Based on these considerations, the conclusion that the data can be modeled as a Markov chain is justified.

In order to infer properties of the depositional mechanism, a matrix of differences of the transition probabilities under the two assumptions is used. Subtracting (20.2.3) from (20.2.2), we get

$$
\hat{P} - \hat{P}^0 = \begin{array}{c}
 \\
1 \\
2 \\
3 \\
4
\end{array}
\begin{matrix}
\begin{array}{cccc}
1 & 2 & 3 & 4
\end{array} \\
\begin{bmatrix}
- & -0.08 & 0.42 & -0.34 \\
0.11 & - & -0.25 & 0.14 \\
-0.05 & -0.30 & - & 0.35 \\
-0.08 & 0.21 & -0.15 & -
\end{bmatrix}
\end{matrix}
\tag{20.2.4}
$$

The positive elements of the difference matrix (20.2.4) represent those transitions that have a higher probability of occurrence than one would have expected from an independent trials assumption. Thus a cyclic process can be identified in the depositional mechanism as follows: Conglomerate (1) → coarse to medium sandstone (3) → fine sandstone (4) → pebbly sandstone (2) → conglomerate (1) or fine standstone (4) → etc. For a detailed discussion of the implications of this cyclic nature, the readers are referred to Miall (1973), and for a general discussion of the role of Markov chains in the transition analysis of structural sequences, to Naylor and Woodcock (1977).

Remarks For a complete understanding of the sedimentation process, the discrete state–discrete parameter models considered here are not adequate. There is a need to incorporate the thickness of lithologies in the model. Two approaches have been suggested to remedy this situation: (1) For transition count data, transition epochs are identified at intervals of equal length on the vertical stratigraphic section. The length of the interval should be such that no lithologies are missed while counting transitions. Now the thickness of lithology i can be modeled as having a geometric distribution with parameter $1 - P_{ii}$, where P_{ii} is the transition probability that lithology i is observed at two consecutive transition epochs (see Krumbein, 1968; Krumbein and Dacey, 1969; Dacey and Krumbein, 1970; and Harbaugh and Bonham-Carter, 1970). (2) The embedded Markov chain model used (as described in Miall, 1966 incorporates the distribution of the thickness of lithologies in a semi-Markov process model. For a discussion of this approach, readers are referred to Vistelius and Feigel'son (1965), Potter and Blakely (1967), Schwarzacher (1968, 1972), and Harbaugh and Bonham-Carter (1970).

REFERENCES

Basava, I. V. and Bhat, U. N. (1985). "Chi-Square Tests for Markov Chain Analysis: Some Comments on a Paper by R. N. Hiscott." submitted for publication.

Carr, D. D., Horowitz, A., Hrabar, S. V., Ridge, K. F., Rooney, R., Straw, W. T., and Webb, W. (1966). "Stratigraphic Sections, Bedding Sequences, and Random Processes." *Science* **154**, 1162–1164.

Dacey, M. F., and Krumbein, W. C. (1970). "Markovian Models in Stratigraphic Analysis." *Math. Geol.* **2**(2), 175–191.

Gingerich, P. D. (1969). "Markov Analysis of Cyclic Alluvial Sediments." *J. Sedimentary Petrology* **39**, 330–332.

Goodman, L. A. (1968). "The Analysis of Cross-Classified Data: Independence, Quasi-independence, and Interactions in Contingency Tables with or without Missing Entries." *J. Am. Stat. Assoc.* **63**, 1091–1131.

Harbaugh, J. W., and Bonham-Carter, G. (1970). *Computer Simulation in Geology.* New York: Wiley.

Hiscott, R. N. (1981). "Chi-Square Tests for Markov Chain Analysis." *Math. Geol.* 13(1), 69–80.

Krumbein, W. C. (1967). "FORTRAN IV Computer Programs for Markov Chain Experiments in Geology." Computer Contribution 13. Kansas Geological Survey, University of Kansas, Lawrence.

Krumbein, W. C. (1968). "FORTRAN IV Computer Program for Simulation of Transgression and Regression with Continuous Time Models." Computer Contribution 26. Kansas Geological Survey, University of Kansas, Lawrence.

Krumbein, W. C. and Dacey, M. F. (1969). "Markov Chains and Embedded Markov Chains in Geology." *J. Int. Assoc. Math. Geol.* 1(1), 79–96.

Miall, A. D. (1973). "Markov Chain Analysis Applied to an Ancient Alluvial Plain Succession." *Sedimentology* 20, 347–364.

Naylor, M. A., and Woodcock, N. H. (1977). "Transition Analysis of Structural Sequences." *Geol. Soc. Amer. Bull.* 88, 1488–1492.

Potter, P. E., and Blakely, R. F. (1967). "Generation of a Synthetic Vertical Profile of a Fluvial Sandstone Body." *J. Soc. Petroleum Eng.* 7, 243–251.

Schwarzacher, W. (1968). "Experiments with Variable Sedimentation Rates." Computer Applications in the Earth Sciences: Colloquium on Simulation. Computer Contribution 22. Kansas Geological Survey, University of Kansas, Lawrence, pp. 19–21.

Schwarzacher, W. (1969). "The Use of Markov Chains in the Study of Sedimentary Cycles." *J. Int. Assoc. Math. Geol.* 1(1), 17–39.

Schwarzacher, W. (1972). "The Semi-Markov Process as a General Sedimentation Model." In D. F. Merriam, ed., *Mathematical Models of Sedimentary Processes.* New York: Plenum Press.

Vistelius, A. B. (1949). "On the Question of the Mechanism of Formation of Strata." *Doklady Akad. Nauk SSSR* 65, 191–194.

Vistelius, A. B., and Feigel'son, T. S. (1965). "On the Theory of Bed Formation." *Doklady Akad. Nauk SSSR* 164, 158–160.

FOR FURTHER READING

Carr, T. R. (1982). "Log-linear Models, Markov Chains and Cyclic Sedimentation." *J. Sedimentary Petrology* 52, 905–912.

Hiscott, R. N. and Attwood, D. (1982). "Solution of the Chi-square Problem." *Math. Geol.* 14, 683–686.

Krumbein, W. C. (1968). "Statistical Models in Sedimentology." *Sedimentology* 10, 7–23.

Krumbein, W. C. (1975). "Markov Models in the Earth Sciences." In R. B. McCannon, ed., *Concepts in Geostatistics.* New York: Springer-Verlag.

Merriam, D. F., ed. (1972). *Mathematical Models in Sedimentary Processes.* New York: Plenum Press.

Powers, D. W. and Easterling, R. G. (1982). "Improved Methodology for Using Embedded Markov Chains to Describe Cyclical Sediments." *J. Sedimentary Petrology* 52, 913–923.

Chapter 21
TIME SERIES ANALYSIS

21.1 INTRODUCTION

A set of observations generated sequentially in time is called a time series. Such series are available for stock prices, national income, sales figures, and prices of all kinds of goods and merchandise, population figures, and other innumerable sequences based on industrial, economic, and social phenomena. Observations in these series vary due to several factors. Some can be easily detected, such as seasonal variations in prices, variations due to the state of the economy in stock prices, and major calamities in population figures. Assuming that it is possible to separate the contributions due to these detectable causes of variation, there would still remain some variation which can only be attributed to random causes—as in any natural phenomenon. Thus, if we are looking for a mathematical model to represent time series, a stochastic process is the obvious choice. Given the right model of this kind, we may then consider the observed time series as a realization (sample function) of the assumed basic stochastic process.

In Chapters 11–20 the applications of stochastic processes were related to Markov and renewal processes. But in the study of time series it should be recognized that the dependence structure represented by Markov and renewal processes may not be adequate. For instance, food prices in December may not be dependent only on the prices in November; they would normally be the result of the cumulative effect of several factors occurring over several preceding months. Further the prices may not be time-invariant, in the sense that the joint distribution of food prices in the months of November and December may not be the same as the joint distribution of prices in the months of April and May. Thus the stochastic processes that must be used in the study of time series belong to the general class of stationary and nonstationary processes. Some properties of these processes have been discussed in Chapter 9. Here, we shall investigate in more detail the possibilities of using the material presented in Chapter 9 for time series analysis.

In order to minimize the required mathematical background, only simple stationary and nonstationary processes will be discussed. Excellent references for further reading in time series analysis are Jenkins and Watts (1968), Koopmans (1974), Brillinger (1975), Box and Jenkins (1976), Fuller (1976), and Priestly (1981). Elementary introductions to the analysis of time series can be found in Kendall (1976), Granger and Newbold (1977), and Chatfield (1980).

Some examples of time series have been presented graphically in Figures 21.1.1–21.1.3. Figure 21.1.1 represents 200 data points N_t from the Fourdrinier paper-making process collected as follows (Tee and Wu, 1972). At equal intervals of 20 min a portion of the paper product was torn off the reel, and three samples were cut out of this portion. The deviation of the average weight of these three samples from 50 g/m^2 is defined as the observation N_t for the sampling time t.

Figure 21.1.2 presents time series data on several geophysical aspects for four recording stations Belém, Recife, Salvador, and Imbituba off the coast of Brazil (Chatfield and Pepper, 1971). The measured variables are water level (or sea level, in centimeters), atmospheric pressure (in millibars), air and sea temperature (in degrees centigrade), and precipitation and evaporation (in millimeters). The individual observations are obtained by averaging over periods of one month, and the data contain 120 points for each series for the 10 year period 1953–1962.

Figure 21.1.3 represents economic time series from Iceland for the period 1953–1968 (Gudmundsson, 1971). They include total exports and imports, the cost of living index, and data from the banks. Specifically, the following

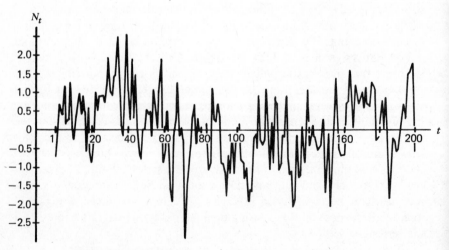

Figure 21.1.1   Time series plot of N_t (Tee and Wu, 1972, p. 483).

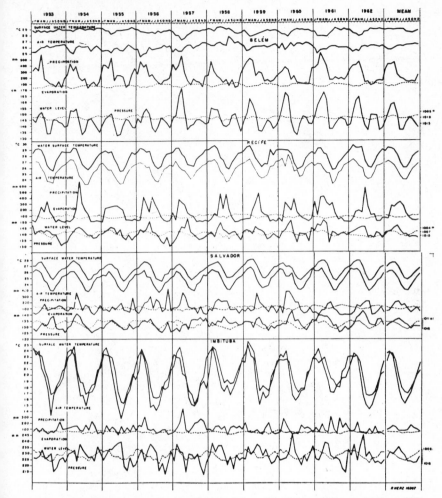

Figure 21.1.2 Time series plot of geophysical data (Chatfield and Pepper, 1971, p. 220).

symbols are used to identify the series: *I*—total imports; *E*—total exports; *N*—notes in circulation; *S*—total bank deposits except current accounts; *C*—total deposits on current accounts; *L*—total credits from the banks, and *P*—the cost of living index.

It may be noted that in the first two examples some sort of averaging has been used already to remove extreme fluctuations in the process. The overall effect of averaging will be discussed later in the chapter. In the series presented in Figure 21.1.3, a logarithmic transformation has already been applied. The role of transformations in time series analysis is also mentioned later in the chapter.

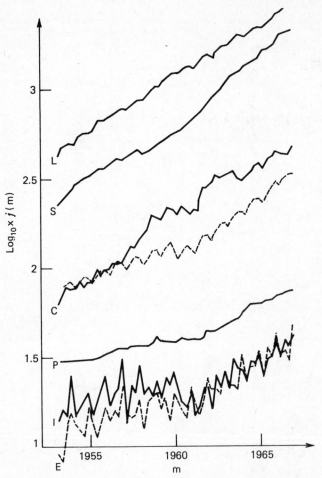

Figure 21.1.3   Time series plot of Icelandic economic variables (Gudmundsson, 1971, p. 399).

Three major objectives of the analysis of time series of the type presented in these examples can be identified:

1. *Describe and explain:*   Using descriptive measures such as means and variances, a general description of the characteristics of the series can be obtained. These characteristics may lead to explaining some of the more obvious factors that contribute to the series.

2. *Forecast future values of the series:*   An appropriate mathematical model is used to forecast the future behavior objectively, after taking into consideration the factors that influence the series.

3. *Control:* If one considers a time series as the result of a phenomenon (or a black box which is usually referred to as a filter) that converts an input process into an output process, looking at the properties of the existing output process, input process can be controlled or the phenomenon can be modified so as to make the output process exhibit desirable properties. In our discussion in this chapter we shall concentrate on the first two objectives of time series analysis.

There are two major approaches in the analysis of time series models. In the time domain analysis covariance and correlation functions are used as diagnostic tools. Knowing the properties of these functions, appropriate models are identified, which are then used in forecasting and control. The second approach is based on the assumption that a time series can be considered to have been made up of several periodic functions, and a harmonic analysis of these functions should lead to the identifications of the properties of the series. Since this technique is based on identifying the frequencies of such periods, the approach is called the analysis in the frequency domain. As described in Chapter 9, the spectral density of the underlying stochastic process is the key tool in this procedure.

A typical time series can be considered to be the result of several components:

1. *Trend:* A long-term movement covering a number of years and giving the general tendency of the series. Growth curves are good examples of time series exhibiting trend. Naturally characteristics of the process depend very much on the time of observation. One method of eliminating the trend factor from a time series is the method of moving averages, and we shall discuss it in some detail in Section 21.4.

2. *Cyclic component:* The result of a periodic movement in definite time periods. For instance, price fluctuations display this factor due to their seasonal variations. Fourier analysis techniques may be employed in the investigation of a known periodic movement like a seasonal variation. Spectral analysis is a method based on the harmonic analysis of the Fourier representation. We shall discuss some of the basic concepts of this technique in Section 21.8.

3. *Oscillatory component (nonrandom):* Further irregular fluctuations due to various unidentifiable causes, after trend and cyclic components have been removed from the data. Some such fluctuations could be deterministic.

4. *Random fluctuations:* Irregular fluctuations due to chance causes alone. Stochastic process representation of the series is justified based on the presence of randomness.

The stochastic models presented in the chapter are aimed at incorporating as many of these modes of fluctuation as necessary and possible. For the sake of simplicity we shall assume, as has been done in most of the investigations on time series, that these factors are additive in nature and independent of each other. In many cases, if they are not additive, proper transformations can be used to make them additive. For instance, if the assumption is that these factors are multiplicative, then the logarithms of the variables should have additive components. Other transformation procedures can be derived to bring the assumptions in line with the simpler procedure.

An important principle to be emphasized in time series modeling is that of parsimony. By a parsimoneous model we mean a model with the smallest number of parameters for an adequate representation of the data. Many times, if appropriate tests are not employed to determine the appropriate number of parameters in the model, one can get bogged down with an unnecessary and unwieldy model which may obscure the basic properties of the process.

21.2 STOCHASTIC MODELS FOR TIME SERIES

We shall restrict our discussion to discrete time series. From a continuous parameter time series, discrete series can be obtained by either sampling at discrete sets of time epochs or by accumulating the variable over a period of time. Let I be the set of integers on the real line (both positive and negative). Suppose $\{ X_n,\ n \in I \}$ is the basic stochastic process representing the time series. As defined in Chapter 9, the process $\{ X_n,\ n \in I \}$ is said to be stationary in the strict sense if and only if

$$F\left(X_{n_1} \leq x_1, X_{n_2} \leq x_2, \ldots, X_{n_k} \leq x_k \right)$$

$$= F\left(X_{n_1+h} \leq x_1, X_{n_2+h} \leq x_2, \ldots, X_{n_k+h} \leq x_k \right)$$

$$h = 0, 1, 2, \ldots; \qquad \text{for all } (n_1, n_2, \ldots, n_k);\ k = 1, 2, \ldots, \in I$$

$$(21.2.1)$$

The process is said to be covariance stationary (also called weakly stationary or stationary in the wide sense) if and only if, for all $n \in I$, $E[X_n]$ and $V[X_n]$ are not functions of n and $\text{Cov}[X_n, X_{n+h}]$ is a function of only h; that is,

$$E[X_n] = \mu \qquad -\infty < \mu < \infty$$

$$V[X_n] = \sigma^2 \qquad 0 \le \sigma^2 < \infty$$

$$\text{Cov}[X_n, X_{n+h}] = \gamma_h \qquad -\infty < \gamma_h < \infty \qquad (21.2.2)$$

The constant time interval h used in the preceding definitions is called the *lag*. Thus, if the observations of the process separated by the same lag have the same covariance and the mean and variance of observations are time-invariant, then the process is said to be covariance stationary. It should be noted that a strictly stationary stochastic process is also covariance stationary, provided the mean value characteristics of (21.2.2) exist.

The converse is true only if mean and covariance functions uniquely determine the joint distributions—as in Gaussian processes. The covariance function γ_h is called the *autocovariance*, and the correlation function $\rho_h = \gamma_h/\sigma^2$ is called the *autocorrelation* at lag h. The autocorrelation function is also called the *correlogram*.

A class of stationary processes suitable as models for a large number of time series can be developed as follows. This development is based on the lucid text of Box and Jenkins (1976) who have provided a general framework for the understanding of the properties of a large class of stationary and nonstationary processes.

Let $\{Y_n\}_{n=-\infty}^{\infty}$ be a sequence of independent random variables distributed normally with zero mean and variance σ^2. This sequence is an example of what engineering scientists and physicists refer to as a white noise process (see Example 9.2.2). It should be noted that with the assumption of normality of Y_n dropped, reference to stationarity in our discussion should be understood as referring only to covariance stationarity. Imagine that the process $\{Y_n, n \in I\}$ passes through a black box and is transformed into a process $\{X_n, n \in I\}$ within the box. Let ψ represent the operator that transforms Y_n to X_n, $n \in I$; we can represent this operation by

$$X_n = \psi Y_n \qquad (21.2.3)$$

Further assume that ψ has the following two properties:

1. It is time-invariant; that is

$$X_n = \psi Y_n \qquad \text{for all } n \in I \qquad (21.2.4)$$

2. If $Y_n = U_n + V_n$, then

$$X_n = \psi Y_n$$

implies

$$X_n = \psi U_n + \psi V_n \qquad (21.2.5)$$

This process of transformation is known as filtering, and the transfer function ψ with properties given here is a linear filter (see Section 9.4).

Suppose the observation X_n of the time series can be represented as a linear function

$$X_n = \mu + \psi_0 Y_n + \psi_1 Y_{n-1} + \psi_2 Y_{n-2} + \cdots \qquad (21.2.6)$$

where we have introduced the constant parameter μ for the sake of generality. Define a backward shift operator B as follows:

$$X_{n-1} = B X_n \qquad (21.2.7)$$

Extending this operation, we may write

$$X_{n-m} = B^m X_n \qquad (21.2.8)$$

Using the operator B, (21.2.6) can be written

$$X_n = \mu + \psi_0 Y_n + \psi_1 B Y_n + \psi_2 B^2 Y_n + \cdots$$

$$= \mu + \psi(B) Y_n \qquad (21.2.9)$$

where

$$\psi(B) = \psi_0 + \psi_1 B + \psi_2 B^2 + \cdots \qquad (21.2.10)$$

Formally the function $\psi(B)$ in (21.2.9), which is called the *transfer function of the filter*, has the properties of the linear filter introduced in (21.2.3)–(21.2.5).

When the sequence of weights $\psi_0, \psi_1, \ldots$ is finite—or if infinite, $\sum_{i=0}^{\infty} \psi_i$ is convergent—the filter is said to be stable, and the resulting process is stationary.

We discuss some special cases of this linear filter model. Without loss of generality, we assume $\mu = 0$ and $\psi_0 = 1$. For, if $\mu > 0$, we can define a modified process $X_n' = X_n - \mu$, $n \in I$, and derive the properties of the process $\{X_n, \ n \in I\}$ through the properties of $\{X_n', \ n \in I\}$. Similar arguments hold for $\psi_0 = 1$.

Consider equation (21.2.6) with $\mu = 0$ and $\psi_0 = 1$, for decreasing values of n.

$$X_n = Y_n + \psi_1 Y_{n-1} + \psi_2 Y_{n-2} + \cdots$$

$$X_{n-1} = Y_{n-1} + \psi_1 Y_{n-2} + \psi_2 Y_{n-3} + \cdots$$

$$X_{n-2} = Y_{n-2} + \psi_1 Y_{n-3} + \psi_2 Y_{n-4} + \cdots \qquad (21.2.11)$$

Eliminating Y_{n-1} between the first two equations of (21.2.11), we get

$$X_n = Y_n + \psi_1(X_{n-1} - \psi_1 Y_{n-2} - \cdots) + \psi_2 Y_{n-2} + \cdots$$

$$= Y_n + \psi_1 X_{n-1} + (\psi_2 - \psi_1^2) Y_{n-2} + (\psi_3 - \psi_1 \psi_2) Y_{n-3} + \cdots$$

$$(21.2.12)$$

Proceeding in this manner, we arrive at an alternate form for (21.2.6) as

$$X_n = Y_n + \sum_{k=1}^{\infty} \eta_k X_{n-k} \qquad (21.2.13)$$

in which X_n is expressed in terms of $X_{n-1}, X_{n-2}, \ldots$ and a random component Y_n. Using the backward shift operator B, we may write (21.2.13) as

$$X_n = Y_n + \sum_{k=1}^{\infty} \eta_k B^k X_n$$

giving

$$Y_n = \left(1 - \sum_{k=1}^{\infty} \eta_k B^k\right) X_n \qquad (21.2.14)$$

which we shall denote as $Y_n = \eta(B) X_n$. Comparing (21.2.14) with (21.2.9), we get the relation between the operator functions

$$\eta(B) = \psi^{-1}(B) \qquad (21.2.15)$$

Both $\psi(B)$ and $\eta(B)$ can be considered to be generating functions of their respective series of coefficients. The properties of the stochastic process

$\{X_n\}$ can be related to the properties of $\psi(B)$ and $\eta(B)$ as follows:

1. The stochastic process $\{X_n\}$ is stationary if $\psi(B)$ converges on or within the unit circle, that is, for $|B| \leq 1$.
2. A stochastic process $\{X_n\}$ is said to be invertible if $\{\eta_k\}$ of the expansion of (21.2.14) form a convergent series. Thus the process is said to be invertible if $\eta(B)$ converges on or within the unit circle.

An important special case of the general model is the *autoregressive* (AR) model. Let X_n be represented as follows, based on the previous p observations and a random component:

$$X_n = \alpha_1 X_{n-1} + \alpha_2 X_{n-2} + \cdots + \alpha_p X_{n-p} + Y_n \qquad (21.2.16)$$

Recall the form of regression equations of statistical theory. Equation (21.2.16) is an AR model of order p [AR(p)]. Iterating successively with (21.2.16), for X_n we can derive the infinite series in Y_k, $k = \ldots -1$, $0, 1, 2, \ldots, n$ as

$$X_n = \psi(B)Y_n \qquad (21.2.17)$$

showing that the AR model is in fact a special case of the general model (21.2.9).

Rewriting (21.2.16), we also get

$$\left(1 - \alpha_1 B - \alpha_2 B^2 - \cdots - \alpha_p B^p\right) X_n = Y_n \qquad (21.2.18)$$

which we may denote as

$$Y_n = \phi(B) X_n \qquad (21.2.19)$$

Comparing (21.2.17) with (21.2.19), we get

$$\psi(B) = \phi^{-1}(B) \qquad (21.2.20)$$

Clearly for the stationarity of the series the convergence of the function $\psi(B) = \phi^{-1}(B)$ is essential. Let $\eta_1, \eta_2, \ldots, \eta_p$ be the reciprocals of the zeros of the polynomial $\phi(B)$ so that we may write

$$\phi(B) = \prod_{k=1}^{p} (1 - \eta_k B)$$

Expanding $X_n = \phi^{-1}(B)Y_n$ in partial fractions, we get

$$X_n = \sum_{k=1}^{p} \frac{A_k}{1 - \eta_k B} Y_n \qquad (21.2.21)$$

In order to have $\phi^{-1}(B)$ convergent for $|B| \le 1$, $|\eta_k| < 1$ for $k = 1, 2, \ldots, p$; consequently the zeros of the polynomial $\phi(B)$ must lie outside the unit circle. This condition can also be seen by considering the variances in (21.2.21). We have

$$X_n = \sum_{k=1}^{p} A_k(1 - \eta_k B)^{-1} Y_n$$

$$= \sum_{k=1}^{p} \left(A_k \sum_{j=0}^{\infty} \eta_k^j Y_{n-k} \right)$$

$$= \sum_{j=0}^{\infty} \left(\sum_{k=1}^{p} A_k \eta_k^j Y_{n-k} \right)$$

$$\sigma_x^2 = \sum_{j=0}^{\infty} \left(\sum_{k=1}^{p} A_k \eta_k^j \right)^2 \sigma_y^2$$

For stationarity σ_x^2 should be finite, and hence $|\eta_k| < 1$, $k = 1, 2, \ldots, p$. Hence the zeros of $\phi(B)$ should lie outside the unit circle. It can be easily shown that this condition remains the same, even when not all zeros of $\phi(B)$ are distinct.

Another important special case of the general linear filter model (21.2.9) is the *moving average* (MA) process. Consider the representation

$$X_n = Y_n - \beta_1 Y_{n-1} - \beta_2 Y_{n-2} - \cdots - \beta_q Y_{n-q} \qquad (21.2.22)$$

This is a moving average process of order q [MA(q)]. If we define a MA operator

$$\theta(B) = 1 - \beta_1 B - \beta_2 B^2 - \cdots - \beta_q B^q \qquad (21.2.23)$$

the model can be put in the compact form

$$X_n = \theta(B)Y_n \qquad (21.2.24)$$

Combining the AR(p) and MA(q) processes, one gets *mixed autoregressive-moving average* [ARMA(p,q)] models. Thus combining (21.2.19) and (21.2.24), we write

$$\phi(B)X_n = \theta(B)Y_n \qquad (21.2.25)$$

or in detail,

$$X_n = \alpha_1 X_{n-1} + \alpha_2 X_{n-2} + \cdots + \alpha_p X_{n-p}$$

$$+ Y_n - \beta_1 Y_{n-1} - \beta_2 Y_{n-2} - \cdots - \beta_q Y_{n-q}$$

In order to exhibit nonstationary properties in a process, a generalized autoregressive operator $\varphi(B)$ which incorporates a stationary operator $\varphi_S(B)$ as well as a nonstationary operator $\varphi_N(B)$ in the form

$$\varphi(B) = \varphi_S(B)\varphi_N(B)$$

may be introduced. Box and Jenkins (1976, Chap. 4) have introduced a special case of this general model with $\varphi_S(B) = \phi(B)$, as defined in (21.2.19) and $\varphi_N(B) = (1 - B)^d$, $d = 0, 1, 2, \ldots$ to account for a class of nonstationary processes with autoregressive and moving average features. A model corresponding to (21.2.25) may now be written as

$$\varphi(B)X_n = \theta(B)Y_n \qquad (21.2.26)$$

where

$$\varphi(B) = \phi(B)(1 - B)^d \qquad d = 0, 1, 2, \ldots \qquad (21.2.27)$$

This model is known as the *autoregressive integrated moving average process of order* (p, d, q) [ARIMA (p, d, q)].

The covariance functions of AR(p) and MA(q) models can be derived following the methods of Chapter 9. We may recall the property that the MA(q) process is covariance stationary, whereas the AR(p) process may not be so. The covariance function of the general linear filter model

$$X_n = \psi(B)Y_n$$

can be shown to be

$$\gamma_h = \sigma_y^2 \sum_{k=0}^{\infty} \psi_k \psi_{k+h} \qquad (21.2.28)$$

where we have assumed Y_n to be identically distributed uncorrelated random variables with variance σ_y^2. In particular, the variance

$$\gamma_0 = \sigma_y^2 \sum_{k=0}^{\infty} \psi_k^2 \qquad (21.2.29)$$

gives the convergence of $\sum_{k=0}^{\infty} \psi_k^2$ as a necessary condition for the finiteness of the variance of the process $\{X_n\}$.

A convenient procedure for the determination of the covariance function γ_h is obtained by using a generating function for γ_h. Let

$$\gamma(B) = \sum_{n=-\infty}^{\infty} \gamma_h B^h \qquad (21.2.30)$$

Multiplying (21.2.28) by B^h and summing, we get after simplification

$$\gamma(B) = \sigma_y^2 \psi(B)\psi(B^{-1}) \qquad (21.2.31)$$

where $\psi(B^{-1})$ can be interpreted as the generating function for a forward shift operator defined by

$$B^{-k}X_n = X_{n+k} \qquad k = 1, 2, \ldots$$

Writing $B^{-k} = F^k$ and the corresponding generating function as $\psi(F)$, (21.2.31) can be written in the form

$$\gamma(B) = \sigma_y^2 \psi(B)\psi(F) \qquad (21.2.32)$$

The spectral density of the general linear filter model follows directly from the discussion given in Chapter 9. In fact Example 9.4.3 is a special case of this general model. Following the arguments of that example, we get [with $\mu = 0$ in (21.2.9)]

$$B(\lambda) = \psi(e^{-i\lambda}) \qquad (21.2.33)$$

and the spectral density of the general model is given by

$$f(\lambda) = \frac{\sigma_y^2}{2\pi} |\psi(e^{-i\lambda})|^2 \qquad (21.2.34)$$

Some special forms of $\psi(B)$ will be discussed in the next few sections.

In Sections 21.3–21.6 we shall identify basic properties of the models introduced here. Our treatment follows generally that of Box and Jenkins

(1976), which provides a comprehensive analysis and major extensions and applications of these models. Even though many of the individual models have existed before, Box and Jenkins have brought all of them under a general framework for the analysis of time series in the time domain. A practitioner's guide to the Box-Jenkins approach has been given by Anderson (1976).

21.3   THE AUTOREGRESSIVE MODEL

In Section 9.2 we gave a first-order autoregressive process as an example of a covariance stationary stochastic process. An autoregressive process of order p ($p \geq 1$) has been introduced in (21.2.16) as a special case of the linear filter model. Following the discussion of Section 9.2, a pth-order autoregressive process can be defined as a covariance stationary stochastic process that satisfies the difference equation

$$X_n = \alpha_1 X_{n-1} + \alpha_2 X_{n-2} + \cdots + \alpha_p X_{n-p} + Y_n \qquad (21.3.1)$$

white noise process

where $\{Y_n\}$, $n = -1, 0, 1, \ldots$ is a sequence of uncorrelated normal random variables with zero mean and constant variance σ_y^2.

Consider the special case $p = 1$. We now write the autoregressive scheme as

$$X_n = \alpha X_{n-1} + Y_n \qquad n = \ldots -1, 0, 1, \ldots \qquad (21.3.2)$$

with $|\alpha| < 1$. As illustrated in (9.2.21)–(9.2.23), then we get

$$X_n = \sum_{i=0}^{\infty} \alpha^i Y_{n-i}$$

$$E(X_n) = 0$$

$$V(X_n) = \frac{1}{1 - \alpha^2} \sigma_y^2$$

$$= \sigma_x^2 \qquad (21.3.3)$$

and

$$\text{Cov}(X_n, X_{n+h}) = \alpha^h \sigma_x^2 \qquad (21.3.4)$$

For a spectral analysis of the AR(1) process the reader is referred to Example 9.4.2, which can also be derived as a special case of (21.2.34). This method will be illustrated for the AR(2) case.

Let the second-order autoregressive process be defined by the difference equation

$$X_n = \alpha_1 X_{n-1} + \alpha_2 X_{n-2} + Y_n \qquad (21.3.5)$$

In order to determine the conditions for the stationarity of this process, multiply (21.3.5) by X_{n-h}, and take expectations. We have

$$\gamma_h = \alpha_1 \gamma_{h-1} + \alpha_2 \gamma_{h-2} \qquad h \geq 1$$

$$\rho_h = \alpha_1 \rho_{h-1} + \alpha_2 \rho_{h-2} \qquad (21.3.6)$$

where we have assumed constant variance for $\{Y_n\}$. Suppose $\rho_0 = 1$. Then

$$\rho_1 = \alpha_1 + \alpha_2 \rho_{-1}$$

which gives (noting $\rho_1 = \rho_{-1}$ for a stationary process)

$$\rho_1 = \frac{\alpha_1}{1 - \alpha_2} \qquad (21.3.7)$$

Substituting this value in (21.3.6) for $h = 2$,

$$\rho_2 = \frac{\alpha_1^2}{1 - \alpha_2} + \alpha_2 \qquad (21.3.8)$$

From (21.3.5) we also have

$$\sigma_x^2 = \alpha_1 \gamma_1 + \alpha_2 \gamma_2 + \sigma_y^2$$

that is,

$$\sigma_x^2 [1 - \alpha_1 \rho_1 - \alpha_2 \rho_2] = \sigma_y^2$$

Substituting values for ρ_1 and ρ_2, we then get

$$\sigma_x^2 \left[1 - \frac{\alpha_1^2}{1 - \alpha_2} - \alpha_2 \left(\frac{\alpha_1^2}{1 - \alpha_2} + \alpha_2 \right) \right] = \sigma_y^2$$

that is,

$$\sigma_x^2 = \frac{(1 - \alpha_2) \sigma_y^2}{(1 + \alpha_2)(1 - \alpha_1 - \alpha_2)(1 + \alpha_1 - \alpha_2)} \qquad (21.3.9)$$

In order to have the coefficient of σ_y^2 in (21.3.9) positive, the following conditions should hold:

$$-1 < \alpha_2 < 1 \qquad \alpha_1 + \alpha_2 < 1 \qquad \alpha_2 - \alpha_1 < 1 \qquad (21.3.10)$$

which give the conditions for stationarity of the AR(2) process.

Noting that the autoregressive process $\{ X_n \}$ can be represented in the form (21.2.17), we can easily conclude that

$$E[X_n] = 0$$

Now let us consider the correlation function ρ_h for the process. As demonstrated before, ρ_h satisfies the difference equation (21.3.6). Using the forward shift operator F defined in (21.2.31), it can be written in the form

$$\left(F^2 - \alpha_1 F - \alpha_2 \right)\rho_h = 0 \qquad h = 2, 3, 4, \ldots \qquad (21.3.11)$$

This is a homogeneous difference equation of order 2 which has the general solution

$$\rho_h = \lambda \xi_1^h + \mu \xi_2^h \qquad (21.3.12)$$

where ξ_1 and ξ_2 are the solutions of the operator equation known as the characteristic equation

$$F^2 - \alpha_1 F - \alpha_2 = 0$$

We have

$$\xi_1 = \frac{\alpha_1 + \sqrt{\alpha_1^2 + 4\alpha_2}}{2}$$

$$\xi_2 = \frac{\alpha_1 - \sqrt{\alpha_1^2 + 4\alpha_2}}{2} \qquad (21.3.13)$$

Clearly

$$\xi_1 + \xi_2 = \alpha_1 \quad \text{and} \quad \xi_1 \xi_2 = -\alpha_2 \qquad (21.3.14)$$

The coefficients λ and μ of (21.3.12) are determined using the initial conditions for ρ_h, $(h = 0, 1)$. We have

$$\rho_0 = 1 = \lambda + \mu$$

$$\rho_1 = \frac{\alpha_1}{1 - \alpha_2} = \lambda \xi_1 + \mu \xi_2 \qquad (21.3.15)$$

Solving for λ and μ and using results (21.3.14), we get

$$\lambda = \frac{\xi_1(1 - \xi_2^2)}{(\xi_1 - \xi_2)(1 + \xi_1\xi_2)}$$

$$\mu = \frac{-\xi_2(1 - \xi_1^2)}{(\xi_1 - \xi_2)(1 + \xi_1\xi_2)} \tag{21.3.16}$$

Thus we finally get

$$\rho_h = \frac{(1 - \xi_2^2)\xi_1^{h+1}}{(\xi_1 - \xi_2)(1 + \xi_1\xi_2)} + \frac{(1 - \xi_1)^2\xi_2^{h+1}}{(\xi_2 - \xi_1)(1 + \xi_1\xi_2)} \tag{21.3.17}$$

It may be verified that under conditions (21.3.10), $|\xi_1| < 1$ and $|\xi_2| < 1$. Therefore when they are real, ρ_h is clearly a sum of two decreasing exponentials.

When $\alpha_1^2 + 4\alpha_2 < 0$, the roots ξ_1 and ξ_2 given by (21.3.13) are complex. Let

$$\xi_1 = pe^{i\theta} \quad \text{and} \quad \xi_2 = pe^{-i\theta}$$

with

$$\alpha_1 = 2p\cos\theta \quad \text{and} \quad -\alpha_2 = p^2 \tag{21.3.18}$$

and hence

$$\cos\theta = \frac{\alpha_1}{2\sqrt{-\alpha_2}}$$

The correlation function ρ_h of (21.3.17) can be written in the form

$$\rho_h = \frac{(\xi_1^{h+1} - \xi_2^{h+1}) - (\xi_1\xi_2)^2[\xi_1^{h-1} - \xi_2^{h-1}]}{(\xi_1 - \xi_2)(1 + \xi_1\xi_2)} \tag{21.3.19}$$

Also we have

$$e^{i\theta} - e^{-i\theta} = 2i\sin\theta$$

$$\xi_1^n - \xi_2^n = p^n(e^{ni\theta} - e^{-ni\theta}) = 2p^n i\sin\theta$$

and

$$\xi_1\xi_2 = p^2$$

Using these results in (21.3.19), we get

$$\rho_h = \frac{2p^{h+1}i\sin(h+1)\theta - 2p^{h+3}i\sin(h-1)\theta}{(1+p)^2 2pi\sin\theta}$$

$$= p^h\left[\frac{\sin(h+1)\theta - p^2\sin(h-1)\theta}{(1+p^2)\sin\theta}\right]$$

$$= p^h\left[\frac{(1-p^2)\sin h\theta\cos\theta + (1+p^2)\cos h\theta\sin\theta}{(1+p^2)\sin\theta}\right]$$

$$= p^h\left(\frac{\sin h\theta}{\tan\psi} + \cos h\theta\right)$$

$$= p^h\left(\frac{\sin h\theta\cos\psi + \cos h\theta\sin\psi}{\sin\psi}\right)$$

$$= p^h\frac{\sin(h\theta + \psi)}{\sin\psi} \tag{21.3.20}$$

where we have written

$$\frac{1+p^2}{1-p^2}\tan\theta = \tan\psi$$

The expression (21.3.20) indicates that the correlation function is a damped sine wave.

The spectral density of the AR(2) process is obtained by the direct application of the result (21.2.34). Equation (21.3.5) can be rearranged in linear filter representation as

$$X_n = \left(\frac{1}{1 - \alpha_1 B - \alpha_2 B^2}\right)Y_n \tag{21.3.21}$$

Now using equation (21.2.34), the spectral density of AR(2) is obtained as

$$f(\lambda) = \frac{\sigma_y^2}{2\pi}\frac{1}{|1 - \alpha_1 e^{-i\lambda} - \alpha_2 e^{-2i\lambda}|^2} \tag{21.3.22}$$

But it should be noted that

$$\left| 1 - \alpha_1 e^{-i\lambda} - \alpha_2 e^{-2i\lambda} \right|^2$$

$$= \left(1 - \alpha_1 e^{-i\lambda} - \alpha_2 e^{-2i\lambda} \right)\left(1 - \alpha_1 e^{i\lambda} - \alpha_2 e^{2i\lambda} \right)$$

$$= 1 + \alpha_1^2 + \alpha_2^2 - 2\alpha_1(1 - \alpha_2)\cos\lambda - 2\alpha_2\cos 2\lambda \quad (21.3.23)$$

In simplifying (21.3.23), we have used the representation

$$\cos\lambda = \frac{e^{i\lambda} + e^{-i\lambda}}{2}$$

Thus in place of (21.3.22) we get

$$f(\lambda) = \left(\frac{\sigma_y^2}{2\pi} \right) \frac{1}{1 + \alpha_1^2 + \alpha_2^2 - 2\alpha_1(1 - \alpha_2)\cos\lambda - 2\alpha_2\cos 2\lambda}$$

$$(21.3.24)$$

The properties of $f(\lambda)$ can now be identified by drawing its graph as a function of $\lambda(-\pi, \pi)$ for appropriate weights α_1 and α_2.

Finally consider the pth-order autoregression model [AR(p)]:

$$X_n = \alpha_1 X_{n-1} + \alpha_2 X_{n-2} + \cdots + \alpha_p X_{n-p} + Y_n \quad (21.3.25)$$

Multiplying by X_{n-h} and taking expectations, we get

$$\gamma_h = \alpha_1\gamma_{h-1} + \alpha_2\gamma_{h-2} + \cdots + \alpha_p\gamma_{h-p} \qquad h > 0 \quad (21.3.26)$$

Dividing throughout by γ_0, (21.3.26) gives

$$\rho_h = \alpha_1\rho_{h-1} + \alpha_2\rho_{h-2} + \cdots + \alpha_p\rho_{h-p} \qquad h > 0 \quad (21.3.27)$$

For $h = 1, 2, \ldots, p$, equations (21.3.27) are known as *Yule-Walker equations* which can be used for the estimation of parameters α_j, $j = 1, 2, \ldots, p$ in terms of the estimates of autocorrelations. For instance, for $p = 2$ we have

$$\rho_1 = \alpha_1\rho_0 + \alpha_2\rho_1$$

$$\rho_2 = \alpha_1\rho_1 + \alpha_2\rho_0$$

Noting that $\rho_0 = 1$ and solving for α_1 and α_2, we get

$$\alpha_1 = \frac{\rho_1(1 - \rho_2)}{1 - \rho_1^2} \qquad \alpha_2 = \frac{\rho_2 - \rho_1^2}{1 - \rho_1^2}$$

Using the estimates r_h for ρ_h

$$\hat{\alpha}_1 = \frac{r_1(1 - r_2)}{1 - r_1^2} \qquad \hat{\alpha}_2 = \frac{r_2 - r_1^2}{1 - r_1^2}$$

Equations (21.3.27) can be used to determine ρ_h in a recursive manner. Alternately, noting that (21.3.27) is a difference equation, it can be solved in the manner illustrated in AR(2) case to show that the correlation function ρ_h is made up of a mixture of damped exponentials and/or damped sine wave. (See books on finite differences such as Jordan 1965, and Levy and Lessman, 1961, for solution techniques of pth-order difference equations.) For further discussions on the properties of AR(p) processes, the reader may also refer to Barlett (1955), Cox and Miller (1965), Box and Jenkins (1976), Fuller (1976), and Pandit and Wu (1983).

The spectral density of the general AR(p) model is obtained by extending the expression (21.3.21) in the obvious manner.

Example 21.3.1

Consider the AR(2) model:

$$X_n = 0.5X_{n-1} + 0.3X_{n-2} + Y_n$$

We shall determine ρ_h, $h = 1, 2, 3, \ldots$ using equations (21.3.15)–(21.3.17). We have

$$\alpha_1 = 0.5 \qquad \alpha_2 = 0.3$$

$$\rho_0 = 1.0000$$

$$\rho_1 = \frac{\alpha_1}{1 - \alpha_2} = 0.7143$$

$$\rho_2 = \frac{\alpha_1^2}{1 - \alpha_2} + \alpha_2 = 0.6571$$

for $h > 2$, $\rho_h = 0.5\rho_{h-1} + 0.3\rho_{h-2}$. Thus we get the following values (Fig. 21.3.1):

h	0	1	2	3	4	5	6	7	8
ρ_h	1.000	0.714	0.657	0.543	0.469	0.397	0.339	0.289	0.246
h	9	10	11	12	13	14	15	16	17
ρ_h	0.210	0.179	0.152	0.130	0.110	0.094	0.080	0.068	0.058

From (21.3.42), for the spectral density $f(\lambda)$, we have

$$f_*(\lambda) = \frac{2\pi}{\sigma_y^2} f(\lambda) = \frac{1}{1 + (0.5)^2 + (0.3)^2 - 2(0.5)(0.7)\cos\lambda - (0.6)\cos\lambda}$$

$$= \frac{1}{1.34 - (.07)\cos\lambda - (0.6)\cos 2\lambda}$$

Thus we get the following values (Fig. 21.3.2):

λ	$-\pi$	$-\dfrac{3\pi}{4}$	$-\dfrac{\pi}{2}$	$-\dfrac{\pi}{4}$	0	$\dfrac{\pi}{4}$	$\dfrac{\pi}{2}$	$\dfrac{3\pi}{4}$	π
$f_*(\lambda)$	0.69	0.55	0.52	1.22	25.00	1.22	0.52	0.55	0.69

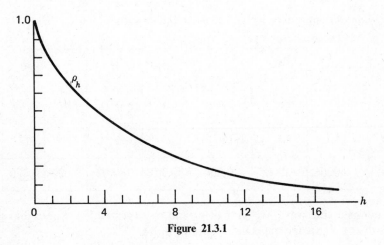

Figure 21.3.1

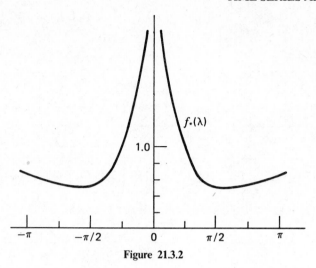

Figure 21.3.2

It is clear that low frequencies dominate in their contribution to the variance.

Example 21.3.2

Consider the AR(2) model:

$$X_n = 0.5X_{n-1} - 0.3X_{n-2} + Y_n$$

Following the same steps as before, we get these values (Fig. 21.3.3):

h	0	1	2	3	4	5	6
ρ_h	1.000	0.385	-0.108	-0.169	-0.052	0.025	0.028
h	7	8	9	10	11	12	13
ρ_h	-0.0066	-0.005	-0.004	-0.001	0.001	0.000	0.000

The damped sine wave behavior of ρ_h can be expected if we note that the roots of the characteristic equation are complex.

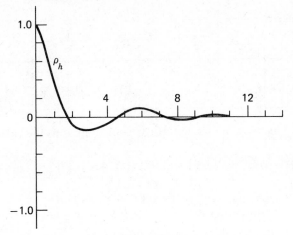

Figure 21.3.3

Similarly for spectral density we get

$$f_*(\lambda) = \frac{2\pi}{\sigma_y^2} f(\lambda) = \frac{1}{1.34 - (1.3)\cos\lambda + (0.6)\cos 2\lambda}$$

Thus we get the following (Fig. 21.3.4):

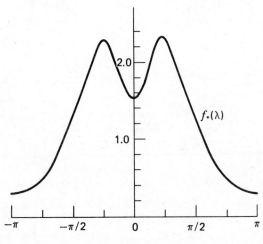

Figure 21.3.4

λ	$-\pi$	$\dfrac{-3\pi}{4}$	$\dfrac{-\pi}{2}$	$-\dfrac{\pi}{4}$	0	$\dfrac{\pi}{4}$	$\dfrac{\pi}{2}$	$\dfrac{3\pi}{4}$	π
$f_*(\lambda)$	0.31	0.44	1.35	2.33	1.56	2.33	1.35	0.44	0.31

Hence low frequencies dominate in their contribution to the variance.

21.4   THE MOVING AVERAGE PROCESS

Consider the moving average process MA(q) defined in (21.2.22), and recall some of its properties derived in Chapter 9. We have

$$X_n = Y_n - \beta_1 Y_{n-1} - \beta_2 Y_{n-2} - \cdots - \beta_q Y_{n-q} \qquad (21.4.1)$$

where β_i, $i = 1, 2, \ldots, q$ are constants and $\{Y_n\}_{n=-\infty}^{\infty}$ is a sequence of uncorrelated normal random variables with

$$E[Y_n] = 0 \qquad V[Y_n] = \sigma_y^2 \qquad n = 0, 1, 2, \ldots$$

From (21.4.1) we get

$$E[X_n] = 0 \qquad V[X_n] = \left(1 + \sum_{m=1}^{q} \beta_m^2\right)\sigma_y^2 \qquad (21.4.2)$$

and

$$\operatorname{Cov}(X_n, X_{n+h}) = \gamma_h = \begin{cases} -\left(\beta_h + \displaystyle\sum_{m=1}^{q-h} \beta_m \beta_{m+h}\right)\sigma_y^2 & h = 1, 2, \ldots, q \\ 0 & h > q \end{cases}$$

$$\times n = 0, 1, 2 \ldots \quad (21.4.3)$$

Consequently the correlation function

$$\rho_h = \begin{cases} \dfrac{-\beta_h + \displaystyle\sum_{m=1}^{q-h} \beta_m \beta_{m+h}}{1 + \displaystyle\sum_{m=1}^{q} \beta_m^2} & h = 1, 2, \ldots, q \\ 0 & h > q \end{cases} \qquad (21.4.4)$$

A significant point to be noted is that the correlation function is zero beyond the order q of the process.

Also the MA(q) process is stationary (covariance stationary if we drop the assumption of normality).

For a moving average process the determination of the parameters β_i, $i = 1, 2, \ldots, q$ using the estimates of autocorrelations ρ_i, $i = 1, 2, \ldots, q$ is not as straightforward as in the case of the AR(p) process. This is because of the nonlinearity of equations (21.4.4). This is illustrated for the case $q = 1$ as follows.

Consider the MA(1) process

$$X_n = Y_n - \beta Y_{n-1} \tag{21.4.5}$$

We have

$$V[X_n] = (1 + \beta^2)\sigma_y^2$$

and

$$\rho_h = \begin{cases} \dfrac{-\beta}{1 + \beta^2} & h = 1 \\ 0 & h \geq 2 \end{cases}$$

This gives for $h = 1$

$$(1 + \beta^2)\rho_1 = -\beta$$

$$\beta^2\rho_1 + \beta + \rho_1 = 0 \tag{21.4.6}$$

Let β_1^* and β_2^* be the roots of equation (21.4.6). We have

$$\beta_1^*\beta_2^* = 1$$

and therefore the roots are reciprocals of each other. The correct value of the root is to be chosen using the invertibility condition stated following equation (21.2.15). Using the arguments employed in deriving (21.2.14), we note that X_n can be written as

$$X_n = Y_n - \beta X_{n-1} - \beta^2 X_{n-2} - \beta^3 X_{n-3} - \cdots \tag{21.4.7}$$

For this expression to be meaningful, one must have $\sum_{k=0}^{\infty} \beta^k < \infty$ (in this case $\eta_k = \beta^k$), that is, $|\beta| < 1$. Thus of the two solutions β_1^* and β_2^* of equation (21.4.6) only one qualifies as a valid solution, which is $|\beta^*| < 1$ (see Box and Jenkins, 1970, pp. 50–51). For a discussion of similar procedures when $q = 2$, the readers are referred to Box and Jenkins (1976, pp. 70–71).

The spectral density of the MA(q) process is obtained by the direct application of the result (21.2.34). Equation (21.4.1) can be written in the linear filter form as

$$X_n = \left(1 - \beta_1 B - \beta_2 B^2 - \cdots - \beta_q B^q\right) Y_n \qquad (21.4.8)$$

Using equation (21.2.34), the spectral density of MA(q) is obtained as

$$f(\lambda) = \frac{\sigma_y^2}{2\pi} |1 - \beta_1 e^{-i\lambda} - \beta_2 e^{-2i\lambda} - \cdots - \beta_q e^{-qi\lambda}|^2 \qquad (21.4.9)$$

Using the simplifications used in (21.3.40)–(21.3.42) for $q = 2$, for MA(q) we get

$$f(\lambda) = \left(\frac{\sigma_y^2}{2\pi}\right)\left[1 + \beta_1^2 + \beta_2^2 - 2\beta_1(1 - \beta_2)\cos\lambda - 2\beta_2\cos 2\lambda\right]$$

$$(21.4.10)$$

These results indicate the dual nature of the AR(p) and MA(q) processes. The duality of the two processes are only to be expected if we note that an AR(p) process can be expressed as an infinite-order moving average process [e.g., (21.2.17)] and MA(q) process can be expressed as an infinite-order autoregressive process.

21.5 MIXED AUTOREGRESSIVE–MOVING AVERAGE PROCESSES

As defined in (21.2.25), the ARMA(p, q) model combines an AR(p) model with an MA(q) model. Thus we write

$$X_n = \alpha_1 X_{n-1} + \alpha_2 X_{n-2} + \cdots + \alpha_p X_{n-p}$$

$$+ Y_n - \beta_1 Y_{n-1} - \beta_2 Y_{n-2} - \cdots - \beta_q Y_{n-q} \qquad (21.5.1)$$

Or using AR(p) and MA(q) operators,

$$\phi(B) X_n = \theta(B) Y_n \qquad (21.5.2)$$

Combining the conditions for the two constituent processes, for stationarity the roots of the equation $\phi(B) = 0$ should lie outside the unit circle, and for invertibility the roots of the equation $\theta(B) = 0$ should lie outside the unit circle.

The determination of the covariance and correlation functions in the general case being cumbersome, we shall illustrate the procedure using an ARMA(1, 1) process. Let

$$X_n = \alpha X_{n-1} + Y_n - \beta Y_n - 1 \qquad (21.5.3)$$

Multiplying (21.5.3) by X_{n-h} and taking expectations, we have

$$E[X_n X_{n-h}] = \alpha E[X_{n-1} X_{n-h}] + E[Y_n X_{n-h}]$$
$$- \beta E[Y_{n-1} X_{n-h}]$$
$$\gamma_h = \alpha \gamma_{h-1} + \gamma_h^{(X,Y)} - \beta \gamma_{h-1}^{(X,Y)} \qquad (21.5.4)$$

where $\gamma_h^{(X,Y)}$ is a cross-covariance function between the X and Y processes. Specifically we get

$$\gamma_0^{(X,Y)} = E[X_n Y_n] = \sigma_y^2 \qquad (21.5.5)$$

To get $\gamma_{-1}^{(X,Y)}$, we multiply (21.5.3) by Y_{n-1} and take expectations:

$$E[Y_{n-1} X_n] = \alpha E[Y_{n-1} X_{n-1}] + E[Y_n Y_{n-1}] - \beta E[Y_{n-1}^2]$$
$$\gamma_{-1}^{(X,Y)} = \alpha \sigma_y^2 + 0 - \beta \sigma_y^2 = (\alpha - \beta)\sigma_y^2 \qquad (21.5.6)$$

Substituting from (21.5.5) and (21.5.6) in (21.5.4), we get

$$\gamma_0 = \alpha \gamma_1 + \sigma_y^2 - \beta(\alpha - \beta)\sigma_y^2$$
$$= \alpha \gamma_1 + [1 + \beta^2 - \alpha\beta]\sigma_y^2$$
$$\gamma_1 = \alpha \gamma_0 - \beta \sigma_y^2$$
$$\gamma_h = \alpha \gamma_{h-1} \qquad h \geq 2 \qquad (21.5.7)$$

Solving for γ_0 and γ_1 between the first two equations of (21.5.7), we get

$$\gamma_0 = \frac{1 + \beta^2 - 2\alpha\beta}{1 - \alpha^2} \sigma_y^2$$
$$\gamma_1 = \frac{(\alpha - \beta)(1 - \alpha\beta)}{1 - \alpha^2} \sigma_y^2$$
$$\gamma_h = \alpha \gamma_{h-1} \qquad h \geq 2 \qquad (21.5.8)$$

Dividing by γ_0, for the correlation function ρ_h we have

$$\rho_1 = \frac{(\alpha - \beta)(1 - \alpha\beta)}{1 + \beta^2 - 2\alpha\beta}$$

$$\rho_h = \alpha\rho_{h-1} \qquad h \geq 2 \tag{21.5.9}$$

indicating exponential decay after ρ_1. The decay is monotonic if $\alpha > 0$ and oscillatory if $\alpha < 0$. Since $-1 < \alpha < 1$ and $-1 < \beta < 1$, the sign of ρ is dependent on the term $\alpha - \beta$.

In the general case a significant result to note is

$$\gamma_h = \alpha_1\gamma_{h-1} + \alpha_2\gamma_{h-2} + \cdots + \alpha_p\gamma_{h-p} \qquad h \geq q + 1 \tag{21.5.10}$$

which is due to the fact that Y_i, $i = 1 = \ldots -1, 0, +1, \ldots$, are uncorrelated random variables. In terms of autocorrelation we have the result

$$\rho_h = \alpha_1\rho_{h-1} + \alpha_2\rho_{h-2} + \cdots + \alpha_p\rho_{h-p} \qquad h \geq q + 1 \tag{21.5.11}$$

The spectral density of the general ARMA(p, q) process is determined by writing (21.5.2) as

$$X_n = \frac{\theta(B)}{\phi(B)} Y_n \tag{21.5.12}$$

and applying (21.2.34). Thus we get

$$f(\lambda) = \frac{\sigma_y^2}{2\pi} \frac{\left|1 - \beta_1 e^{-i\lambda} - \beta_2 e^{-2i\lambda} - \cdots - \beta_q e^{-qi\lambda}\right|^2}{\left|1 - \alpha_1 e^{-i\lambda} - \alpha_2 e^{-2i\lambda} - \cdots - \alpha_p e^{-pi\lambda}\right|^2} \tag{21.5.13}$$

Appropriate simplifications follow in special cases. For example, for the ARMA(1, 1) given by (21.5.3), we get

$$f(\lambda) = \left(\frac{\sigma_y^2}{2\pi}\right) \frac{1 + \beta^2 + 2\beta\cos\lambda}{1 + \alpha^2 + 2\alpha\cos\lambda} \tag{21.5.14}$$

21.6 AUTOREGRESSIVE INTEGRATED MOVING AVERAGE PROCESSES

We have introduced the *autoregressive integrated moving average process* (ARIMA) as a way of representing certain types of nonstationary behavior in the process. For this purpose we define a generalized autoregressive operator $\varphi(B)$ as

$$\varphi(B) = \phi(B)(1 - B)^d \tag{21.6.1}$$

where $d > 0$ and an integer, and $\phi(B)$ is a stationary operator [i.e., when the roots of the characteristic equation $\phi(B) = 0$ lie outside the unit circle]:

$$\phi(B) = 1 - \alpha_1 B - \alpha_2 B^2 - \cdots - \alpha_p B^p \tag{21.6.2}$$

Let $\theta(B)$ be the moving average operator

$$\theta(B) = 1 - \beta_1 B - \beta_2 B^2 - \cdots - \beta_q B^q \tag{21.6.3}$$

Now the ARIMA (p, d, q) model is given by

$$\varphi(B) X_n = \phi(B)(1 - B)^d X_n = \theta(B) Y_n \tag{21.6.4}$$

The operator $1 - B$ is a difference operator since

$$(1 - B) X_n = X_n - X_{n-1} \tag{21.6.5}$$

Therefore the operator $(1 - B)^d$ is an operator that gives the dth difference.

Because of the term $(1 - B)^d$, $d > 0$, the generalized autoregressive operator $\varphi(B)$ is a nonstationary operator. [For (covariance) stationarity $\varphi(B)$ should be such that all roots of the characteristic equation $\varphi(B) = 0$ lie outside the unit circle. In this case d of them lie on the unit circle.] But since $\varphi(B) = \phi(B)(1 - B)^d$, and $\phi(B)$ is a stationary operator, a process represented by $\varphi(B)$ is nonstationary, but its dth difference is stationary. Reversing the procedure, summing a stationary AR(p) process d times, one gets a nonstationary process that can be represented by the operator $\varphi(B)$. Because of this summing procedure (which is equivalent to "integrating") the process bears the ARIMA name.

A discussion of the ARIMA(p, d, q) model along the lines given in the last three sections is beyond the scope of this chapter. Interested readers are referred to Box and Jenkins (1976). Nevertheless, two possible approaches can be identified for purposes of understanding. Writing out $\varphi(B)$, we get

$$\varphi(B) = \phi(B)(1 - B)^d = 1 - \varphi_1 B - \varphi_2 B - \cdots - \varphi_{p+d} B^{p+d}$$

$$\tag{21.6.6}$$

Using this expansion, ARIMA(1, 2, 1) can be written as

$$\varphi(B)X_n = \theta(B)Y_n \tag{21.6.7}$$

where

$$\varphi(B) = 1 - (2 + \alpha)B + (1 + 2\alpha)B^2 - \alpha B^3$$

$$\theta(B) = 1 - \beta B$$

Thus we get the model

$$X_n = (2 + \alpha)X_{n-1} - (1 + 2\alpha)X_{n-2} + \alpha X_{n-3} + Y_n - \beta Y_{n-1}$$

Alternately, one can put equation (21.6.4) in the form

$$X_n = \psi(B)Y_n \tag{21.6.8}$$

by appropriately identifying the form for the coefficients ψ_i, $i = 0, 1, 2, \ldots$. A convenient way of doing this would be to use the operator $\varphi(B)$ on both sides of (21.6.8) and write

$$\varphi(B)X_n = \varphi(B)\psi(B)Y_n$$

$$= \theta(B)Y_n$$

Hence

$$\varphi(B)\psi(B) = \theta(B) \tag{21.6.9}$$

Now the coefficients ψ_i, $i = 0, 1, 2, \ldots$ are obtained by comparing coefficients of like power of B. Since Y_n is a white noise process, the properties of X_n can be obtained by determining the properties of $\psi(B)$.

A special case of the ARIMA(p, d, q) model is the *integrated moving average process* [ARIMA(0, 1, 1) or IMA(0, 1, 1)] which provides the exponential smoothing scheme of forecasting. To get the exponential form, we start with

$$X_n - X_{n-1} = Y_n - \beta Y_{n-1}$$

$$(1 - B)X_n = (1 - \beta B)Y_n \tag{21.6.10}$$

This gives

$$Y_n = \frac{1 - B}{1 - \beta B} X_n$$

$$= \frac{1 - \beta B - (1 - \beta) B}{1 - \beta B} X_n$$

$$= \left[1 - (1 - \beta) \sum_{j=1}^{\infty} \beta^{j-1} B^j \right] X_n \qquad (21.6.11)$$

Rearranging the terms, we then get

$$X_n = (1 - \beta) \sum_{j=1}^{\infty} \beta^{j-1} X_{n-j} + Y_n$$

$$= \sum_{j=1}^{\infty} g_j X_{n-j} + Y_n \qquad (21.6.12)$$

where $\{ g_j, \ j = 1, 2, \ldots \}$ is a geometric distribution. Hence this scheme is known as exponential smoothing in which the next value of the "smoothed" time series is estimated as a weighted sum of all previous observations, while the weights form a geometric distribution.

An alternate form for (21.6.12) is obtained by writing

$$X_n = \frac{1 - \beta B}{1 - B} Y_n$$

$$= \frac{1 - B + (1 - \beta) B}{1 - B} Y_n$$

$$= \left[1 + (1 - \beta) B \sum_{j=0}^{\infty} B^j \right] Y_n$$

$$= Y_n + (1 - \beta) \sum_{j=1}^{\infty} Y_{n-j} \qquad (21.6.13)$$

For a discussion of the role of exponential smoothing in forecasting, readers are referred to Brown (1963) and Box and Jenkins (1976).

It is interesting to note that the exponential smoothing scheme (21.6.10) gives the random walk (if we assume it to have started an infinitely long time before) discussed in earlier chapters if we set $\beta = 0$. From (21.6.12) we get

$$X_n = X_{n-1} + Y_n.$$

An extension of the ARIMA process to include periodic changes such as seasonal fluctuations is realized by using the Tth-backshift operator B^T in place of B in the ARIMA model, where T is the period exhibited by the process. Now the model represented by (21.6.4) takes the form

$$\phi(B^T)(1 - B^T)^d X_n = \theta(B^T)Y_n \qquad (21.6.14)$$

Since all models considered in Sections 21.3–21.5 are special cases of the ARIMA model, this model provides a fairly general framework for purposes of model building. Knowing the specific properties of the special cases, one can search for the right model for representing a given time series. We shall indicate the essential steps of this procedure in the next section.

21.7 TIME SERIES ANALYSIS IN THE TIME DOMAIN

In Sections 21.2–21.6 we presented several stochastic models for time series. In this section we shall briefly discuss how these models can be used in analyzing time series. Since the material to be covered is extensive and it includes varied practical insight, we must caution the reader that we shall only introduce the highlights and leave the rest to be gathered from books such as Anderson (1971), Nelson (1973), Koopmans (1974), Brillinger (1975), Anderson (1976), Box and Jenkins (1976), Fuller (1976), Kendall (1976), Jenkins (1979), Chatfield (1980), and Priestley (1981).

Following Box and Jenkins, the major steps in the analysis of time series include model identification, parameter estimation and model verification. Once the right model is identified, it can be used for forecasting or control purposes.

In contrast with random phenomena to which stochastic models based on Markov and renewal processes are appropriate, basic elements of the structure of time series may not be evident from observations. The only features obvious from the data are likely to be those pertaining to strong periodic fluctuations and dominant nonstationary trends. The time series may be the result of several factors, which we may not be able to separate; nevertheless, the objective of this analysis is to identify the results of these interacting factors through appropriate techniques.

In analyzing a time series, one starts out by plotting observations against time. The time plot is very helpful in identifying features such as trend and periodic fluctuations. While plotting the data, it may be desirable to employ transformations in order to stabilize the variance or to make the data conform to the basic assumptions of the model such as the additivity of components and normality. If trends are indicated, variations due to trends can be accounted for by fitting an appropriate curve or using some kind of a filtering mechanism which may include averaging or differencing. Periodic fluctuations such as seasonal variations can be eliminated many times by applying an appropriate filter to the series. An example of such a filter is the case where the new monthly figure is the old figure minus the average of that monthly figure over adjacent years (Granger and Newbold, 1977, p. 66). Simple averaging and taking differences are other examples of such filters.

The next step is to identify an appropriate model from the general class of ARIMA processes. The Box-Jenkins method of model identification is based on knowing the properties of models pertaining to their characteristics such as autocorrelation and periodic behavior. In this connection the following specific properties are significant:

1. If the correlation function does not approach zero reasonably fast, nonstationarity should be suspected. If the series is stationary, very often after a high short-term value the correlation function tends to get smaller very fast. It should also be noted that the correlation function of alternating series alternate with successive observations on different sides of zero. A trend in the series makes the correlation function stay on one side of the zero until lag values become large. Finally, periodicity in the series will be reflected in the correlation function as well. Noting that spectral analysis is an analysis of variance of the process (see Chapter 9), the properties of a series can also be identified using a plot of its spectral density (also known as the power spectrum). In any case it should be emphasized that considerable experience is needed in the interpretation of the properties of these functions by observing their plots.

2. For an $AR(p)$ process the correlation function is a mixture of damped exponentials and/or damped sine waves.

3. For a $MA(q)$ process the correlation function $\rho_h = 0$ for $h > q$.

4. For an $ARMA(p, q)$ process, as illustrated by equation (21.5.11), if $q < p$, ρ_h will have the characteristics of the correlation function of an $AR(p)$ process regardless of the value of h. If $q \geq p$, there will be $q - p + 1$ initial values $\rho_0, \rho_1, \ldots, \rho_{q-p}$ which do not follow the $AR(p)$ pattern.

Another characteristic that we have not defined so far but can help identify an $AR(p)$ model is the *partial autocorrelation*. Let α_{kj} denote the jth coefficient in an autoregressive process of order k (i.e., α_j of the $AR(k)$

model is written as α_{kj} to indicate the order of the process). Now α_{kk} is the last coefficient and is known as the partial autocorrelation function. Equation (21.3.27) can be written as

$$\rho_j = \alpha_{k1}\rho_{j-1} + \alpha_{k2}\rho_{j-2} + \cdots + \alpha_{kk}\rho_{j-k} \qquad j = 1, 2, \ldots, k$$

These Yule-Walker equations can be written down in a matrix form as follows:

$$\begin{bmatrix} 1 & \rho_1 & \rho_2 & \cdots & \rho_{k-1} \\ \rho_1 & 1 & \rho_1 & \cdots & \rho_{k-2} \\ \vdots & \vdots & \vdots & & \vdots \\ \rho_{k-1} & \rho_{k-2} & \rho_{k-3} & \cdots & 1 \end{bmatrix} \begin{bmatrix} \alpha_{k1} \\ \alpha_{k2} \\ \vdots \\ \alpha_{kk} \end{bmatrix} \begin{bmatrix} \rho_1 \\ \rho_2 \\ \vdots \\ \rho_k \end{bmatrix}$$

$$P_k \alpha_k = \rho_k \tag{21.7.1}$$

Solving these equations for $k = 1, 2, \ldots,$

$$\alpha_{11} = \rho_1$$

$$\alpha_{22} = \frac{\begin{vmatrix} 1 & \rho_1 \\ \rho_1 & \rho_2 \end{vmatrix}}{\begin{vmatrix} 1 & \rho_1 \\ \rho_1 & 1 \end{vmatrix}}$$

$$\alpha_{33} = \frac{\begin{vmatrix} 1 & \rho_1 & \rho_1 \\ \rho_1 & 1 & \rho_2 \\ \rho_2 & \rho_1 & \rho_3 \end{vmatrix}}{\begin{vmatrix} 1 & \rho_1 & \rho_2 \\ \rho_1 & 1 & \rho_1 \\ \rho_2 & \rho_1 & 1 \end{vmatrix}}$$

and so on. Thus the denominator of the expression for α_{kk} is the determinant $|P_k|$, and the numerator is $|P_k^*|$, where the matrix P_k^* has the same elements as P_k except the last column which is replaced by ρ_k. The Box-Jenkins procedure for model identification uses the fact that if the process is actually AR(p), then α_{kk} is nonzero for $k = p$ and $= 0$ for $k > p$.

Other model identification procedures include the use of residual variance plots proposed by Whittle (1963), Akaike's (1969) FPE criterion which refines the residual variance plot criterion, Akaike's (1974) AIC criterion which is based on information theoretic concepts, Akaike's (1978) BIC criterion which is a Bayesian modification of his AIC criterion, the use of an inverse autocorrelation function proposed by Cleveland (1972), and Parzen's (1974) CAT criterion which uses the integrated relative mean square error of the spectral estimate. Some of them are specific only to autoregressive models, and others are more generally applicable. For more information on these procedures, the readers are referred to Priestly (1981, pp. 370–380, 601) who provides a concise description. Also see Gray, Kelley, and McIntire (1978) for the R- and S-array method and Woodward and Gray (1981) for a model identification method based on a generalized partial autocorrelation function.

The next problem is one of estimation of model parameters and model characteristics. The model identification procedures described before are based on autocorrelations, partial autocorrelations, and residual variance, all of which will have to be estimated from time series data. The ergodic theorems given in Chapter 9 justify the use of time averages in place of ensemble averages in estimation. Thus estimates are usually based on sample observations, say N of them, $(x_1, x_2, \ldots, x_N)$ of a stochastic process. Some of the specific estimates used in time series analysis are as follows:

Mean:

$$\bar{x} = \frac{1}{N} \sum_{n=1}^{N} x_n \qquad (21.7.2)$$

Variance:

$$c_0 = \hat{\sigma}_x^2 = \frac{1}{N} \sum_{n=1}^{N} (x_n - \bar{x})^2 \qquad (21.7.3)$$

Autocovariance with lag h:

$$c_h = \frac{1}{N} \sum_{n=1}^{N-h} (x_n - \bar{x})(x_{n+h} - \bar{x}) \qquad h = 0, 1, 2, \ldots \qquad (21.7.4)$$

Autocorrelation with lag h:

$$r_h = \frac{c_h}{c_0} \qquad h = 0, 1, 2, \ldots \qquad (21.7.5)$$

Standard error of the autocorrelation for a stationary Gaussian process

(Bartlett, 1946):

$$V[r_h] \simeq \frac{1}{N} \sum_{k=-\infty}^{\infty} \left\{ \rho_k^2 + \rho_{k+h}\rho_{k-h} - 4\rho_h\rho_k\rho_{k-h} + 2\rho_h^2\rho_k^2 \right\}$$

(21.7.6)

If it is known that the autocorrelations ρ_h die out at lags $h > q$, then the approximation (21.7.6) yields (the large-lag standard error)

$$V[r_h] \simeq \frac{1}{N} \left\{ 1 + 2 \sum_{k=1}^{q} \rho_k^2 \right\} \qquad h > q \qquad (21.7.7)$$

Large-lag approximation to the covariance between r_h and r_{h+m} (Barlett, 1946):

$$\text{Cov}(r_h, r_{h+m}) \simeq \frac{1}{N} \sum_{k=-\infty}^{\infty} \rho_k\rho_{k+m} \qquad (21.7.8)$$

Standard error of partial autocorrelation for large lags (Quenouille, 1949):

$$V[\hat{\alpha}_{kk}] \simeq \frac{1}{N} \qquad (21.7.9)$$

The method of estimation of model parameters depends on the nature of the model. For instance, for an autoregressive model the least squares method is appropriate. Another method would be to use the Yule-Walker equations (21.7.1) along with estimates of autocorrelations for $\rho_1, \rho_2, \ldots$.

Estimating the parameters of a moving average process is not straightforward. Box and Jenkins (1970) have given a procedure using the estimates of ρ_h in equations (21.4.4) and solving the q nonlinear equations for the unknown $\beta_1, \beta_2, \ldots, \beta_q$. As indicated in the discussion of the MA(1) case following equation (21.4.4), the equations may generate multiple solutions. Nevertheless, these solutions provide initial estimates for further improvement by other methods.

As one may imagine, parameter estimation in more general models is much more complicated. For appropriate solution techniques, the readers are referred to Box and Jenkins (1976) and Priestley (1981).

Model verification is generally carried out using an analysis of the residuals, which are the differences between the observations and the fitted values. Box and Jenkins (1976, Chap. 8) discuss this problem in great detail. For a short discussion of other available approaches, reference can also be made to Chatfield (1980, pp. 76–79).

The final topic we shall discuss in this section is forecasting (also known as prediction in the literature). The problem involves forecasting future values of a time series having a set of observations from it up to the present time. The various models discussed in the last few sections can be conveniently used for this purpose. For instance, the exponential smoothing scheme presented in (21.6.12) is a one-step-ahead forecast function, since the forecast value $\hat{X}_n$ of X_n can be obtained as [noting that $E(Y_n) = 0$]

$$\hat{X}_n = \sum_{j=1}^{\infty} (1 - \beta)\beta^{j-1} x_{n-j} \qquad (21.7.10)$$

where $\beta < 1$ and $x_{n-1}, x_{n-2}, \ldots$, are observed values of the series.

In the general ARIMA model, consider observations x_j, $j = \ldots$, $(n - 1), n$, and suppose we wish to forecast the values of the time series after a lead time l; that is, we wish to determine the properties of X_{n+l}. Using the ARIMA model in the form given by equation (21.6.8),

$$X_n = \psi(B)Y_n \qquad (21.7.11)$$

we have

$$X_{n+l} = \sum_{j=-\infty}^{n+l} \psi_{n+l-j} Y_j = \sum_{j=0}^{\infty} \psi_j Y_{n+l-j}$$

where $\psi_0 = 1$. This equation can be rewritten as

$$X_{n+l} = (Y_{n+l} + \psi_1 Y_{n+l-1} + \cdots + \psi_{l-1} Y_{n+1})$$

$$+ (\psi_l Y_n + \psi_{l+1} Y_{n-1} + \cdots) \qquad (21.7.12)$$

Suppose we wish to make a forecast of X_{n+l}, which we shall denote as $\hat{X}_{n+l}$, using a linear function of the present and past observations $x_n, x_{n-1}, \ldots$. By virtue of (21.7.11), $\hat{X}_{n+l}$ will also be a linear function of $Y_n, Y_{n-1}, \ldots$ which we may represent as

$$\hat{X}_{n+l} = \psi_l^* Y_n + \psi_{l+1}^* Y_{n-1} + \cdots \qquad (21.7.13)$$

where $\psi_l^*, \psi_{l+1}^*, \ldots$ are unknown. Comparing (21.7.13) with (21.7.12), the mean square error of the forecast is obtained as

$$E[X_{n+l} - \hat{X}_{n+l}]^2 = \left[1 + \psi_1^2 + \cdots + \psi_{l-1}^2\right]\sigma_y^2$$

$$+ \sum_{j=0}^{\infty} \left(\psi_{l+j} - \psi_{l+j}^*\right)\sigma_j^2 \qquad (21.7.14)$$

which is minimized when $\psi_{l+j}^* = \psi_{l+j}$. Thus the forecast resulting in the smallest mean square error is given by

$$\hat{X}_{n+l} = \psi_l Y_n + \psi_{l-1} Y_{n-1} + \cdots \qquad (21.7.15)$$

which can be considered to be the conditional expectation of X_{n+l}, given $X_n, X_{n-1}, \ldots$. Comparing (21.7.15) with (21.7.12), the term

$$Y_{n+l} + \psi_1 Y_{n+l-1} + \cdots + \psi_{l-1} Y_{n+1}$$

can be considered to be the error in the forecast.

The properties of the minimum mean square error forecasts have been investigated by Box and Jenkins (1976). They also discuss problems such as the calculation of forecast weights and updating forecasts for models coming under the ARIMA class.

It can be easily seen that the exponential smoothing scheme considered earlier belongs to the class of minimum mean square error forecasts.

For further information on this topic, in addition to Box and Jenkins (1976), the following references may be suggested: Brown (1963), Nelson (1973), Granger and Newbold (1977), Montgomery and Johnson (1977), Jenkins (1979), Chatfield (1980, Chap. 5), and Pandit and Wu (1983).

21.8   SPECTRAL ANALYSIS OF TIME SERIES DATA

So far in this chapter we have presented stochastic models for time series, their properties, and procedures for using them. We may call this the parametric approach to time series analysis. A nonparametric approach would be to start with the single assumption that a covariance stationary stochastic process can be represented as a sum of periodic functions with random coefficients, as illustrated in Section 9.4. This leads us to spectral analysis.

The spectral density (or spectral distribution) provides the distribution of the variance in the process corresponding to various frequencies of the underlying periodic functions. To adopt spectral analysis of observed time series data, one should have an appropriate estimator of the spectral density and its distribution properties. In this section we shall provide only an introduction to this topic. For further reading, references can be made to Jenkins and Watts (1968), Anderson (1971), Koopmans (1974), Brillinger (1975), Fuller (1976), and Priestley (1981).

In Chapter 9 we have noted that the spectral density $f(\lambda)$ can be written as [(9.4.10) and (9.4.11)].

$$f(\lambda) = \frac{1}{2\pi} \sum_{h=-\infty}^{\infty} \gamma_h e^{-i\lambda h}$$

$$= \frac{1}{2\pi}\left[\gamma_0 + 2\sum_{h=1}^{\infty} \gamma_h \cos\lambda h\right] \qquad (21.8.1)$$

where γ_h is the covariance function of the process at lag h. Using ergodic theorems (Section 9.3), an estimator for $f(\lambda)$ can be identified as

$$\hat{f}(\lambda) = I_N(\lambda) = \frac{1}{2\pi} \sum_{h=-\infty}^{\infty} c_h e^{-i\lambda h}$$

$$= \frac{1}{2\pi}\left[c_0 + 2\sum_{h=1}^{N-1} c_h \cos\lambda h\right] \qquad (21.8.2)$$

where the c_h, $h = 0, 1, 2, \ldots$ are the sample autocovariances defined in (21.7.3) and (21.7.4) based on N observations.

Now consider N observations $(x_1, x_2, \ldots, x_N)$ from a zero-mean stationary Gaussian process. Let

$$z_\nu^{(N)} = \frac{1}{\sqrt{2\pi N}} \sum_{n=1}^{N} x_n e^{-i\lambda_\nu n} \qquad (21.8.3)$$

where $\lambda_\nu = 2\pi\nu/N$, $-[(N-1)/2] \le \nu \le [N/2]$. Here we have denoted by $[u]$ the largest integer less than or equal to u. Equation (21.8.3) defines the normalized finite Fourier transform of the data [$\sqrt{2\pi/N}\, z_\nu^{(N)}$ is the finite Fourier transform].

Linear combinations of normal variates are also normal. Therefore $\{z_\nu^{(N)}\}$ have multivariate complex normal distributions with zero means. Thus the variance of $z_\nu^{(N)}$ is given by $E|z_\nu^{(N)}|^2$ [note that $|z_\nu^{(N)}|^2 = z_\nu^{(N)}\bar{z}_\nu^{(N)}$]. It is possible to show (see Koopmans, 1974, p. 259) that if the spectral density function of the stochastic process $\{X_n, n = 0, \pm 1, \ldots\}$ is continuous, then for sufficiently large N,

$$E|z_\nu^{(N)}|^2 \simeq f(\lambda_\nu) \tag{21.8.4}$$

When $f(\lambda)$ is smooth in the vicinity of λ_ν, the approximation is good even for moderate values of N.

Equation (21.8.4) justifies the use of $|z_\nu^{(N)}|^2$ as an estimator of the spectral density function. It is known as the *periodogram* of the time series. For frequencies λ, $-\pi < \lambda < \pi$, define

$$I_N(\lambda) = |z^{(N)}(\lambda)|^2$$

where

$$z^{(N)}(\lambda) = \frac{1}{\sqrt{2\pi N}} \sum_{n=1}^{N} e^{-i\lambda n} \tag{21.8.5}$$

It is easy to show that the $I_N(\lambda)$ given by this equation is in fact the $I_N(\lambda)$ of equation (21.8.2). We have from (21.8.5)

$$I_N(\lambda) = \frac{1}{2\pi N} \left(\sum_{n=1}^{N} x_n e^{-i\lambda n} \right) \left(\sum_{r=1}^{N} x_r e^{i\lambda r} \right)$$

$$= \frac{1}{2\pi N} \left[\sum_{n=1}^{N} \sum_{r=1}^{N} x_n x_r \cos(n - r)\lambda \right] \tag{21.8.6}$$

In order to transform the double summation over n and r into a summation over the diagonal points of the lattice, we change variables in (21.8.6) from n and r into n and $h = n - r$. Then we get

$$I_N(\lambda) = \frac{1}{2\pi} \sum_{h=-(N-1)}^{N-1} \left\{ \frac{1}{N} \sum_{n=1}^{N-|h|} x_n x_{n+h} \right\} \cos \lambda h$$

$$= \frac{1}{2\pi} \sum_{n=-(N-1)}^{N-1} c_h \cos \lambda h$$

$$= \frac{1}{2\pi} \left\{ c_0 + 2 \sum_{h=1}^{N-1} c_h \cos \lambda h \right\}$$

Due to the normality of $z_\nu^{(N)}$ and equation (21.8.4), we also have the following result (see Koopmans, 1974, p. 265):

$$\lim_{N \to \infty} V(I_N(\lambda)) = \begin{cases} f^2(\lambda) & \lambda \neq 0, \pi \\ 2f^2(0) & \lambda = 0 \\ 2f^2(\pi) & \lambda = \pi \end{cases} \qquad (21.8.7)$$

The implication of this variance result is that $I_N(\lambda)$ is not a consistent estimator of the spectral density (for consistency the variance of the estimator $\to 0$ as $N \to \infty$). The reason for this situation is the fact that for large values of the lag h as compared to N the autocovariances c_h are obtained as averages of relatively few values regardless of the value of N. In order to eliminate the inconsistency without losing the estimator's unbiasedness, a weighted covariance estimator $\omega_M(h)c_h$ is used in (21.8.2) in place of c_h. The weight function $\omega_M(h)$ is known as the *lag window* and has the following properties (for $M < N$, $h = 0 \pm 1, \pm 2, \dots$):

1. $0 \leq \omega_M(h) \leq \omega_M(0) = 1$.
2. $\omega_M(h)$ is symmetric in h.
3. $\omega_M(h) = 0$ for $|h| > M$.

Thus incorporating the lag window, equation (21.8.2) may be written as

$$\hat{f}(\lambda) = \frac{1}{2\pi} \sum_{h=-\infty}^{\infty} \omega_M(h) c_h e^{-i\lambda h} \qquad (21.8.8)$$

showing that $\hat{f}(\lambda)$ is the Fourier transform of $\omega_M(h)c_h$. Let $W_M(\lambda)$ be the Fourier transform of $\omega_M(h)$; that is,

$$W_M(\lambda) = \frac{1}{2\pi} \sum_{h=-\infty}^{\infty} \omega_M(h) e^{-i\lambda h} \qquad (21.8.9)$$

Since $I_n(\lambda)$ is the Fourier transform of c_h, using the convolution property of Fourier transforms, we may write

$$\hat{f}(\lambda) = \int_{-\pi}^{\pi} W_M(\lambda - \mu) I_N(\mu) \, d\mu \qquad (21.8.10)$$

The function $W_M(\lambda)$ is known as the *spectral window*. The reason for the use of the term 'window' in characterizing $\omega_M(h)$ and $W_M(\lambda)$ is that we may think of $W_M(\lambda)$ as being effectively zero outside a small interval, say

$(-\varepsilon, \varepsilon)$, and that the integral (21.8.10) may be regarded as giving a view of $I_N(\mu)$ through a narrow "window" from $\lambda - \varepsilon$ to $\lambda + \varepsilon$ (Priestley, 1981, p. 436).

Various forms for the function $\omega_M(h)$ have been suggested in the literature. For a discussion of such lag windows and their properties, the readers are referred to Priestley (1981, pp. 437–449).

Distribution properties of spectral estimators are obtained based on the distribution properties of the statistic $I_N(\lambda)$ and the impact of the lag window on it. For a discussion of these properties, the readers are referred to Koopmans (1974), Fuller (1976), and Priestley (1981).

Carrying out spectral analysis in practice requires attention to several factors not mentioned in our discussion. For instance, sampling observations at discrete time points from a continuous parameter stochastic process may generate a time series with properties that have no resemblance to those

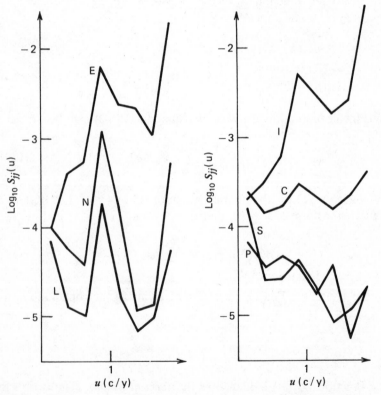

Figure 21.8.1 Power spectra of Icelandic series for economic variablels (Gudmundsson, 1971, p. 401).

of the original process. Therefore the selection of the right length of the time interval between observations is important. For an understanding of such problems, the readers are well advised to refer to some of the books mentioned earlier before starting on a spectral analysis of time series data.

In order to set up an appropriate mathematical model for a time series such as one belonging to the ARIMA class, it is necessary to have some understanding of the behavior of the underlying stochastic process. The analysis of the properties of the covariance and correlation functions is one approach. Another approach is through spectral analysis. In most cases carrying out spectral analysis becomes necessary in order to zero in on the model as well as to verify it after its selection. Spectral analysis is like drawing a histogram which is an invaluable tool of data analysis.

As an example we give in Figure 21.8.1 the power spectra (spectral density functions) of the economic time series from Iceland (Gudmundsson, 1971) presented in Figure 21.1.3. The difference in the periodic behavior of basic processes underlying series C (total bank deposits on current accounts) and I (total imports), on the one hand, and S (total bank deposits except current accounts) and P (cost of living index), on the other, is very significant. In C and I high frequencies dominate representing fast fluctuations in the series, whereas in S and P low frequencies dominate indicating a slow movement.

REFERENCES

Akaike, H. (1969). "Fitting Autoregressive Models for Prediction." *Ann. Inst. Stat. Math.* **21**, 243–247.

Akaike, H. (1974). "A New Look at the Statistical Model Identification." *IEEE Trans. Auto. Control* **AC-19**, 716–722.

Akaike, H. (1978). "A Bayesian Analysis of the Minimum AIC Procedure." *Ann. Inst. Stat. Math.* **30A**, 9–14.

Anderson, O. D. (1976). *Time Series Analysis and Forecasting: The Box-Jenkins Approach*. London: Butterworths.

Anderson, T. W. (1971). *The Statistical Analysis of Time Series*. New York: Wiley.

Bartlett, M. S. (1946). "On the Theoretical Specifications of Sampling Properties of Autocorrelated Time Series." *J. Roy. Stat. Soc. Supp.* **8**, 27–41.

Bartlett, M. S. (1955). *An Introduction to Stochastic Processes*. Cambridge: Cambridge University Press, Chap. 9.

Box, G. E. P., and Jenkins, G. M. (1976). *Time Series Analysis: Forecasting and Control*. 2nd ed. San Francisco: Holden-Day.

Brillinger, D. R. (1975). *Time Series: Data Analysis and Theory*. New York: Holt, Rinehart and Winston.

Brown, R. G. (1963). *Smoothing, Forecasting and Prediction of Discrete Time Series*. Englewood Cliffs, N.J.: Prentice-Hall.

Chatfield, C. (1980). *The Analysis of Time Series: An Introduction*. 2nd ed. New York: Chapman and Hall.

Chatfield, C., and Pepper, P. M. G. (1971). "Time Series Analysis: An Example from Geophysical Data." *Appl. Stat.* **20**, 217–238.

Cleveland, W. S. (1972). "The Inverse Autocorrelations of a Time Series and Their Applications." *Technometrics* **14**, 277–293.

Cox, D. R., and Miller, H. D. (1965). *The Theory of Stochastic Processes*. London: Chapman and Hall.

Fuller, W. A. (1976). *Introduction to Statistical Time Series*. New York: Wiley.

Granger, C. W. J., and Newbold, P. (1977). *Forecasting Economic Time Series*. New York: Academic Press.

Gray, H. L., Kelley, G. D., and McIntire, D. D. (1978). "A New Approach to ARMA Modeling." *Comm. Stat.* **B7**, 1–77.

Gudmundsson, G. (1971). "Time Series Analysis of Imports, Exports and Other Economic Variables." *J. Roy. Stat. Soc.* **A134**, 383–412.

Jenkins, G. M. (1979). *Practical Experiences with Modelling and Forecasting Time Series*. Jersey: Gwilym Jenkins and Partners.

Jenkins, G. M. and Watts, D. G. (1968). *Spectral Analysis and its Applications*. San Francisco: Holden-Day.

Jordan, C. (1965). *Calculus of Finite Differences*. 3rd ed. New York: Chelsea.

Kendall, M. G. (1976). *Time Series*. 2nd ed. London: Griffin.

Koopmans, L. H. (1974). *The Spectral Analysis of Time Series*. New York: Academic Press.

Levy, H., and Lessman, F. (1961). *Finite Difference Equations*. New York: Macmillan.

Montgomery, D. C., and Johnson, L. A. (1977). *Forecasting and Time Series Analysis*. New York: McGraw-Hill.

Nelson, C. R. (1973). *Applied Time Series Analysis for Managerial Forecasting*. San Francisco: Holden-Day.

Pandit, S. M., and Wu, S. M. (1983). *Time Series and Systems Analysis with Applications*. New York: Wiley.

Parzen, E. (1974). "Some Recent Advances in Time Series Modelling." *IEEE Trans. Auto. Control* **AC-19**, 723–729.

Priestley, M. B. (1981). *Spectral Analysis and Time Series*. Vol. 1: *Univariate Series*. New York: Academic Press.

Quenouille, M. H. (1949). "Approximate Tests of Correlation in Time Series." *J. Roy. Stat. Soc.* **B11**, 68–84.

Tee, L. H., and Wu, S. M. (1972). "An Application of Stochastic and Dynamic Models for the Control of Papermaking Process." *Technometrics* **14**, 481–496.

Whittle, P. (1963). *Prediction and Regulation by Least Square Methods*. New York: Van Nostrand.

Woodward, W. A., and Gray, H. L. (1981). "On the Relationship Between the *S* Array and the Box-Jenkins Method of ARMA Model Identification." *J. Amer. Stat. Assoc.*. **76**, 579–587.

FOR FURTHER READING

Anderson, A. (1982). "An Empirical Examination of Box-Jenkins Forecasting." *J. Roy. Stat. Soc.* **A145**, 472–478.

Bhansali, R. J. (1979). "A Mixed Spectrum Analysis of the Lynx Data." *J. Roy. Stat. Soc.* **A142**, 199–209.

Brown, R. G., and Meyer, R. F. (1961). "The Fundamental Theorem of Exponential Smoothing." *Oper. Res.* **9**, 673–685.

Burman, J. P. (1980). "Seasonal Adjustment by Signal Extraction." *J. Roy. Stat. Soc.* **A143**, 321–337.

Findley, D. F., ed. (1978). *Applied Time Series Analysis*. New York: Academic Press.

Grenander, U., and Rosenblatt, M. (1957). *Statistical Analysis of Stationary Time Series*. New York: Wiley.

Hannan, E. J. (1960). *Time Series Analysis*. London: Methuen.

Hannan, E. J. (1970). *Multiple Time Series*. New York: Wiley.

Malinvaud, E. (1966). *Statistical Methods of Econometrics*. Chicago: Rand McNally, Pt. 4.

Parzen, E. (1967). *Time Series Analysis Papers*. San Francisco: Holden-Day.

Rosenblatt, M., ed. (1963). *Time Series Analysis*. Proc. Symposium on Time Series Analysis, Brown University, June 11–14, 1962. New York: Wiley.

Slutsky, E. (1927). "The Summation of Random Causes as a Source of Cyclical Processes" (in Russian). *Problems of Economic Conditions* **3**, 1. English Translation, *Econometrica* **5**, 105 (1937).

Taylor, S. J. (1980). "Conjectured Models for Trends in Financial Prices, Tests and Forecasts." *J. Roy. Stat. Soc.* **A143**, 338–362.

Tintner, G. (1940). *The Variate Difference Method*. Bloomington, Ind.: Principia Press.

Tintner, G. (1952). *Econometrics*. New York: Wiley, Pt. 3.

Wallis, K. F. (1982). "Seasonal Adjustment and Revision of Current Data: Linear Filters for the *X*-11 Method." *J. Roy. Stat. Soc.* **A145**, 74–85.

Wold, H. O. A., ed. (1954). *A Study in the Analysis of Stationary Time Series*. 2nd ed. Stockholm: Almquist and Wisell.

Wold, H. O. A. (1965). *Bibliography on Time Series and Stochastic Processes*. International Statistical Institute. Cambridge, Mass.: The MIT Press.

Yaglom, A. M. (1962). *An Introduction to the Theory of Stationary Random Functions*. Englewood Cliffs, N.J.: Prentice-Hall (English trans.).

Appendix A

SOME FUNDAMENTALS OF MATRIX THEORY

A.1 INTRODUCTION

Matrix theory has extensive applications in the study of stochastic processes as well as many other areas. We would like to present here a few definitions and properties of some fundamental and pertinent results required for understanding the techniques employed in this book. The coverage is by no means complete, and we have omitted the proofs of the properties. If the reader is interested in a more extensive coverage of the subject, a few references are given at the end of this appendix.

DEFINITION A.1.1 *The rectangular array*

$$
P = \begin{bmatrix}
p_{11} & p_{12} & \cdots & p_{1n} \\
p_{21} & p_{22} & \cdots & p_{2n} \\
\vdots & \vdots & & \vdots \\
p_{m1} & p_{m2} & \cdots & p_{mn}
\end{bmatrix}
$$

is called a matrix. *It has m rows and n columns, and is usually denoted by P, $\|p_{ij}\|$, or $[p_{ij}]$. The elements p_{ij}, $i = 1, 2, \ldots, m$, $j = 1, 2, \ldots, n$ are assumed to be real numbers. We shall use the notation $P_{m \times n} = \|p_{ij}\|$ to denote a matrix with m rows and n columns. When the dimensions (number of rows or columns) are not relevant, we shall just use P.*

DEFINITION A.1.2 *We call a $P_{1 \times n}$ matrix a* row vector, *$P_{n \times 1}$ matrix a* column vector, *and $P_{1 \times 1}$ matrix a* scalar.

A.2 MATRIX OPERATIONS

Let $P_{m_1 \times n_1} = \|p_{ij}\|$ and $Q_{m_2 \times n_2} = \|q_{ij}\|$ be given matrices we now define some basic matrix operations.

DEFINITION A.2.1 (1) *If* $m_1 = m_2$ *and* $n_1 = n_2$, *then matrix addition* (*subtraction*) *is defined as*

$$P \pm Q = R$$

where the elements r_{ij} *of* R *are given by* $r_{ij} = p_{ij} \pm q_{ij}$ *for all* i, j. (2) *If* $n_1 = m_2$, *then matrix multiplication is defined by* $P \cdot Q = PQ = R$ *where the elements of* R *are given by*

$$r_{ij} = \sum_{l=1}^{n} p_{il} \cdot q_{lj} \qquad \text{for all } i, j$$

(3) *If* k *is a scalar, then*

$$kP = Pk = \|kp_{ij}\|$$

(4) *If* $m_1 = m_2$ *and* $n_1 = n_2$, *then*

$$P = Q \qquad iff \quad p_{ij} = q_{ij} \text{ for each } i, j$$

$$P \leq Q \qquad iff \quad p_{ij} \leq q_{ij} \text{ for each } i, j$$

$$P < Q \qquad iff \quad p_{ij} < q_{ij} \text{ for each } i, j$$

DEFINITION A.2.2 *Let* $P_{m \times n}$ *be given as*

$$P = \begin{bmatrix} p_{11} & p_{12} & \cdots & p_{1n} \\ p_{21} & p_{22} & \cdots & p_{2n} \\ \vdots & \vdots & & \vdots \\ p_{m1} & p_{m2} & \cdots & p_{mn} \end{bmatrix}$$

Then the matrix

$$P^T = \begin{bmatrix} p_{11} & p_{21} & \cdots & p_{m1} \\ p_{12} & p_{22} & \cdots & p_{m2} \\ \vdots & \vdots & & \vdots \\ p_{1n} & p_{2n} & \cdots & p_{mn} \end{bmatrix}$$

is called the transpose of P. *Clearly* P^T *has* n *rows and* m *columns.*

There are several types of matrices whose structure and properties have allowed them to have special names. If $P_{m \times n}$ has all entries zero, we say that it is a *null matrix* and denote it by $\varnothing_{m \times n}$. If the number of rows of P equal the number of columns, then we say that P is a *square matrix*. There are some square matrices that are of interest.

DEFINITION A.2.3 *If $P_{n \times n}$ is a square matrix, then* (1) P is upper (lower) triangular *iff for each i, $p_{ij} = 0$ for $j < i$ ($p_{ij} = 0$ for $j > i$);* (2) P *is the* identity matrix, *denoted by I iff*

$$P = I = \begin{bmatrix} 1 & 0 & 0 & \cdots & 0 \\ 0 & 1 & 0 & \cdots & 0 \\ \vdots & \vdots & \vdots & & \vdots \\ 0 & 0 & 0 & \cdots & 1 \end{bmatrix}$$

(3) P *is* idempotent *iff $P \cdot P = P^2 = P$;* (4) *let $Q_{n \times n}$ be another square matrix, then P is the inverse of Q, denoted by Q^{-1} iff $PQ = QP = I$;* (5) *P is a* diagonal matrix *iff $p_{ij} = 0$ when $i \neq j$;* (6) *P is* symmetric *iff $P = P^T$;* (7) *a square matrix $P_{n \times n} = \| p_{ij} \|$ is called a* positive matrix *iff $p_{ij} \geq 0$ for $i, j = 1, 2, \ldots, n$ and at least one $p_{ij} > 0$.*

In the theory of stochastic processes the following matrix occurs frequently:

DEFINITION A.2.4 *Let $P_{n \times n}$; then P is said to be a* stochastic matrix *iff for each i and j, $0 \leq p_{ij} \leq 1$ and for all i and $\sum_{j=1}^{n} p_{ij} = 1$. If in addition we have $\sum_{i=1}^{n} p_{ij} = 1$ for all j, we say P is* doubly stochastic.

We give next some basic properties of matrix operations. For ease of notation and clarity we shall assume that such operations are well defined.

PROPERTY A.2.1 *Let P, Q, and R be matrices, k a scalar, and $\varnothing$ the null matrix; we then have* (1) $P + Q = Q + P$; (2) $P + [Q + R] = [P + Q] + R$; (3) $k[P + Q] = kP + kQ = [P + Q]k$; (4) *there exists an R such that* $P + R = Q$; (5) $P[Q + R] = PQ + PR$; (6) $[Q + R]P = QP + RP$; (7) $P(QR) = (PQ)R$; (8) $P + \varnothing = \varnothing + P = P$; (9) $P - P = \varnothing$; (10) $P \cdot \varnothing = \varnothing \cdot P = \varnothing$; (11) *in general, the following are not true:* (a) $PQ = QP$, (b) *if $PQ = \varnothing$, then $P = \varnothing$ or $Q = \varnothing$,* (c) *if $PQ = PR$, then $Q = R$.*

PROPERTY A.2.2 *Let P, P^T, Q, and Q^T be given; then* (1) *if $R = P^T$, then $R^T = P$;* (2) $(kP)^T = kP^T$; (3) $(P + Q)^T = P^T + Q^T$; (4) *if $R = PQ$, then $R^T = Q^T P^T$;* (5) $I^T = I$.

A.3   DETERMINANTS

In order to study the theory of matrices, we must investigate the structure of a matrix. This can be done by looking at what a determinant is and developing some aspects of the theory of determinants. We will first define what we mean by a determinant, a cofactor, and the rank of a matrix, after which we give some properties of a determinant.

DEFINITION A.3.1   *The determinant of a square matrix P, denoted by $|P|$, is defined as $|P| = \Sigma(\pm)p_{1i}p_{2j} \cdots p_{nr}$, where the sum is taken over all permutations of the second subscript. A term has a plus sign if $(i, j, \ldots, r)$ is an even permutation of $(1, 2, \ldots, n)$ and a minus sign if it is an odd permutation.*

DEFINITION A.3.2   *In $P_{m \times n}$ let $R_{l \times l}$ be a submatrix of P; then $|R|$ is an lth-order minor of P (note that the order refers to the number of rows in a square matrix).*

DEFINITION A.3.3   *Let $P_{m \times n}$ be given; then we say the rank of P, denoted $r(P)$, is l, if there is a submatrix of P that has a nonzero lth-order minor of P and every minor of P of order greater than l is zero.*

DEFINITION A.3.4   *Let $P_{n \times n}$ be a square matrix; we say β_{ij} is the cofactor of the element p_{ij} of P iff β_{ij} equals $(-1)^{i+j}$ times the determinant of the submatrix obtained from P by deleting the ith row and jth column of P.*

The properties of a determinant given next follow from its definition. We suppress the number of rows and columns in each matrix for notational convenience. Also we assume that all operations are well defined.

PROPERTY A.3.1   *Let P, Q, and R be n-order square matrices; then* (1) $|P| = \Sigma_{j=1}^{n} p_{ij}\beta_{ij}$; (2) *if* $Q = P^T$, *then* $|Q| = |P|$; (3) *if* $R = P \cdot Q$, *then* $|R| = |P| \cdot |Q|$; (4) *in general, it is not true that if* $R = P + Q$, *then* $|R| = |P| + |Q|$; (5) *if every element in a row (column) of P is zero, then* $|P| = 0$; (6) *if Q is obtained from P by multiplying one row (column) of P by the scalar k, then* $|Q| = k|P|$, *if* $Q = kP$, *then* $|Q| = k^n|P|$; (7) *if Q is obtained from P by interchanging any two adjacent rows (columns), then* $|Q| = -|P|$; (8) *if two rows (columns) of P are equal, then* $|P| = 0$; (9) *adding a multiple of the kth row to the lth row does not change the value of the determinant of a matrix P*; (10) *if* $|P| \neq 0$, *then P is said to be* nonsingular, *and otherwise, P is said to be* singular.

We should note that the last property of the determinant is really the definition of a nonsingular matrix. We shall examine it more closely when we discuss bases, independence, and inverse of a matrix.

A.4   ADJOINT AND INVERSE

Given a matrix P, there are many matrices that may be formed from P, its determinant, and cofactors. One such matrix is called the adjoint of P, denoted by adj P, and it is defined as follows:

DEFINITION A.4.1.   *Let P be a nth-order square matrix, with β_{ij} the cofactor of the element p_{ij}; then the* adjoint *of P is given by*

$$\text{adj } P = \begin{bmatrix} \beta_{11} & \beta_{21} & \cdots & \beta_{n1} \\ \beta_{12} & \beta_{22} & \cdots & \beta_{n2} \\ \vdots & \vdots & & \vdots \\ \beta_{1n} & \beta_{2n} & \cdots & \beta_{nn} \end{bmatrix}$$

We should point out that the cofactors of the kth row of P are the elements of the kth column of the adj P. Thus β_{ij} is the element in the jth row and ith column of adj P. Property A.4.1 lists the properties of an adjoint.

PROPERTY A.4.1   *Let P, Q, and R be nth-order square matrices*: (1) *if $|P| \neq 0$, then $|\text{adj } P| = |P|^{n-1}$, if $|P| = 0$, then P (adj P) = (adj P)P = $\varnothing$; (2) if rank $(P) = n - 1$, then rank (adj P) = 1, if rank $(P) < n - 1$, then adj $P = 0$; (3) if $R = P \cdot Q$, then adj R = (adj P) $\cdot$ (adj Q); (4) P(adj P) = (adj P)P = $|P|I_{n \times n}$.*

Associated with a nonsingular square matrix P is another matrix which, when multiplied by P, gives the identity matrix. More formally, we have Definition A.4.1.

DEFINITION A.4.1   *If P and Q are nth-order square matrices such that $PQ = QP = I$, then Q is the* inverse *of P and is denoted by P^{-1}. Similarly we can say that P is the inverse of Q.*

We can show that matrix P has an inverse P^{-1} if and only if it is nonsingular (i.e., $|P| \neq 0$). Since $|P| \neq 0$ and P(adj P) = (adj P)$P = |P|I$, we can easily see that $P^{-1} = \text{adj } P/|P|$.

PROPERTY A.4.2   *Let P, Q, and R be n-order nonsingular matrices. Then (1) P^{-1} is unique; (2) if $PQ = PR$, then $Q = R$ (note all that is required here is $|P| \neq 0$); (3) if $R = PQ$, then $R^{-1} = Q^{-1}P^{-1}$; (4) if $R = P^T$, then $R^{-1} = Q^T$ where $Q = P^{-1}$; (5) if $R = P^{-1}$, then $R^{-1} = P$; (6) if $p_{ij}^{(n)}$ is the (i, j)th element of P^n, the nth power of P, and for each i, j, $\lim_{n \to \infty} p_{ij}^{(n)} = 0$, then*

$I - P$ *is nonsingular and*

$$(I - P)^{-1} = \sum_{n=0}^{\infty} P^n$$

A.5 PARTITIONING

Suppose we can partition the matrices P and Q in the following manner:

$$P_{m \times n} = \begin{bmatrix} P_{11} & P_{12} \\ P_{21} & P_{22} \end{bmatrix} \qquad Q_{m \times n} = \begin{bmatrix} Q_{11} & Q_{12} \\ Q_{21} & Q_{22} \end{bmatrix}$$

where the submatrices P_{ij} and Q_{ij} have m_i rows and n_j columns ($i, j = 1, 2$). We should point out that $m_1 + m_2 = m$ and $n_1 + n_2 = n$. With these assumptions, matrix addition and subtraction may be defined as

$$P \pm Q = \begin{bmatrix} P_{11} \pm Q_{11} & P_{12} \pm Q_{12} \\ P_{21} \pm Q_{21} & P_{22} \pm Q_{22} \end{bmatrix}$$

The notion of partitioning of a matrix may be extended from these examples to any general partitioning as long as the operations are well defined. Partitioning of a matrix plays a stronger role when determining the inverse of a nonsingular matrix $P_{n \times n}$. Suppose

$$P_{n \times n} = \begin{bmatrix} P_{11} & P_{12} \\ P_{21} & P_{22} \end{bmatrix} \quad \text{and} \quad |P| \neq 0$$

where P_{11} is an $s \times s$ matrix, P_{12} is an $s \times m$ matrix, P_{21} is an $m \times s$ matrix, and P_{22} is an $m \times m$ matrix, such that $n = m + s$. Since $|P| \neq 0$, we know P^{-1} exists and suppose it is given as

$$P^{-1} = \begin{bmatrix} Q & R \\ S & T \end{bmatrix}$$

where $Q_{s \times s}$, $R_{s \times m}$, $S_{m \times s}$, and $T_{m \times m}$. Then we have

$$Q = \left(P_{11} - P_{12} P_{22}^{-1} P_{21} \right)^{-1}$$

$$R = -Q P_{12} P_{22}^{-1}$$

$$S = -P_{22}^{-1} P_{21} Q$$

$$T = P_{22}^{-1} - P_{22}^{-1} P_{21} R$$

In particular, when

$$P = \begin{bmatrix} I & P_{12} \\ 0 & P_{22} \end{bmatrix} \quad \text{and} \quad |P| \neq 0$$

we have

$$P^{-1} = \begin{bmatrix} I & -P_{12}P_{22}^{-1} \\ 0 & P_{22}^{-1} \end{bmatrix}$$

A.6   SYSTEM OF LINEAR EQUATIONS

Consider the system of equations

$$p_{11}x_1 + p_{12}x_2 + \cdots + p_{1n}x_n = \pi_1$$

$$p_{21}x_1 + p_{22}x_2 + \cdots + p_{2n}x_n = \pi_2$$

$$\vdots$$

$$p_{m1}x_1 + p_{m2}x_2 + \cdots + p_{mn}x_n = \pi_m$$

This system of m equations in n unknowns is said to be *consistent* if there exists an $(x_1, x_2, \ldots, x_n)$ that satisfies all the equations; otherwise, the system is said to be *inconsistent*. If $\pi_1 = \pi_2 = \cdots = \pi_m = 0$, the system is called *homogeneous*. We may transform this system into matrix notation:

$$\begin{bmatrix} p_{11} & p_{12} & \cdots & p_{1n} \\ p_{21} & p_{22} & \cdots & p_{2n} \\ \vdots & \vdots & & \vdots \\ p_{m1} & p_{m2} & \cdots & p_{mn} \end{bmatrix} \begin{bmatrix} x_1 \\ x_2 \\ \vdots \\ x_n \end{bmatrix} = \begin{bmatrix} \pi_1 \\ \pi_2 \\ \vdots \\ \pi_m \end{bmatrix}$$

or

$$PX = \Pi$$

Again, the theory of systems of linear equations is extensive. We will give only a few of the more fundamental results. Let us form the matrix $P\Pi$ by augmenting P with the column vector Π. The matrix $P\Pi$ is called the *augmented matrix.* For this system of equations we have the following properties:

PROPERTY A.6.1 *Suppose we had the system of equations just given; then* (1) *the system of equations are consistent iff* $r(P) = r(P\Pi)$; (2) *if* X^0 *is a solution to the system of equations, it need not be unique;* (3) *if* $m = n$, $\Pi \neq \varnothing$ *and* $|P| \neq 0$, *then the system has a unique solution* (*if* $\Pi \neq \varnothing$, *the system is called nonhomogeneous*); (4) *if* $\Pi = \varnothing$, *then* $X = \varnothing$ *is called the trivial solution to the system;* (5) *if the system is homogeneous* ($\Pi = \varnothing$) *and* $m = n$, *the system has a nontrivial solution iff* $r(P) < n$ (*or* $|P| = 0$).

A.7 VECTOR SPACE

We will use the notation $X = (x_1, x_2, \ldots, x_n)$ to represent a point in euclidean n space E^n, where for each i, x_i is a real number. A vector Y in E^n is said to be a *linear combination* of $X^1, X^2, \ldots, X^m$, where X^i is in E^n if there exist scalars $c_1, c_2, \ldots, c_m$ such that $Y = c_1 X^1 + c_2 X^2 + \cdots + c_m X^m$. A set of vectors $X^1, X^2, \ldots, X^m$ is said to be *dependent* if there exist scalars $c_1, c_2, \ldots, c_m$ not all zero such that $c_1 X^1 + c_2 X^2 + \cdots + c_m X^m = 0$. Otherwise, $X^1, X^2, \ldots, X^m$ are said to be *independent*. The basic properties of independent and dependent vectors are given next.

PROPERTY A.7.1 *Let* $X^1, X^2, \ldots, X^m$ *be a set of vectors; then* (1) *if* X^i *and* X^j *are dependent for some* i, j ($= 1, 2, \ldots, m$), *then* $X^1, X^2, \ldots, X^m$ *are dependent;* (2) *if* $X^1, X^2, \ldots, X^m$ *are independent, then any subset of these vectors has independent vectors;* (3) *if* $X^i = \varnothing$, *the zero vector for some* $i = 1, 2, \ldots, m$, *then the vectors in the set are dependent;* (4) *the vectors in the preceding set are independent iff vectors in each of their subsets are independent;* (5) *if* $m \leq n$, *then if we consider the rows of matrix* P *as vectors, they are dependent iff* $r(P) < m$.

Let $S = \{ X^1, X^2, \ldots, X^m \}$ for each i, $X^i \in E^n$ be a set of vectors. We say W, $W \subseteq E^n$, is *spanned* by S iff every vector in W can be expressed as a linear combination of vectors from S. A *basis* for W is a set of independent vectors that spans W. In particular, if $W = E^n$ and

$$e^1 = (1, 0, 0, \ldots, 0)$$

$$e^2 = (0, 1, 0, \ldots, 0)$$

$$\vdots$$

$$e^n = (0, 0, 0, \ldots, 1)$$

then $S = \{e^1, e^2, \ldots, e^n\}$ is the standard basis for E^n. Some fundamental properties of E^n related to vectors that span a subset of E^n and the basis for a subset of E^n are given next.

PROPERTY A.7.2 *Let $S = \{X^1, X^2, \ldots, X^m\}$ where for each i $X^i \in E^n$; then (1) if S is a basis for E^n, then $m = n$; (2) any two bases of E^n have the same number of elements in the bases, and that number is called the* dimension *of E^n, denoted $\dim(E^n)$; (3) if $m > n$, then the vectors in the set S are dependent; (4) if S is a set of independent vectors and $X^{m+1} \in E^n$ is not in the space spanned by S, then the set $S^1 = \{X^1, X^2, \ldots, X^m, X^{m+1}\}$ is independent; (5) if $m < n$ and S forms an independent set of vectors that form a basis for $W \subseteq E^n$, then S is a part of a basis for E^n; (6) if $W \subseteq E^n$, then $\dim(W) \leq \dim(E^n)$; (7) if $P_{n \times n}$ and the rows of P are independent, then P^{-1} exists; (8) let $S^1 = \{X^1, X^2, \ldots, X^n\}$ and $S^2 = \{Y^1, Y^2, \ldots, Y^n\}$ be two bases for E^n such that $X = \sum_{i=1}^n k_i X^i$ and $Y = \sum_{i=1}^n l_i Y^i$ where k_i and l_i are scalars for each i—then there exists a nonsingular matrix P such that $X = P \cdot Y$.*

By a polynomial in λ of degree n over E^1 we mean an expression of the form

$$c(\lambda) = a_n \lambda^n + a_{n-1} \lambda^{n-1} + \cdots + a_1 \lambda + a_1 \lambda^0$$

where $a_i \in E^1$ for each i.

A.8   EIGENVALUES AND EIGENVECTORS

Suppose the matrix P is a square nth-order matrix, and there exists a column vector $X, X \neq 0$ such that

$$PX = \lambda X$$

where $\lambda \in E^1$. Rearranging, we get

$$(\lambda I - P)X = 0$$

We note that $c(\lambda) = |\lambda I - P|$ is a polynomial in λ of degree n over E^1.

DEFINITION A.8.1 *The polynomial $c(\lambda) = |\lambda I - P|$ is called the* characteristic polynomial *of the matrix P. The equation $|\lambda I - P| = 0$ is called the* characteristic equation *of P, and the roots $(\lambda_1, \lambda_2, \ldots, \lambda_n)$ of $c(\lambda)$ are called the* eigenvalues (latent roots *or* characteristic roots) *of P. A column vector $X_i, X_i \neq \varnothing$ which satisfies $PX_i = \lambda_i X_i$ is called the* right eigenvector *of P,* *belonging to the eigenvalue* λ_i, *and a row vector* Y_i, $Y_i \neq \emptyset$ *which satisfies* $Y_i P = \lambda_i Y_i$ *is called the* left eigenvector *of* P, *belonging to the eigenvalue* λ_i (*eigenvectors are also called* latent vectors *or* characteristic vectors).

Various types of matrices possess several important properties as related to the concept of eigenvalues and eigenvectors.

PROPERTY A.8.1.   *Let* $c(\lambda)$ *be given as in Definition A.8.1; then* (1) *the following are equivalent*: (*a*) $\lambda = (\lambda_1, \lambda_2, \ldots, \lambda_n)$ *are eigenvalues of* P, (*b*) $Q = \lambda I - P$ *is singular*, (*c*) $|\lambda I - P| = 0$; (2) *the eigenvalues of* P *need not be all distinct or real numbers*; (3) *let* $P_{n \times n}$, *and let* $c(\lambda)$ *be its characteristic polynomial; then* $c(P) = 0$ (*this is the* Cayley-Hamilton Theorem); (4) *if* $\lambda_1, \lambda_2, \ldots, \lambda_k$ *are distinct eigenvalues of* P *and if* $X^1, X^2, \ldots, X^k$ *are the eigenvectors associated with the eigenvalues, then* $X^1, X^2, \ldots, X^k$ *are independent*; (5) *the eigenvalues of* P *and* P^T *are the same*; (6) *if* $\lambda_1, \lambda_2, \ldots, \lambda_n$ *are the eigenvalues of* $P_{n \times n}$ *and* k *is a scalar, then* $(k\lambda_1, k\lambda_2, \ldots, k\lambda_n)$ *and* $(\lambda_1 - k, \lambda_2 - k, \ldots, \lambda_n - k)$ *are the eigenvalues of* kP *and* $P - kI$; (7) *the eigenvector* X_i *belonging to the eigenvalue* λ_i *is unique apart from a nonzero scalar multiplier*; (8) *if* λ *is a nonzero eigenvalue of the nonsingular matrix* P, *then* $|P| \cdot \lambda^{-1}$ *is the eigenvalue of* adj P; (9) *if* P *is symmetric, then* (*a*) *the eigenvalues of* P *are real*, (*b*) *if* $\lambda_i \neq \lambda_j$ *and* X *and* Y *are the eigenvectors associated with* λ_i *and* λ_j, *then* $\sum_{i=1}^{n} x_i \cdot y_i = 0$; (10) *if* $Tr(P) = \sum_{i=1}^{n} p_{ii}$ *and* $\lambda_1, \lambda_2, \ldots, \lambda_n$ *are the eigenvalues of* $P_{n \times n}$, *then* $Tr(P) = \sum_{i=1}^{n} \lambda_i$, *where* $Tr(P)$ *is called the* trace *of the matrix* $P_{n \times n}$.

Suppose $P_{n \times n}$, and if we can permute the rows and columns of P such that we obtain a matrix of the form

$$\begin{bmatrix} P_1 & 0 \\ R & Q \end{bmatrix}$$

where P_1 is an $r \times r$ matrix, then P is called *reducible*; otherwise, P is called *irreducible*.

Theorem A.8.1   (Perron-Frobenius Theorem): *If* P *is positive* ($\geq \emptyset$) *and irreducible, then* P *has a real positive eigenvalue* λ_1, *with the properties* (1) *there exists a vector* $X > \emptyset$ *such that* $PX = \lambda_1 X$; (2) *if* λ *is any other eigenvalue of* P, *the* $|\lambda| \leq \lambda_1$; (3) λ_1 *increases when any element of* P *increases*; (4) λ_1 *is a simple root of the characteristic equation* $|\lambda I - P| = 0$; (5) $\lambda_1 \leq max_i(\sum_k p_{ik})$, $\lambda_1 \leq max_k(\sum_i p_{ik})$; (6) *if* $P_{n \times n}$ *has exactly* t *eigenvalues with modulus equal to* λ_1, *then these numbers are all different and are the roots of* $\lambda^t - \lambda_1^t = 0$, *and, in general, the set of* n *eigenvalues, when plotted as points in the complex* λ *plane, is invariant under a rotation of the plane* *through the angle $2\pi/t$ but not through smaller angles. When $t > 1$, then P can be reduced to the following cyclic form by a permutation applied to both rows and columns:*

$$P = \begin{bmatrix} 0 & P_{12} & 0 & \cdots & 0 \\ 0 & 0 & P_{23} & \cdots & 0 \\ \vdots & \vdots & \vdots & & \vdots \\ 0 & 0 & 0 & \cdots & P_{t-1,t} \\ P_{t1} & 0 & 0 & \cdots & 0 \end{bmatrix}$$

where the submatrices on the main diagonal are square.

A.9 SIMILAR MATRICES

DEFINITION A.9.1 *Two matrices $P_{n \times n}$ and $Q_{n \times n}$ are called* similar *if there exists an $n \times n$ matrix R with $|R| \neq 0$ for which*

$$R^{-1}PR = Q$$

This also implies

$$RQR^{-1} = P$$

If Q is a diagonal matrix ($q_{ij} = 0$, $i \neq j$) and the first relation holds, then P is said to be diagonalized.

Suppose $P_{n \times n}$ has n distinct eigenvalues. Let R be given by

$$R = (X_1, X_2, \ldots, X_n)$$

where X_i is the right eigenvector belonging to the eigenvalue λ_i ($i = 1, 2, \ldots, n$). Then

$$PR = RQ$$

$$Q = \begin{bmatrix} \lambda_1 & 0 & \cdots & 0 \\ 0 & \lambda_2 & \cdots & 0 \\ \vdots & \vdots & & \vdots \\ 0 & 0 & \cdots & \lambda_n \end{bmatrix}$$

Similarly it can be shown that R^{-1} is made up of left eigenvectors belonging to the eigenvalues $\lambda_1, \lambda_2, \ldots, \lambda_n$ for rows (after choosing suitable multiplicative constants). When Q is diagonal, as in the preceding matrix, we also have

$$P^m = (RQR^{-1})(RQR^{-1}) \cdots (RQR^{-1})$$

$$= RQ^m R^{-1}$$

$$= R \begin{bmatrix} \lambda_1^m & 0 & \cdots & 0 \\ 0 & \lambda_2^m & \cdots & 0 \\ \vdots & \vdots & & \vdots \\ 0 & 0 & \cdots & \lambda_n^m \end{bmatrix} R^{-1}$$

Suppose the eigenvalues of $P_{n \times n}$ are not distinct. Let $\lambda_1, \lambda_2, \ldots, \lambda_k$ be the distinct eigenvalues of $P_{n \times n}$, λ_i with multiplicity n_i $(i = 1, 2, \ldots, k)$, such that $n_1 + n_2 + \cdots + n_k = n$. Then generalizing the diagonalizing property given before, it can be shown that there exists a matrix $R_{n \times n}$ with $|R| \neq 0$,

$$RPR^{-1} = \begin{bmatrix} J_{n_1}(\lambda_1) & 0 & \cdots & 0 \\ 0 & J_{n_2}(\lambda_2) & \cdots & 0 \\ \vdots & \vdots & & \vdots \\ 0 & 0 & \cdots & J_{n_k}(\lambda_k) \end{bmatrix}$$

where $J_{n_i}(\lambda_i)$ is an $n_i \times n_i$ square matrix of the form

$$J_{n_i}(\lambda_i) = \begin{bmatrix} \lambda_i & 1 & 0 & \cdots & 0 \\ 0 & \lambda_i & 1 & \cdots & 0 \\ \vdots & \vdots & \vdots & & \vdots \\ 0 & 0 & 0 & \cdots & \lambda_i \end{bmatrix}$$

$$= \lambda_i I + M$$

where M is an $n_i \times n_i$ matrix with ones as the first superdiagonal elements and zeros elsewhere. Clearly $M^{n_i} = 0$ and for $m \geq n_i$,

$$J_{n_i}^m(\lambda_i) = \lambda_i^m I = \binom{m}{1} \lambda_i^{m-1} M + \cdots + \binom{m}{n_i - 1} \lambda_i^{m-n_i+1} M^{n_i-1}$$

The generalized diagonal form of P given here is known as the *Jordan canonical form* of P.

Two other properties of irreducible positive matrices may prove to be useful. These are given next, preceded by the definition of primitive matrices.

DEFINITION A.9.2 *If in a positive irreducible matrix there is only one eigenvalue λ_1 with the largest modulus, then such a matrix is called* primitive.

PROPERTY A.9.1 *If the irreducible positive matrix P has exactly t eigenvalues of maximum modulus λ_1, then there exists a $R_{n \times n}$ matrix with $|R| \neq 0$, such that*

$$RPR^{-1} = \begin{bmatrix} \Lambda & 0 \\ 0 & J \end{bmatrix}$$

where

$$\Lambda = \begin{bmatrix} \lambda_1 \varepsilon_0 & 0 & \cdots & 0 \\ 0 & \lambda_1 \varepsilon_1 & \cdots & 0 \\ \vdots & \vdots & & \vdots \\ 0 & 0 & \cdots & \lambda_1 \varepsilon_{t-1} \end{bmatrix}$$

where $\varepsilon_k = e^{2\pi i k / t}$ $(k = 0, 1, 2, \ldots, t - 1)$, and J is in the Jordan canonical form with all diagonal elements less than λ_1 in modulus. In particular, if $t = 1$ (P primitive),

$$RPR^{-1} = \begin{bmatrix} \lambda_1 & 0 \\ 0 & J \end{bmatrix}$$

PROPERTY A.9.2 *If the irreducible positive matrix P has exactly t eigenvalues of modulus λ_1, and if X_1 and Y_1 are the normalized right and left eigenvectors belonging to λ_1, such that $X_1 Y_1 = 1$, then*

$$\lim_{n \to \infty} \frac{1}{t} \left[\left(\frac{1}{\lambda_1 n} \right) P^n + \left(\frac{1}{\lambda_1 n + 1} \right) P^{n+1} + \cdots + \left(\frac{1}{\lambda_1 n + t - 1} \right) P^{n+t-1} \right]$$

$$= X_1 Y_1$$

and the convergence to the limit is geometric and uniform for all elements. In particular, when $t = 1$ (P primitive),

$$\lim_{n \to \infty} \left(\frac{1}{\lambda_1^n} \right) P^n = X_1 Y_1$$

Denoting $P^n = \| p_{ij}^{(n)} \|$, *this shows*

$$\lim_{n \to \infty} \left(\frac{p_{ij}^{(n)}}{\lambda_1^n} \right) = x_{i1} y_{1j} > 0 \qquad i, j = 1, 2, \ldots, n$$

where

$$X_1 = \begin{bmatrix} x_{11} \\ x_{21} \\ \vdots \\ x_{n1} \end{bmatrix}$$

and

$$Y_1 = (y_{11}, y_{12}, \ldots, y_{1n}).$$

FOR FURTHER READING

Aitken, A. C. (1948). *Determinants and Matrices.* Edinburgh: Oliver and Boyd.

Faddeev, D. K., and Faddeeva, V. N. (1963). *Computational Methods of Linear Algebra.* San Francisco: W. H. Freeman & Co.

Gantmacher, F. R. (1959). *Applications of the Theory of Matrices.* New York: Interscience (English trans.).

Hadley, G. (1964). *Linear Algebra.* Reading, Mass.: Addison-Wesley.

Halmos, P. R. (1958). *Finite-Dimensional Vector Spaces.* New York: Van Nostrand.

Herstein, I. N. (1969). *Topics in Algebra.* Waltham, Mass.: Blaisdell.

Herstein, I. N., and Debreu, G. (1953). "Nonnegative Square Matrices." *Econometrica* **21**, 597–607.

Hoffman, K., and Kunze, R. (1965). *Linear Algebra.* Englewood Cliffs, N.J.: Prentice-Hall.

Schreier, O., and Sperner, E. (1955). *Modern Algebra and Matrix Theory.* New York: Chelsea.

Appendix B

MISCELLANEOUS TOPICS

In this appendix we shall present useful results from several areas that have been cited in the book. For further reading on these topics, references are given at the end of the appendix.

B.1 RIEMANN-STIELTJES INTEGRAL

Let $f(x)$ and $\phi(x)$ be real-valued functions on $[a, b]$, and suppose that $f(x)$ is bounded on $[a, b]$ and $\phi(x)$ is monotonically increasing there. By a partition P of $[a, b]$, we mean a finite sequence of points $x_0, x_1, \ldots, x_n$ such that

$$a = x_0 \leq x_1 \leq x_2 \leq \cdots \leq x_{n-1} \leq x_n = b$$

For any partition P of the closed interval $[a, b]$, we define

$$N_i = \text{least upper bound } f(x) \qquad \text{where } x \in [x_{i-1}, x_i]$$

$$n_i = \text{greatest lower bound } f(x) \qquad \text{where } x \in [x_{i-1}, x_i]$$

$$\Delta\phi_i = \phi(x_i) - \phi(x_{i-1})$$

$$U(P, f, \phi) = \sum_{i=1}^{n} N_i \Delta\phi_i$$

$$L(P, f, \phi) = \sum_{i=1}^{n} n_i \Delta\phi_i$$

Then

$$\overline{\int_a^b} f(x) \, d\phi(x) = \text{greatest lower bound } U(P, f, \phi)$$

$$\underline{\int_a^b} f(x) \, d\phi(x) = \text{least upper bound } L(P, f, \phi)$$

where greatest lower bound and least upper bound are taken over all partitions of $[a, b]$.

We say $f(x)$ is Riemann-Stieltjes integrable with respect to $\phi(x)$ over $[a, b]$ if and only if

$$\underline{\int_a^b} f(x)\, d\phi(x) = \overline{\int_a^b} f(x)\, d\phi(x)$$

When $f(x)$ is Riemann-Stieltjes integrable with respect to $\phi(x)$ over $[a, b]$, we write its integral as

$$\int_a^b f(x)\, d\phi(x)$$

It should be pointed out that one may define the Riemann-Stieltjes integral with respect to a function $\phi(x)$, where $\phi(x)$ is of bounded variation on $[a, b]$. A function $\phi(x)$ is of bounded variation on $[a, b]$ iff

$$V(\phi; a, b) = \text{least upper bound} \sum_{i=1}^{n} |\Delta\phi_i| < +\infty$$

where the least upper bound is taken over all partitions of $[a, b]$. (Rudin, 1964.)

B.2   LAPLACE TRANSFORMS

The proofs of the properties have been omitted, and all operations are assumed to be well defined.

DEFINITION B.2.1   *Let $f(t)$ be a real-valued function in $[0, \infty)$. We define the Laplace transform of $f(t)$ as*

$$L\{f(t)\} = \phi(s) = \int_0^\infty e^{-st} f(t)\, dt \qquad Re(s) > 0$$

If $f(t)$ is piecewise continuous on every interval $[0, N]$ and of exponential order α [i.e., there exist constants M_1, M_2, and α such that for all $t > M_2$ we have $|f(t)| < M_1 e^{\alpha t}$], then it can be shown that $L\{f(t)\} = \phi(s)$ exists. In Section B.1 we defined what is meant by the Riemann-Stieltjes integral; in turn one may also define the Laplace-Stieltjes transform of $F(t)$.

DEFINITION B.2.2   *Let $F(t)$ be a real-valued function; then we define the Laplace-Stieltjes transform of $F(t)$ as*

$$\int_0^\infty e^{-st}\, dF(t) \qquad Re(s) > 0$$

We note that if $F(t)$ is absolutely continuous and its Laplace-Stieltjes transform exists, then $F(t)$ is a differentiable monotonically increasing function and

$$dF(t) = F'(t)\, dt$$

The Laplace-Stieltjes transform of $F(t)$ then equals the Laplace transform for this case.

The following properties apply only to Laplace transforms, although analogous properties hold for Laplace-Stieltjes transforms. Let $L\{f(t)\} = \phi(s)$, and assume that all operations are well defined.

PROPERTY B.2.1 (1) *If* $L\{f_i(t)\} = \phi_i(s)$ *and* $f(t) = \sum_{i=1}^{\infty} \xi_i f_i(t)$, *where* ξ_i *is a constant* $(i = 1, 2, \ldots)$, *then* $\phi(s) = \sum_{i=1}^{\infty} \xi_i \phi_i(s)$. (2) *If* $g(t) = e^{\xi t} f(t)$, *then* $L\{g(t)\} = \phi(s - \xi)$. (3) *If*

$$g(t) = \begin{cases} f(t - \xi) & \text{for } t > \xi \\ 0 & \text{for } t \leq \xi \end{cases}$$

then $L\{g(t)\} = e^{-\xi s} \phi(s)$. (4) *If* $\xi \neq 0$ *and* $g(t) = f(\xi t)$, *then*

$$L\{f(\xi t)\} = \frac{1}{\xi} \phi\left(\frac{s}{\xi}\right)$$

(5) *If* $g(t) = d^n[f(t)]/dt^n = f^{(n)}(t)$, *then*

$$L\{g(t)\} = s^n \phi(s) - s^{n-1} f(0) - s^{n-2} f^{(1)}(0) - \cdots$$

$$- sf^{(n-2)}(0) - f^{(n-1)}(0)$$

Here the continuity at 0 of $f^{(n)}(t)$ *is assumed for each n.* (6) *If* $g(t) = t^n f(t)$, *then* $L\{g(t)\} = (-1)^n \phi^{(n)}(s)$. (7) *When the indicated limit exists, we have*

$$\lim_{s \to \infty} \phi(s) = 0$$

$$\lim_{t \to 0} f(t) = \lim_{s \to \infty} s\phi(s)$$

$$\lim_{t \to \infty} f(t) = \lim_{s \to 0} s\phi(s)$$

(8) *Let* $f(t)$ *be the probability density function of a continuous random variable*

T, then $E(T) = -\phi^{(1)}(0)$. (9) *Let*

$$f(t) = f_1(t) * f_2(t) = \int_{\tau=0}^{t} f_1(\tau) f_2(t - \tau)\, d\tau$$

and $L\{f_i(t)\} = \phi_i(s)$ $(i = 1, 2)$; *then* $\phi(s) = \phi_1(s) \cdot (\phi_2(s)$. (10) *If* $f(t)$ *is such that*

$$\int_0^x f(t)\, dt = 0$$

for all $x > 0$, *then* $f(t)$ *is called a* null function *and* $\phi(s) = 0$.

Perhaps a word about the uniqueness of the Laplace transform of $f(t)$ is in order. Suppose $f_2(t)$ is a null function and $f(t) = f_1(t) + f_2(t)$; then by properties (1) and (10) we have

$$\phi(s) = \phi_1(s) + \phi_2(s) = \phi_1(s) = L\{f_1(t)\}$$

One can see that several different functions may have the same Laplace transforms, but if we do not consider null functions, the Laplace transform of a function is unique.

Finally, we give two theorems that are useful in limiting operations dealing with transforms.

Theorem B.2.1 (An Abelian Theorem): *If for some nonnegative number* ξ (≥ 0) *we have*

$$\lim_{t \to \infty} \frac{F(t)}{t^\xi} = \frac{C}{\Gamma(\xi + 1)}$$

and

$$\psi(s) = \int_0^\infty e^{-st}\, dF(t)$$

exists for $\mathrm{Re}(s) > 0$, *then*

$$\lim_{s \to 0^+} s^\xi \psi(s) = C$$

Theorem B.2.2 (A Tauberian Theorem): *If* $F(t)$ *is nondecreasing and*

$$\psi(s) = \int_0^\infty e^{-st}\, dF(t)$$

exists for $\text{Re}(s) > 0$, *and if for some constant* ξ (≥ 0)

$$\lim_{s \to 0} s^{\xi} \psi(s) = C$$

then

$$\lim_{t \to \infty} \frac{F(t)}{t^{\xi}} = \frac{C}{\Gamma(\xi + 1)}$$

(Widder, 1946.)

B.3   GENERATING FUNCTIONS

Analogous to the transform of a function is that of a transform of a sequence of real numbers $\{a_n\}_{n=0}^{\infty}$. This is commonly called a Z-transform or generating function of $\{a_n\}_{n=0}^{\infty}$.

DEFINITION B.3.1.   *Let* $\{a_n\}_{n=0}^{\infty}$ *be a sequence of real numbers. If*

$$A(z) = \sum_{n=0}^{\infty} a_n z^n$$

exists, then $A(z)$ *is called the* generating function *of the sequence* $\{a_n\}_{n=0}^{\infty}$.

Since the series $A(z)$ converges to a unique number, the generating function of a sequence of real numbers is unique. The similarity between generating functions and Laplace transforms is obvious and is further exemplified by the properties of generating functions. We again assume all operations are well defined and let the generating functions of $\{a_n\}_{n=0}^{\infty}$ and $\{b_n\}_{n=0}^{\infty}$ be $A(z)$ and $B(z)$, respectively.

PROPERTY B.3.1   (1) *If* $c_n = \xi_1 a_n + \xi_2 b_n$ *for each n, where* ξ_1 *and* ξ_2 *are constants, then* $C(z) = \sum_{n=0}^{\infty} c_n z^n = \xi_1 A(z) + \xi_2 B(z)$. (2) *If* $b_n = a_{n+k}$, *then* $B(z) = z^{-k} A(z) - \sum_{n=0}^{k-1} b_r z^{r-k}$. (3) *If* $a_n = n^k$ *and* $b_n = n^{k-1}$ *for* $k \geq 1$, *then* $A(z) = z B^{(1)}(z) = z \, dB(z)/dz$. (4) *If* $c_n = \sum_{r=0}^{n} a_r b_{n-r}$, *then* $C(z) = \sum_{n=0}^{\infty} c_n z^n = A(z) \cdot B(z)$. (5) *Let X be a nonnegative, discrete random variable, and let* $P(X = n) = p_n$ *and* $P(X > n) = q_n$; *if* $P(z) = \sum_{n=0}^{\infty} p_n z^n$ *and* $Q(z) = \sum_{n=0}^{\infty} q_n z^n$, *then* (a) $Q(z) = [1 - P(z)]/(1 - z)$, (b) $E(X) = P^{(1)}(1) = Q(1)$, (c) $V(X) = P^{(2)}(1) + P^{(1)}(1) - [P^{(1)}(1)]^2 = 2Q^{(1)}(1) + Q(1) - [Q(1)]^2$; *note that when* $p_n = P(X = n)$, *we may write* $P(z) = E[z^X]$.

Finally, we give three theorems that are useful in analyzing stochastic systems.

Theorem B.3.1   (Abel's Theorem): *If* $\lim_{n \to \infty} a_n = a$, *then*

$$\lim_{z \to 1^-} \left\{ (1 - z) \sum_{n=0}^{\infty} a_n z^n \right\} = a$$

Theorem B.3.2   (Tauber's Theorem): *If* $\lim_{z \to 1^-} (1 - z) \sum_{n=0}^{\infty} a_n z^n = a$ *and* $\lim_{n \to \infty} n(a_n - a_{n-1}) = 0$, *then*

$$\lim_{n \to \infty} a_n = a$$

Theorem B.3.3   *Let* $\{a_n\}_{n=0}^{\infty}$ *be a nonnegative sequence of real numbers whose generating function is*

$$A(z) = \sum_{n=0}^{\infty} a_n z_n \qquad |z| < 1$$

The following hold (for a and c real constants):

1. $$\sum_{n=0}^{\infty} a_n = a \qquad iff \ \lim_{z \to 1^-} A(z) = a$$

2. $$\lim_{m \to \infty} \left\{ \frac{1}{m} \sum_{n=0}^{m} a_n \right\} = c \qquad iff \ \lim_{z \to 1^-} \left\{ (1 - z) A(z) \right\} = c$$

(Beightler et al., 1961; Feller, 1968; Hardy 1949; Whittaker and Watson, 1927.)

B.4   A LEMMA FROM NUMBER THEORY

LEMMA   *Let* $n_1, n_2, \ldots, n_k$ *be positive integers with the greatest common divisor* (g.c.d.) *d. Then there exist a positive integer M and nonnegative integers* c_j *(j = 1, 2, \ldots, k) such that for all* $m \geq M$

$$md = \sum_{j=1}^{k} c_j n_j \qquad (B.4.1)$$

As we have not been able to locate the proof of this lemma anywhere, its proof is as follows.

Proof   Let A be the set $\{a_0, a_1, \ldots\}$, where $a_i = b_1^{(i)} n_1 + b_2^{(i)} n_2 + \cdots + b_k^{(i)} n_k$, and where in turn $b_j^{(i)}$ are integers in $(-\infty, \infty)$, $(j = 1, 2, \ldots, k)$. Let $d' = \inf\{a_i | a_i > 0$ and $\in A\}$. Since d is the g.c.d. of

$n_1, n_2, \ldots, n_k$, d divides a_i, $a_i \in A$. Hence d divides d', and therefore

$$d' > d. \tag{B.4.2}$$

Consider any element $a_i \in A$. We may write

$$a_i = \delta d' + l \tag{B.4.3}$$

where δ is an integer in $(-\infty, \infty)$ and $0 \le l < d'$ and l is an integer. Let

$$a_i = \sum_{j=1}^{k} b_j^{(i)} n_j$$

and

$$d' = \sum_{j=1}^{k} b_j' n_j$$

where $b_j^{(i)}$ and b_j' are integers in $(-\infty, \infty)$. From (B.4.3) we have

$$\sum_{j=1}^{k} b_j^{(i)} n_j - \delta \sum_{j=1}^{k} b_j' n_j = l$$

giving

$$\sum_{j=1}^{k} \left(b_j^{(i)} - \delta b_j' \right) n_j = l$$

Now $b_j^{(i)} - \delta b_j'$ are also integers in $(-\infty, \infty)$. Writing $b_j^{(i)} - \delta b_j' = p_j$, we have

$$l = \sum_{j=1}^{k} p_j n_j$$

and therefore $l \in A$. But d' is the smallest positive integer in A and $0 \le l < d'$. This is not possible unless $l = 0$. Thus

$$a_i = \delta d' \tag{B.4.4}$$

It should be noted that $n_1, n_2, \ldots, n_k \in A_i$, and hence d' divides $n_1, n_2, \ldots, n_k$. But d is the g.c.d., and hence

$$d' \le d \tag{B.4.5}$$

Combining (B.4.2) and (B.4.5), we get

$$d' = d \tag{B.4.6}$$

This also shows that $d \in A$, and it can be written as

$$d = \sum_{i=1}^{r} b_i^+ n_i - \sum_{i=r+1}^{k} b_i^- n_i$$

$$= N_1 - N_2 \tag{B.4.7}$$

where $b_i^+, b_i^- \geq 0$. If $N_2 = 0$,

$$md = N_1 d = \sum_{i=1}^{k} c_i n_i \tag{B.4.8}$$

where $c_i = b_i^+$ $(i = 1, \ldots, r)$, $c_i = 0$ $(i = r + 1, \ldots, n)$. If $N_2 > 0$, let

$$M = \frac{N_2^2}{d} \tag{B.4.9}$$

Clearly M is an integer. Consider any positive integer $m \geq M$, and write $m = M + \delta(N_2/d) + l'$, where δ and l' are nonnegative integers and $0 \leq l' < N_2/d$. We have

$$md = N_2^2 + \delta N_2 + l'd$$

$$= N_2^2 + \delta N_2 + l'(N_1 - N_2)$$

$$= N_2(N_2 + \delta - l') + N_1 l'$$

$$= \sum_{i=1}^{k} c_i n_i \tag{B.4.10}$$

where c_i are nonnegative integers. Note that $N_2 + \delta - l'$ is a positive integer since $l' < N_2$. (Karlin and Taylor, 1975, p. 77.) $\square$

B.5 FUBINI-TONELLI THEOREM AND LEBESGUE CONVERGENCE THEOREM

Interchanging the variables of integration and commuting the operations of taking a limit and integrating are common practice in the areas of probability theory and stochastic processes. The justification for doing this is given in Theorems B.5.1 and B.5.2.

Theorem B.5.1 (Fubini-Tonelli Theorem): *Let X and Y be subsets of R^+ = $R \cup \{-\infty, +\infty\}$ and $f: X \times Y \to R$; then*

$$\int_X \left[\int_Y f(x, y)\, dy \right] dx = \int_Y \left[\int_X f(x, y)\, dx \right] dy \qquad \text{(B.5.1)}$$

provided that either one of the iterated integrals in (B.5.1) *exists.*

Theorem B.5.2 (Lebesgue Convergence Theorem): *Let $\{f_n(x)\}_{n=0}^{\infty}$ be a sequence of continuous functions on $[a, b]$, and let $g(x)$ be integrable on $[a, b]$. If $|f_n(x)| \leq g(x)$ for all $x \in [a, b]$ and for all n, and $\lim_{n \to \infty} f_n(x) = f(x)$, then*

$$\lim_{n \to \infty} \int_a^b f_n(x)\, dx = \int_a^b \lim_{n \to \infty} f_n(x)\, dx = \int_a^b f(x)\, dx$$

We have not stated these theorems in their usual measure-theoretic setting; thus the hypothesis has been somewhat strengthened. The corresponding operations dealing with summations instead of integrals follow similar considerations. (Royden, 1966.)

B.6 PARTIAL FRACTION EXPANSION

A method of expansion useful in integration and inversion of transforms is given in Theorem B.6.1.

Theorem B.6.1 *Let $P(x)/Q(x)$ be a rational function of the real variable x, consisting of the ratio of polynomials $P(x)$ and $Q(x)$ which are of real coefficients and of degrees less than n, and n (≥ 1), respectively. Let $Q(x)$ have only real roots $\xi_1, \xi_2, \ldots, \xi_r$, with multiplicities $m_1, m_2, \ldots, m_r$, such that $m_1 + m_2 + \cdots m_r = n$. Then*

$$\frac{P(x)}{Q(x)} = \frac{A_{11}}{x - \xi_1} + \frac{A_{12}}{(x - \xi_1)^2} + \cdots + \frac{A_{1m_1}}{(x - \xi_1)^{m_1}}$$

$$+ \frac{A_{21}}{x - \xi_2} + \frac{A_{22}}{(x - \xi_2)^2} + \cdots + \frac{A_{2m_2}}{(x - \xi_2)^{m_2}}$$

$$\vdots$$

$$+ \frac{A_{r1}}{x - \xi_r} + \frac{A_{r2}}{(x - \xi_r)^2} + \cdots + \frac{A_{rm_r}}{(x - \xi_r)^{m_r}}$$

where

$$A_{i,m_i-j} = \frac{1}{j!} \lim_{x \to \xi_i} \frac{d^j}{dx^j}\left[(x - \xi_i)^{m_i}\frac{P(x)}{Q(x)}\right] \qquad j = 0, 1, 2, \ldots, m_i - 1$$

(Dettman, 1965.)

REFERENCES

Beightler, C. S., Mitten, L. G., and Nemhauser, G. L. (1961). "A Short Table of Z-Transforms and Generating Functions." *ORSA* **9**, 574–578.

Dettman, J. W. (1965). *Applied Complex Variables*. New York: Macmillan.

Feller, W. (1968). *An Introduction to Probability Theory and Its Applications*. Vol. 1. 3rd ed. New York: Wiley, Chap. 11.

Hardy, G. (1949). *Divergent Series*. Oxford: Oxford University Press.

Karlin, S. and Taylor, H. M. (1975). *A First Course in Stochastic Processes*. 2nd ed. New York: Academic Press.

Muth, E. J. (1977). *Transform Methods with Applications to Engineering and Operations Research*. Englewood Cliffs, N.J.: Prentice-Hall.

Royden, H. L. (1966). *Real Analysis*. New York: Macmillan.

Rudin, W. (1964). *Principles of Mathematical Analysis*. New York: McGraw-Hill.

Whittaker, E. T., and Watson, G. N. (1927). *A Course of Modern Analysis*. 4th ed. Cambridge, Cambridge University Press.

Widder, D. V. (1946). *The Laplace Transform*. Princeton: Princeton University Press.

ANSWERS TO EXERCISES

CHAPTER 2

Modeling Exercises

1. State space: $\{A, B, C\}$; parameter space (months): $\{0, 1, 2, \ldots\}$. For a population size N, the number of persons buying brand X is Np_X, where p_X is the probability a person buys brand X ($X = A, B, C$).

2. State space: $\{X: 0 < X < \infty\}$; parameter space: $\{0, 1, 2, \ldots\}$ or $\{T: 0 \le T < \infty\}$. X_{n+1} should depend only on X_n and $(I_n - C_n)$ but not on X_i and $(I_i - C_i)$, $i = 1, 2, \ldots, n - 1$, in the discrete parameter case.

3. Inventory level, observation epoch, demand process, mode of replenishment of inventory.

4. $P(D_n = k) = \binom{n}{k} p^k q^{n-k}$
   $P(D_{n+1} = k) = qP(D_n = k) + pP(D_n = k - 1)$.

5. Jobs—State space: $\{0, 1, 2, \ldots\}$; parameter space: $\{T: 0 \le T < \infty\}$. Job length-State space: $\{X: 0 \le T < \infty\}$; parameter space: $\{T: 0 \le T < \infty\}$. Job arrival process, distribution of job length, processing discipline, number of processors used.

6. $f_{X|T}(x) = f(x)/F(T)$; $g(x) = [1 - F(T)]\sum_{n=1}^{\infty} [f(x)]^{n*}$ where $F(T)$ is the distribution function of gap length and $n*$ signifies the n-fold convolution of $f(x)$ with itself.

7. The modes of births, deaths, and changes in population size occur due to immigration and emigration (rates); dependency characteristics of these rates on population size and time.

8. $P[X(t) = k] = \binom{N}{k} [1 - F(T)]^k [F(T)]^{N-k}$, ($k = 0, 1, 2, \ldots, N$).

9. $[f(x)]^{n*}$, where $n*$ signifies the n-fold convolution of $f(x)$ with itself; $F_n(T) - F_{n+1}(T)$.

$$e^{-\lambda t} \frac{\lambda^n t^{n-1}}{(n-1)!}; \quad e^{-\lambda t} \frac{(\lambda t)^n}{n!}.$$

10. Replace $f(x)$ in 9 above by $f(x)*g(x)$.

11. (a) $P[S_n^{(N, N_1)} = k] = \dfrac{\dbinom{N_1}{k}\dbinom{N - N_1}{n - k}}{\dbinom{N}{n}}$.

12. (a) $0\ (j \le i - 1)$, $q\ (j = i)$, $p\ (j = i + 1)$, $0\ (j > i + 1)$.

   (b) $0\ (j \le i - 1)$, $\dfrac{N - N_1 + n + i}{N - n}\ (j = i)$, $\dfrac{N_1 - i}{N - n}\ (j = i + 1)$, $0\ (j > i + 1)$.

Review Exercises

2. (i) $[(1 - p)/p]^{1/2}$,   (ii) $[(1 - p)/np]^{1/2}$,   (iii) $(1 - p)^{1/2}$,
   (iv) $[n(1 - p)]^{-1/2}$,   (v) $\lambda^{-1/2}$,   (vi) 1,   (vii) $n^{-1/2}$.

8. $p\mu_s e^{-\mu_s t} + (1 - p)\mu_1 e^{-\mu_1 t}$; $p/\mu_s + (1 - p)/\mu_1$;
   $p(2 - p)/\mu_s^2 + (1 - p^2)/\mu_1^2 - 2p(1 - p)/\mu_s\mu_1$.

10. (i) $1 - p + pz$,   (ii) $(1 - p + pz)^n$,   (iii) $pz[1 - (1 - p)z]^{-1}$,
    (iv) $p[1 - (1 - p)z]^{-n}$,   (v) $e^{-\lambda(1 - z)}$.

11. (i) $\dfrac{\lambda}{\theta + \lambda}$ (ii) $\dfrac{e^{-\theta a} - e^{\theta b}}{\theta(b - a)}$ (iii) $e^{-\theta b}$.

12. p^n.

13. $\sum\limits_{r=k}^{n} \dbinom{n}{r} p^r (1 - p)^{n-r}$.

14. $[1 - (1 - p_1)^2][1 - (1 - p_2)^3]$.

15. $p_3^2 + p_1 p_2 (2 - p_1)(1 - p_3^2)$.

16. $p_2(3p_1^2 - 2p_1^3)$.

17. 0.3232, 0.6667, 0.0101.

18. 0.9851, 0.0149.

19. 0.4044.

20. 0.6065.

21. 0.3679.

24. e^{-x}, e^{-y}, No.

32. (b) \$60,000, $24 \times 10^8 (\$^2)$.

33. (a) Poisson with parameter λp.

    (b) 0.665.

CHAPTER 3

2. $S:\{0, 1, \ldots, a, > a\};\ P(X_k = j) = p_j = \begin{pmatrix} N \\ j+1 \end{pmatrix} p^{j+1} q^{N-j-1}.$

$$P = \begin{bmatrix} 1 & 0 & 0 & \cdots & 0 & \cdots & 0 & 0 & 0 \\ p_{-1} & p_0 & p_1 & \cdots & p_{N-1} & \cdots & 0 & 0 & 0 \\ \vdots & \vdots & \vdots & & \vdots & & \vdots & \vdots & \vdots \\ 0 & 0 & 0 & \cdots & 0 & \cdots & p_{-1} & p_0 & 1 - p_0 - p_{-1} \\ 0 & 0 & 0 & \cdots & 0 & \cdots & 0 & 0 & 1 \end{bmatrix}$$

States 0 and $> a$ are absorbing and others transient.

3. $S:\{0, 1, \ldots, N\}.$

$$P = \begin{bmatrix} 0 & 1 & 0 & 0 & \cdots & 0 & 0 & 0 \\ \dfrac{1}{N} & 0 & \dfrac{N-1}{N} & 0 & \cdots & 0 & 0 & 0 \\ 0 & \dfrac{2}{N} & 0 & \dfrac{N-2}{N} & \cdots & 0 & 0 & 0 \\ \vdots & \vdots & \vdots & \vdots & & \vdots & \vdots & \vdots \\ 0 & 0 & 0 & 0 & & \dfrac{N-1}{N} & 0 & \dfrac{1}{N} \\ 0 & 0 & 0 & 0 & & 0 & 1 & 0 \end{bmatrix}$$

Markov chain is irreducible and periodic (period = 2).

4. (a)

$$P = \begin{bmatrix} p & q & 0 & 0 \\ 0 & 0 & p & q \\ p & q & 0 & 0 \\ 0 & 0 & p & q \end{bmatrix}$$

(b) No.

5. $S:\{HHH, HHT, HTH, HTT, THH, THT, TTH, TTT\}$

Top half of the transition probability matrix:

$$\begin{bmatrix} p & q & 0 & 0 & 0 & 0 & 0 & 0 \\ 0 & 0 & p & q & 0 & 0 & 0 & 0 \\ 0 & 0 & 0 & 0 & p & q & 0 & 0 \\ 0 & 0 & 0 & 0 & 0 & 0 & p & q \end{bmatrix}$$

The bottom half is identical with the top half.

6. S: $\{0, 1, 2, \dots\}$

$$P = \begin{bmatrix} q & p & 0 & 0 & \cdots \\ q & 0 & p & 0 & \cdots \\ q & 0 & 0 & p & \cdots \\ \cdot & \cdot & \cdot & \cdot & \\ \cdot & \cdot & \cdot & \cdot & \\ \cdot & \cdot & \cdot & \cdot & \cdots \end{bmatrix}$$

T = recurrence time; $P(T = n) = p^{n-1}q$, $n = 1, 2, \dots$. $E(T) = 1/q$.

$\sum_{n=1}^{\infty} P(T = n) = 1$. Markov chain is irreducible and positive recurrent.

7. Population size N. S: $\{0, 1, 2, \dots, N\}$.

$$P = \begin{bmatrix} 1 & 0 & 0 & 0 & \cdots & 0 \\ \dfrac{1}{N} & 0 & \dfrac{N-1}{N} & 0 & \cdots & 0 \\ \dfrac{2}{N} & 0 & 0 & \dfrac{N-2}{N} & \cdots & 0 \\ \vdots & \vdots & \vdots & \vdots & & \vdots \\ \dfrac{N-1}{N} & 0 & 0 & 0 & \cdots & \dfrac{1}{N} \\ 1 & 0 & 0 & 0 & \cdots & 0 \end{bmatrix}$$

8.

$$P = \begin{bmatrix} q & p & 0 & 0 & 0 & 0 \\ 0 & 0 & q & 0 & p & 0 \\ 0 & 0 & 0 & q & 0 & p \\ q & p & 0 & 0 & 0 & 0 \\ 0 & 0 & 0 & 0 & 0 & 1 \\ 0 & 1 & 0 & 0 & 0 & 0 \end{bmatrix}$$

9. S: $\{0, 1, 2, \dots\}$

$$P = \begin{bmatrix} q & p & & & \\ rq & pr + sq & ps & & \\ & rq & pr + sq & ps & \\ & & \cdot & & \\ & & & \cdot & \\ & & & & \cdot \end{bmatrix}$$

10. S: $\{0, 1, 2 \ldots, N\}$

$X_{n+1} = 0$ if $X_n = 0$; $X_{n+1} = \min\{N, X_n + J_{n+1} - 1\}$, if $X_n > 0$ where J_{n+1} is the number of new jobs arriving during nth service. Let $\alpha_N = \sum\limits_{N}^{\infty} a_r$.

$$P = \begin{bmatrix} 1 & 0 & 0 & \cdots & 0 & 0 \\ a_0 & a_1 & a_2 & \cdots & a_{N-1} & \alpha_N \\ 0 & a_0 & a_1 & \cdots & a_{N-2} & \alpha_{N-1} \\ \vdots & \vdots & \vdots & & \vdots & \vdots \\ 0 & 0 & 0 & \cdots & a_0 & \alpha_1 \end{bmatrix}$$

State 0 is absorbing and others are transient.

11. S: $\{0, 1, 2, \ldots\}$

$$P = \begin{bmatrix} q & p & & & \\ q & 0 & p & & \\ & q & 0 & p & \\ & & q & 0 & p \\ & & & & \ddots \end{bmatrix}$$

12. S: $\{W, D, S, L\}$

$$P = \begin{bmatrix} 1.0 & 0 & 0 & 0 \\ 0.3 & 0.4 & 0.3 & 0 \\ 0 & 0.2 & 0.5 & 0.3 \\ 0 & 0 & 0 & 1.0 \end{bmatrix}$$

States W and L are absorbing and states D and S are transient.

13. S: {Freshman, Sophomore, Junior, Senior, Graduate, Dropout}

$$P = \begin{bmatrix} 0.10 & 0.65 & & & & 0.25 \\ & 0.10 & 0.80 & & & 0.10 \\ & & 0.05 & 0.92 & & 0.03 \\ & & & 0.04 & 0.95 & 0.01 \\ & & & & 1.00 & \\ & & & & & 1.00 \end{bmatrix}$$

14. S: $\{3, 4, 5, 6, 7\}$

$$P = \begin{bmatrix} 0.1 & 0 & 0 & 0 & 0.9 \\ 0.1 & 0.1 & 0 & 0 & 0.8 \\ 0.2 & 0.1 & 0.1 & 0 & 0.6 \\ 0.1 & 0.2 & 0.1 & 0.1 & 0.5 \\ 0.1 & 0.1 & 0.2 & 0.1 & 0.5 \end{bmatrix}$$

15. $S = \{$Mark, Tom, Joe, Dick, Harry, Sam$\}$.

$$\begin{bmatrix} 1 & 0 & 0 & 0 & 0 & 0 \\ 0 & 0 & 1 & 0 & 0 & 0 \\ 0 & 1 & 0 & 0 & 0 & 0 \\ \frac{1}{4} & \frac{1}{4} & 0 & 0 & \frac{1}{4} & \frac{1}{4} \\ 0 & \frac{1}{4} & \frac{1}{4} & \frac{1}{4} & 0 & \frac{1}{4} \\ \frac{1}{4} & \frac{1}{4} & 0 & \frac{1}{4} & \frac{1}{4} & 0 \end{bmatrix}$$

$\{$Mark$\}$: Absorbing; $\{$Tom, Joe$\}$: Recurrent, periodic (2); $\{$Dick, Harry, Sam$\}$: Transient, aperiodic.

17. S: $\{E, A, B, C, AB, AC, BC, ABC\}$

$$\begin{bmatrix} 1 & 0 & 0 & 0 & 0 & 0 & 0 & 0 \\ 0 & 1 & 0 & 0 & 0 & 0 & 0 & 0 \\ 0 & 0 & 1 & 0 & 0 & 0 & 0 & 0 \\ 0 & 0 & 0 & 1 & 0 & 0 & 0 & 0 \\ \frac{4}{15} & \frac{3}{15} & \frac{6}{15} & 0 & \frac{2}{15} & 0 & 0 & 0 \\ \frac{6}{25} & \frac{4}{25} & 0 & \frac{9}{25} & 0 & \frac{6}{25} & 0 & 0 \\ \frac{6}{15} & 0 & \frac{4}{15} & \frac{3}{15} & 0 & 0 & \frac{2}{15} & 0 \\ \frac{12}{75} & \frac{8}{75} & \frac{6}{75} & \frac{18}{75} & \frac{12}{75} & \frac{4}{75} & \frac{9}{75} & \frac{6}{75} \end{bmatrix}$$

18. (a) No brand switching. Should try to alter pattern through advertising, etc.

(b) Most of the customers are unhappy with the existing brands. Easy to influence them through other imaginative means.

(c) B is not favored at all. Phase out B, or introduce new product in its place or improve its image.

(d) Same as (b) with a longer periodicity.

19. (a) $\{3\}$: absorbing; $\{1,6\}$: Recurrent, periodic (2); $\{2\}$: transient; $\{4,5\}$: Transient, aperiodic.

(b) $\{1,4,5\}$: recurrent, aperiodic; $\{2,3\}$: transient aperiodic.

20. (a) $\{1,4\}$: recurrent, aperiodic; $\{2,5\}$: recurrent, aperiodic; $\{3,6\}$: transient, aperiodic.

(b) $\{2\}$: absorbing; $\{6\}$: absorbing; $\{1,4,7,8\}$: recurrent, aperiodic; $\{3,9,10\}$: transient aperiodic; $\{5\}$: transient.

CHAPTER 4

4. $\binom{N}{n}\frac{1}{2}N.$

5. (a) For $n \geq 2$, P^n has identical rows $[p^2\ pq\ pq\ q^2]$.

   (b) For $n \geq 3$, P^n has identical rows
   $[p^3\ p^2q\ p^2q\ pq^2\ p^2q\ pq^2\ pq^2\ q^3].$

   (c) For $n \geq k$, P^n has identical rows with elements $p^i q^j$ $(i + j = k)$ where i is the number of H's and j is the number of T's in the corresponding state.

6. $(1 + p + p^2)/p^3.$

7. $\nu = \pi_{10}^{-1} = 1 + q + 2p + q^2 + pq + q^3/p$
   $\pi_{20} = q^3\nu/p,\ \pi_{11} = q\nu,\ \pi_{12} = q^2\nu,\ \pi_{01} = p\nu,$
   $\pi_{02} = (pq + p)\nu.$

8. $\pi_0 = \left[1 + \dfrac{p}{(rq)^K}\dfrac{(rq)^K - (ps)^K}{rq - ps}\right]^{-1}.$

   $\pi_n = \dfrac{1}{s}\left(\dfrac{ps}{rq}\right)^n \pi_0.$

9. 11.48 time intervals.

10. See Section 13.5.

11. See Section 13.5.

12. $\pi_0 = \dfrac{1 - \rho}{1 - \rho^{N+1}},\ \pi_n = \rho^n\pi_0,$ where $\rho = p/q.$

13. (a) $\left(\dfrac{1}{m}, \dfrac{1}{m}, \ldots, \dfrac{1}{m}\right).$

    (b) For $K = 8$. $S = \{0, 1, 2 \ldots, 7\}$

$$P = \begin{bmatrix}
0 & \frac{1}{6} & \frac{1}{6} & \frac{1}{6} & \frac{1}{6} & \frac{1}{6} & \frac{1}{6} & 0 \\
0 & 0 & \frac{1}{6} & \frac{1}{6} & \frac{1}{6} & \frac{1}{6} & \frac{1}{6} & \frac{1}{6} \\
\frac{1}{6} & 0 & 0 & \frac{1}{6} & \frac{1}{6} & \frac{1}{6} & \frac{1}{6} & \frac{1}{6} \\
\frac{1}{6} & \frac{1}{6} & 0 & 0 & \frac{1}{6} & \frac{1}{6} & \frac{1}{6} & \frac{1}{6} \\
\frac{1}{6} & \frac{1}{6} & \frac{1}{6} & 0 & 0 & \frac{1}{6} & \frac{1}{6} & \frac{1}{6} \\
\frac{1}{6} & \frac{1}{6} & \frac{1}{6} & \frac{1}{6} & 0 & 0 & \frac{1}{6} & \frac{1}{6} \\
\frac{1}{6} & \frac{1}{6} & \frac{1}{6} & \frac{1}{6} & \frac{1}{6} & 0 & 0 & \frac{1}{6} \\
\frac{1}{6} & \frac{1}{6} & \frac{1}{6} & \frac{1}{6} & \frac{1}{6} & \frac{1}{6} & 0 & 0
\end{bmatrix}$$

$$\pi_0 = \frac{1}{K}.$$

14. $\dfrac{(q/p)^i-(q/p)^N}{1-(q/p)^N}$

15. See Section 17.2.

16. 0.8027; 0.2000; 4 minutes.

17. 55%. Should aim at a loss/gain ratio of 1/3.

18. 0.2899; 4.58.

19. 0.54, 0.37, 3.19.

20. 0.834, $\frac{2}{3}$ year.

23. 3/11, 3n/11 out of n periods, 2.66.

24. 72.

25. 22.22.

26. (a) 0.625, 3.333 (b) 0.250, 3.333.

27. (0.114, 0.095, 0.138, 0.065, 0.588), 1.596, 0.3082.

28. 3.72.

30. (i) 2, (ii) $\frac{2}{3}$, (iii) yes.

32. 0.646, 0.68.

33. (1) $\begin{pmatrix} 1.2 & 0.4 & 0.4 \\ 0.4 & 1.2 & 0.4 \\ 0.4 & 0.4 & 1.2 \end{pmatrix}$; (2) 1/3; (3) $\frac{1}{6},\frac{1}{2}$; (4) 2, 2.

34. (0.090, 0.360, 0.225, 0.180, 0.145); 11.11.

35. Once every 14 years.

36. (2.07, 3.62, 4.14); 4 years and 1.5 years.

37. S: $\{0,1,2,3,4\}$

$$P = \begin{bmatrix} .10 & .90 & 0 & 0 & 0 \\ .15 & 0 & .85 & 0 & 0 \\ .35 & 0 & 0 & .65 & 0 \\ .40 & 0 & 0 & 0 & .60 \\ 1 & 0 & 0 & 0 & 0 \end{bmatrix}$$

(0.30, 0.27, 0.21, 0.14, 0.08) (3.45, 2.73, 2.04, 1.60, 1.00) weeks; a 4 week replacement policy recommended.

38. See Section 17.4.

CHAPTER 5

1. 1/36.

3. (a) [{1,2,4},3] not lumpable; [{1,2},3,4] lumpable.

  (b) [1,{2,3},{4,5}] lumpable; [1,{2,3,4,5}] not lumpable.

4. (a) $\begin{pmatrix} .6 & .4 & 0 \\ 0 & 1 & 0 \\ .7 & 0 & .3 \end{pmatrix}$ (b) $\begin{pmatrix} 1 & 0 & 0 \\ 0 & .6 & .4 \\ .2 & 0 & .8 \end{pmatrix}$

8. (a) $\left[\dfrac{a}{a+b} + \dfrac{b}{a+b}(1 - a - b)^n \right](1 - \alpha) \Big/$

$\qquad \left[\dfrac{a}{a+b} + \left(\dfrac{b}{a+b} - \alpha \right)(1 - a - b)^n \right].$

(b) $\left[\dfrac{1}{2} + \dfrac{1}{2}(1 - 2q)^n \right](1 - \alpha) \Big/ \left[\dfrac{1}{2} + \left(\dfrac{1}{2} - \alpha \right)(1 - 2q)^n \right].$

9. (a) $\begin{pmatrix} 0.176 & 0.824 \\ 0.455 & 0.545 \end{pmatrix}.$ (b) Yes; χ^2 statistic with 2 $d.f. = 0.3285$.

10. 1.4; 1.9 minutes.

12. χ^2 statistic with 1 $d.f. = 3.33$; can be considered to be an independent trials process with about 7% level of significance.

14. Likelihood ratio statistic with 24 $d.f. = 57.62$; impact of the introduction of promotional game significant.

CHAPTER 6

4. $\pi_o = \left(1 + \sum\limits_{n=1}^{\infty} p_o p_1 \cdots p_{n-1} \right)^{-1}$, $\pi_n = p_0 p_1 \cdots p_{n-1} \pi_0.$

5. $\pi_0 = 1 - p/q$, $\pi_n = \dfrac{p^n (1 - q)^{n-1}}{q^n (1 - p)^n}(1 - p/q).$

7. $\pi_j = (1 - p/q)(p/q)^j$, $j = 0, 1, 2\ldots$, when $p/q < 1$.

8. $\pi_0 = 1 + \sum\limits_{m=1}^{\infty} \left(\dfrac{a_{01} a_{12} a_{23} \cdots a_{m-1, m}}{a_{10} a_{21} a_{32} \cdots a_{m, m-1}} \right)^{-1}$,

$\qquad \pi_n = \dfrac{a_{01} a_{12} \cdots a_{n-1, n}}{a_{10} a_{21} \cdots a_{n, n-1}} \pi_0.$

9. $\pi_j = (1 - p/q)(p/q)^j$, $j = 0, 1, 2\ldots$, where $p/q < 1$.

10. Periodic chain with period 2. Stationary probabilities are

$$p_0 = \frac{1}{2}(1 - p/q), \quad p_n = \frac{1}{2q}(1 - p/q)(p/q)^{n-1}, \quad n = 1, 2, \ldots .$$

11. $\pi_j = (1 - p/q)(p/q)^j$, $j = 0, 1, 2\ldots$ when $p/q < 1$;
$(p/q)/(1 - p/q)$.

14. (a) p^n, $qp^n(1 - p^n)$, 1.

(b) $(q + 2p)^n$, $pq(q + 2p)^{n-1}[1 - (q + 2p)^n]$, $\min(1, r/p)$.

(c) $1, \dfrac{n}{8}, 1.$

(d) $1, n, 1.$

(e) $\left(\dfrac{11}{8}\right)^{n}, \dfrac{47}{64}\left(\dfrac{11}{8}\right)^{n-1}\left[1 - \left(\dfrac{11}{8}\right)^{n}\right], 0.3027.$

15.   (a) $P(X_3 = 0) = q(1 + p + p^2),\ P(X_3 = 1) = p^3$

      (b) $P(X_3 = 0) = q(1 + p + p^2),\ P(X_3 = 2) = p^3$

      (c)

X_3	0	1	2	3	4
Prob	.4183168	.2983296	.1853824	.0690432	.0223360
X_3	5	6	7	8	
Prob	.0052992	.0011264	.0001536	.0000128	

17. $S:\{-B, -B+1, \ldots, 0, 1, 2, \ldots\}$

$$P = \begin{bmatrix} \alpha_0 & a_1 & a_2 & a_3 & \cdots \\ \alpha_{-1} & a_0 & a_1 & a_2 & \cdots \\ \alpha_{-2} & a_{-1} & a_0 & a_1 & \cdots \\ \cdot & \cdot & \cdot & \cdot & \cdots \\ \cdot & \cdot & \cdot & \cdot & \cdots \end{bmatrix}$$

where $a_j = P(P_t - D_t = j);\ \alpha_j = P(P_t - D_t \le j).$

18.

$$Q_{n+1} = \begin{cases} Q_n + X_{n+1} - 1 & \text{if } Q_n > 0 \\ X_{n+1} & \text{if } Q_n = 0 \end{cases}$$

where X_n is the number of customers arriving during the nth service period. $S: \{0, 1, 2, \ldots\}$

$$P = \begin{bmatrix} a_0 & a_1 & a_2 & a_3 & \cdots \\ a_0 & a_1 & a_2 & a_3 & \cdots \\ & a_0 & a_1 & a_2 & \cdots \\ \bigcirc & & a_0 & a_1 & \cdots \end{bmatrix}$$

19.

$$Q_{n+1} = \begin{cases} Q_n + 1 - X_{n+1} & \text{if } Q_n + 1 - X_{n+1} > 0 \\ 0 & \text{if } Q_n + 1 - X_{n+1} \le 0 \end{cases}$$

where X_n, $n = 1, 2 \ldots$ are *i.i.d* random variables distributed as b_j, $j = 0, 1, 2 \ldots$.

$S: \{0, 1, 2, \dots\}$

$$P = \begin{bmatrix} \beta_1 & b_0 & & \bigcirc \\ \beta_2 & b_1 & b_0 & \\ \beta_3 & b_2 & b_1 & b_0 \\ \vdots & \vdots & \vdots & \vdots \end{bmatrix}$$

where $\beta_j = \sum_{r=j}^{\infty} b_r$.

20.

$$Q_{n+1} = \begin{cases} Q_n + X_{n+1} - s & \text{if } Q_n > s \\ X_{n+1} & \text{if } Q_n \le s \end{cases}$$

where X_n is the number of customers arriving during $((n-1)\sigma, n\sigma]$.
$S: \{0, 1, 2, \dots, s, s+1, \cdots\}$

$$\begin{bmatrix} a_0 & a_1 & a_2 & a_3 & \cdots \\ a_0 & a_1 & a_2 & a_3 & \cdots \\ \vdots & \vdots & \vdots & \vdots & \\ a_0 & a_1 & a_2 & a_3 & \cdots \\ & a_0 & a_1 & a_2 & \cdots \\ \bigcirc & & a_0 & a_1 & \cdots \end{bmatrix}$$

CHAPTER 7

1. (a) 0.041   (b) 0.206.

3. $\dfrac{r}{n+1}T, \dfrac{r(n-r+1)}{(n+1)^2(n+3)}T^2$.

5. (a) $\sum_{1}^{3} \dfrac{1}{\mu_i}$ (b) $\dfrac{1}{\mu_1} + \dfrac{\alpha}{\mu_2} + \dfrac{1}{\mu_3}$ (= E, say); $\dfrac{1}{\mu_1 E}, \dfrac{\alpha}{\mu_2 E}, \dfrac{1}{\mu_3 E}$.

6. $\dfrac{\lambda}{\lambda + s\mu}$.

8. $F_{Z_1}(z) = 1 - e^{-(\lambda_1 + \lambda_2)z}$; $F_{Z_2}(z) = 1 - e^{-\lambda_1 z} - e^{-\lambda_2 z} + e^{-(\lambda_1 + \lambda_2)z}$.

10. Poisson with mean 175.5.

11. $N(t) = \dfrac{N(0)}{1 - N(0)\lambda t}$.

15. (2)
$$\begin{bmatrix} -\lambda_0 & \lambda_0 & 0 & 0 & 0 & \cdots \\ 0 & -\lambda_1 & \lambda_1 & 0 & 0 & \cdots \\ 0 & 0 & -\lambda_2 & \lambda_2 & 0 & \cdots \\ \cdot & \cdot & \cdot & \cdot & \cdot & \cdots \end{bmatrix},$$

(4)
$$\begin{bmatrix} -\lambda_0 & \lambda_0 & 0 & 0 & \cdots \\ \mu_1 & -(\lambda_1 + \mu_1) & \lambda_1 & 0 & \cdots \\ 0 & \mu_2 & -(\lambda_2 + \mu_2) & \lambda_2 & \cdots \\ \cdot & \cdot & \cdot & \cdot & \cdots \end{bmatrix},$$

(5) $\begin{bmatrix} -\lambda & \lambda \\ \mu & -\mu \end{bmatrix}$.

16. See Section 11.5.

20. 0.32, 1.52 hr.

21. Justified only if computer is used more than 5333 hours per year.

22. (a) 0.3679, (b) 0.0803, (c) 1/3, (d) 40 mins, (e) 4 secs.

23. (a) $p_0 = \left(1 + \dfrac{\mu_0}{\alpha_3\mu_1} + \dfrac{\alpha_2\mu_0}{\alpha_3\mu_2} + \dfrac{\mu_0}{\mu_3}\right)^{-1}$, $p_1 = \dfrac{\mu_0}{\alpha_3\mu_1}p_0$,

$p_2 = \dfrac{\alpha_2\mu_0}{\alpha_3\mu_2}p_0$, $p_3 = \dfrac{\mu_0}{\mu_3}p_0$.

(b) Use an additional equation: $P_4'(t) = \alpha_4\mu_1 P_1(t) + \beta_4\mu_3 P_3(t)$.

24. Poisson with mean λ/μ.

25. $p_n = (1 - \rho)\rho^{n+k}$, $n = -k, -k+1, \ldots, 0, 1, \ldots$.

26. See Chapter 11.

27. (a) $p_n = (1 - \rho)\rho^n$ where $\rho = \lambda/(1 - \alpha)\mu$, $n = 0, 1, 2, \ldots$.

28. $p_{00} = (1 + \rho_a + \rho_b + \rho_a\rho_b)^{-1}$, $p_{10} = \rho_1 p_{00}$, $p_{01} = \rho_b p_{00}$, $p_{11} = \rho_a\rho_b p_{00}$, where $\rho_a = \lambda_a/\mu_a$, $\rho_b = \lambda_b/\mu_b$ and state 0–working, state 1–not working;

$$p_{00} = \dfrac{\gamma + \rho_a}{(\gamma + \rho_a + \rho_a\gamma)(a + \rho_b + \rho_a\rho_b) + \rho_a^2},$$

$$p_{10} = \dfrac{\rho_a\gamma}{\gamma + \rho_a}p_{00}, \qquad p_{01} = \dfrac{\rho_b(\gamma + \rho_a + \rho_a\gamma)}{\gamma + \rho_a}p_{00},$$

$$p_{11} = \dfrac{\rho_a\rho_b(\gamma + \rho_a + \rho_a\gamma) + \rho_a^2}{\gamma + \rho_a}p_{00}$$

where $\gamma = \lambda_a/\lambda_b$.

29. Capital $\to 0$ if $\lambda/\mu \le 1$ and $\alpha/\beta \le 1$ and $\to \infty$ otherwise.

30. Same as 29.

31. No; with one every 25 mins, traffic intensity $\rho = 5/6 < 1$; but with one every 40 mins, $\rho = 4/3 > 1$.

32. $\lambda p/\mu < 1$; $p_n = (1 - \rho)\rho^n$, $n = 0, 1, 2, \ldots$, where $\rho = \lambda p/\mu$.

33. S: $\{(xy) = 00, 10, 01, 11, 20, 02\}$ where $x =$ number of priority jobs and $y =$ number of nonpriority jobs. $(\alpha\lambda = \lambda_1, (1 - \alpha)\lambda = \lambda_2)$

$$\begin{bmatrix} -\lambda & \lambda_1 & \lambda_2 & 0 & 0 & 0 \\ \mu_1 & -(\mu_1 + \lambda) & 0 & \lambda_2 & \lambda_1 & 0 \\ \mu_2 & 0 & -(\mu_1 + \lambda) & \lambda_1 & 0 & \lambda_2 \\ 0 & 0 & \mu_1 & -\mu_1 & 0 & 0 \\ 0 & \mu_1 & 0 & 0 & -\mu_1 & 0 \\ 0 & 0 & \mu_2 & 0 & 0 & -\mu_2 \end{bmatrix}$$

34. (a) $s = 2$: $p_{ij} = \rho_1^i \rho_2^j p_{00}$, $i, j = 0, 1, 2, i + j \le 2$.

$s = 3$: $p_{ij} = \rho_1^i \rho_2^j p_{00}$, $i, j = 0, 1, 2, 3, i + j \le 3$.

(b) $p_{ij} = \rho_1^i \rho_2^j p_{00}$, $i, j = 0, 1, 2, \ldots, s$, $i + j \le s$.

(c) $p_{i_1, i_2, \ldots, i_k} = \rho_1^{i_1} \rho_2^{i_2} \cdots \rho_k^{i_k} p_{00\ldots0}$, $i_1, i_2, \ldots, i_k = 0, 1, \ldots, s$,

$$\sum_{r=1}^{k} i_r \le s.$$

35. $P_0'(t) = \mu P_1(t)$.

$P_1'(t) = -(\lambda + \mu)P_1(t) + \mu P_2(t)$.

$P_n'(t) = -(\lambda + \mu)P_n(t) + \lambda P_{n-1}(t) + \mu P_{n+1}(t)$ $(n > 1)$.

$P_i(0) = 1$ if $i = 1$ and $= 0$, otherwise.

$[P_n(t)) = P(\text{Number of customers in the system at time } t = n)]$.

36. Let $P_{nm}(t) = P(\text{Number of customers in the system at time } t = n,$

Number of departures until $t = m$).

$P_{00}'(t) = -\lambda P_{00}(t)$.

$P_{i0}'(t) = -(\lambda + \mu)P_{i0}(t) + \lambda P_{i-1,0}(t)$.

$P_{0m}'(t) = \lambda P_{0m}(t) + \mu P_{1, m-1}(t)$ $(m < K)$.

$P_{nm}'(t) = -(\lambda + \mu)P_{nm}(t) + \lambda P_{n-1, m}(t) + \mu P_{n+1, m-1}(t)$

$(m < K)$.

$P_{nK}'(t) = \mu P_{n+1, K-1}(t)$.

37. $\dfrac{3\rho}{2(1 - \rho)}$.

38. See Section 15.6.

39. Reliability $R(t) = \dfrac{1}{\xi_1 - \xi_2}[\xi_1 e^{-(\lambda_2 - \xi_2)t} - \xi_2 e^{-(\lambda_2 - \xi_1)t}]$.

$\xi_1, \xi_2 = \frac{1}{2}[-(3\lambda + \mu) \pm (\lambda^2 + \mu^2 + 6\lambda\mu)^{1/2}]$

Mean time to failure $E(L) = \dfrac{\lambda_2 + 3\lambda + \mu}{\lambda_2(\lambda_2 + 3\lambda + \mu) + 2\lambda^2}$.

Long term availability $= \dfrac{1 + 2\rho}{(1 + \rho)^2(1 + \rho_2)}$

where $\rho = \lambda/\mu$ and $\rho_2 = \lambda_2/\mu_2$.

40. Reliability $R(t)$ is the same as for Ex. 39 with

$$\xi_1, \xi_2 = \frac{1}{2}\left[-(5\lambda + \mu) \pm (\lambda^2 + \mu^2 + 10\lambda\mu)^{1/2}\right].$$

Mean time to failure $E(L) = \dfrac{\lambda_2 + 5\lambda + \mu}{\lambda_2(\lambda_2 + 5\lambda + \mu) + 6\lambda^2}$

Long term availability $= \left(\dfrac{1 + 3\rho_1}{1 + 3\rho_1 + 3\rho_1^2}\right)\left(\dfrac{1}{1 + \rho_2}\right)$

where $\rho_i = \lambda_i/\mu_i$ $(i = 1, 2)$.

41. Reliability

$$R(t) = \dfrac{1}{\xi_1 - \xi_2}\left[(\xi_1 + \lambda + \mu + 2\lambda\alpha)e^{\xi_1 t} - (\xi_2 + \lambda + \mu + 2\lambda\alpha)e^{\xi_2 t}\right]$$

where $\xi_1, \xi_2 = \frac{1}{2}[-(3\lambda + \mu) \pm (\lambda^2 + \mu^2 - 2\lambda\mu + 8\lambda\mu\alpha)^{1/2}]$.

Mean time to failure $E(L) = \dfrac{\lambda + \mu + 2\lambda\alpha}{2\lambda(\lambda + \mu - \mu\alpha)}$.

42. m_b = Number of registered births during the observation interval T.
m_d = Number of deaths.
T_n = Total time the population size was n, $n = 1, 2, \ldots, s$.
$\hat{\lambda} = \dfrac{m_b}{T - T_s}, \hat{\mu} = \dfrac{m_d}{\displaystyle\sum_{n=1}^{s} nT_n}$.

43. $\hat{\lambda} = 0.25$, $\hat{\mu} = 0.30$.

44. (a) m_i = Number of times state i has occurred, $i = 0, 1$.
T_i = Total time the process has occupied state i, $i = 0, 1$.
$\hat{\lambda} = m_0/T_0$, $\hat{\mu} = m_1/T_1$.

(b) $-2\ln\Lambda =$
$$2\left[m_0\ln\frac{m_0}{\lambda^0 T_0} + m_1\ln\frac{m_1}{\mu^0 T_1} - \left(m_0 - \lambda^0 T_0\right) - \left(m_1 - \mu^0 T_1\right)\right]$$

(c) $-2\ln\Lambda = 1.16782$, which under H_0 is χ^2 with 2 $d.f.$ No reason to assume that the sample observations have not come from a normal workday.

CHAPTER 8

1. $F(z) = \dfrac{pz}{1 - (1 - p)z}$, $P(z) = 1 + \dfrac{pz}{1 - z}$.

2. $\displaystyle\sum_{r=1}^{n} e^{-r\lambda}\frac{(r\lambda)^n}{n!}$.

3. (a)
0	1	2	3	4
166.7	133.3	100.0	66.7	33.3

(b)
0	1	2	3	4	5	6
159.0	134.0	96.0	59.0	30.8	14.0	5.6

7	8
1.9	0.6

(c)
0	1	2	3	4	5	6
150	105	73.5	51.0	36.0	25.5	18.0

7	8	9			
12.0	9.0	6.0	·	·	·

4.   Same.

5. (a) $e^{-\lambda t}\dfrac{(\lambda t)^n}{n!}$; λt; λ.

(b)   1 if $nb < t \le (n + 1)b$,   0 otherwise; $\left[\dfrac{t}{b}\right]$; $\dfrac{1}{b}$.

6. (i)   (a) $\dfrac{\theta\left[\lambda_2 + \alpha(\lambda_1 - \lambda_2)\right] + \lambda_1\lambda_2}{\theta\left[\theta + \lambda_1 - \alpha(\lambda_1 - \lambda_2)\right]}$

(b) $\dfrac{\lambda_1\lambda_2}{\theta\left[\lambda_1 - \alpha(\lambda_1 - \lambda_2)\right]}$,

(ii)   (a) $\dfrac{\lambda^2}{\theta(\theta + 2\lambda)}$ (b) $\dfrac{\lambda}{2\theta}$.

11. $s_n^{(k)} = u_{n-k}(1 - F_k), \quad \lim_{n \to \infty} s_n^{(k)} = \dfrac{1 - F_k}{\mu},$

$r_n^{(k)} = f_1^{(n+k)} + \displaystyle\sum_{m=1}^{n} u_m f^{(n-m+k)}, \quad \lim_{n \to \infty} r_n^{(k)} = \dfrac{1 - F_k}{\mu},$

$r_n^k = pq^{k-1}, \ n = 1, 2, \ldots, \ k = 1, 2 \ldots,$ when $f_n^{(r)} = pq^{r-1},$

$r = 1, 2, \ldots .$

16. $\dfrac{1}{\mu + \lambda}[\mu e^{-\lambda t} + \lambda e^{-\mu t}] - 1.$

19. (a) $T[1 - F(T)] + \displaystyle\int_0^T xf(x)\, dx.$

(b) $\dfrac{C_1 F(T) + C_2[1 - F(T)]}{T[1 - F(T)] + \displaystyle\int_0^T xf(x)\, dx}.$

(c) $\dfrac{\lambda\left[C_1 - (C_1 - C_2)e^{-\lambda T}\right]}{1 - e^{-\lambda T}}.$

(d) $\dfrac{\mu(1 - e^{-\lambda T})}{\lambda + \mu(1 - e^{-\lambda T})}.$

CHAPTER 9

1. $\gamma(m, n) = \sigma^2 \quad n = m; \qquad \rho(m, n) = 1 \quad n = m$
    $ = 0 \quad n \neq m \qquad = 0 \quad n \neq m.$

2. (a) $\gamma(m, n) = [\min(m, n)]p(1 - p); \quad \rho(m, n) = [\min(m, n)]/\sqrt{mn};$
       no.

(b) $\gamma(s, t) = [\min(s, t)]\sigma^2; \ \rho(s, t) = [\min(s, t)]/\sqrt{st};$ no.

5. $\gamma(m, n) = [\min(m, n)]\sigma^2;$ no.

7. $\gamma(m, n) = \dfrac{ab}{(a + b)^2}(1 - a - b)^{n-m}; \ \rho(m, n) = (1 - a - b)^{n-m}.$

11. (a) $\gamma(1) = -0.25\sigma_y^2, \ \gamma(2) = -0.5\sigma_y^2$

(b) $\gamma(1) = 0.5\sigma_y^2, \ \gamma(2) = -0.26\sigma_y^2, \ \gamma(3) = -0.1\sigma_y^2$

12. (a) $\dfrac{(0.8)^h}{0.36}\sigma_y^2,$ (b) $\dfrac{(0.4)^h}{0.84}\sigma_y^2,$ (c) $\dfrac{(-0.5)^h}{0.75}\sigma_y^2$

18. $f(\lambda) = \dfrac{\sigma^2}{4\pi}e^{-\lambda^2/4} \qquad -\infty < \lambda < \infty.$

19. $f(\lambda) = \dfrac{\sigma^2}{\pi}\dfrac{\alpha}{\lambda^2 + \alpha^2} \qquad -\infty < \lambda < \infty$

AUTHOR INDEX

673

SUBJECT INDEX

DATE DUE

APR 1 3 '87			
AUG 9 '89			
AUG 9 '89			
DEC 0 9 '89			
MAY 0 3 '89			
MAY 1 5 '94			
DEC 0 5 '97			
120205			